W. Kahle · H. Leonhardt · W. Pl

Color Atlas and Textbook of Human Anatomy

in 3 Volumes

Volume 3:

Nervous System and Sensory Organs

by Werner Kahle

Translated by H. L. and A. D. Dayan

3rd revised edition

178 color plates with 578 drawings
by Gerhard Spitzer

1986
Georg Thieme Verlag Stuttgart · New York
Thieme Inc. New York

Prof. Dr. med. *Werner Kahle*
Neurologisches Institut
(Edinger Institut) der Universität Frankfurt/Main, FRG

Prof. Dr. med. *Helmut Leonhardt*
Direktor des Anatomischen Instituts der Universität Kiel, FRG

Univ.-Prof. Dr. med. univ. *Werner Platzer*
Vorstand des Anatomischen Instituts der Universität Innsbruck, Austria

Gerhard Spitzer,
Frankfurt/Main, FRG

Hedi L. Dayan, M.B., and
Anthony D. Dayan, M.D.,
Beckenham, Kent, UK

Library of Congress Cataloguing in Publication Data

Kahle, W. (Werner)
Color atlas and textbook of human anatomy.

Translation of: Taschenatlas der Anatomie.
Includes bibliographies and indexes.

Contents: v. 1. Locomotor system / by Werner Platzer – v. 2. Internal organs / by Helmut Leonhardt – – v. 3. Nervous system and sensory organs / Werner Kahle.

1. Anatomy, Human-Atlases.
I. Leonhardt, Helmut.
II. Platzer, Werner. III. Title.
[DNLM: 1. Anatomy-atlases.
QS 17 K12t]
QM25.K3413 1986
611'.022'2 86-5679

1st German edition 1976
2nd German edition 1978
3rd German edition 1979
4th German edition 1984
5th German edition 1986

1st English edition 1978
2nd English edition 1984

1st Dutch edition 1978
2nd Dutch edition 1981
1st French edition 1979
1st Greek edition 1985
1st Italian edition 1979
1st Japanese edition 1979
2nd Japanese edition 1981
3rd Japanese edition 1984
1st Spanish edition 1977

Typesetting by Druckhaus Dörr, (Linotype System 5 [202])
Printed in West Germany by Druckhaus Dörr, D-7140 Ludwigsburg

ISBN 3-13-533503-8 (Georg Thieme Verlag, Stuttgart)
ISBN 0-86577-251 -7 (Thieme Inc., New York)
1 2 3 4 5 6

Foreword

This pocket atlas is designed to provide a plain and clear compendium of the essential facts of human anatomy for the student of medicine. It also demonstrates the basic knowledge of the subject for students of related disciplines and for the interested layman. For all students preparation for their examinations and practice requires repetition of visual experiences. Text and illustrations in this book have been deliberately juxtaposed to provide visual demonstration of the topics of anatomy.

The pocket atlas is divided according to organ systems into three volumes: Volume 1 deals with the locomotor system, Volume 2 with the internal organs and skin and Volume 3 with the nervous system and the organs of the special senses. The topographic relationships of the peripheral pathways of nerves and vessels are considered in Volume 1, in so far as they are closely related to the locomotor system; Volume 2 systematically describes the distribution of the vessels. The floor of the pelvis (pelvic cavity), which has a close functional relationship with the organs of the lesser pelvis, and the relevant topography are incorporated in Volume 2. The developmental anatomy (embryology) of the teeth is briefly mentioned in Volume 2 because it aids unterstanding of the eruption of the teeth. The common embryological origins of the male and female genital organs are also discussed because it helps to explain their structure in the adult, as well as their not infrequent variants and malformations. Certain problems connected with pregnancy and childbirth are mentioned in the chapter on the female reproductive organs. But these do not cover all the knowledge of embryology required by students. The notes on physiology and biochemistry are deliberately brief and only serve to provide better understanding of structural details. Reference should be made to textbooks of physiology and biochemistry. Finally, it must be emphasized that no pocket atlas can replace a major textbook or the opportunity to examine macroscopic dissections and microscopic preparations.

The reference list mentions textbooks and original papers as a guide to the more advanced literature, and it also cites clinical textbooks of relevance to the study of anatomy.

Those who require less detailed knowledge of the structure of the human body will find clear illustrations, too, of the anatomic bases of the more important methods of medical examination. To help the nonmedical reader, everyday English terms for the major organs and their parts have been supplied as far as feasible; these terms are also listed in the index.

Frankfurt/Main, Kiel, Innsbruck

The Editors

Foreword to the 3rd English Edition of Volume 3

In this revision of the third edition a number of illustrations have been changed, Latin terminology has been brought into line with current international nomenclature and the results of recent research have been included in the text. The index has been completely revised. As far as possible, the Latin terms have been supplemented by English names, so that even the nonmedical reader may rapidly find important key words.

I wish to thank all the readers whose suggestions have helped to improve the text. I am particularly grateful to my colleagues Prof. *Leonhardt* for his important suggestions, and to Prof. *Platzer* for reading the text and for his many original preparations which have served as models for the illustrations. Above all I thank the publisher Dr. h. c. *G. Hauff* and his colleagues for generously enabling me to make all the changes I wanted.

Frankfurt, August 1985 *Werner Kahle*

Vol. 1: Locomotor System by W. Platzer

Vol. 2: Internal Organs by H. Leonhardt

Vol. 1: Locomotor System by W. Platzer

Vol. 2: Internal Organs by H. Leonhardt

Nervous System

Development and Classification of the Nervous System

The nervous system transmits messages. In primitive forms of life this function is performed by the sensory cells themselves **ABC 1**. They are excited by stimuli from their environment and the impulses are conducted via their process to a muscle cell **ABC 2.** This is the simplest reaction to stimuli from outside. Sensory cells with their own processes are found in man only in the olfactory epithelium. In more highly differentiated organisms there is inserted between the sensory cell and the muscle cell an additional cell, which transmits the messages ('impulses'): the nerve cell **BC 3.** It is able to transmit the stimulus to a number of muscle cells or other nerve cells and thus is formed a *nerve network* **C.** Even in the human body there occurs a diffuse network of this type and all the viscera, blood vessels and glands are innervated by it. It is called the **vegetative,** *visceral* or *autonomic,* **nervous system,** and can be divided into two antagonistic parts, the **sympathetic** and **parasympathetic** systems, which together are responsible for preserving a constant internal environment, homeostasis.

Vertebrates, in addition to the vegetative system, have a **somatic system,** which consists of the central nervous system **CNS** (brain and spinal cord) and the peripheral nervous system **PNS** (nerves which supply the head, trunk and limbs). Together they serve conscious perception, voluntary movement and the processing of messages, *integration.*

The CNS develops from the *medullary plate* of the ectoderm **D 4,** which becomes the *neural groove* **D 5** and then the *neural tube* **D 6.** Finally, the neural tube differentiates into the spinal cord **D 7** and the brain **D 8.**

Functional Circuits

The nervous system, the organism and the environment are functionally interrelated. Stimuli from the environment **E 9** are transmitted by the sensory cells **E 10** over **sensory (afferent) nerves E 11** to the CNS **E 12.** The CNS responds by sending instructions via **motor (efferent) nerves E 13** to the muscles **E 14.** The control and regulation of the muscular response **E 15** is achieved by sensory cells in the muscles, which provide a *feedback* through sensory nerves **E 16** to the central nervous system. This afferent pathway does not transmit stimuli from outside, *exteroception,* but only signals from the interior of the body, *proprioceptive stimuli.* Thus, there is both **exteroceptive** and **proprioceptive sensibility.**

However, the organism does not only respond to its surroundings but it also acts spontaneously on them, and for this, too, there is a corresponding functional circuit. The action, instigated by the CNS via the efferent nerves **F 17,** is registered by the sense organs, and information is then returned to the CNS via the afferent nerves *(reafference).* Depending upon whether the result corresponds to what was desired or not, additionai impulses are sent by the CNS to increase or decrease the action. A great number of these circuits form the basis of nervous activity.

Just as we distinguish exteroceptive sensibility of the skin and mucous membranes from proprioceptive sensibility of the muscle and tendon receptors and the vegetative sensory innervation of the viscera, so motor activity can be divided into actions concerned with the environmental oikotropic somatomotor system (striated voluntary muscle) and the internal or idiotropic visceromotor system (smooth visceral muscle).

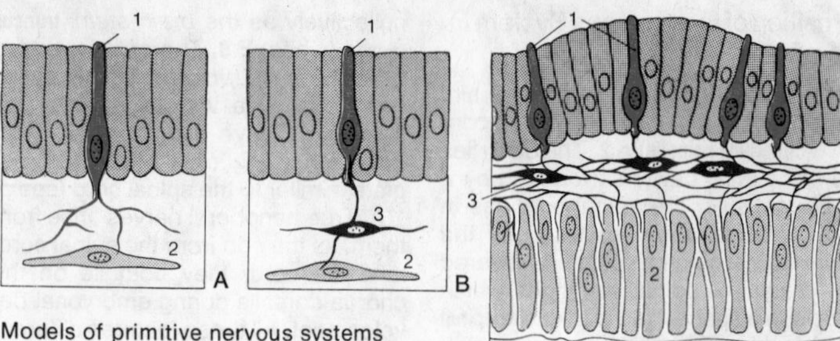

Models of primitive nervous systems
(after Parker and Bethe)

A Sensory cell with a process to a muscle cell

B Nerve cell as connection between a sensory and muscle cell

C Diffuse nerve network

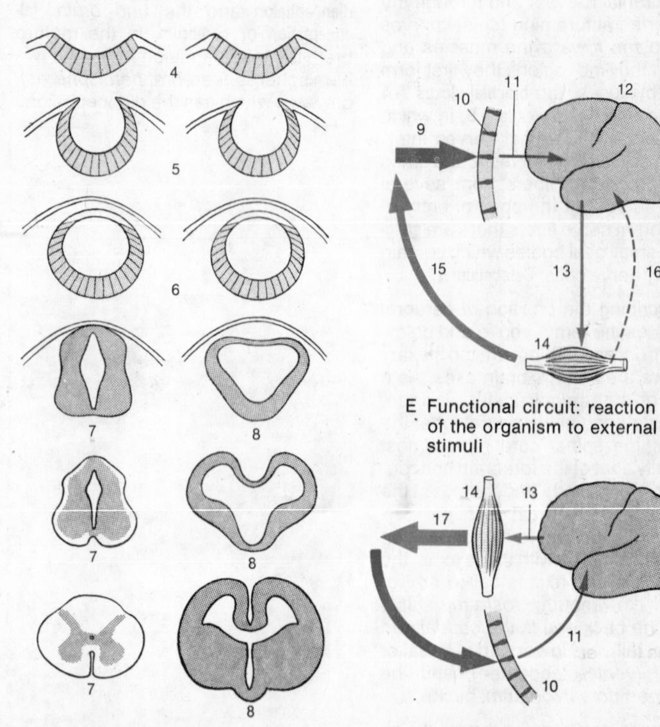

D Embryonal development of the central nervous system. Spinal cord on the left, brain on the right

E Functional circuit: reaction of the organism to external stimuli

F Functional circuit: action of the organism on its environment

Position of the Nervous System in the Body

The CNS is divided into the brain, **(encephalon) A1** and the spinal cord (SC) **(medulla spinalis) A2.** The brain lies in the cranial cavity surrounded by a bony capsule; the spinal cord lies in the vertebral canal enclosed by the bony spinal column. Both are covered by cranial or spinal meninges which enclose a space filled with **cerebrospinal fluid.** In this way the CNS is surrounded on all sides by bony walls and a fluid cushion.

The peripheral nerves pass through holes (foramina) in the base of the skull (cranial nerves) and through the intervertebral foramina (spinal nerves **A3**), to run toward the muscles and skin. In the limb regions they first form nerve plexuses, the **brachial plexus A4** and the **lumbosacral plexus A5,** in which the fibers of the spinal nerves intermingle so that the nerves to the limbs come to contain fibers from several spinal nerves. At the point of entry of the afferent nerve fibers there are **ganglia A6,** small oval bodies which contain sensory nerve cells, **Cerebellum A7.**

In describing the position of cerebral structures the terms above and below and in front and behind, are too inexact as there are various brain axes. As a result of the upright posture in man, there is a bend in the neural tube: the axis of the spinal cord runs almost vertically, that of the forebrain horizontally **(Forel's axis,** red), and the axis of the lower parts of the brain obliquely (**Meynert's axis,** blue). The terms for position are in accordance with these axes: the front end of the axis is called oral or rostral *(os,* mouth; *rostrum,* ship's bow), the back end is the *caudal* part *(cauda,* tail), the lower part is basal or *ventral (venter,* abdomen), and the upper part *dorsal (dorsum,* back).

The lower parts of the brain, which merge into the spinal cord, are known collectively as the *brain stem,* **truncus cerebri** (white) **B8.** The anterior part is called the *forebrain,* **prosencephalon** (grey) **B9.** The various parts of the brain stem have a uniform architecture, consisting of a basal and alar plate, similar to the spinal cord (see p. 13). True peripheral nerves arise from them, as they do from the spinal cord. Like the cord, they both lie on the chorda dorsalis during embryonal development. All these features differentiate the brain stem from the forebrain. This classification differs from the official one, in which the diencephalon is regarded as part of the brain stem.

The forebrain consists of two parts, the **diencephalon** and the *end brain,* **telencephalon** or **cerebrum.** In the mature brain the telencephalon forms the two hemispheres *(cerebral hemispheres),* between which lies the diencephalon.

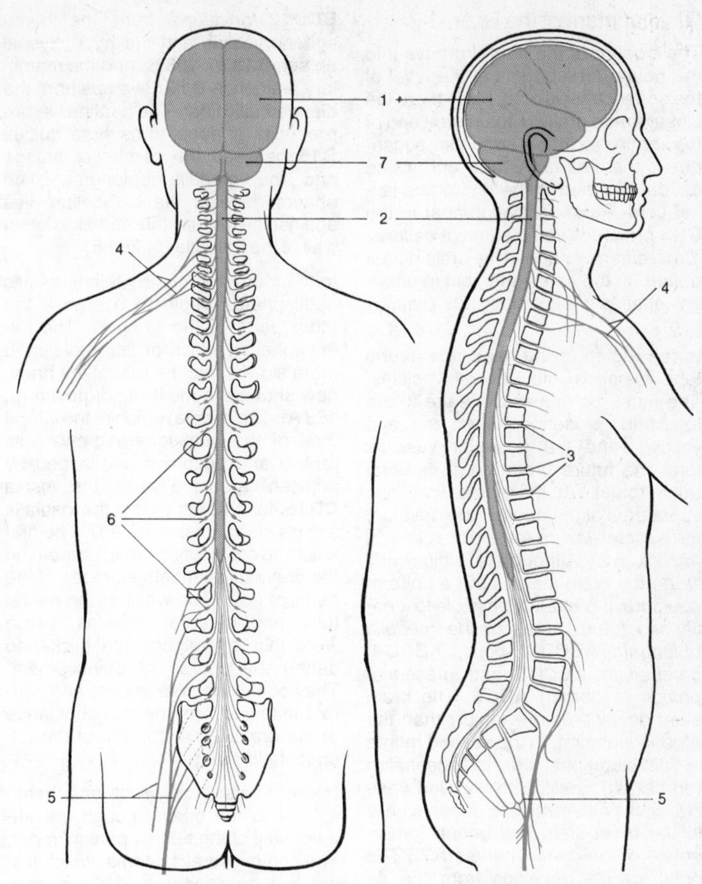

A Position of the central nervous system within the body

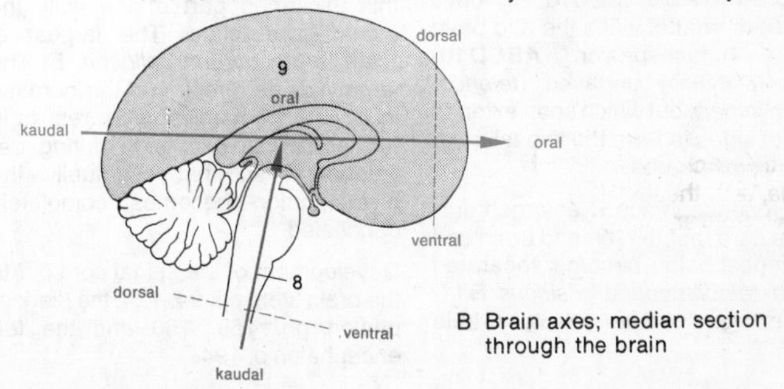

B Brain axes; median section through the brain

Development of the Brain

The closure of the neural groove into the neural tube begins at the level of the upper cervical cord. From there the closure runs orally to the rostral end of the brain *(oral neuroporus,* subsequently *lamina terminalis)* and in the caudal direction to the end of the spinal cord. Further development in the CNS proceeds in the same directions. Thus, different parts of the brain do not mature at the same time, but in different phases *(heterochronous maturation).*

In the region of the head, the neural tube expands into several vesicles. The most rostral vesicle is the future forebrain, prosencephalon (red and yellow), and the posterior vesicles form the future brain stem, cerebral trunk (blue). At the same time two curvatures of the neural tube develop, the *parietal flexure* **A1** and the *cervical flexure* **A2**. Although at this early stage the brain stem is still a uniform structure, it is already possible to identify the future areas of the *medulla oblongata* **ABCD3,** *pons,* **ABCD4,** *cerebellum* **ABCD5** and *mesencephalon* (midbrain) **ABC6.** The brain stem develops more rapidly than the prosencephalon. In the second month of development, the telencephalon (end brain) is still a thin-walled vesicle, whilst differentiation of nerve cells in the brain stem has begun (emergence of the cranial nerves **A7**). The optic vesicle develops from the diencephalon **AB8** (p. 316 A); optic put **A9**. In front of it lies the end brain vesicle (telencephalon **ABCD10)** which is initially unpaired *(telencephalon impar),* but which soon extends on both sides to form the two telencephalic hemispheres.

The prosencephalon **B** enlarges during the third month. The end brain and the diencephalon become separated by the *telodiencephalic sulcus* **B11.** The anlage of the olfactory bulb **BCD12** develops from the hemispheric vesicle and the hypophyseal anlage **B13** (p. 188 B) and the mamillary eminence **B14** develops from the diencephalic floor. The pontine flexure produces a deep transverse sulcus **B15** between the cerebellar anlage and the medulla oblongata. The undersurface of the cerebellum lies against the thin, membrane-like dorsal wall of the medulla (p. 263 **E**).

In the fourth month, the telencephalic hemispheres begin to overgrow the other parts of the brain **C.** The telencephalon, which at first developed more slowly than the rest of the brain, now shows the most rapid growth (p. 158 A). The middle region of the lateral face of the hemisphere grows less rapidly and becomes overlapped by adjacent areas. This is the *insula* **CD16.** In the sixth month the insula is still visible on the surface **D.** The first sulci and convolutions (gyri) appear on the previously smooth surfaces of the hemispheres. The walls of the neural tube and cerebral vesicles, which were thin at first become thickened during the course of development. They contain the nerve cells and neural tracts and form the true substance of the brain. Development of the telencephalon p. 194.

Nerve fibers grow from one hemisphere to the other through the anterior wall of the telencephalon impar. The commissural systems, which join the two hemispheres, develop from this thickened part of the wall, the commissural plate. The largest of them is the *corpus callosum* **E.** The largely caudal direction of the increase in size of the hemispheres, results in further *caudal extension* during development so that eventually the diencephalon becomes completely concealed.

Development of the spinal cord p. 56, the brain stem pp. 94, 124, the diencephalon pp. 158, 160 and the telencephalon p. 194.

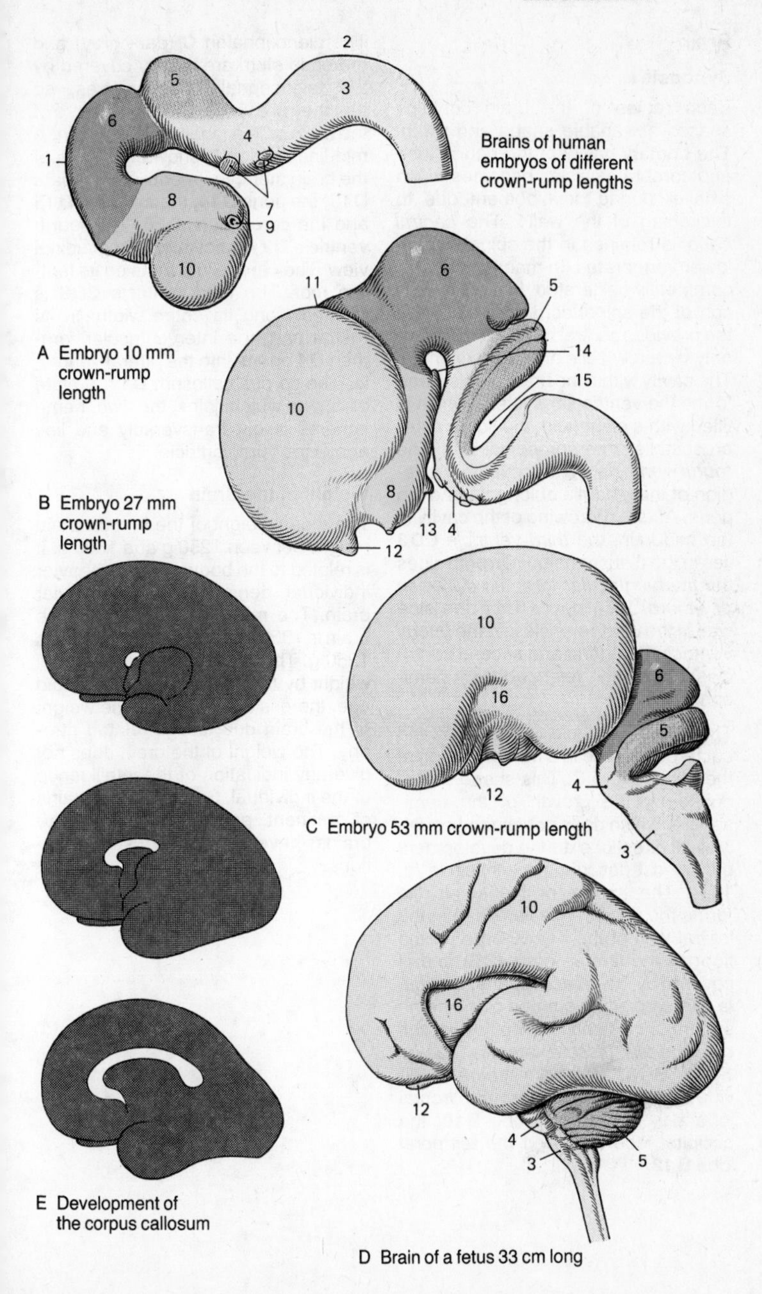

Brains of human embryos of different crown-rump lengths

A Embryo 10 mm crown-rump length

B Embryo 27 mm crown-rump length

C Embryo 53 mm crown-rump length

D Brain of a fetus 33 cm long

E Development of the corpus callosum

Brain

Synopsis

Each region of the brain contains spaces of variable shape and width. The primary cavity of the neural tube and cerebral vesicles becomes much smaller during development due to thickening of the walls. The *central canal* is retained in the spinal cord in lower vertebrates. In man the canal is completely obliterated. In a cross section of the spinal cord the position of the previous central canal **A1** is shown only by a few cells of its former lining. The cavity within the brain persists and forms the ventricular system, which is filled with a clear fluid, the liquor cerebrospinalis or cerebrospinal fluid. The *fourth ventricle* **AD2** occurs in the region of the medulla oblongata and the pons. After a narrowing of the cavity in the midbrain, the *third ventricle* **CD3** lies in the diencephalon. On both sides the *interventricular foramen (foramen of Monro)* **DE4** opens from the side wall of the third ventricle into the *lateral ventricle* **CE5** (first and second ventricles) of both *telencephalic* hemispheres.

The lateral ventricle is curved **E** and cut through twice in a frontal section of the hemisphere **C.** This shape is produced by the growth of the hemisphere, which does not expand equally in all directions during development but almost describes a semicircle (p. 194). The middle of the semicircle forms the *insula.* This lies deep in the lateral wall of the hemisphere on the floor of the lateral fossa **(C6)** and is covered by the adjacent parts, *opercula* **C7,** so that the surface of the hemisphere shows only a deep fissure *sulcus lateralis (fissura lateralis, fissura Sylvii)* **BC8.** The hemisphere is divided into several lobes: the frontal lobe **B9,** the parietal lobe **B10,** the occipital lobe **B11** and the temporal lobe **B12.**

The diencephalon **C** (dark grey) and the brain stem are largely covered by the telencephalic hemispheres, so that they are visible only at the base of the brain or in a longitudinal section. A mid-line section **D** shows the parts of the brain stem, the medulla oblongata **D13,** the pons **D14,** the midbrain **D15** and the cerebellum **D16.** The fourth ventricle **D2** is shown in a longitudinal view. The cerebellum rests on its tent-like roof. The third ventricle **CD3** is opened along its entire width. In its rostral part, the interventricular foramen **D4** opens into the lateral ventricle. The corpus callosum **D17,** a plate of fibers which joins the two hemispheres is cut transversely and lies above the third ventricle.

Weight of the Brain

The mean weight of the human brain ranges between 1250 g and 1600 g. It is related to the body weight: a heavier individual generally has a heavier brain. The mean weight of the male brain is 1350 g and that of the female is 1250 g. The brain has attained its full weight by the age of 20 years. In old age, there is a decrease in the weight of the brain due to age-related atrophy. The weight of the brain does not give any indication of the intelligence of the individual. Studies of the brains of eminent people (so-called elite brains) have shown the usual variations.

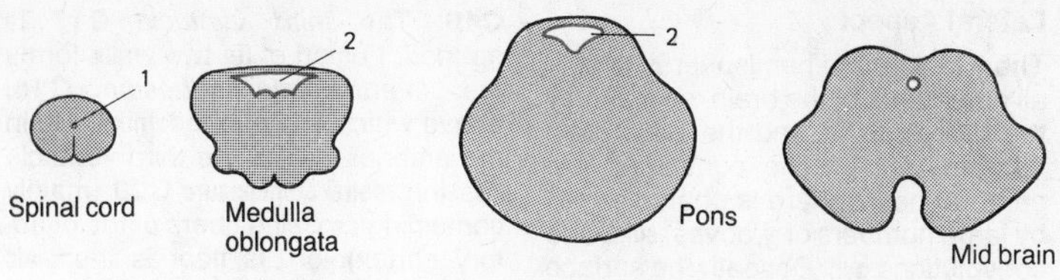

A Sections through the spinal cord and brain stem in true relative sizes

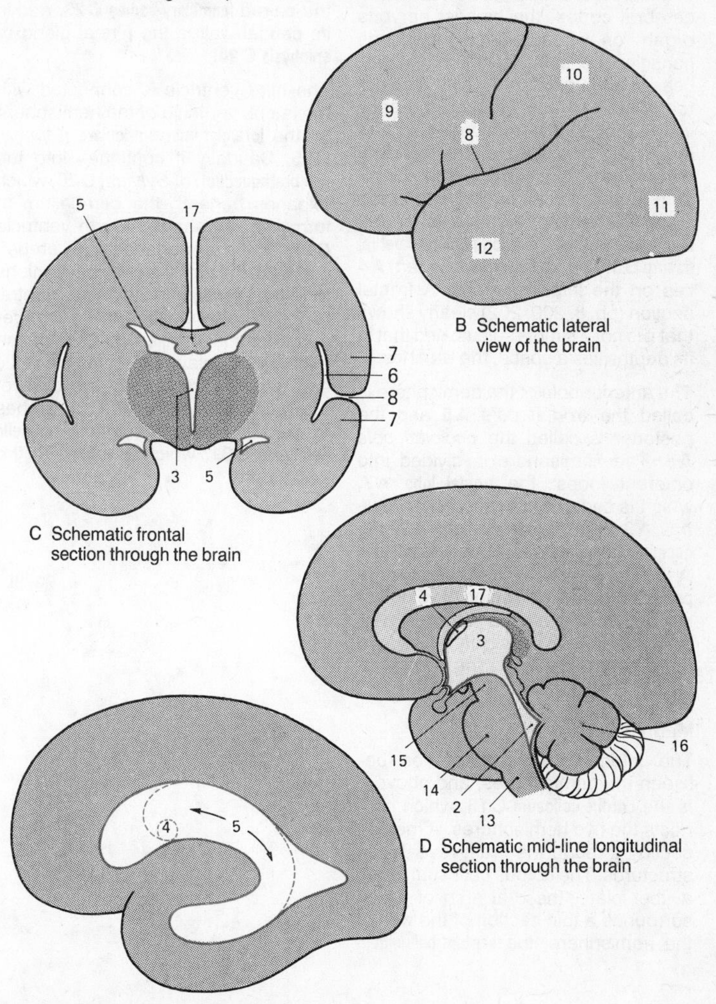

B Schematic lateral view of the brain

C Schematic frontal section through the brain

D Schematic mid-line longitudinal section through the brain

Schematic paramedian longitudinal section through the brain

Lateral Aspect

The two cerebral hemispheres overlie all other parts of the brain, so that only the **cerebellum A1** and the *brain stem* **A2** are visible. The surface of the cerebral hemisphere is characterized by large numbers of grooves, **sulci,** and convolutions **gyri.** Beneath the surface of the relief of the convolutions lies the cerebral cortex, the highest nervous organ, on whose integrity depends consciousness, memory, thought processes and voluntary activity. The extent of the cerebral cortex is greatly increased by the formation of sulci and gyri. Only one third of the cerebral cortex lies on the surface of the hemisphere and two thirds lie deep in the sulci. The hemispheres are separated by a deep furrow, the **longitudinal cerebral fissure B3**. The **lateral sulcus** (*sylvian)* **A4** lies on the lateral surface. A frontal section (pp. 8, 200, 202) clearly shows that it is not a simple sulcus, and that in its depths lies a space, the **lateral fossa.**

The anterior pole of the hemisphere is called the *frontal pole* **A5** and the posterior is called the *occipital pole* **A6.** The hemisphere is divided into different lobes: the **frontal lobe A7,** which is separated by the *central sulcus* **A8** from the **parietal lobe A9,** the **occipital lobe A10,** and the **temporal lobe A11.** The central sulcus separates the **prencetral gyrus A12** (region of voluntary motor control) from the **postcentral gyrus A13** (region of sensibility). The two together are known as the *central region.*

Median Section

The interbrain **diencephalon** lies between the hemispheres, and above it is the **corpus callosum C15,** which connects the two hemispheres. A median section through the brain reveals the structures. The corpus callosum forms a fiber plate, the oral arch of which surrounds a thin section of the wall of the hemisphere, the **septum pellucidum C16.** The *third ventricle* **C17** is opened. Fusion of its two walls forms the *interthalamic adhesion* **C18,** above which arches the **fornix C19.** In the anterior wall of the third ventricle lies the **rostral commissure C20** (mainly containing crossing fibers of the olfactory cortex), on its floor is the **optic chiasm C21**, the **hypophysis C22,** and the paired **mamillary bodies C23,** and in its caudal wall is the pineal gland or **epiphysis C24**.

The third ventricle is connected with the lateral ventricle of the hemisphere by the **foramen interventriculare** *(Monroi)* **C25.** Caudally it continues into the **cerebral aqueduct** (of *Sylvius)* **C26,** which expands beneath the cerebellum to form the tent-shaped fourth ventricle **C27.** On the cut surface of the cerebellum **C28,** the furrows and convolutions form the so-called *arbor vitae.* Rostral to the cerebellum lies the quadrigeminal lamina, **lamina tecti C29** of the midbrain (a relay station for the optic and acoustic tracts). On the base of the brain stem the **pons C30** arches forward and leads over into the **medulla oblongata C31,** which is joined to the spinal cord.

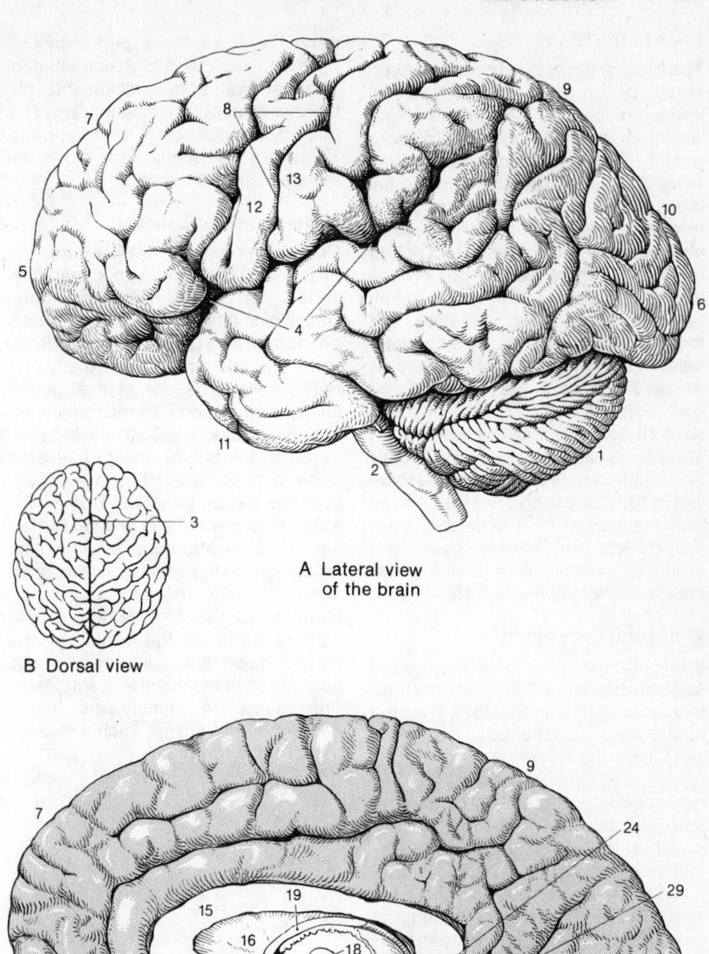

A Lateral view of the brain

B Dorsal view

C

Median section through the brain, medial surface of the right hemisphere

Base of the Brain

The base of the brain affords a general survey of the brain stem, the ventral surface of the frontal **A1** and temporal **A2** lobes, and the floor of the diencephalon. The *longitudinal cerebral fissure* **A3** separates the two frontal lobes, on whose basal surface lie bilaterally the *olfactory lobes* with the **olfactory bulb A4** and the **olfactory tract A5.** In the **olfactory trigone A6** the tract divides into two *olfactory striae,* which delimit the **anterior perforated substance A7** through which enter many blood vessels. At the *chiasm* **A8** where the **optic nerves A9** cross, the base of the diencephalon begins with the **hypophysis A10** and the **mamillary bodies A11.** The **pons A12** arches forward caudally and is continuous with the **medulla oblongata A13.** Many cranial nerves merge from the brain stem. The cerebellum is divided into the medial, deep-lying *cerebellar vermis* **A14** and the two *cerebellar hemispheres* **A15.**

White and Grey Matter

If the brain is cut into slices, the cut surfaces show the white and grey matter, the **substantia alba and grisea.** The grey matter consists of collections of nerve cells and the white matter of fiber tracts, i.e. the processes of the neurons which appear light because of their whitish covering, the *myelin sheaths.* In the spinal cord the grey matter lies centrally **B16** and is surrounded by white matter. In the brain stem **B17** and the diencephalon, the grey and white matter are variously distributed: the grey areas are called **nuclei.** In the telencephalon **B18** the grey matter lies at the outer margin and forms the cerebral **cortex,** while the white matter lies internally; this arrangement is the opposite of that in the spinal cord.

The arrangement in the spinal cord represents a primitive condition found in fishes and amphibia, where the nerve cells are in a periventricular position, even in the telencephalon. The cerebral cortex represents the highest level of organization, which is fully developed only in mammals. There are transitional formations between nucleus and cortex.

Developmental Zones

During embryonal development, the neural tube is divided into longitudinal zones: the ventral half of the lateral wall, which differentiates early, is called the **basal plate C19** and is considered to be the site or origin of the motor nerve cells. The dorsal half of the lateral wall, which differentiates later, is called the **alar plate C20,** and is considered as the site of origin of sensory nerve cells. Between the alar and basal plates lies an area **C21** from which autonomic nerve cells are said to arise. Thus, a structural plan can be recognized in the spinal cord and brain stem, knowledge of which aids understanding of the organization of the various parts of the brain. In the diencephalon and telencephalon the derivatives of the basal and alar plates are difficult to identify and many authors do not accept such a classification of the forebrain.

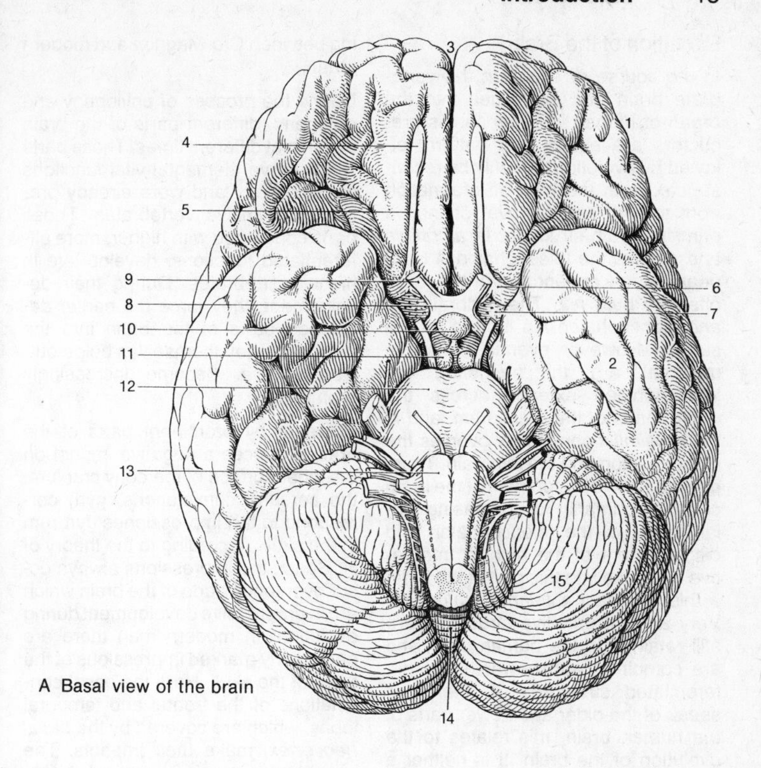

A Basal view of the brain

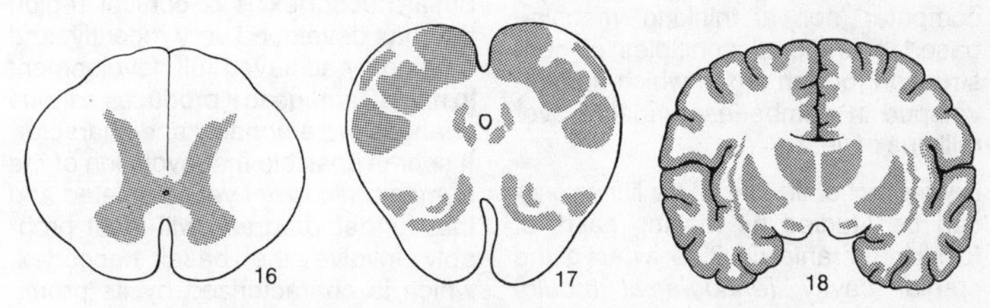

B Distribution of white and grey matter

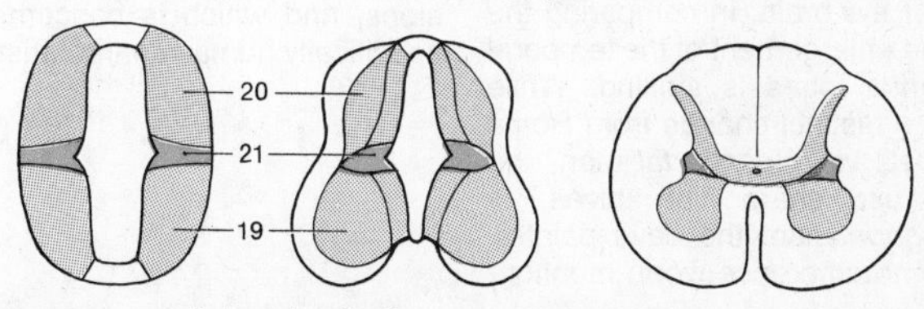

C Longitudinal zones of the CNS

Evolution of the Brain

In the course of evolution the vertebrate brain has developed into the organ of human intelligence. As precursors are extinct, the pattern followed by evolution can only be reconstructed with the help of information from species which have retained a primitive brain structure. In *amphibia* and *reptiles* the telecephalon **A1** appears as an appendage of the large olfactory bulb **A2**. The midbrain **A3** and diencephalon **A4** lie free on the surface. However, even in a primitive mammal, e.g. the *hedgehog*, the telencephalon extends across the rostral parts of the brain stem, and in the *prosimii* it completely overlies the diencephalon and the midbrain. Thus, phylogenetic development of the brain consists primarily of increasing enlargement of the telencephalon and displacement of the highest integrative functions into this part of the brain – this is in fact a *telencephalization*. Very ancient primitive structures are still retained in the human brain and are combined with newer highly differentiated structures. If, then, we speak of the older and newer parts of the human brain, this relates to the evolution of the brain. It is neither a computer nor a thinking machine based on rational principles of construction, but an organ which has developed in numberless variants over millions of years.

Development of the form of the human brain can be studied by making casts of fossil intracranial cavities: a cast of the cranial cavity *(endocranial mould)* constitutes a gross impression of the shape of the brain. In comparing the casts the enlargement of the temporal and frontal lobes is striking. While there is a distinct change from *Homo pekinensis* via *Neandertal* man, the first to use sharp flint knives, to *Cro-Magnon* man, the cave painter, there is no difference worth mentioning between Cro-Magnon and modern man.

During the process of phylogeny and ontogeny, different parts of the brain develop at different times. Those parts which serve elemental vital functions develop early and were already present in primitive vertebrates. Those parts concerned with higher, more differentiated functions, develop late in higher vertebrates. During their development, they force the earlier developed parts of the brain into the depths as they themselves bulge outwards, i. e. become increasingly prominent.

Pressure by prominent parts of the brain produces a negative imprint on the inner surface of the bony cranium: the cerebral convolutions (gyri) correspond to the impressiones gyrorum in the skull. According to the theory of H. Spatz, the impressions always occur over those parts of the brain which are in progressive development during evolution. In modern man there are particularly marked impressions of the base of the skull. Here the basal convolutions of the frontal and temporal lobes, which are covered by the *basal neocortex,* make their imprints. The basal neocortex is a cortical region that has developed very recently, and it has only achieved full development in man. Damage to it produces serious changes in personality and character. It seems possible that evolution of the human brain is not yet completed and that further progress will most probably involve the basal neocortex, which is characterized by its prominence and its ability to form impressions, and which is concerned with specifically human characteristics.

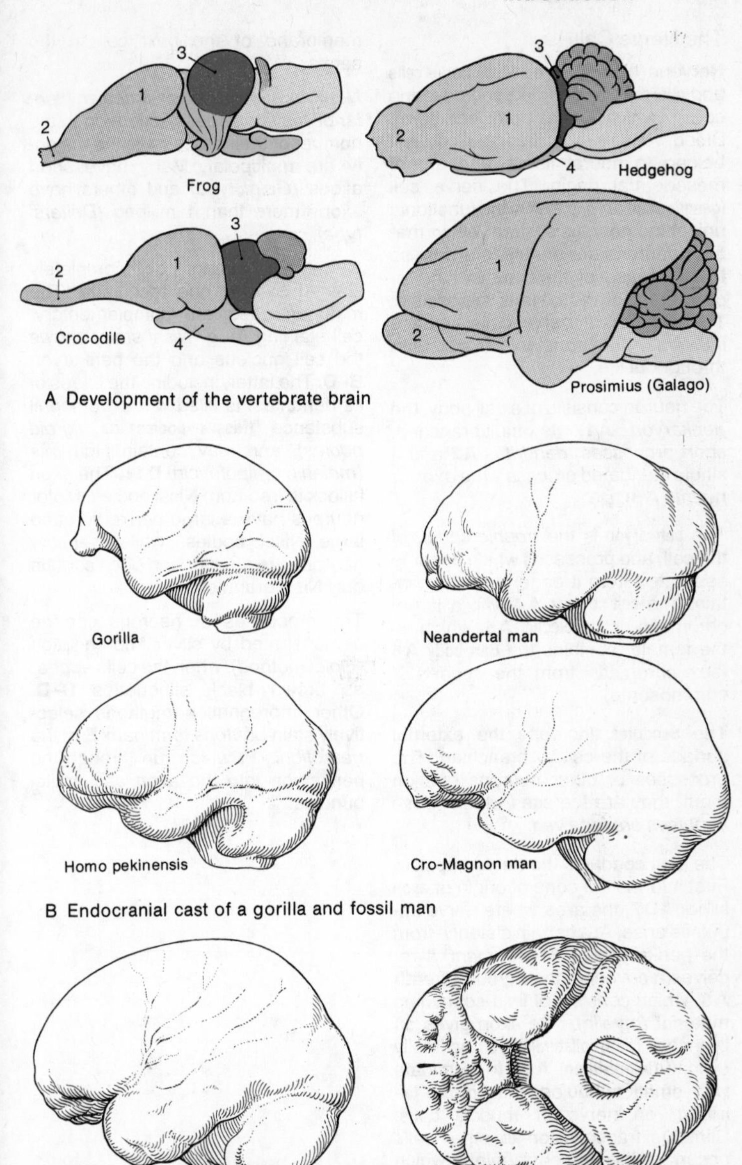

A Development of the vertebrate brain

B Endocranial cast of a gorilla and fossil man

C Endocranial cast of homo sapiens; lateral and basal view

The Nerve Cell

Nervous tissue consists of **nerve cells** and **glial cells** (supporting and covering cells), which arise from ectoderm. Blood vessels and meninges do not belong to neural tissue and are of mesodermal origin. The nerve cell **(ganglion cell** or **neuron) A** is the functional unit of the nervous system. When mature, neurons are unable to divide, so that increase of their number or replacement of old cells is impossible. The number of nerve cells in each individual is constant from birth throughout life.

The neuron consists of a cell body, the *perikaryon* **A1**, its multibranched, short processes, *dendrites* **A2** and a single elongated process – the *axon* or *neurite* **ABCD3.**

The **perikaryon** is the *trophic center* of the cell, and processes which become separated from it degenerate. It contains the **cell nucleus A4** with a large, chromatin-rich **nucleolus A5,** which, in the female, contains the **Barr body A6** (sex chromatin from the second X chromosome).

The **dendrites,** increase the external surface of the cell by branching. The processes of other neurons end on them: they are the sites where *nerve impulses are received.*

The **axon** conducts the nerve impulse. First it forms the cone of origin or **axon hillock AD7,** the area where nerve impulses arise. At a certain distance from the perikaryon *(initial segment)* it receives a covering, the **myelin sheath A8,** which consists of lipid-containing material *(myelin).* The axon gives off branches **(axon collaterals) A9** and finally divides **(telodendron) A10** to terminate with small endbulbs, the **boutons terminaux,** on nerve or muscle cells. Stimulus transmission to other cells occurs at the bouton terminal, which forms a synapse with the surface membrane of the next cell in the series.

Neurons are classed as *unipolar, bipolar* or *multipolar* cells according to the number of their processes; the majority are multipolar. Many have short axons *(Golgi type)* and others have axons more than 1 m long *(Deiters' type).*

A neuron cannot be completely stained by any one technique. The methods used are complementary; cell staining (e. g. *Nissl stain)* shows the cell nucleus and the perikaryon **B–D.** The latter, including the bases of its dendrites, is filled with chromophil substance **(Nissl's bodies** or *tigroid bodies),* and may contain pigments *(melanin* or *lipofuscin)* **D11.** The axon hillock is free from Nissl bodies. Motor neurons have a large perikaryon and large Nissl bodies, whilst sensory neurons are smaller and often contain only Nissl granules.

The processes of neurons can be demonstrated by silver impregnation *(Golgi method),* when the cells appear as browny-black silhouettes **B–D.** Other impregnation methods selectively stain *boutons terminaux* **E** or the *neurofibrils* **F,** which run through the perikaryon into the axon as parallel bundles.

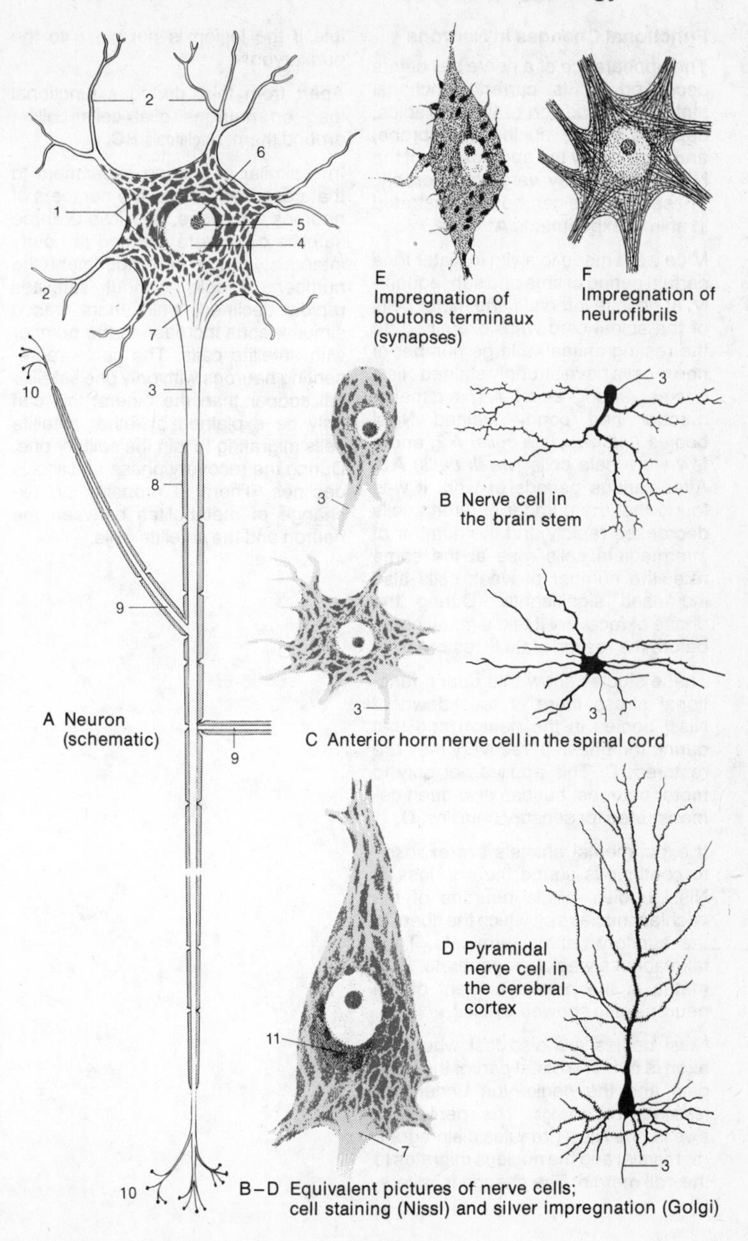

B–D Equivalent pictures of nerve cells; cell staining (Nissl) and silver impregnation (Golgi)

Functional Changes in Neurons

The appearance of a nerve cell differs according to its current functional state. The condition of the nucleolus, deposits on the nuclear membrane, and particularly the appearance of the Nissl bodies may vary considerably. These changes can be demonstrated in animal experiments **A:**

Mice were made to swim in water for a certain period of time and subsequently, at definite intervals, the motor cells of the spinal cord were examined. In the resting animal, a large number of nerve cells have strongly stained Nissl bodies *(strong cells)* **A1,** a smaller number had poorly stained Nissl bodies *(intermediate cells)* **A2,** and a few were pale cells *(weak cells)* **A3.** After various periods of work it was found that the number of strong cells decreased rapidly and the number of intermediate cells rose at the same rate; the number of weak cells also increased significantly. During the phase of recovery it takes many hours before the resting state is restored.

These studies show that during functional stress there is breakdown of Nissl bodies in the neuron and that during the phase of recovery they are restored, **C.** This applies not only to motor neurons, but has also been demonstrated for sensory neurons, **D.**

If experimental animals are exposed to continuous sound there is loss of Nissl bodies in the neurons of the cochlear nucleus on which the fibers of the auditory nerves terminate. They take about a week to regenerate. Estimates of the protein content of the neurons also showed a dificit.

Nissl bodies are also lost when the axon is cut. The distal part of the axon dies and the perikaryon undergoes **retrograde cell change.** The perikaryon swells, the Nissl granules disintegrate (tigrolysis) and the nucleus migrates to the cell margin. The change is reversible if the lesion is not close to the perikaryon.

Apart from this, during a functional load on neurons glial cells collect around them, **satellitosis BC.**

In a similar swimming experiment to that described above, the numbers of neurons with none, one, two or three satellite cells were counted at hourly intervals: during the experiment the number of cells without satellites rapidly declined, while there was a simultaneous increase in the number with satellite cells. The curve representing neurons with only one satellite fell sooner than the others; this can only be explained by other satellite cells migrating to join the solitary one. During the recovery phase satellitosis declines. There is probably an exchange of metabolites between the neuron and the satellite cells.

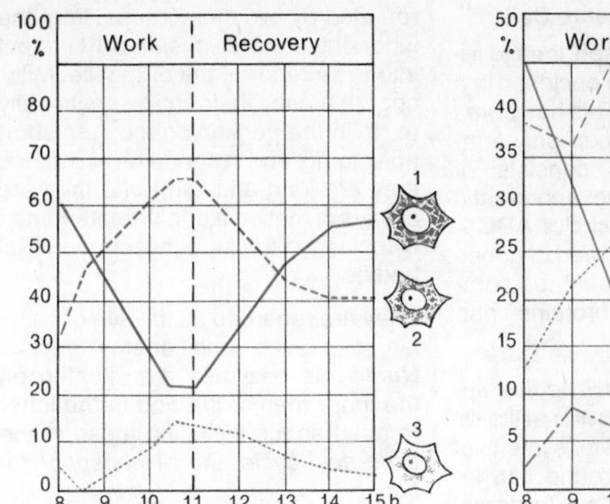

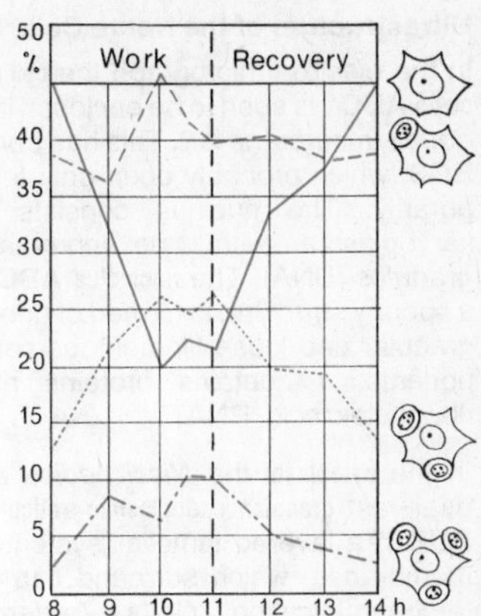

A Decrease of Nissl granules in anterior horn cells in the treading experiment (mouse) (after Kulenkampff)

B Increase in the number of satellite cells around anterior horn cells in the treading experiment (mouse) (after Kulenkampff)

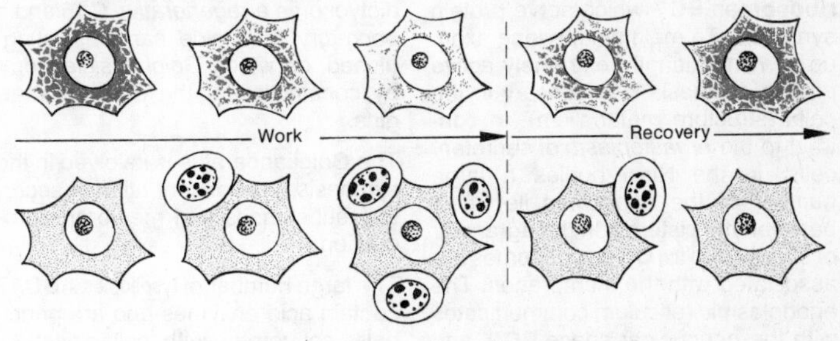

C Changes in nerve cells under stress

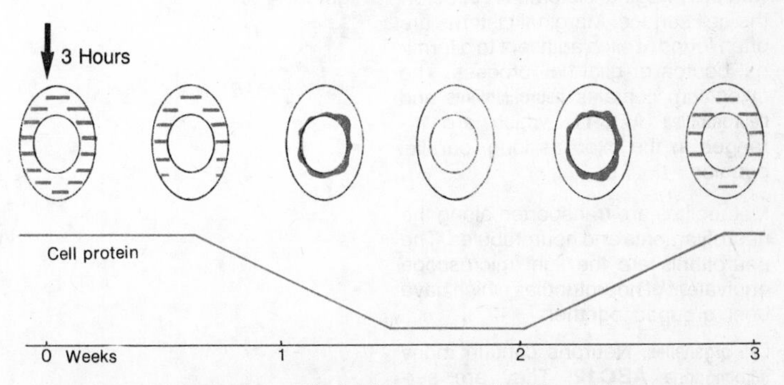

D Decrease and restoration of Nissl substance in cells of the cochlear nucleus after exposure to a continuous sound (guinea pig) (after Hamberger and Hyden)

Ultrastructure of the Nerve Cell

In the electron micrograph the **cell nucleus ABC 1** is seen to be enclosed in a *double membrane* **A 2.** This has *pores* **BC 3,** which probably open only temporarily. The nucleus consists of karyoplasma with fine *chromatin granules* (DNA). The **nucleolus ABC 4,** a spongy structure, is formed of dense granular and loose filamentous components; it contains proteins and ribonucleic acid, RNA.

In the **cytoplasm** the *Nissl bodies* appear as **granular endoplasmic reticulum ABC 5,** a layered lamellar system of membranes, which surround flat, intercommunicating clefts *(cisterns)* **BC 6.** They have been shown to contain *cholinesterases* and various other substances. To the outside of the membrane adhere small granules, the **ribosomes BC 7** which serve protein synthesis. To maintain the long axon, up to 1 m in length, extremely active protein synthesis is necessary in the cells (structure metabolism). In contrast to the *ergastoplasm* of secretory cells, in the Nissl bodies a large number of the *ribosomes* lie freely between the cisterns. In the agranular or **smooth reticulum C 8** no ribosomes are associated with the membranes. The endoplasmic reticulum communicates with the *perinuclear space* **BC 9,** and with the *marginal cisterns* **A 10,** below the cell surface. Marginal cisterns are often found at sites adjacent to a terminal bouton or glial cell process. The cytoplasm contains **neurofilaments** and **neurotubules ABC 11,** which are arranged in the axon as long, parallel bundles.

Metabolites are transported along the neurofilaments and neurotubules. The neurofibrils are the light microscope equivalent of neurotubules which have been grouped together.

Cell organelles. Neurons contain many **mitochondria ABC 12.** They are surrounded by two membranes, from the inner one of which *cristae* **C 13** project into the interior of the organelle. Mitochondria alter their shape continually (e. g. in the perikaryon they are short and plump, and in dendrites and axons they are long and slim) and they are always in motion along predetermined cytoplasmic tracks between the Nissl bodies.

They are regarded as the site of cellular respiration and energy supply. Numerous enzymes are situated on the inner membrane and in the inner (matrix) space including those of the *citric acid cycle* and of *phosphorylation.*

The **Golgy apparatus** consists of a number of **dictyosomes ABC 14,** multilayered, noncommunicating cisterns surrounded by a membrane. On the dictyosome a *regenerative* **C 15** and a *secretory* **C 16** side can be distinguished, on which *Golgi vesicles* form by constriction of the cisternal margins.

The Golgi apparatus is involved in the synthesis and concentration of secretory substances and membrane production.

The large number of **lysosomes ABC 17** contain acid enzymes and are principally concerned with cell digestion. Pigment **A 18.**

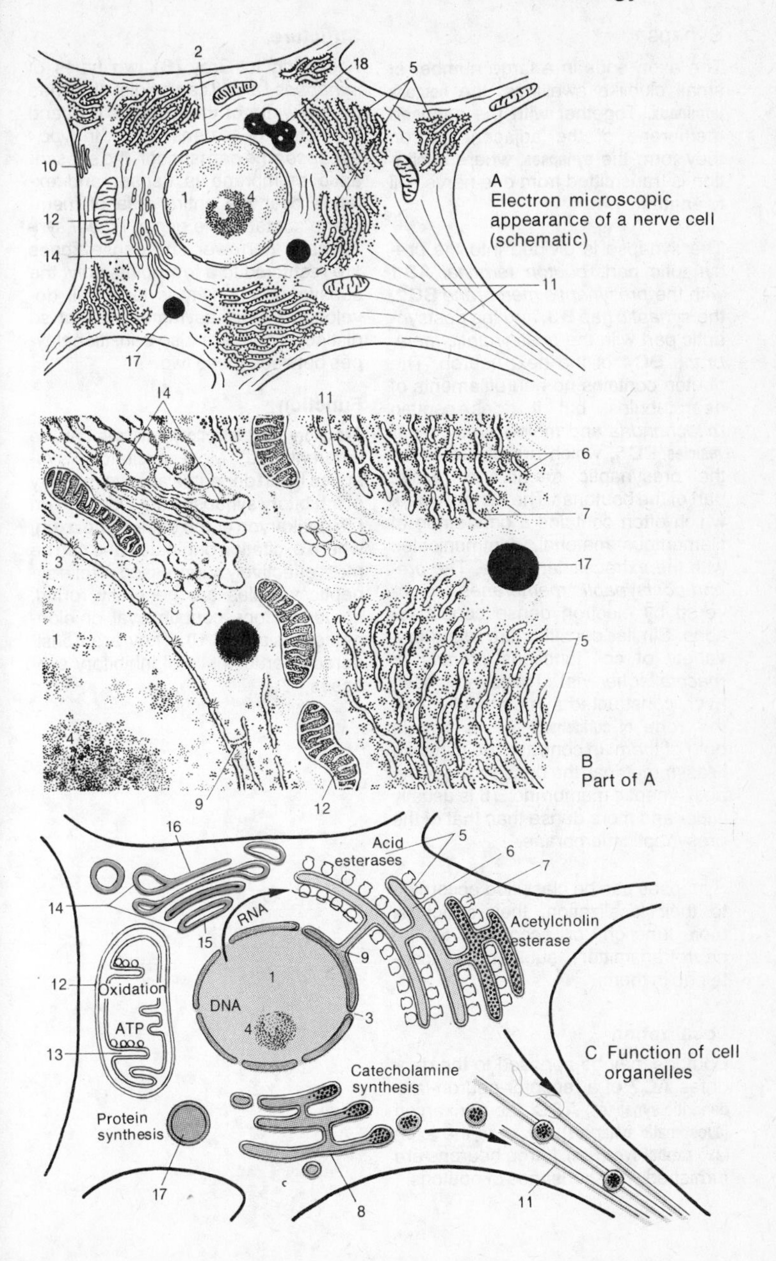

A
Electron microscopic appearance of a nerve cell (schematic)
B
Part of A
C Function of cell organelles
Acid esterases
RNA
DNA
Acetylcholin esterase
Oxidation
ATP
Catecholamine synthesis
Protein synthesis

Synapse

The axon ends in a large number of small clublike swellings, the **boutons terminaux.** Together with the apposed membrane of the adjacent neuron they form the **synapses**, where excitation is transmitted from one nerve cell to another.

The synapse is divided into the presynaptic part, *bouton terminal* **AB1** with the *presynaptic membrane* **BC2,** the *synaptic gap* **B3,** and the postsynaptic part with the *postsynaptic membrane* **BC4** of the next neuron. The bouton contains no neurofilaments of neurotubules, but it does contain *mitochondria* and mainly small, clear **vesicles BC5,** which are clustered on the presynaptic membrane (active part of the boutons). The *synaptic gap,* which often contains a dark stripe of filamentous material, communicates with the extracellular space. The *pre- and postsynaptic membranes* are covered by electron dense condensations. Similar densities are found in a variety of cell junctions *(zonula or macula adherens).* These are, however, constructed symmetrically, i. e. the zone **of condensation** is similar in both of them: In contrast, the synapse is asymmetrical, the dense zone of the postsynaptic membrane **B6** is usually wider and more dense than that of the presynaptic membrane.

Synapses can be classified according to their localization, their structure, their function, or according to the neurotransmitter substances contained in them.

Localization

Boutons may be apposed to the dendrites **AC7** of a receptor neuron **(axodendritic synapses) A8C,** the perikaryon **(axosomatic synapses) A9,** or to the axon **(axo-axonal synapses).** Large neurons are furnished with thousands of boutons.

Structure

According to *Gray* **(B)** two types of *synapses I* and *II* can be distinguished by the width of the synaptic gap and the nature of its dense zone. In *type I synapses* the gap is wider, the subsynaptic membrane is denser and extends over the entire area of membrane contact. The gap in *type II synapses* is narrower and dense zones are restricted to a few points only; the subsynaptic density is less well developed so the asymmetry is not so distinct. There are also transitional types between these two.

Function

Excitatory and **inhibitory synapses** can be differentiated. Most excitatory synapses lie on dendrites and the majority of inhibitory synapses are localized on the perikaryon or the base of the axon, where excitation originates and can be most effectively suppressed. While synaptic vesicles are generally round, some boutons contain oval or elongated vesicles **C10.** They are considered characteristic of inhibitory synapses.

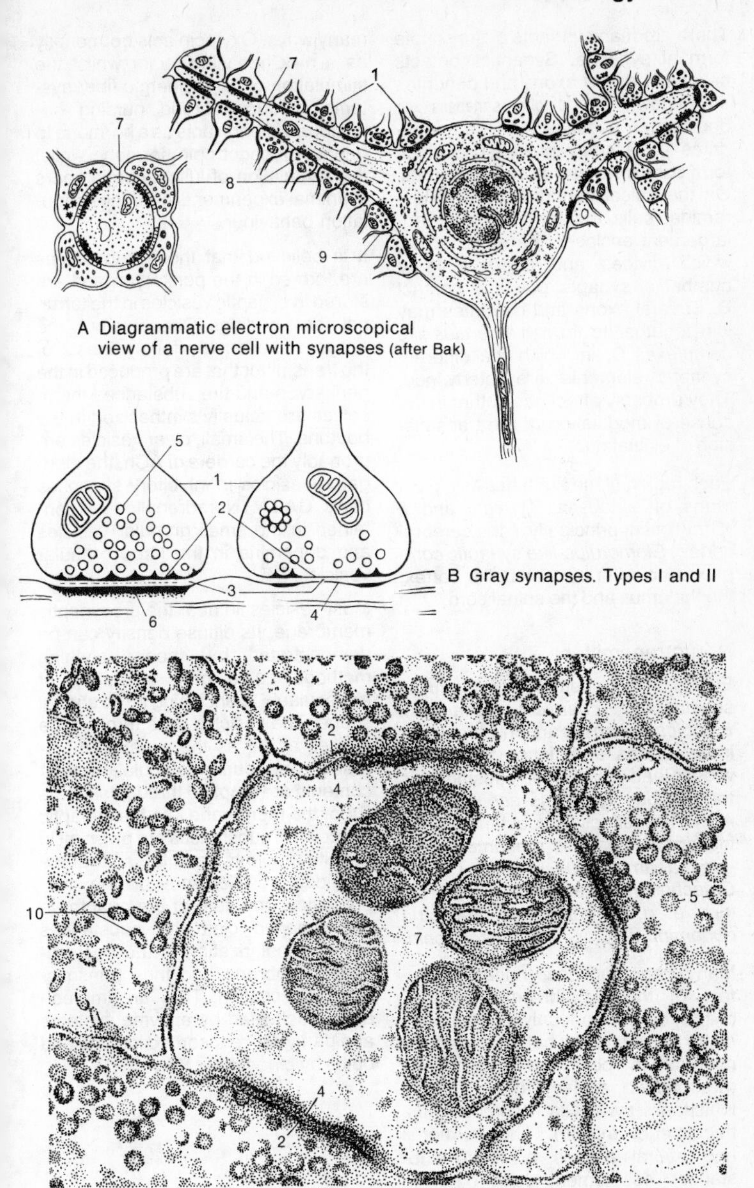

A Diagrammatic electron microscopical view of a nerve cell with synapses (after Bak)

B Gray synapses. Types I and II

C Transverse section of a dendrite surrounded by synapses (after Uchizono)

There are many variants of the simple form of synapse. Synaptic contacts between parallel axons and dendrites are known as **parallel synapses** or boutons en passage **A1.** Many dendrites have spiny processes, which form **spinous synapses A2** with boutons. On the apical dendrites of large pyramidal cells the terminal axon enlargement encloses the entire spine, which divides and bears a large number of synaptic points of contact **B.** Several axons and dendrites may join together to form *glomerulus-like complexes* **C,** in which the different synaptic elements are intertwined. They probably affect each other in the sense of modulation of the transmission of excitation.

Each region of the brain has its typical forms of synapses. *Types I* and *II (Gray)* occur principally in the cerebral cortex. *Glomerulus-like synaptic complexes* occur in the cerebellar cortex, the thalamus and the spinal cord.

Neurotransmitter

Excitation is transmitted by chemical substances *(chemical synapses). Electrical synapses* are only present in invertebrates and fishes. The most widespread transmitter substance in the nervous system is *acetylcholine,* ACh. The effective substance in inhibitory synapses is presumed to be *gamma amino butyric acid,* GABA. *Catecholamines* also act as transmitters, e. g. *noradrenaline* (NA) and *dopamine* (DA), as well as *serotonin.*

Many neuropeptides (releasing factors of the hypothalamus Vol. 2, p. 153) do not work only as hormones in the blood stream, but also as transmitters in excitatory synapses (neurotensin, cholecystokinin) and in inhibitory synapses (somatostatin, thyroliberin, motilin). The hormonal and neural actions of these substances supplement each other in many ways. Oxytocin acts hormonally as a milk releasing factor whilst the stimulation of oxytocinergic fiber systems produces brood nursing behaviour. Luliberin acts as a hormone to release gonadotropic hormone whilst the stimulation of luliberinergic fibers in animal experiments produces copulation behaviour.

It is believed that these substances are formed in the perikaryon and are stored in synaptic vesicles in the terminal nerve endings. Often it is only the enzymes necessary for synthesis of the transmitter that are produced in the perikaryon and the substances themselves are actually synthesized in the boutons. The small, clear vesicles are probably the carriers of ACh, the elongated vesicles in inhibitory synapses carry GABA. Noradrenaline is contained in the small granular vesicles and dopamine in the large granular vesicles, **DE.**

Most vesicles lie near the presynaptic membrane. Its diffuse density can be demonstrated by special staining methods as a grid consisting of trabeculae **F3,** enclosing a hexagonal space. The vesicles pass through the spaces as far as the synaptic membrane, and during stimulation their contents are emptied through *stomata* **F4** of the membrane into the synaptic space. The substances are given off in definite amounts (quanta), so that the individual vesicles may be the morphological correlate of the quanta. In the synaptic cleft the substances cause depolarization of the postsynaptic membrane and thus transmission of excitation. They are immediately inactivated by enzyme systems and partly are reabsorbed into the end bulbs by *pinocytosis* **F5**.

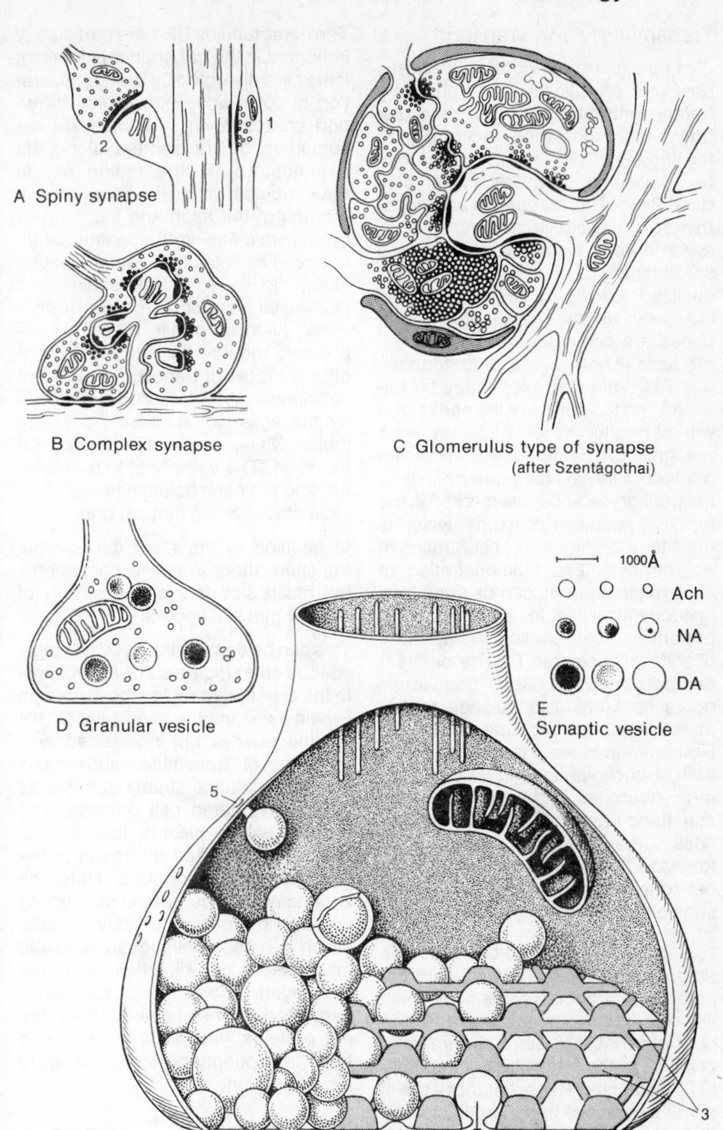

A Spiny synapse

B Complex synapse

C Glomerulus type of synapse (after Szentágothai)

D Granular vesicle

E Synaptic vesicle

F Model of a synapse (after Akert, Pfenninger, Sandri and Moor)

Transmitter, Axon Transport

Just as a gland cell normally secretes only one specific material, so every neuron only produces one particular transmitter substance. According to the substance produced, neurons may be classified into **cholinergic, catecholaminergic** (*noradrenergic* and *dopaminergic),* **serotoninergic** and peptidergic neurons. Catecholaminergic and serotoninergic neurons can be visualized directly by fluorescence microscopy as the transmitter substances are converted into fluorescent products when exposed to formalin gas **AB.** Thus the axons may be followed and the outline of the perikaryon with its nonfluorescent nucleus can be recognized. Fluorescence is least marked in the axons, more definite in the perikaryon and most marked in the terminal swellings of axons, which is the site of the highest concentration of neurotransmitters. Demonstration of cholinergic neurons can be done by a histochemical test for an enzyme important in acetylcholine metabolism, acetylcholinesterase **C.** The different neuropeptides of the peptidergic nerve cells are distinguished by immunohistochemical reactions D. Neuropeptides were demonstrated in both catecholaminergic and cholinergic neurones and some neurones contained two different neuropeptides. It may, therefore, be necessary to revise the concept that each nerve cell only produces one transmitter substance.

Neurotransmitters are produced in the *smooth endoplasmic reticulum* of the perikaryon, from the cisterns of which vesicles become detached. It is also possible that the *Golgi apparatus* is involved in the production of transmitter substances. The substances travel either by inactive transport, in storage form as molecular particles, or as vesicles in the axoplasm, as far as the boutons terminaux.

The neurotubules **DE1** are particularly important in the transport mechanism. If they are disrupted by the administration of colchicine, intra-axonal transport breaks down. Materials are assumed to be transported along the neurotubules. In the region of the neurotubules, viscosity is less than in the rest of the axon and this permits cytoplasmic flow in the centrifugal direction. The velocity of flow is greatest along the walls of the tubules and decreases as one moves further away, (velocity profile **E2**). This hypothesis may be used to explain the different rates of transport of different substances. ATP, which is hydrolysed by the action of ATPase **E3** in the tubule wall provides the energy for transport. The velocity of intra-axonal transport of catecholamines has been estimated at 5 to 6 mm per hour.

In addition to the rapid intra-axonal transport, there is also a continuous, but much slower axoplasmic flow of about 1 mm in 24 hours.

This can be demonstrated by placing a ligature on a single neuron **F.** Proximal to the constricted region the axoplasm is held back and the axon swells. This plasma flow is not concerned with transport of transmitter substances but with continual supply of nutrients for the elongated cell process. The velocity of the plasma flow corresponds to the speed of growth of the axon during development. Thus, the axon is not a type of rigid conducting wire, but more like a plasma column, which is being continuously renewed in a centrifugal direction from the perikaryon. There is also centripetal intra-axonal transport, and in this manner proteins, viruses and toxins can travel from peripheral nerve endings to the perikaryon.

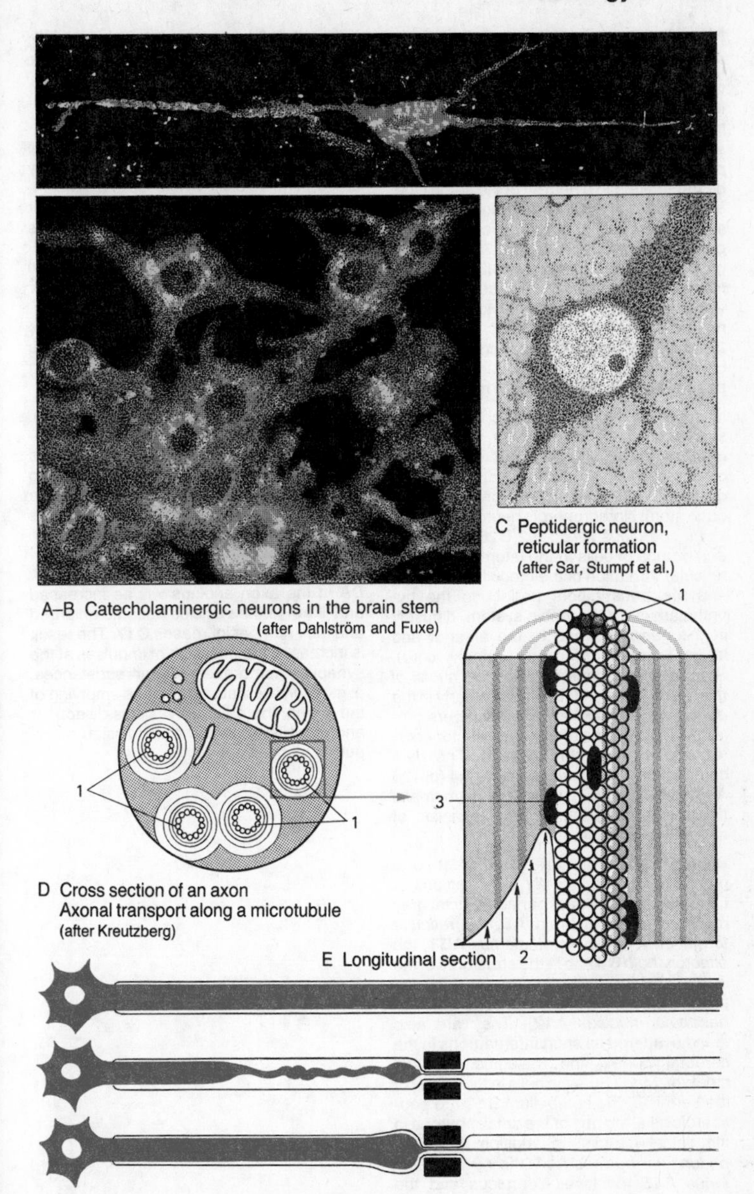

A–B Catecholaminergic neurons in the brain stem (after Dahlström and Fuxe)

C Peptidergic neuron, reticular formation (after Sar, Stumpf et al.)

D Cross section of an axon
Axonal transport along a microtubule (after Kreutzberg)

E Longitudinal section

F Blockage of the axoplasm after axon ligation (after Weiss and Hiscoe)

Neuron Systems

Groups of neurons that possess the same transmitter substances and whose axons form discrete fiber bundles are named after their neurotransmitter as *cholinergic, noradrenergic, dopaminergic, serotoninergic* or *peptidergic systems.* The nerve impulse can be transmitted to a neuron of the same type, as well as to one with a different transmitter substance. In the neuronal chain of the *parasympathetic system* (p. 270), *cholinergic* neurons conduct impulses from the central nervous system to peripheral ganglia, where they are again changed to cholinergic neurons. In the *sympathetic system* (p. 270), neurons in the spinal cord are also cholinergic, but in peripheral ganglia there is transmission to *noradrenergic* nerve cells.

It is not yet fully known which transmitter substances are involved in the central neurons, particularly in the nerve cells of the cerebral cortex. Noradrenergic, dopaminergic and serotoninergic neurons lie in the brain stem. Noradrenergic neurons form the *locus coeruleus* **A1** (p. 92 B28, 124 D18) and groups of cells in the lateral part of the reticular formation of the medulla oblongata and the pons (fibers project to the hypothalamus, to the limbic system, diffusely in the neocortex and to the anterior and lateral horns of the spinal cord). Serotoninergic neurons lie in the *nuclei of the raphe* **A2** (p. 100 B28) particularly in the *dorsal nucleus of the raphe* **A3** (fibers project to the hypothalamus, the olfactory cortex and to the limbic system). The pars compacta of the *substantia nigra* **A4** (p. 126 A17, 128 **AB1**) from which the *nigrostriatal fibers* extend to the striatum consists of dopaminergic neurons.

Peptidergic neurons are found most commonly in phylogenetically older regions of the brain: in the central periaqueductal grey matter of the mid brain **A5**, the *reticular formation* **A6**, the *hypothalamus* **AB7**, the *olfactory bulb* **B8** and in the structures of the limbic system (*cingulate gyrus* **B9,** *hippocampus* **B10**, *amygdaloid body* **B11** and *habenular nucleus* **A12**). They are also found scattered as short interneurons in the cerebral cortex, the thalamus and the striatum. Many Purkinje cells in the cortex of the vermis of the cerebellum **B13** are peptidergic as are many of the small neurons in the spinal ganglia. In addition the *interpeduncular nucleus* **A14**, the *nucleus solitarius* **A15** the locus coeruleus and the nuclei of the raphe are rich in peptidergic nerve cells. The most important peptidergic fiber systems are the *fornix,* the *terminal stria* and the medial forebrain bundle, the *medial telencephalic fascicle.*

The structure, breakdown and storage of neurotransmitters may be influenced by drugs; in nerve cells an excess or a deficiency may be produced and this leads to motor or psychiatric changes. Certain compounds *(neuroleptics)* produce sedation or tranquillization, whilst others *(stimulating amines)* produce an alert consciousness. Still other substances, e. g. LSD, cause hallucinations. With some of these substances the changes produced in the neurons may be demonstrated by fluorescence or electron microscopy.

Neurotropic drugs attack the formation, transport or storage of the transmitter substances. For example, in the neurons of the *substantia nigra,* the dopamine-containing vesicles **C16** move through the axons to the boutons, where they lie in precise amounts as stored dopamine. If through the administration of a certain drug the enzyme responsible for the breakdown of DA is inhibited, DA in the axon endings will be increased and likewise the vesicles become enlarged and their number increases **C17.** The result is increased transmission of impulses at the synaptic membranes. Other substances, e. g. *reserpine,* cause complete emptying of the stores, the granular vesicles disappear and there is no longer transmission of impulses at the synapse.

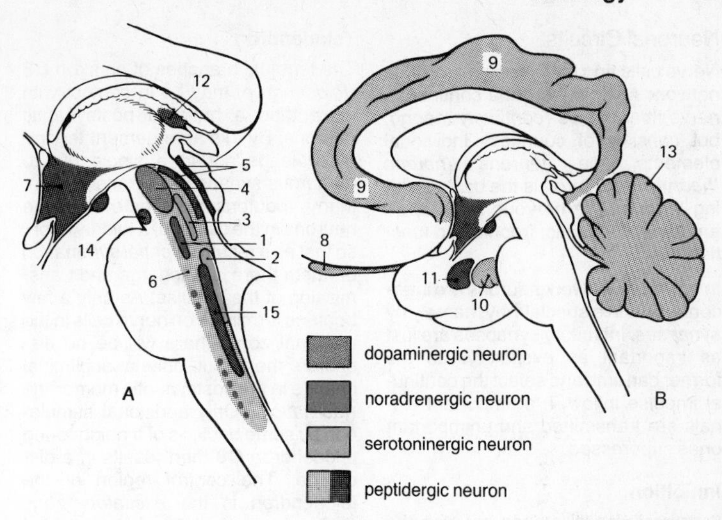

AB Monoaminergic and peptidergic cell groups in the brain

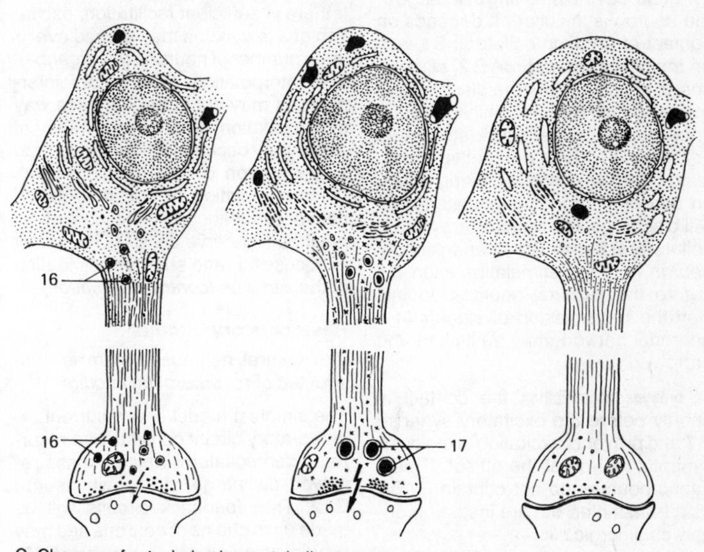

C Changes of catecholamine metabolism (after Hassler und Bak)

normal metabolism

inhibition of catecholamine breakdown

emptying of catecholamine store

Neuronal Circuits

Nerve cells and their processes form a network **A,** which is not a continuous nerve fiber plexus *(continuity theory),* but consists of countless individual elements, the neurons *(neuron theory).* The neuron is the basic building block of the nervous system, an anatomical, genetic, trophic and functional unit.

In the nerve network neurons are interconnected in a special way, namely by *synapses.* Inhibitory synapses are just as important as excitatory; as the former can limit and select the continual impulse inflow, i. e. important signals are transmitted and unimportant ones suppressed.

Inhibition

Postsynaptic inhibition does not stop the transmission of an impulse across a synapse, but the resulting discharge of the neuron is inhibited. It depends on contact of an axon collateral **B1** with an inhibitory *interneuron* **B2**, stimulation of which inhibits the discharge of the next cell **B3.** An inhibitory influence may be exerted on the stimulated cell itself **B4**. Such a negative recoupling (feedback) has a braking effect, so that only strongly stimulated cells will discharge and transmit a signal. A collateral **B6** may run to an inhibitory neuron from the stimulating axon **B5** before the terminal boutons. In this way the transmission of signals in a neuronal network may be limited and narrowed down.

In **presynaptic inhibition** the contact is shortly before the excitatory synapse **B7** and prevents excitation of the postsynaptic neuron at the outset. Presynaptic boutons do not contain GABA as a transmitter, but are instead probably cholinergic.

Telodendron

The terminal branches of an axon **C8** *(telodendron)* may be in contact with more than a hundred postsynaptic neurons. By this arrangement the impulse leads not to a single but to numerous synapses for transmission. Many boutons terminate on the neurons in the center of a telodendron, so that as a result of *spatial summation* on them there is discharge and transmission of the impulse. As only a few boutons terminate on nerve cells in the marginal zone, there will be no discharge there but only a subliminal change in the postsynaptic membrane *(facilitation).* Only additional stimulation by some boutons of a neighboring telodendron **C9** then results in a discharge. The central region of the telodendron is the *excitatory zone* **C10,** the marginal area is the zone of *facilitation* **C11.**

If there is sufficient facilitation, excitation of a few axons may spread over a large number of neurons *(divergence).* The interpolation of inhibitory interneurons may act in the opposite way and excitation of a large number of axons may become concentrated onto a few neuron chains *(convergence).* Often stimulation of one neuron will cause simultaneous inhibition of all neighboring nerve cells **D.** The result is 'focussing' and sharper delineation of the stimulus *(contrast formation).*

Reverberatory Circuits

The central nervous system is constructed of *reverberating circuits.*

The simplest model of a recurrent reverberatory circuit consists of a recurrent axon collateral which causes renewed discharge of its own neuron **E12.** True feedback circuits will be made up to chains of neurons and may be very complex in their structure **E13.**

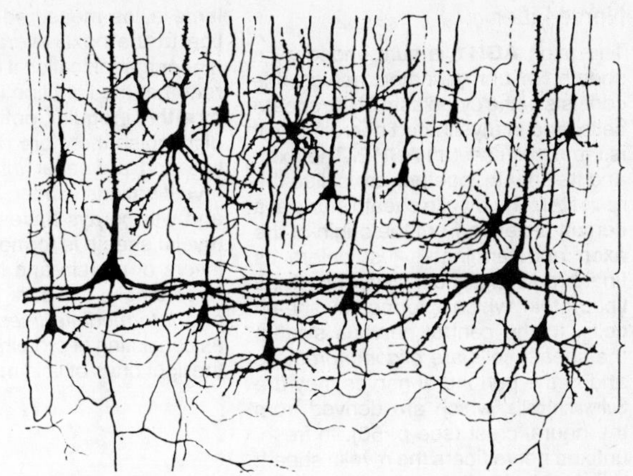

A Neuronal network in the cerebral cortex: silver impregnation (after Cajal)

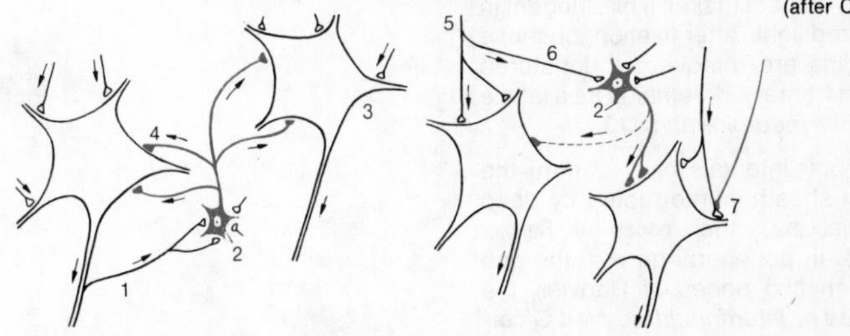

B Synaptic inhibition of nerve cells by interneurons (after Eccles)

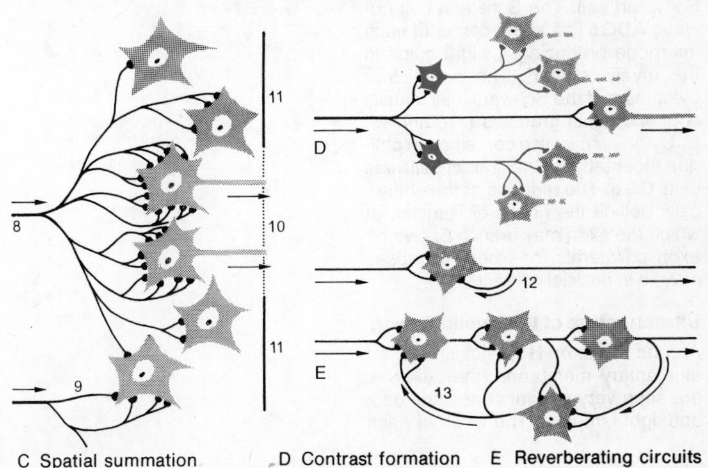

C Spatial summation D Contrast formation E Reverberating circuits

Nerve Fiber

The axon **AGH 1** is surrounded by a sheath. In unmyelinated nerves this consists of the cytoplasm of the sheath cells, and in myelinated nerve fibers it is the *myelin sheath* **ABH 2.** The axon and its sheath together are called the **nerve fiber.** The myelin sheath starts at a certain distance from the origin of the axon and terminates just before its final branching. It consists of **myelin**, a lipoprotein, which is formed by sheath cells. In the central nervous system the sheath cells are *oligodendrocytes* and in the peripheral nerves they are **Schwann cells,** which are derived from the neural crest (see p. 56). In fresh, unfixed nerve fibers the myelin sheath is highly refractile and structureless. Its lipid content makes it birefringent in polarized light. After fixation, because the lipids are removed, a denatured protein framework remains as a lattice structure *(neurokeratin)* **D 3.**

At regular intervals of (1–3 mm) the myelin sheath is interrupted by deep constrictions, the **nodes of Ranvier ABG 4.** In peripheral nerves, the gap between two nodes of Ranvier, the **internodal** or *interannular* **segment G** corresponds to the extent of one Schwann cell. The Schwann cell nucleus **ADG 5** lies at the center of each internode producing a slight bulge in the myelin sheath. The perinuclear cytoplasm of the Schwann cell usually contains small granules (π-*granules)* **E.** Cytoplasm is also contained in oblique indentations, the **Schmidt-Lanterman clefts CG 6.** The margins of the sheath cells delimit the nodes of Ranvier, at which the axon may branch **F,** give off axon collaterals, or where synapses may lie in parallel contact.

Ultrastructure of the Myelin Sheath

Around the *axon* **H 1,** enclosed by an elementary membrane, the **axolemma,** are seen very regular concentric dark and light lamellae. The width of each lamella, as measured from one dark line to the next, averages 120 Å. At higher magnification it becomes apparent that the light line is again divided by a thin irregular, dotted line **H 7.** We distinguish therefore a dense primary line and a weaker *intermediate line.* Investigations under polarized light and with x-rays have shown that the myelin sheath is formed of alternating layers of protein and lipid molecules. Accordingly, the dark layers *(primary* and *intermediate lines)* are regarded as consisting of protein molecules and the light ones of lipid molecules.

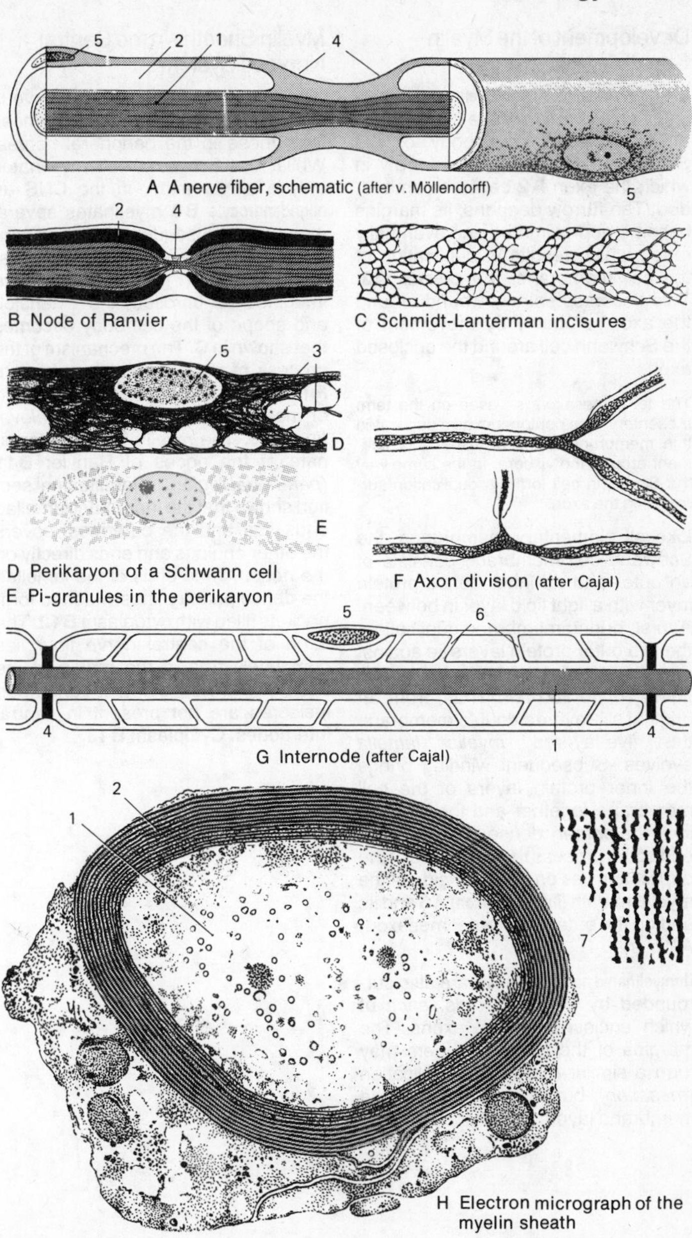

A A nerve fiber, schematic (after v. Möllendorff)

B Node of Ranvier

C Schmidt-Lanterman incisures

D Perikaryon of a Schwann cell

E Pi-granules in the perikaryon

F Axon division (after Cajal)

G Internode (after Cajal)

H Electron micrograph of the myelin sheath

Development of the Myelin Sheath

The **development of the myelin sheath** provides an indication of the construction of its lamellae. The body of the *Schwann cell* **A1** forms a furrow in which the axon **A2** becomes embedded. The furrow deepens, its margins become approximated and eventually meet, which results in duplication of the cell membrane, the **mesaxon A3.** This becomes spirally wound around the axon, probably by movement of the Schwann cell around the enclosed axon.

The term *mesaxon* is based on the term mesentery. The peritoneum as a duplicated thin membrane forms a suspensory ligament around the viscera. In the same way the Schwann cell forms a duplication surrounding the axon.

Like all elementary membranes, the Schwann cell membrane consists of an outer and an inner dense protein layer with a light lipid layer in between. At first, during membrane duplication, the two outer protein layers lie against one another and then fuse to form an *intermediate line* **A4.** Thus, from an original six-layered double membrane the five-layered *myelin lamella* evolves. Subsequent winding brings the inner protein layers of the cell membrane together and they fuse to form the *primary* dense *line* **A5.** At the end of the process the beginning of the duplication lies on the inner side of the myelin sheath **(inner mesaxon) A6** and its end on the outer side (outer mesaxon) **A7.**

Unmyelinated nerve fibers A8 are also surrounded by Schwann cells, each of which encloses several axons. The margins of the furrows in them may form a similar membrane duplication *(mesaxon),* but without fusion of the membrane layers.

Myelin Sheaths in the Central Nervous System

Myelin sheaths in the central nervous system **B,** show essential differences from those in the peripheral nerves. Whilst the Schwann cell myelinates only a single axon, in the CNS an **oligodendrocyte B9** myelinates several axons and subsequently will also connect several internodes by protoplasmic bridges. If the internodes are imagined as unrolled, the extension and shape of the cell body becomes that shown in **C.** The mechanism of the process of myelination is not known. From the protoplasmic bridge the outer mesaxon forms an *external elevation* **B10.** The myelin lamellae terminate at the nodes of Ranvier **B11** *(paranodal region).* A longitudinal section shows that the innermost lamellae ends first and the outermost covers the other endings and ends directly on the node. At the ends of the lamellae the dense primary lines widen to form pockets filled with cytoplasm **B12.** The axon of the central nerve fiber lies completely free in the region of the node of Ranvier. Schmidt-Lanterman incisures are not present in central internodes. Cytoplasm **B13.**

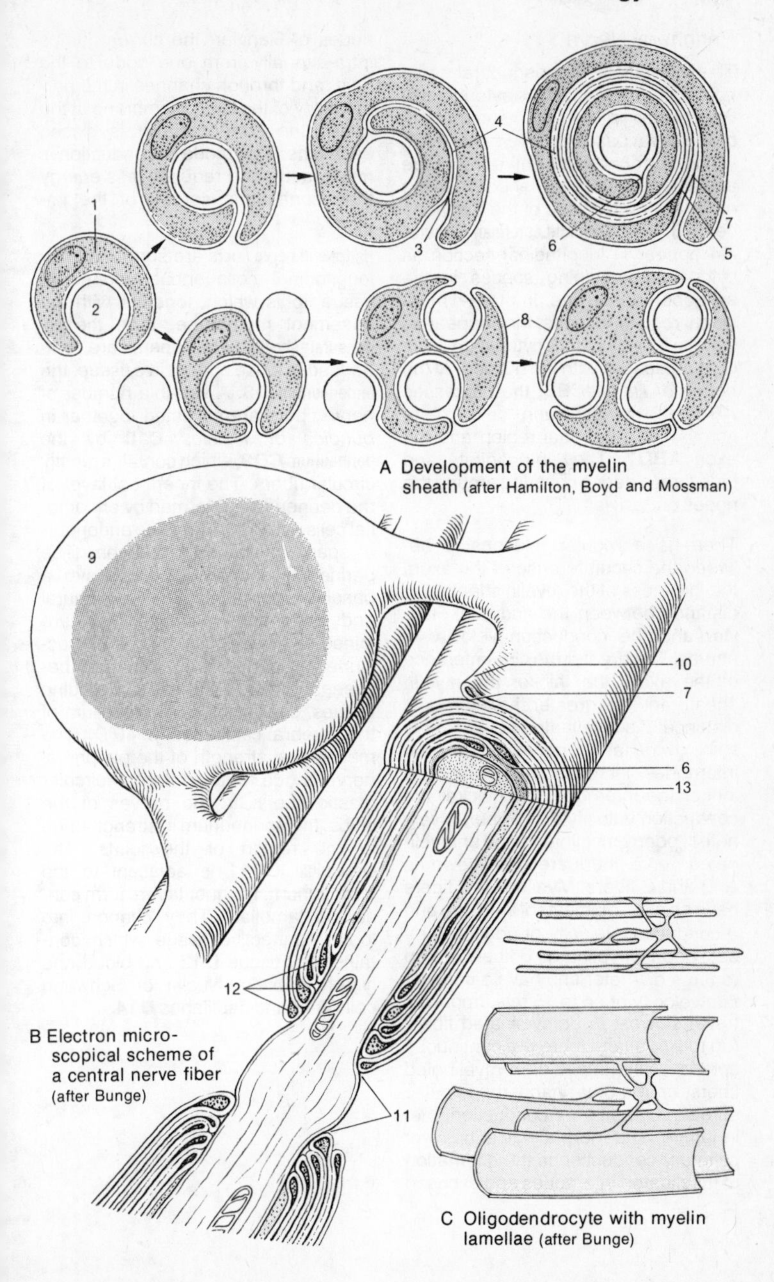

A Development of the myelin sheath (after Hamilton, Boyd and Mossman)

B Electron microscopical scheme of a central nerve fiber (after Bunge)

C Oligodendrocyte with myelin lamellae (after Bunge)

Peripheral Nerve

The **myelin sheath** of a peripheral nerve fiber is surrounded by the cytoplasm of the **Schwann cell A1.** On its cell membrane lies a *basement membrane* **A2,** which surrounds the entire internode and acts as a strict barrier between the nerve fibers. Nucleus of the Schwann cell **A3.** The **Schmidt-Lanterman incisures A4** appear in longitudinal section as cytoplasm-containing spaces in the split-open dense line. In three-dimensional reconstructions, they appear as spirals in which the cytoplasm is in communication with the outside. At the *nodes of Ranvier* **B5,** the processes **AB6** of the Schwann cell extend across the paranodal region and the axon **ABD7.** They interdigitate and thus form a dense covering around the nodes of Ranvier.

There is a regular relationship between the circumference of the axon, the thickness of the myelin sheath, the distance between the nodes of Ranvier and the conduction velocity in nerves. The greater the circumference of the axon, the thicker the myelin sheath and the greater the internodal distance. If a myelinated nerve fiber is still growing, e. g. a nerve in a limb, the internodes increase in length. The longer the internode, the quicker the conduction velocity in the fiber. Myelinated, poorly myelinated and unmyelinated nerve fibers are also known as A, B and C fibers. Myelinated A fibers have an axon diameter of 3–20 μm and a conduction velocity of up to 120 m/sec, poorly myelinated B fibers are up to 3 μm diameter and have a conduction velocity of up to 15 m/s. Impulses travel slowest in unmyelinated fibers (up to 2 m/s) and there is a continuous spread of the impulse. In myelinated fibers, on the other hand, the impulse spreads by saltatory conduction, i. e. in jumps. The morphological basis of saltatory conduction is the alternation of myelinated internodes and exposed nodes of Ranvier; the current jumps intra-axonally from one node to the next, and through changes in the permeability of the axon membrane at the nodes the current circuit is closed each time. This mode of conduction is much faster and requires less energy than continuous spread of the impulse.

Peripheral nerve fibers are surrounded by longitudinal collagenous connective tissue fibrils which, together with the basement membrane, form the **endoneural sheath.** The fibers are embedded in loose connective tissue, the **endoneurium D8.** A variable number of nerve fibers are collected together in *bundles* or *fasicles* **C10** by the **perineurium CD9,** which contains mostly circular fibers. The innermost layer of the perineurium is formed by endothelial cells which surround the endoneural space in several thin layers. The perineural endothelial cells have a basal membrane on their perineural and endoneural surfaces and are joined to one another by zonulae occludentes. They form a barrier between the nerve and the surrounding tissues, similar to the endothelium of the cerebral capillaries (p. 40 D). The mechanical strength of the peripheral nerve is due to its content of circular elastic fibers. In the nerves of the limbs, the perineurium is strengthened in the region of the joints. The **epineurium CD11** is adjacent to the perineurium. Its inner layers form concentric lamellae. They change into loose connective tissue which contains fatty tissue **D12** and blood and lymph vessels. Nuclei of Schwann cells **D13** and capillaries **D14.**

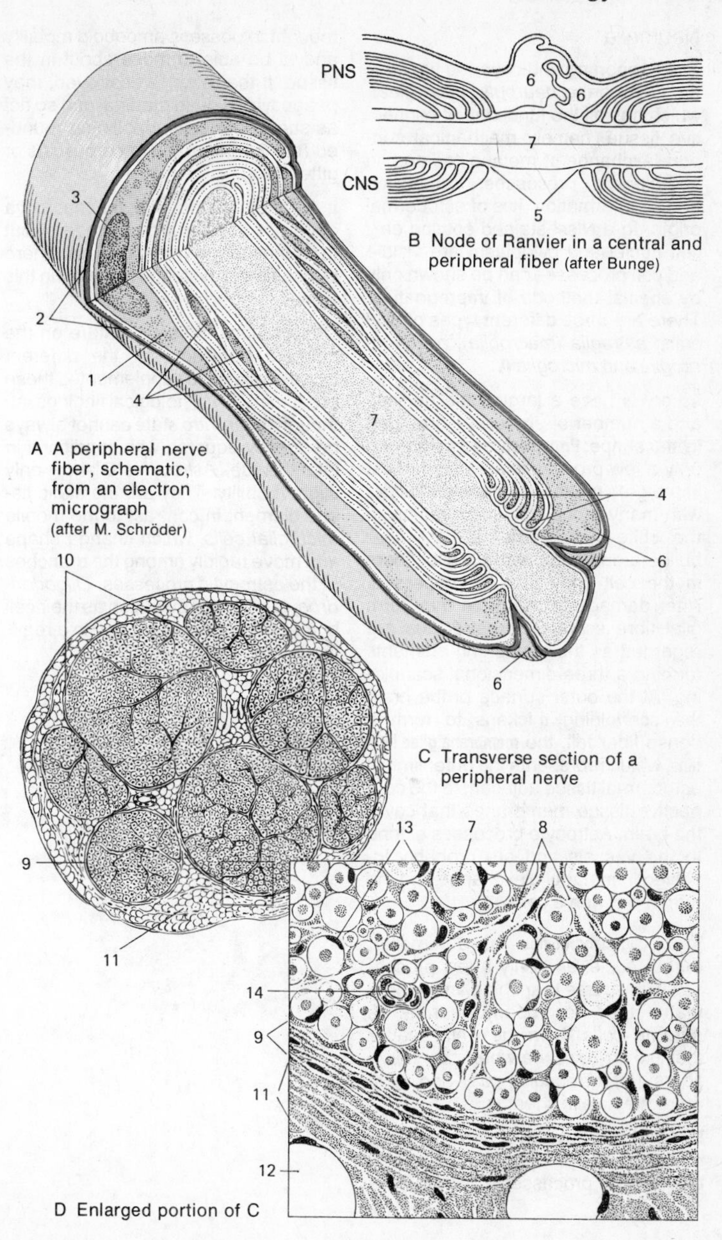

A A peripheral nerve fiber, schematic, from an electron micrograph (after M. Schröder)

B Node of Ranvier in a central and peripheral fiber (after Bunge)

C Transverse section of a peripheral nerve

D Enlarged portion of C

Neuroglia

The supporting and covering tissue of the CNS is the *neuroglia* (glia: glue), which has all the functions of connective tissue, namely mechanical support, exchange of metabolites and, in pathological processes, catabolism and scar formation. It is of ectodermal origin. In a Nissl-stained section only cell nuclei and cytoplasm are visible and cell processes can be shown only by special methods of impregnation. There are three different types of glial cells: *astroglia (macroglia), oligodendroglia* and *microglia* **A.**

Astrocytes have a large, pale nucleus and a number of processes arranged in star shape. **Protoplasmic astrocytes** with only a few processes are commonest in the grey matter, and **fibrous astrocytes** with many processes occur mainly in the white matter. They are the fiber builders and comprise the glial fibers in the cell body and its processes. After damage to the brain they form glial fibre scars. The astrocytes are regarded as the supporting elements forming a three-dimensional scaffolding. At the outer surface of the brain the scaffolding thickens to form a dense fiber felt, the **membrana gliae limitans,** which represents the outer limit of ectodermal tissue adjacent to the connective tissue membranes that cover the brain. Astrocyte processes extend to the vessels and are probably involved in metabolic exchange and the nutrition of nerve cells.

The **oligodendrocytes** have smaller, darker nuclei and only a few processes with sparse branching. In the grey matter they accompany neurons as *satellite cells* **B.** In the white matter they lie in rows between the nerve fibers *(interfascicular glia)* and form there the myelin sheaths.

Microglial cells have oval or rod-shaped nuclei and short, thick, many-branched processes. They are thought to possess amoeboid mobility and to be able to move about in the tissue. If the tissue is destroyed, they phagocytose dead material and so act as scavenger cells, becoming rounded (compound granular corpuscles or gitter cells).

It is often stated that the microglia does not develop from ectoderm but from mesoderm (*mesoglia),* but there is insufficient evidence to maintain this view.

Observations in tissue culture on the different behaviour of the different types of glia have supplemented these predominantly histological findings, although the *in vitro* state cannot always be directly equated with conditions in intact tissue. *Astrocytes* **C** show only slight mobility. They are the static tissue element in contrast to the mobile *microglial cells,* which change shape and move rapidly among the branches of the astrocytic processes. *Oligodendrocytes* **D** show pulsation as their cell bodies contract and enlarge in a regular rhythm.

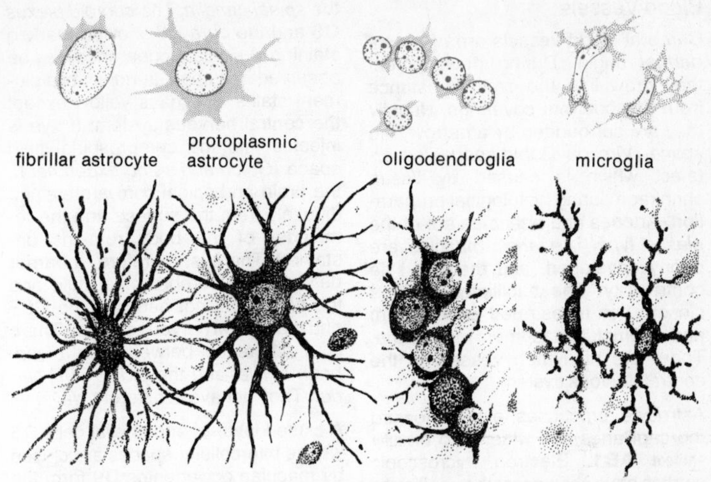

A Equivalent pictures of neuroglia; above Nissl staining, below silver impregnation

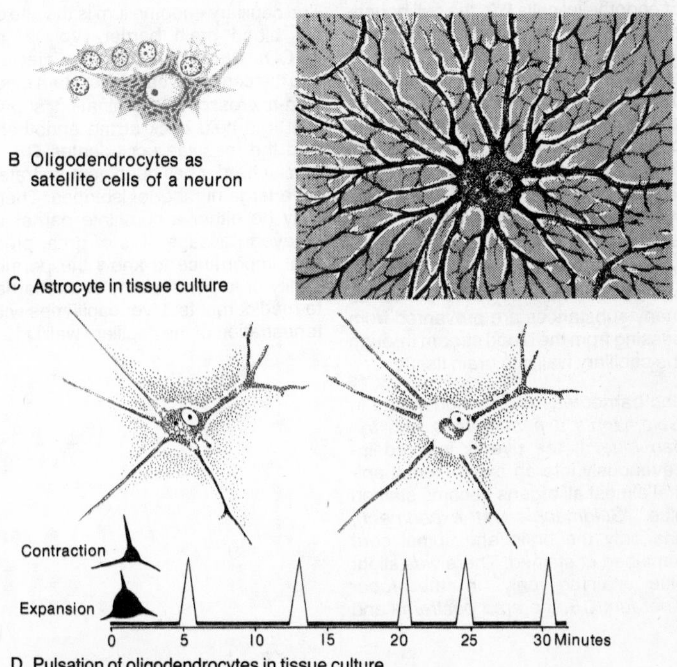

B Oligodendrocytes as satellite cells of a neuron

C Astrocyte in tissue culture

D Pulsation of oligodendrocytes in tissue culture (after Lumsden and Pomerat)

Blood Vessels

Cerebral blood vessels are of *mesodermal* origin. During development they grow into the brain substance from mesodermal coverings. Usually they are surrounded by a narrow free space *(Virchow-Robin space),* an artefact which is caused by tissue shrinkage during histological preparation. Arteries and arterioles are of the *elastic type,* i. e. their muscles are poorly developed, and there is little contractility. The capillaries have a closed, non-fenestrated endothelium and a complete basement membrane. There are no lymph vessels in the central nervous system.

Astrocyte processes extend toward the capillaries and widen into **vascular endfeet AB1.** Electron microscopic studies have shown that the capillaries are completely surrounded by vascular end feet. The capillary wall consists of endothelial cells **B2,** the cell boundaries of which overlap one another like roof tiles and are held together by *maculae occludentes.* Then follows the basement membrane **B3** and the astrocyte covering **B4,** which is comparable to the *membrana gliae limitans:* both of them separate the ectodermal tissue of the CNS from the adjacent mesodermal tissue.

Brain tissue is effectively separated from the rest of the body by the **blood-brain barrier.** It is a selective barrier as many substances are prevented from passing from the blood stream through the capillary wall into brain tissue.

The barrier was first demonstrated in *Goldmann's experiments* **C** with *trypan blue.* If the dye is injected intravenously into an experimental animal almost all organs become stained blue *(Goldmann's 1st experiment),* and only the brain and spinal cord remained unstained. There was slight blue staining only in the *tuber cinereum* **C5,** the *area postrema* and the *spinal ganglia.* The *choroid plexus* **C6** and the *dura* **C7** showed marked staining. A similar appearance may be observed in human jaundice: bile pigment stains all organs yellow except the central nervous system. If dye is injected into the cerebrospinal fluid space *(Goldmann's 2nd experiment),* the brain and spinal cord are stained diffusely from the surface and the remainder of the body remains unstained. There is, therefore, a barrier between the cerebrospinal fluid and the blood, but not between the CSF and the central nervous system. There is a difference between the *blood-brain barrier* and the *blood-CSF barrier.* They behave in different ways.

A closed layer of endothelial cells **D8** whose intercellular spaces are closed by maculae occludentes **D9** form the wall of the cerebral capillaries **D10.** Basal membrane **D11.**

The capillary endothelium is the site of the blood brain barrier (Vol. 2, p. 40 CD). When peroxidase is injected into the cerebral vascular system electron-microscopic examination shows that it is held back at the endothelia and the maculae occludentes **D8.** A similar barrier has been demonstrated for a large number of isotopes. There may be either a complete barrier or delayed passage. It is of great practical importance to know the permeability or impermeability of the barrier to medicaments. Liver capillaries with fenestration of the capillary wall **D12.**

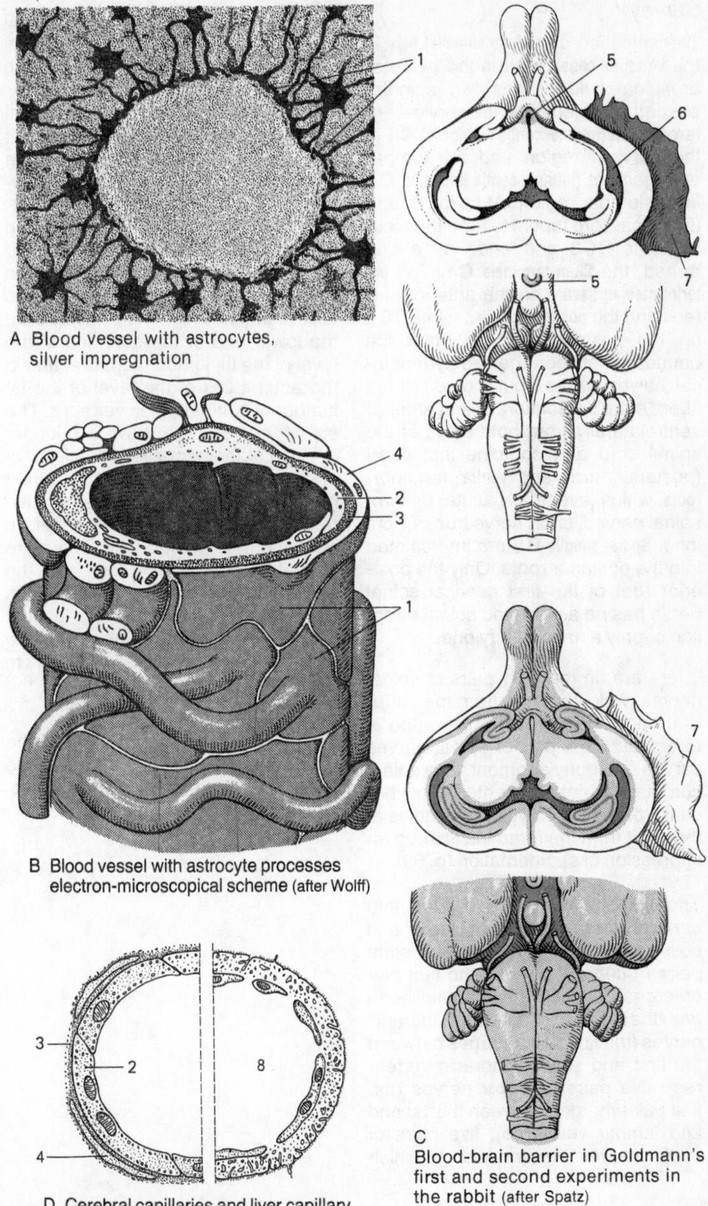

A Blood vessel with astrocytes, silver impregnation

B Blood vessel with astrocyte processes electron-microscopical scheme (after Wolff)

C Blood-brain barrier in Goldmann's first and second experiments in the rabbit (after Spatz)

D Cerebral capillaries and liver capillary, electron-microscopical scheme

Survey

The spinal cord (**medulla spinalis**) lies in the **vertebral canal** suorrunded by cerebrospinal fluid. It has two spindle-shaped enlargements, the cervical enlargement (**intumescentia cervicalis**) **C1** in the cervical region and the lumbar enlargement (**intumescentia lumbalis**) **C2** in the lumbar region. At its lower end the spinal cord tapers to form the **conus medullaris BC3** and ends as a fine thread, the **filum terminale C4.** The **anterior median fissure** on the anterior surface and the **posterior median sulcus BC5** on the posterior surface mark the boundary between the two symmetrical halves of the spinal cord. Nerve fibers enter dorsolaterally and emerge ventrolaterally from both sides of the spinal cord and combine into **dorsal** (posterior) **roots** and **ventral** (anterior) **roots** which eventually unite to form spinal nerves, short nerve trunks 1 cm long. **Spinal ganglia B6** are intercalated into the posterior roots. Only the posterior root of the first cervical spinal nerve has no associated spinal ganglion or only a rudimentary one.

There are, in man, 31 pairs of spinal nerves, which emerge from the vertebral canal through the intervertebral foramina. Each pair of spinal nerves serves one body segment. The spinal cord itself is not segmented. Only because the nerve fibers form bundles at their exit from the foramina is there an impression of segmentation (p. 60).

The spinal nerves are divided into *cervical, thoracic, lumbar, sacral* and *coccygeal nerves.* There are eight pairs of cervical nerves (the first pair emerges between the occipital bone and the atlas); 12 pairs of thoracic nerves (the first pair emerges between the first and second thoracic vertebrae); five pairs of **lumbar** nerves (the first pair emerges between the 1st and 2nd lumbar vertebrae); five pairs of sacral nerves the first pair of which emerges from the uppermost sacral foramina; and one pair of coccygeal nerves, which emerges between the first and second coccygeal vertebrae. Originally the spinal cord and vertebral canal were the same length, so that each spinal nerve emerged from the foramen lying at its own level. However, during development the vertebral column grows much more in length than the spinal cord, so that the lower end of the cord, in comparison with the surrounding vertebrae comes to lie higher and higher. In the newborn the lower end of the cord lies at the level of the III. lumbar vertebra, and in the adult it lies at the level of the Ist lumbar or XIIth thoracic vertebra. The spinal nerves can now no longer emerge at the same vertebral level of origin from the cord, but instead their roots run downward for a certain distance in the vertebral canal before reaching their point of exit. The more caudal the origin of a root from the spinal cord the longer is its course in the vertebral canal. Thus, the level at which a spinal nerve emerges from the vertebral column in the adult ceases to be at the corresponding level of the spinal cord.

From the conus medullaris on, the vertebral canal contains only a dense mass of descending spinal roots, known as the **cauda equina B7.**

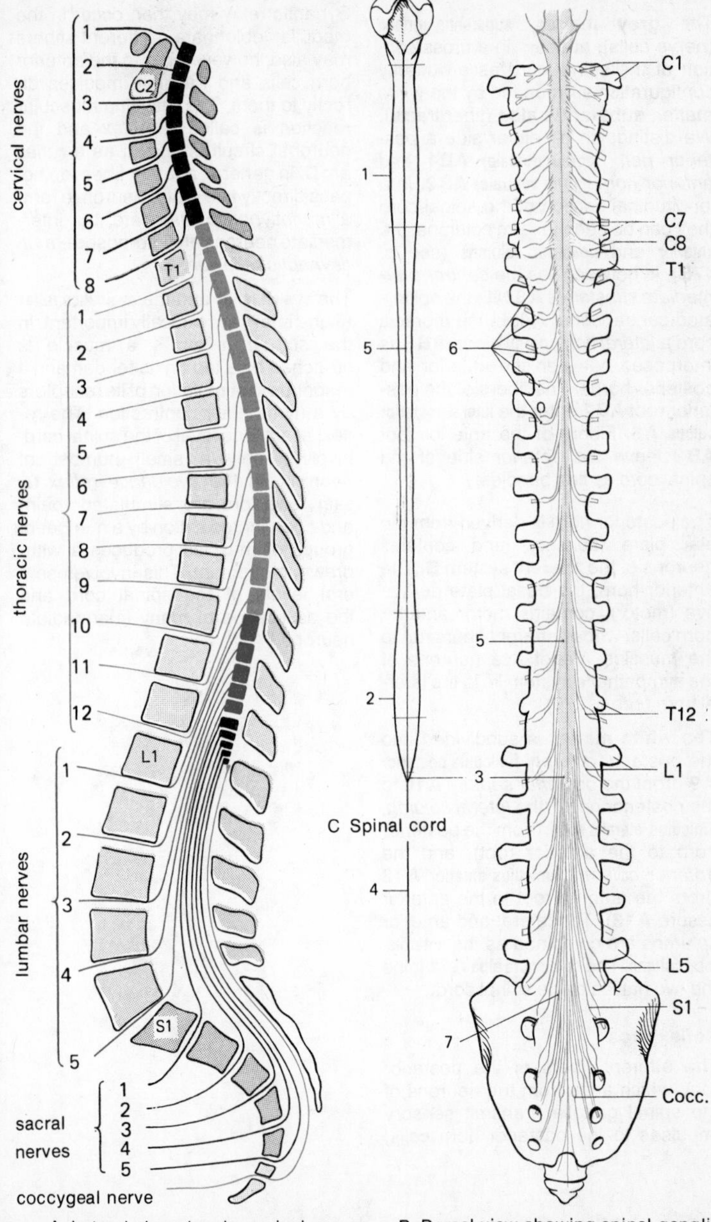

A Lateral view showing spinal nerves B Dorsal view showing spinal ganglia

Structure, Reflexes

The **grey matter, substantia grisea** (nerve cells), appears in a cross-section of the spinal cord as a butterfly configuration surrounded by the *white matter,* **substantia alba** (fiber tracts). We distinguish on either side a *posterior horn* **(cornu dorsale) AB1** and *anterior horn* **(cornu ventrale) AB2.** In a longitudinal section of the spinal cord they can be seen to form columns, the **anterior** and **posterior columns** (see p. 47**B**). In between lies the *central intermediate substance* **A3** with the obliterated central canal **A4.** In the thoracic cord a *lateral horn,* **cornu laterale AB5,** is interposed between the anterior and posterior horns. The fibers of the posterior root **AB7** enter the **lateral posterior sulcus A6.** Fibers of the anterior root **AB8** leave the anterior side of the spinal cord as fine bundles.

The posterior horn is derived from the alar plate (sensory) and contains neurons of the afferent system **B.** The anterior horn, the basal plate derivative (motor), contains motor anterior horn cells, whose efferent fibers run to the muscles. Vegetative neurons of the sympathetic system lie in the lateral horn (see p. 270).

The **white matter** is subdivided into the *posterior column,* **funiculus posterior A9** (from the *posterior septum* **A10** to the posterior horn), the *lateral column,* **funiculus lateralis A11** (from the posterior horn to the anterior root), and the *anterior column,* **funiculus anterior A12** (from the anterior root to the anterior fissure **A13**). The lateral and anterior columns are combined as the **anterolateral column.** The **comissura alba A14** joins the two halves of the spinal cord.

Reflex Arcs

The afferent fibers of the posterior root, which arise from the neurons of the spinal ganglia, transmit sensory impulses to the posterior horn cells, from which signals pass to the brain **C.** Synaptic relay may then occur in the medulla oblongata. Afferent fibers may also, however, run to the anterior horn cells and transmit impulses directly to them. The resulting muscular reaction is called a *reflex* and the neuronal circuit is known as a *reflex arc* **D.** In general, afferent fibers do not pass directly to a motor neuron to form a *monosynaptic reflex* arc, but intermediate neurons are interposed – *multisynaptic reflex arc* **E.**

The **stretch reflex** and the **avoidance reflex** (flight reflex) are clinically important. In the *stretch reflex* **F,** a muscle is stretched by a tap on its tendon and it responds to stimulation of its receptors by a momentary contraction. The reflex, at any one level of the spinal cord, involves only a small number of neurons. In the *avoidance reflex* **G,** skin receptors are stimulated (pain) and coordinated action by a number of groups of muscles produces a withdrawal movement. This involves several levels of the spinal cord and the assistance of many intermediate neurons.

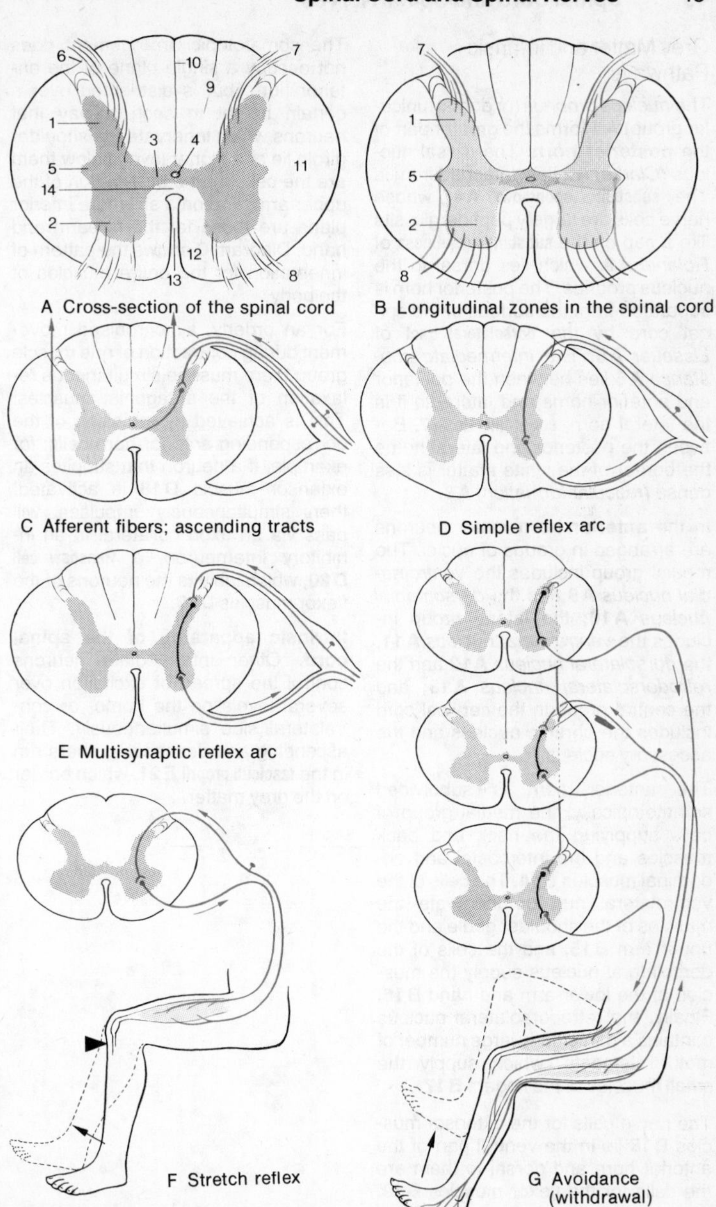

A Cross-section of the spinal cord

B Longitudinal zones in the spinal cord

C Afferent fibers; ascending tracts

D Simple reflex arc

E Multisynaptic reflex arc

F Stretch reflex

G Avoidance (withdrawal) reflex

Grey Matter and Intrinsic Pathways

The *nucleus proprius* (or dorsal funicular group) **A1** forms the greater part of the **posterior horn.** The dorsal nucleus *(Clarke)* **A2** lies detached from it. The **substantia spongiosa A4**, whose nerve cells are largely peptidergic, sits like a cap on the **substantia gelatinosa** of *Rolandi* **A3,** which lies dorsal to the nucleus proprius. The posterior horn is separated from the surface of the spinal cord by the **dorsolateral tract** of *Lissauer* **A5.** The *intermediate substance* **A6** lies between the posterior and anterior horns and lateral to it is the lateral horn, **cornu laterale A7.** Between the posterior and lateral horns the border of the white matter is less dense *(reticular formation)* **A8.**

In the **anterior horn** motor neurons are arranged in groups of nuclei. The medial group includes the *ventromedial nucleus* **A9** and the *dorsomedial nucleus* **A10;** the lateral group includes the *ventrolateral nucleus* **A11,** the *dorsolateral nucleus* **A12** and the *retrodorsolateral nucleus* **A13;** and the central group in the cervical cord includes the phrenic nucleus and the accessory nucleus.

The anterior horn is subdivided somatotopically, the medial group of cells supplying the neck and back muscles and the intercostal and abdominal muscles **B14.** The cells of the ventrolateral nucleus innervate the muscles of the shoulder girdle and the upper arm **B15,** and the cells of the dorsolateral nucleus supply the muscles of the lower arm and hand **B16.** Finally, the retrodorsolateral nucleus contains a particularly large number of motor elements, which supply the small muscles of the fingers **B17.**

The nerve cells for the extensor muscles **B18** lie in the ventral part of the anterior horn and dorsal to them are the cells for the flexor muscles **B19.**

The somatotopic arrangement does not occupy a single plane of the anterior horn, but is distributed over a certain height in such a way that neurons which innervate the shoulder girdle lie at a higher level; below them are the cells for the innervation of the upper arm, and on a still more inferior plane are those for the forearm and hand. Diagram **C** shows the pattern of innervation for the entire muscles of the body.

For an orderly, synchronised movement during contraction of one muscle group there must be simultaneous relaxation of the antagonist muscles. This is achieved by inhibition of the corresponding anterior horn cells: for example, if a neuron that supplies an extensor muscle **D18** is activated, then simultaneously impulses will pass via an axon collateral to an inhibitory interneuron, a **Renshaw cell D20,** which inhibits the neurons of the flexor muscles **D19.**

Intrinsic apparatus of the spinal cord. Other intermediate neurons control the spread of excitation over several levels on the homo- or contralateral side simultaneously. Their ascending and descending fibers run in the **fasciculi proprii E21,** which border on the grey matter.

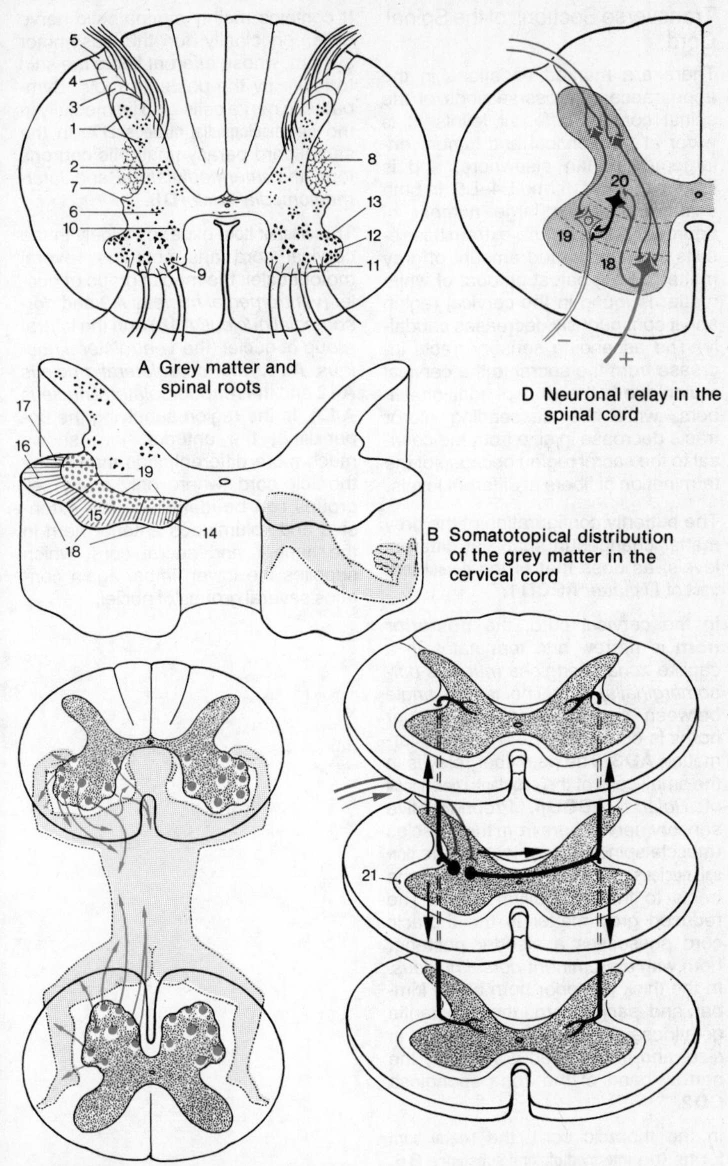

A Grey matter and spinal roots

B Somatotopical distribution of the grey matter in the cervical cord

C Summary of somatotopical organization of the grey matter (after Bossy)

D Neuronal relay in the spinal cord

E Fasciculi proprii

Transverse Sections of the Spinal Cord

There are marked variations in the appearance of cross-sections of the spinal cord at different levels. It is wider at the cervical and lumbar enlargements than elsewhere, and is largest at C4–C5 and L4–L5. In both enlargements the large number of neurons supplying the extremities results in an increased amount of grey matter. The greatest amount of white matter is found in the cervical region and it continuously decreases caudally. The ascending sensory tracts increase from the sacral to the cervical region due to the entry of additional fibers, while the descending motor tracts decrease in size from the cervical to the sacral region because of the termination of fibers at different levels.

The butterfly configuration of the grey matter changes in shape at different levels, as does that of the **dorsolateral tract** of *Lissauer* **ABCD1.**

In the cervical cord, the **posterior horn** is narrow and terminates in a caplike *zona spongiosa (nucleus dorsomarginalis)* **A2.** The lateral angle between the posterior and anterior horns is occupied by the reticular formation **AD3.** The pain fiber relay is in the small cells of the **substantia gelatinosa** of *Rolandi* **ABCD4.** Proprioceptive sensory nerve fibers from the muscles (muscle spindles) terminate in the **dorsal nucleus** *(Clarke)* **AB5** where the tracts to the cerebellum begin. The reduced grey matter in the thoracic cord possesses a slender posterior horn with a prominent *dorsal nucleus.* In the thick posterior horn of the lumbar and sacral cord, the substantia gelatinosa **CD4** is much increased in size, and is bordered dorsally by the narrow band of the zona spongiosa **CD2.**

In the thoracic cord, the **lateral horn** forms the **intermediolateral substance B6.** It contains mainly sympathetic nerve cells, principally for the vasomotor system, whose efferent fibers are said to leave by the posterior roots. Sympathetic nerve cells also lie medially in the **intermediomedial nucleus B7.** In the sacral cord parasympathetic neurons form the *intermediolateral* and *intermediomedial nuclei* **D8.**

The **anterior horn** expands widely in the cervical cord and contains several motor nuclei; the medial group of nuclei *(ventromedial nucleus* **A9** and *dorsomedial nucleus* **A10)** and the lateral group of nuclei (the *ventrolateral nucleus* **A11,** the *dorsolateral nucleus* **A12** and the *retrodorsolateral nucleus* **A13).** In the region supplying the upper limb, the anterior horn shows much more differentiation than in the thoracic cord, where only a few cell groups can be identified. The extensive and voluminous anterior horn in the lumbar and sacral cord, which supplies the lower limbs, again contains several groups of nuclei.

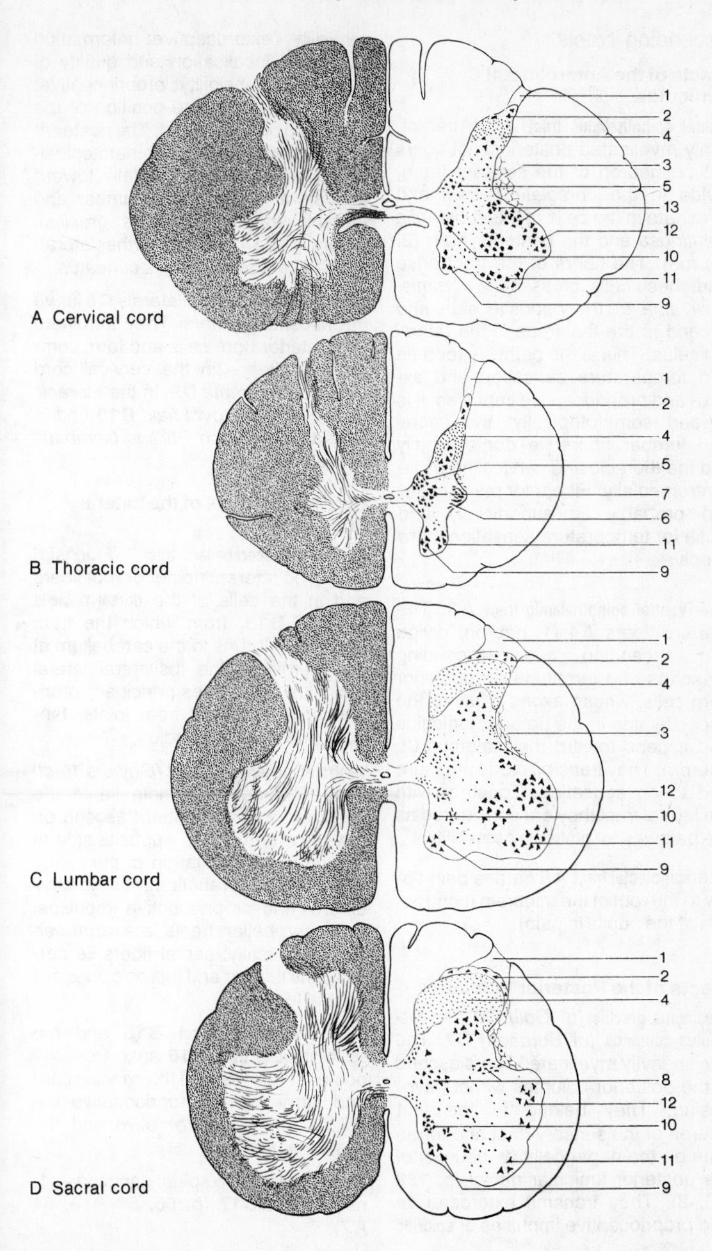
1
2
4
3
5
13
12
10
11
9
A Cervical cord
1
2
4
5
7
6
11
9
B Thoracic cord
1
2
4
3
12
10
11
9
C Lumbar cord
1
2
4
8
12
10
11
9
D Sacral cord

Ascending Tracts

Tracts of the Anterolateral Funiculus

Lateral spinothalamic tract A1. Afferent, thinly myelinated posterior root fibers **A2** (1. neuron of the sensory tract), divide in the *dorsolateral tract* and terminate in the cells of the *substantia gelatinosa* and the posterior horn (2. neuron). The fibers of the tract arise from these cells, cross in the *commissura alba* to the opposite side and ascend to the thalamus in the *lateral* funiculus. This is the pathway for pain and temperature sensation and extero- and proprioceptive impulses. It is divided somatotopically; the sacral and lumbar fibers lie dorsolaterally and the thoracic and cervical fibers lie ventromedially. Fibers for pain sensation probably lie superficially and those for temperature sensation more deeply.

The **ventral spinothalamic tract A3.** The afferent fibers **A4** (1. neuron) divide into ascending and descending branches and terminate on posterior horn cells, whose axons cross to the opposite side in the anterior funiculus and ascend toward the thalamus (2. neuron). They transmit crude pressure and touch sensation. Together with the lateral tract they are considered as the pathway for **protopathic sensibility.**

The **spinotectal tract A5** carries pain fibers to the roof of the midbrain (contraction of the pupils in pain).

Tracts of the Posterior Funiculus

Fasciculus gracilis (of *Goll)* **C6** and **fasciculus cuneatus** (of *Burdach)* **C7.** The thick heavily myelinated fibers ascend in the posterior columns without synapsing. They belong to the first neuron of the sensory tract and terminate on the nerve cells (2. neuron) of the posterior funicular nuclei (p. 132 B5, 8). They transmit exteroceptive and proprioceptive impulses of **epicritic sensibility** (exteroceptive: information about the localisation and quality of cutaneous sensibility; proprioceptive: information about the position of the limbs and body posture). The posterior columns are arranged somatotopically: the sacral fibers lie medially, toward the lateral side are the lumbar and thoracic tracts *(fasciculus gracilis).* Fibers from T1 to C2 lie further laterally and form the *fasciculus cuneatus.*

Short descending collaterals **C8** leave the ascending fibers. They terminate on posterior horn cells and form compact bundles – in the cervical cord *Schultze's comma* **D9,** in the *thoracic cord Flechsig's oval field* **D10** and in the sacral cord the *Phillipe-Gombault triangle* **D11.**

Cerebellar Tracts of the Lateral Funiculus

Dorsal spinocerebellar tract *(Flechsig)* **B12.** The afferent posterior root fibers end in the cells of the **dorsal nucleus** *(Clarke)* **B13,** from which the tract originates. It runs to the cerebellum at the margin of the ipsilateral lateral funiculus and carries principally proprioceptive impulses from joints, tendons and muscle spindles.

Ventral spinocerebellar tract *(Gowers' tract)* **B14.** The cells of origin lie in the posterior horn. The fibers ascend on the same and on the opposite side at the ventrolateral margin of the spinal cord to the cerebellum. They carry extero- and proprioceptive impulses. Both cerebellar tracts are arranged somatotopically: sacral fibers lie dorsally and lumbar and thoracic ones are ventral to them.

The **spino-olivary tract B15** and the **spinovestibular tract B16** arise from the posterior horn cells of the cervical cord and carry mainly proprioceptive impulses to the interior olive and the vestibular nuclei.

Nerve cells in the spinal ganglion – (1. neuron) **ABC17** (p. 56, 58 A13, 64 A7).

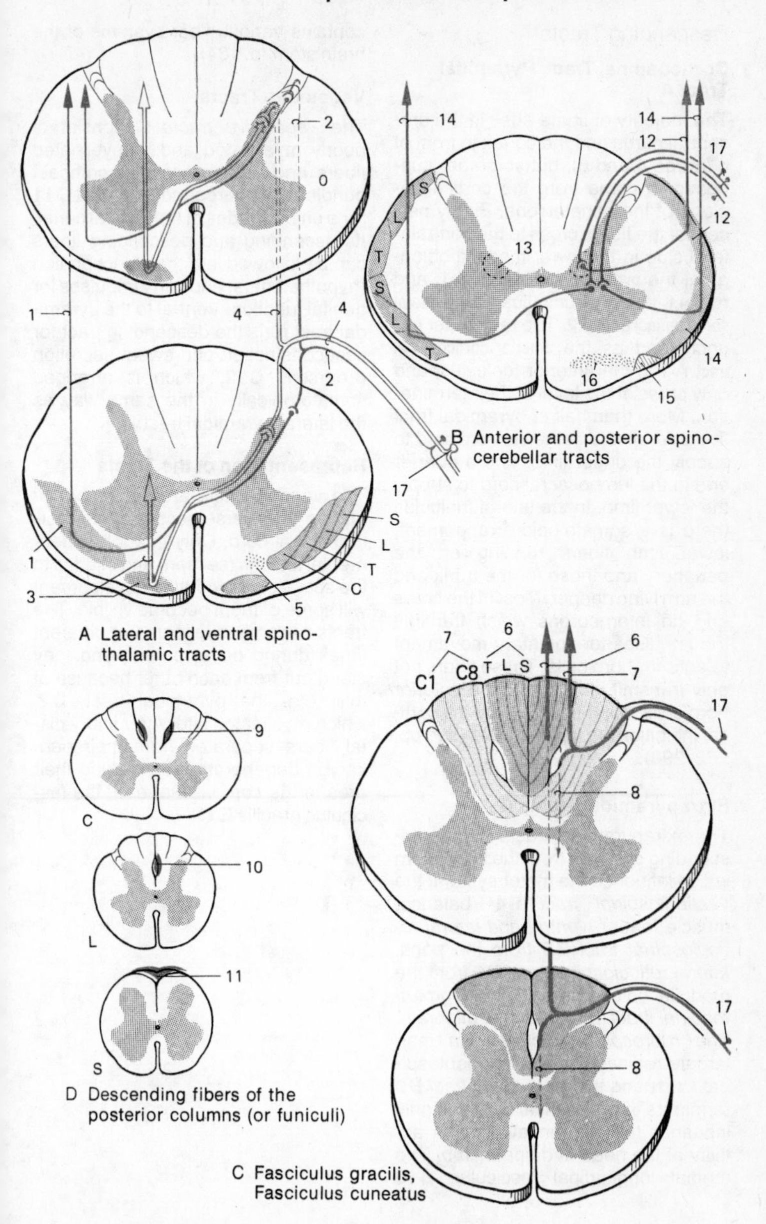

A Lateral and ventral spinothalamic tracts

B Anterior and posterior spinocerebellar tracts

C Fasciculus gracilis, Fasciculus cuneatus

D Descending fibers of the posterior columns (or funiculi)

Descending Tracts

Corticospinal Tract, Pyramidal Tract A

The majority of fibers arise in the precentral gyrus and the cortex in front of it (areas 4 and 6), but some are supposed to come from the cortical regions of the parietal lobe. Eighty percent of the fibers cross to the contralateral side in the lower medulla oblongata, the **pyramidal decussation A1,** and run in the lateral funiculus as the **lateral corticospinal tract A2.** The remainder run uncrossed as the **anterior corticospinal tract A3** in the anterior funiculus, and only cross at the level of their termination. More than half of pyramidal tract fibers terminate in the cervical cord to supply the upper limb, and a quarter end in the lumbosacral cord to supply the lower limb. In the lateral funiculus there is a somatotopic arrangement, lower limb fibers running on the periphery and those for the trunk and the arm lying deeper. Most of the fibers end on interneurons which transmit the impulses for voluntary movement to anterior horn cells. These fibers not only transmit impulses to the anterior horn cells, but they also transmit cortical inhibition via interneurons (see pp. 282, 290).

Extrapyramidal Tracts B

The extrapyramidal tracts include descending systems from the brain stem which influence the motor system: the *vestibulospinal tract* **B4** (balance, muscle tonus), *ventral and lateral reticulospinal tract* **B5** from the pons, *lateral reticulospinal tract* **B6** from the medulla oblongata and the *tegmentospinal tract* **B7** from the midbrain. The *rubrospinal tract* **B8** (in man largely replaced by the tegmentospinal tract) and the *tectospinal tract* **B9** terminate in the cervical cord and only influence the differentiated motor activity of the head and upper limb. The medial longitudinal fasciculus **B10** contains various fiber systems of the brain stem (p. 134).

Vegetative Tracts

The *vegetative tracts* **C** consist of poorly myelinated and unmyelinated fibers and only rarely form compact bundles. The parependymal tract **C11** runs on both sides of the central canal. Its ascending and descending fibers can be followed into the diencephalon (hypothalamus) and carry impulses for genital function. Ventral to the pyramidal tract runs the descending tract for vasoconstriction and sweat secretion (Foerster) **C12,** which is arranged somatotopically in the same way as the lateral pyramidal tract.

Representation of the Tracts

The various tracts are not recognisable in transverse section of the normal spinal cord. Only through experimental lesions (section) and injuries to the spinal cord, or during development will some of them become visible. The tracts become myelinated at different times during development, and they stand out from each other because of this, e.g. the pyramidal tract, **D2,** which myelinates late. After injury distal fibers separated from their perikaryon degenerate thus making their area in the cord visible, e. g. the fasciculus gracilis **E13.**

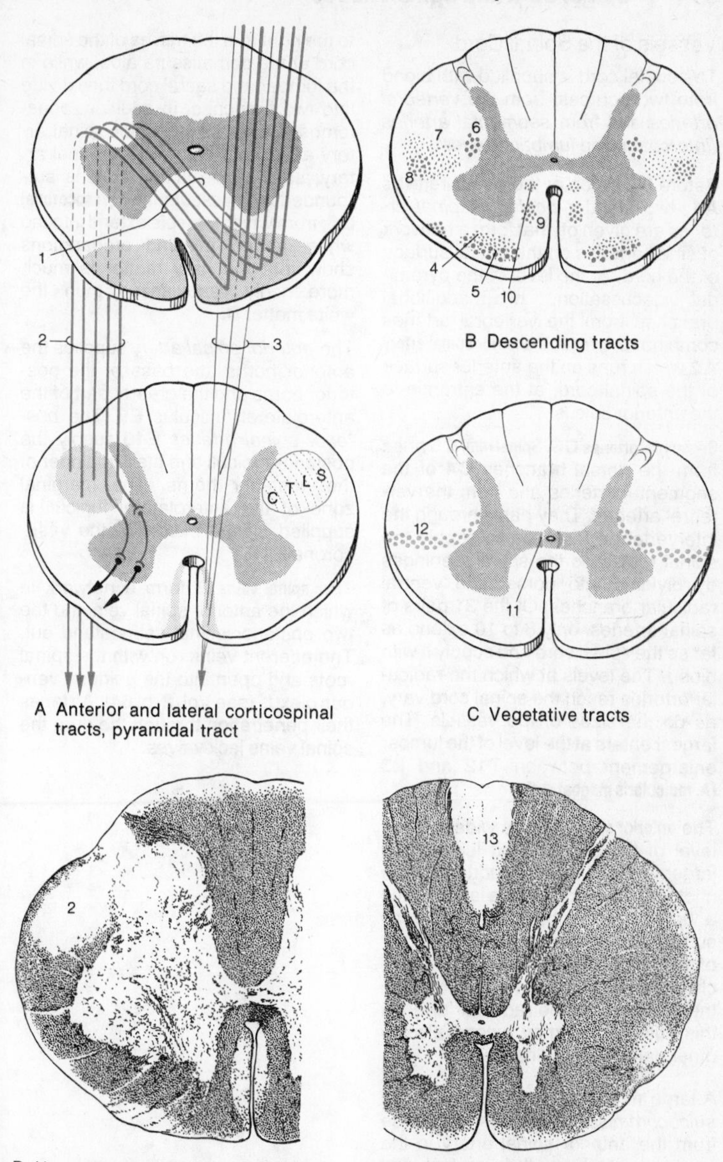

A Anterior and lateral corticospinal tracts, pyramidal tract

B Descending tracts

C Vegetative tracts

D Unmyelinated pyramidal tract in an infant

E Degeneration of the fasciculus gracilis after injury to the spinal cord

Vessels of the Spinal Cord

The spinal cord is supplied with blood from two sources, from the *vertebral arteries* and from *segmental arteries (intercostal* and *lumbar arteries).*

Before the union of the **vertebral arteries A1,** two slender *posterior spinal arteries* are given off that form a network of small arteries on the dorsal surface of the cord. At the level of the pyramidal decussation, two additional branches from the vertebral arteries combine to form the **anterior spinal artery A2** which runs on the anterior surface of the spinal cord, at the entrance to the anterior sulcus.

Segmental arteries C3. Spinal rami C5 arise from the dorsal branches **C4** of the segmental arteries and from the vertebral arteries. They pass through the intervertebral foramina to supply the spinal roots and the spinal meninges by dividing into dorsal and ventral radicular branches. Of the 31 pairs of spinal arteries, only 8 to 10 extend as far as the spinal cord and supply it with blood. The levels at which the radicular arteries reach the spinal cord vary, as do the sizes of the vessels. The largest enters at the level of the lumbar enlargement between T12 and L3 **(A. radicularis magna) A6.**

The **anterior spinal artery** is widest at the level of the cervical and lumbar enlargement and is much reduced in the midthoracic region. As the latter is also a border region between the areas supplied by radicular arteries, this part of the spinal cord is most vulnerable to circulatory disturbances. According to the variations in the radicular arteries this vulnerability may extend also to other parts of the spinal cord.

A large number of small arteries, the *sulcocommissural arteries,* **D7,** arise from the anterior spinal artery in the anterior sulcus. In the cervical and thoracic cord they branch alternately to the right and left halves of the spinal cord at the commissura alba, while in the lumbar and sacral cord they divide into two branches. In addition, anastomoses with the posterior spinal artery arise from the anterior spinal artery, so that the spinal cord is surrounded by a vascular ring **(vasocorona) D8,** from which vessels extend into the white matter. Injected preparations show that the grey matter is much more richly vascularized than the white matter **D.**

The *anterior spinal artery* supplies the anterior horns, the base of the posterior horns and the greater part of the anterolateral funiculus **E9.** The *posterior spinal arteries* **E10** supply the posterior funiculi and the remainder of the posterior horns. The marginal zone of the anterolateral funiculi is supplied by the plexus of the vasocorona **E11.**

The **spinal veins B** form a network in which the anterior spinal vein and the two posterior spinal veins stand out. The efferent veins run with the spinal roots and open into the epidural venous plexus (see Vol. 2, p. 64). As far as their penetration through the dura the spinal veins lack valves.

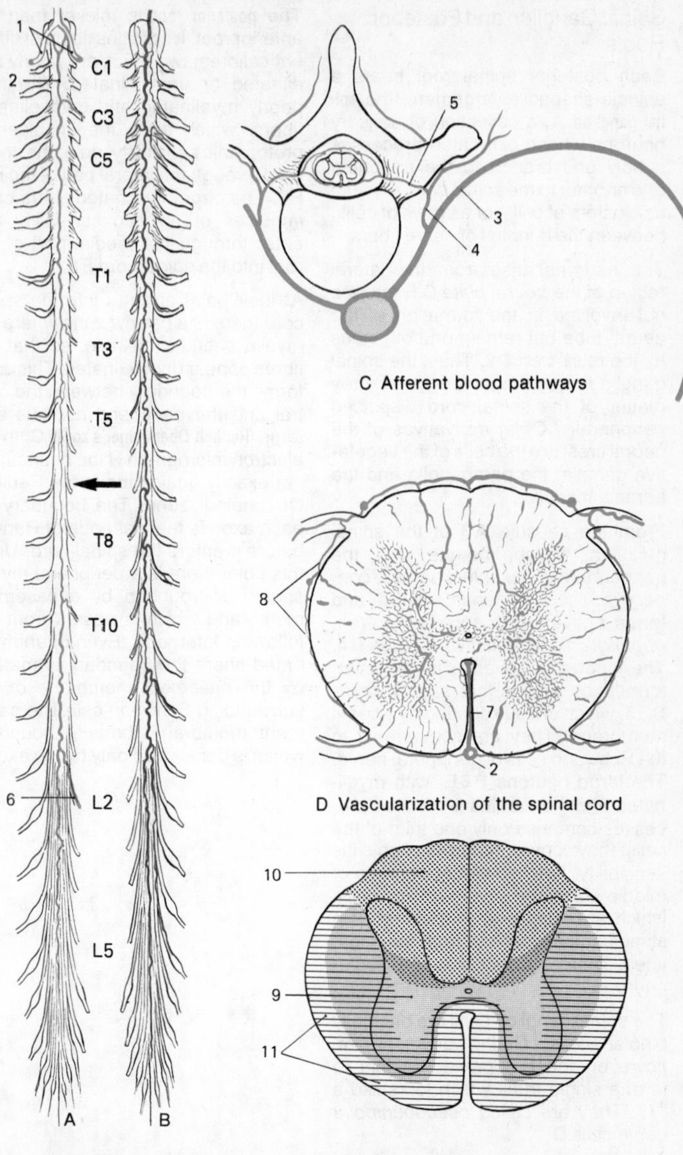

Arteries and veins of the spinal cord

C Afferent blood pathways

D Vascularization of the spinal cord

E Areas supplied by the arteries of the spinal cord (after Gillilan)

Spinal Ganglion and Posterior Roots

Each posterior spinal root bears a spindle-shaped enlargement, the **spinal ganglion A,** a collection of sensory neurons whose bifurcated processes supply one branch to the periphery and another to the spinal cord. They lie as clusters of cells, or as rows of cells, between the bundles of nerve fibers.

The neurons arise from the lateral region of the *neural plate* **C1,** and are not involved in the formation of the neural tube but remain on both sides as the **neural crest C2.** Thus, the spinal ganglia may be regarded as the grey matter of the spinal cord displaced peripherally. Other derivatives of the neural crest are the cells of the vegetative ganglia, the paraganglia and the adrenal medulla.

From the capsule **A3** of the spinal ganglion, which merges into the perineurium of the spinal nerves, connective tissue extends inward and forms a covering for each neuron *(endoganglionic connective tissue)* **B4.** The innermost covering is, however, formed by ectodermal *satellite cells* **BE5,** which are surrounded by a basal membrane. They are comparable to the Schwann cells of peripheral nerve. The large neurons **B6E,** with myelinated and glomerulus-like coiled processes comprise only one third of the cells; they convey impulses of epicritic sensibility. The remainder consists of middle-sized and small ganglion cells, which are supposed to transmit pain stimuli and sensations from the intestines. There are also some multipolar neurons.

The spinal ganglion cells are originally bipolar cells. During development, however, the two processes join to form a single trunk which forks like a "T". They are called *pseudounipolar* nerve cells **D.**

The **posterior root** is thicker than the anterior root. It contains fibers of different calibres; two thirds are poorly myelinated or unmyelinated. The fine, poorly myelinated and unmyelinated fibers, which carry the impulses of protopathic sensibility, enter the spinal cord through the lateral part of the root **F7.** The large myelinated fibers carry impulses of epicritic sensibility and enter through the medial part of the root into the spinal cord **F8.**

At their point of entry into the spinal cord there is a narrow zone where the myelin sheaths are thin, so that the fibers appear unmyelinated. This zone forms the boundary between the central and the peripheral nervous systems **(Redlich-Obersteiner's zone) G.** In the electron micrograph **H** the boundary is not exactly analogous to the Redlich-Obersteiner zone. The boundary for each axon is the last node of Ranvier before it enters the spinal cord. Up to this point there is a peripheral myelin sheath, surrounded by a basement membrane which is absent from the following internode. Even in unmyelinated fibers the boundary is marked by the basement membrane of the surrounding Schwann cell. The basement membrane forms a boundary which is penetrated only by the axon.

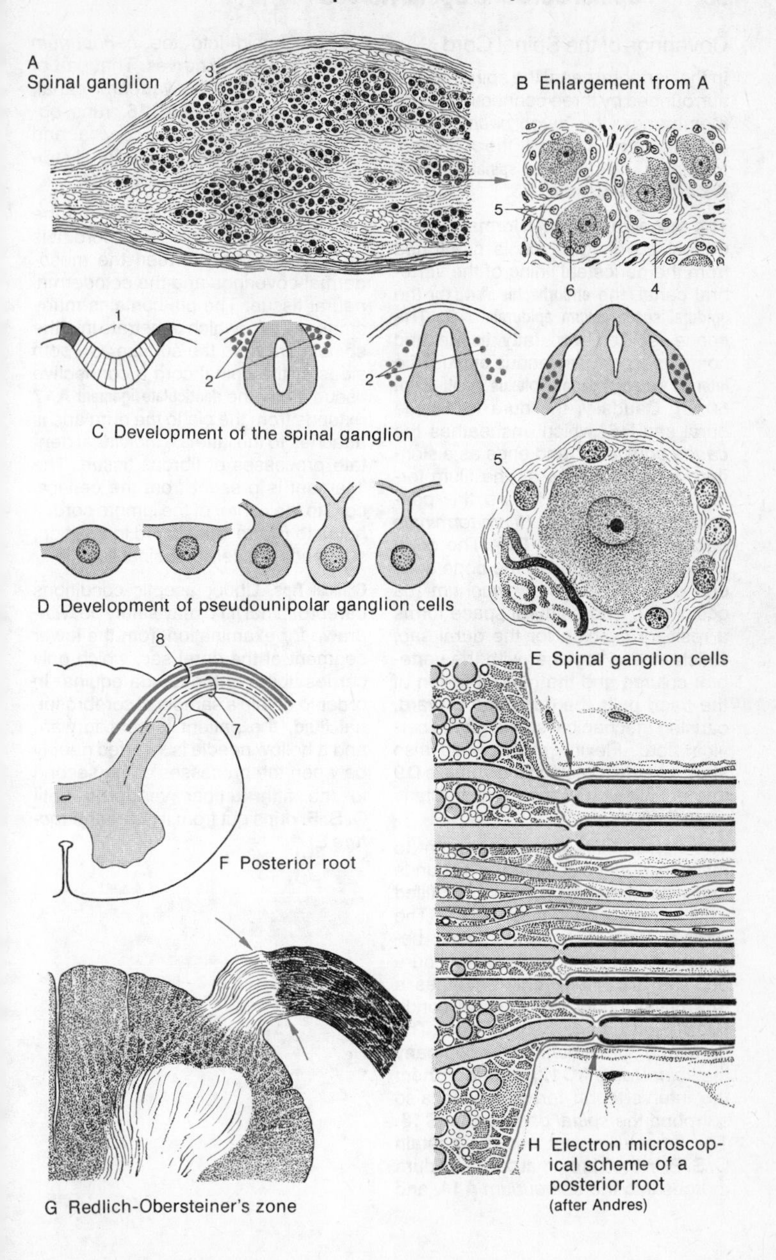
A
Spinal ganglion
3
B Enlargement from A
5
6
4
1
2
2
C Development of the spinal ganglion
D Development of pseudounipolar ganglion cells
5
E Spinal ganglion cells
8
7
F Posterior root
G Redlich-Obersteiner's zone
H Electron microscopical scheme of a posterior root
(after Andres)

Coverings of the Spinal Cord

In the vertebral canal the spinal cord is surrounded by three connective tissue membranes: the *tough pachymeninx* or **dura mater A1** and the *soft leptomeninx,* divided into **spinal arachnoid A2** and **spinal pia mater A3.**

The spinal **dura mater** forms the outermost covering which is separated from the periosteal lining of the vertebral canal, the **endorhachis A4,** by the **epidural space cavum epidurale A5.** The space is filled with fatty tissue and contains a profuse venous plexus, the **internal vertebral venous plexus** (Vol. 2, p. 60 C). Caudally, the dura forms the dural sac **B6,** which ensheathes the *cauda equina* **B7** and ends as a slender strand that runs as the filum terminale externa as far as the periosteum of the coccyx (*filum terminale durae matris spinalis* **B8**). The dural sac is only attached to the bone at its oral end, the *foramen magnum* (os occipitale). The epidural space forms a resilient cushion for the dural sac, which moves together with the vertebral column and the head. Flexion of the head pulls the dural sac upward, causing mechanical stress on the cervical cord. Flexion of the head also stretches the roots and the vessels **D9** and dorsiflexion of the head shortens them **D10.**

The **arachnoid** lies closely adherent to the inner surface of the *dura.* It bounds the **subarachnoid space,** which is filled with *cerebrospinal fluid* **AC11.** The **subdural space,** a capillary space between the inner surface of the dura and the arachnoid, only becomes a true cavity under pathological conditions (e. g. subdural hemorrhage). The dura and the arachnoid accompany the *spinal roots* **AC12,** enter with them the intervertebral foramina and also surround the *spinal ganglion* **ABC13.** The funnel-like root pockets contain C. S. F. in their proximal part. The dura merges into the *epineurium* **A14,** and the arachnoid into the *perineurium* **A15** of the spinal nerves. That part of the root leaving the vertebral canal, the *radicular nerve* **AC16,** runs obliquely downward in the cervical and lumbosacral regions and obliquely upward in the midthoracic region **C.**

The **spinal pia mater** lies directly on the outer glial layer of the spinal cord. It is the border-zone between the mesodermal coverings and the ectodermal neural tissue. The pia contains many small vessels, which penetrate into the spinal cord from the surface. On both sides of the spinal cord a connective tissue plate, the **denticulate ligament A17** extends from the pia to the dura and is attached to the latter by individual dentate processes of fibrous tissue. The ligament is present from the cervical cord to the center of the lumbar cord. It helps to fix the spinal cord in position, suspended in cerebrospinal fluid.

Clinical Tips. Under aseptic conditions cerebrospinal fluid may safely be withdrawn for examination from the lower segment of the dural sac, which only carries fibers of the cauda equina. In order to obtain a sample of cerebrospinal fluid, the patient is bent forward and a hollow needle is inserted deeply between the processes of the second to the fifth lumbar vertebrae until C. S. F. drips out from it (*lumbar puncture* **E**).

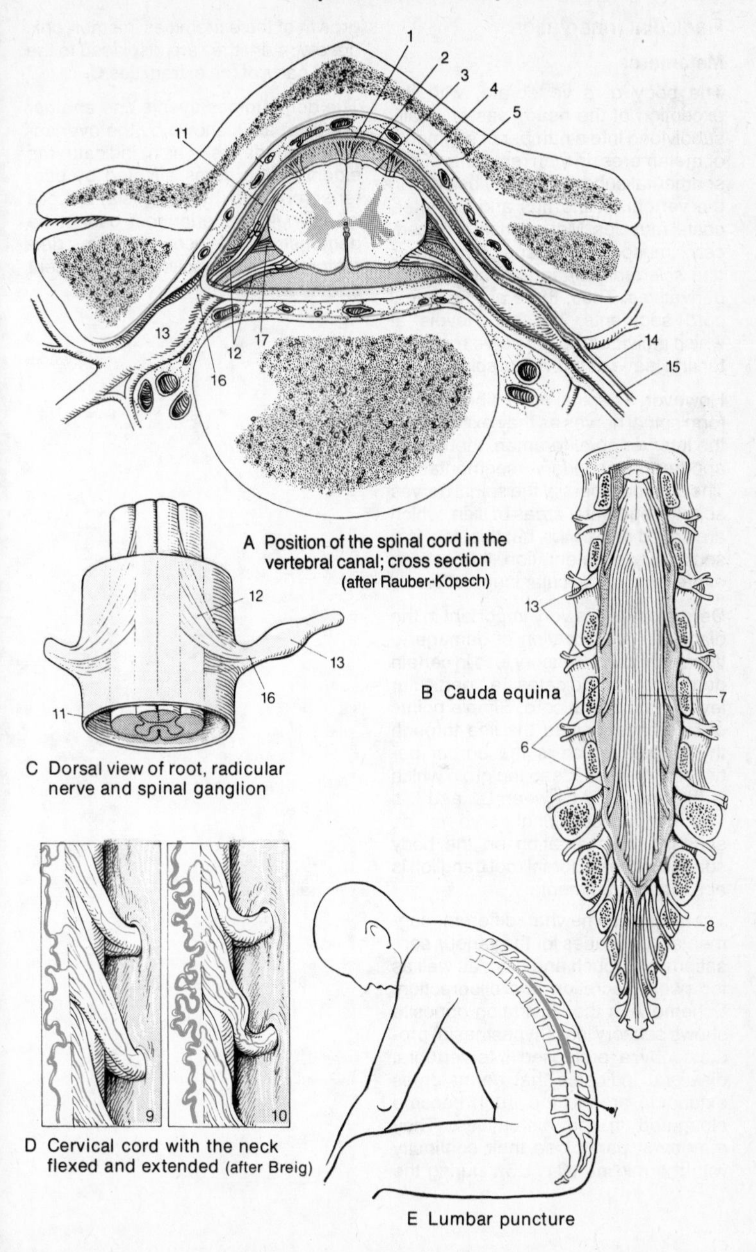

A Position of the spinal cord in the vertebral canal; cross section (after Rauber-Kopsch)

B Cauda equina

C Dorsal view of root, radicular nerve and spinal ganglion

D Cervical cord with the neck flexed and extended (after Breig)

E Lumbar puncture

Radicular Innervation

Metameres

The body of a vertebrate, with the exception of the head, was originally subdivided into a number of segments or metameres. In man remnants of this segmental subdivision may be seen in the vertebrae, the ribs and the intercostal muscles. Metamerism only concerns mesodermal tissues (*myotomes* and *sclerotomes*) and not ectodermal derivatives. Thus, there are no spinal cord segments, but only levels at which individual spinal nerve roots enter and leave as individual spinal roots.

However, as spinal nerve fibers join to form spinal nerves as they exit through the intervertebral foramen, there is an apparent secondary segmentation. The sensory fibers of the spinal nerves supply segmental areas of skin, which are called **dermatomes,** but this, too, is a secondary segmentation. It is only an expression of radicular innervation.

Dermatomes are very important in the diagnosis and location of damage to the spinal cord. Sensory loss in certain dermatomes indicates a particular level in the spinal cord. Simple points of reference include the line through the nipples which is the border between T4 and T5, and the groin which is the boundary between L1 and L2. The first cervical spinal nerve has no sensory representation on the body surface since its dorsal root ganglion is absent or is rudimentary.

There are somewhat different segmental boundaries for the various sensations i. e. touch and pain, as well as for sweat secretion and piloerection. Scheme **A** in the illustration opposite shows sensory loss (hypesthesia) produced by a prolapsed intervertebral disk and indicates that dermatomes extending around the trunk become elongated in the extremities. They may even partly lose their continuity with the midline (C7, L5). During the growth of the extremities in embryonic development they are displaced to the distal parts of the extremities **C.**

The dermatomes overlie one another like tiles, as is shown by the overlapping boundaries which indicate the extent of the areas affected by posterior root pain (hyperalgesia) **B.** Loss of a single posterior root cannot be demonstrated since its own dermatome is also supplied from adjacent dermatomes.

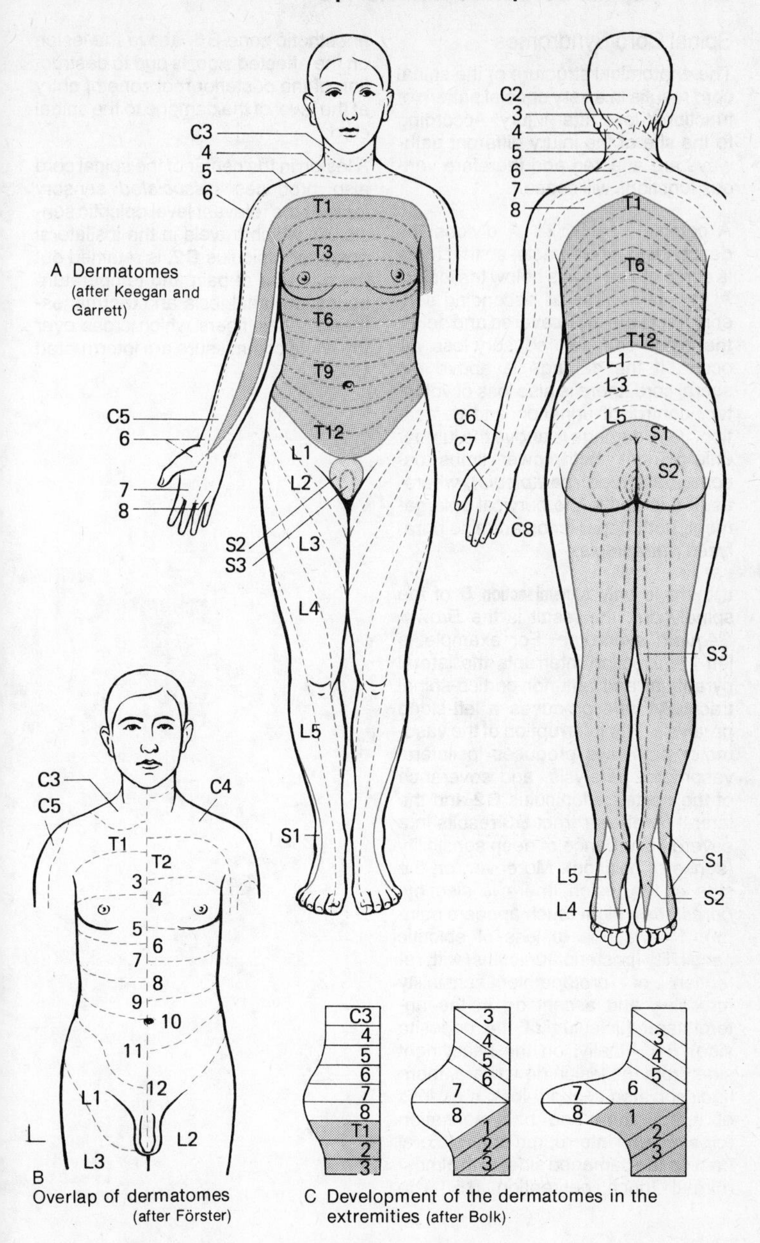

A Dermatomes (after Keegan and Garrett)

B Overlap of dermatomes (after Förster)

C Development of the dermatomes in the extremities (after Bolk)

Spinal Cord Syndromes

The anatomical structure of the spinal cord results in a very special pattern of functional loss after injury. According to the site of the injury different pathways are affected and therefore various functions will cease.

A complete **transection A** divides the descending motor tracts, so that there is complete paralysis below the lesion. At the same time all ascending sensory tracts are also severed and below the lesion complete sensory loss will occur. If the damage is above the sacral cord, there is also loss of voluntary control of urination and defecation. If the damage is above the lumbar enlargement, both lower limbs become paralyzed *(paraplegia),* whereas, if it is above the cervical enlargement, both upper limbs are also paralyzed *(tetraplegia).*

If there is only a **hemisection B** of the spinal cord, the result is the *Brown-Séquard syndrome.* For example, a left hemisection interrupts the lateral pyramidal and anterior cortico-spinal tracts **B1** and produces a left-sided paralysis. The interruption of the vasomotor pathways produces ipsilateral vasomotor paralysis, and severance of the posterior funiculus **B2** and the lateral cerebellar tract **B3** results in a severe disturbance of deep sensibility (sense of position). Moreover, on the side of the lesion, there is also *hyperesthesia* (light touch appears painful). This is due to loss of epicritic sensibility (posterior funiculus) with retention of protopathic sensibility (crossing and ascending in the anterolateral funiculus of the opposite side) **B4.** Finally, on the intact right side, from the lesion downward, there is *dissociated* sensory loss, i. e., loss of temperature and pain sensation (crossed and interrupted anterolateral tract on the damaged side) with almost normal touch perception **B5.** The anesthetic zone **B6,** above the lesion on the affected side, is due to destruction of the posterior root zone of entry at the level of the damage to the spinal cord.

A lesion in the center of the spinal cord also produces *dissociated* sensory loss. At the relevant level epicritic sensibility, which travels in the ipsilateral posterior funiculus **C2,** is retained but there is loss of pain and temperature sensation (analgesia and thermanesthesia): their fibers which cross over the white commissure are interrrupted **C5.**

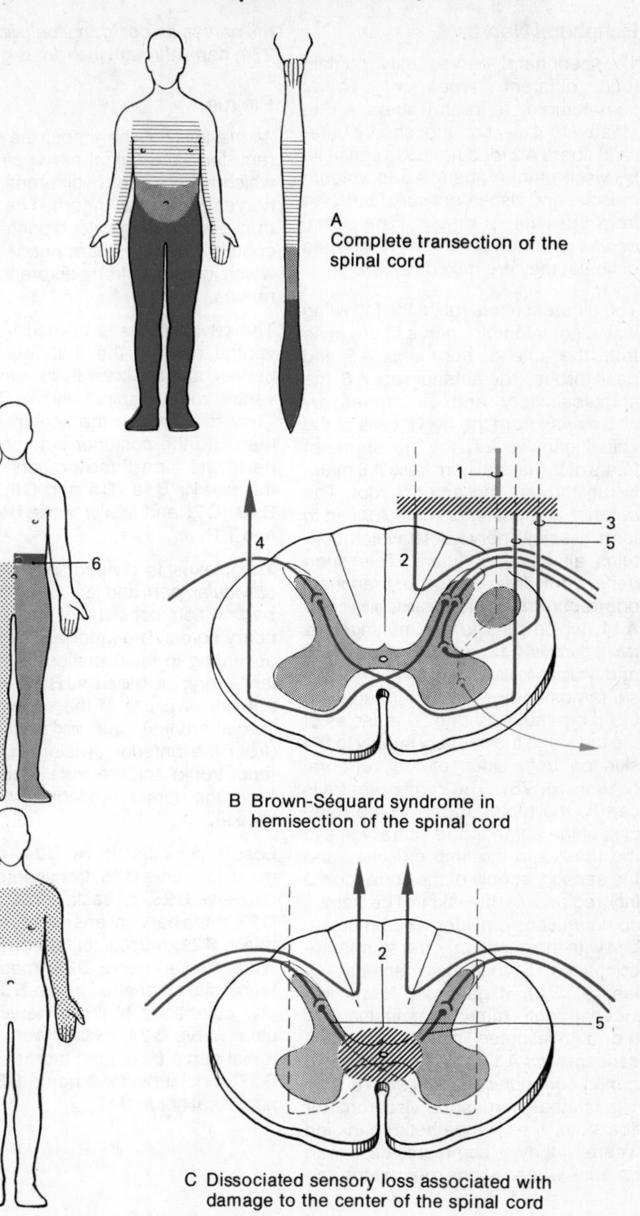

A
Complete transection of the spinal cord

B Brown-Séquard syndrome in hemisection of the spinal cord

C Dissociated sensory loss associated with damage to the center of the spinal cord

Peripheral Nerves

The peripheral nerves may contain four different types of fibers: somatomotor (efferent) fibers **A1** to striated muscle, somatosensory (afferent) fibers **A2** for cutaneous sensibility, visceromotor fibers **A3** to smooth muscle, and viscerosensory fibers **A4** from the internal organs. The spinal nerves generally contain several types of fibers; they are *mixed nerves.*

The different fibers run in the following ways: somatomotor nerve fibers arise from the anterior horn cells **A5** and pass through the anterior root **A6,** the somatosensory and viscerosensory fibers arise from the nerve cells of the spinal ganglia **A7,** the visceromotor fibers of the lateral horn cells **A8** mainly run through the anterior root. The anterior and posterior roots **A9** join to form a spinal nerve **A10** which contains all types of fibers. This short nerve trunk divides into four branches: one recurrent branch, **ramus meningeus A11,** which runs to the meninges, a **ramus dorsalis A12,** a **ramus ventralis A13** and a **ramus communicans A14.** The dorsal ramus carries the motor supply to the deep (autochthonous) muscles of the back and the sensory supply to the skin on both sides of the vertebral column (p. 78). The ramus ventralis carries the motor supply to the muscles of the anterior and lateral walls of the trunk and the limb muscles, and the sensory supply of the corresponding regions of the skin. The ramus communicans provides a communication with the ganglion of the sympathetic chain **A15** (vegetative nervous system p. 270). It generally forms two independent rami communicantes, the unmyelinated *ramus communicans griseus* **A17** and the myelinated *ramus communicans albus* **A16.** The ramus albus carries the visceromotor fibers to the sympathetic ganglion where they communicate with neurons whose axons re-enter the spinal nerves as *postganglionic fibers* (p. 274) partly through the ramus griseus.

Plexus

At the level of the limbs, the ventral rami form the spinal nerve plexus in which fibers from different spinal nerves are exchanged. The nerve trunks which then extend peripherally, contain newly rearranged fibers, which originate from different spinal nerves.

The **cervical plexus** is formed from the ventral rami of the first four spinal nerves, and the **brachial plexus** from the ventral rami of spinal nerves C5–T1. They run through the scalenus foramen into the posterior triangle of the neck and form three primary trunks, the **superior B18 (C5** and **C6**), **medial B19 (C7)** and **inferior trunks B20 (C8** and T1).

The plexus is divided into a *supraclavicular part* and an *infraclavicular part,* which consists of three secondary cords. The latter are described according to their relation to the axillary artery: the **lateral cord B21** (from the anterior divisions of the superior and medial trunks), the **medial cord B22** (from the anterior division of the inferior trunk) and the **posterior cord B23** (from the dorsal divisions of all three trunks).

Lesser occipital nerve **B24,** greater auricular nerve **B25,** transverse cervical nerve **B26,** supraclavicular nerves **B27,** deep cervical ansa **B28,** phrenic nerve **B29,** musculocutaneous nerve **B30,** median nerve **B31,** medial antebrachial cutaneous nerve **B32,** medial cutaneous brachial nerve **B33,** ulnar nerve **B34,** axillary nerve **B35,** radial nerve **B36,** long thoracic nerve **B37,** medial pectoral nerve **B38,** hypoglossal nerve **B39.**

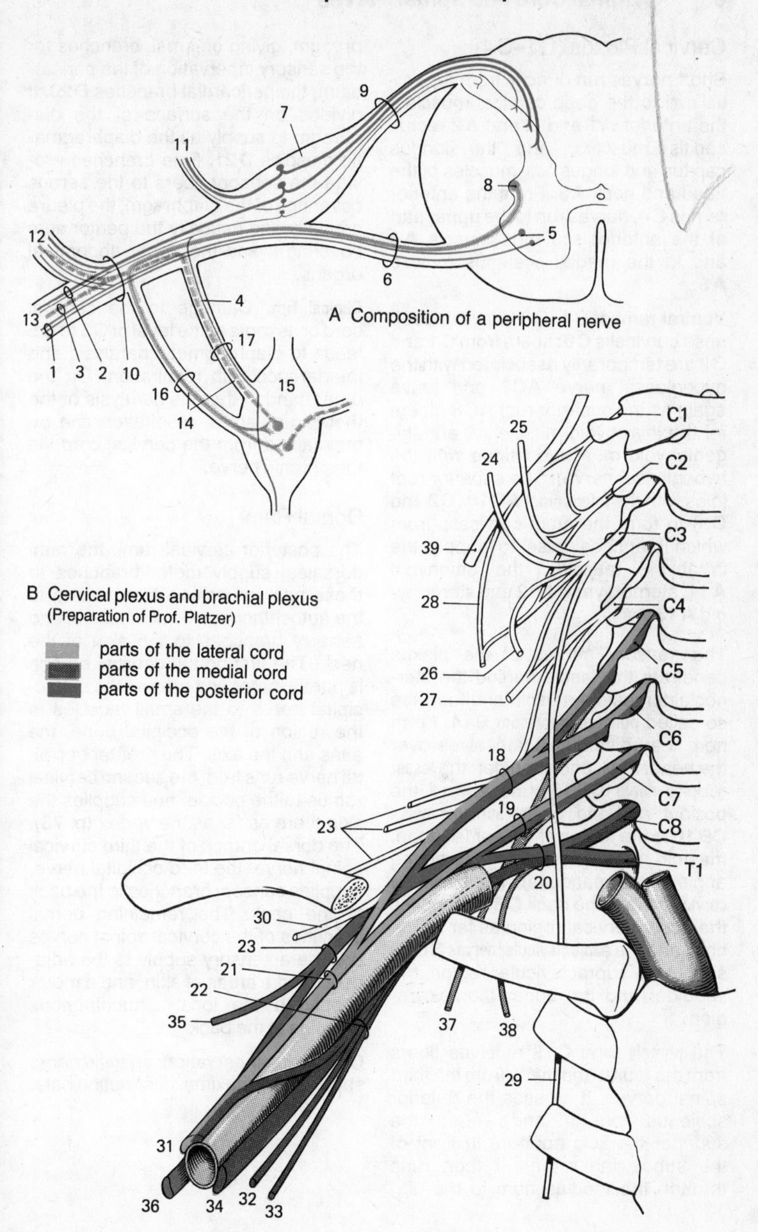

A Composition of a peripheral nerve

B Cervical plexus and brachial plexus
(Preparation of Prof. Platzer)

Cervical Plexus (C1–C4)

Short nerves run directly from the **ventral rami** to the deep cervical muscles: the anterior **A1** and lateral **A2** rectus capitis muscles, and the longus capitus and longus colli muscles of the head and neck **A3.** From the anterior ramus C4, nerves run to the upper part of the anterior scalenus muscle **A4** and to the medial scalenus muscle **A5.**

Ventral rami of C1–C3 form the deep ansa cervicalis **C6:** fibers from C1 and C2 are temporarily associated with the hypoglossal nerve **AC7** and leave again as the superior root **AC8** (fibers for the thyrohyoid muscle **A9** and the geniohyoid muscle continue with the hypoglossal nerve). The superior root joins with the **inferior root AC10** (C2 and C3) to form the ansa cervicalis from which branches arise to supply the infrahyoid muscles: the omohyoid **A11,** sternothyroid **A12** and sternohyoid **A13.**

The sensory nerves of the plexus penetrate the fascia behind the sternocleidomastoid muscle and form the so-called **punctum nervosum B14.** From here they distribute themselves over the head, neck and shoulder: the **lesser occipital nerve CB15** runs toward the occiput and the **greater auricular nerve CB16** to the region of the ear (earlobe, mastoid process and region of the angle of the mandible). The **transverse cervical nerve** of the neck **CB17** supplies the upper cervical region as far as the chin, and the **supraclavicular nerves BC18** supply the supraclavicular region, the shoulder and the upper thoracic region.

The **phrenic nerve C19** receives fibers from the fourth and many from the third spinal nerves. It crosses the anterior scalenus muscle and enters the superior thoracic aperture in front of the subclavian artery. It then runs through the mediastinum to the diaphragm, giving off small branches for the sensory innervation of the pericardium, the pericardial branches **D20.** It divides on the surface of the diaphragm to supply all the diaphragmatic muscles **D21.** Fine branches provide the sensory fibers to the serous coverings of the diaphragm, the pleura cranially and caudally the peritoneum covering it and the upper abdominal organs.

Clinical Tips. Damage to the cervical cord or its roots at the level of C3 to C5 leads to diaphragmatic paralysis and interference with respiration. On the other hand, if there is paralysis of the thoracic muscles, respiration can be maintained from the cervical cord via the phrenic nerve.

Dorsal Rami

The posterior cervical rami, the rami dorsales, supply motor branches to those neck muscles which belong to the autochthonous back muscles, and sensory branches to the skin of the neck. The first dorsal cervical branch is purely motor and runs as the suboccipital nerve to the small muscles in the region of the occipital bone, the atlas and the axis. The greater occipital nerve runs from the second cervical ramus to the occiput and supplies the skin there as far as the vertex (p. 78). The dorsal branch of the third cervical spinal nerve, the third occipital nerve, supplies sensory branches to the back of the neck. The remaining dorsal branches of the cervical spinal nerves provide a sensory supply to the adjacent caudal areas of skin, and a motor supply to the local autochthonous muscles of the back.

Cutaneous innervation: specific nerve supply dark, maximal distribution pale.

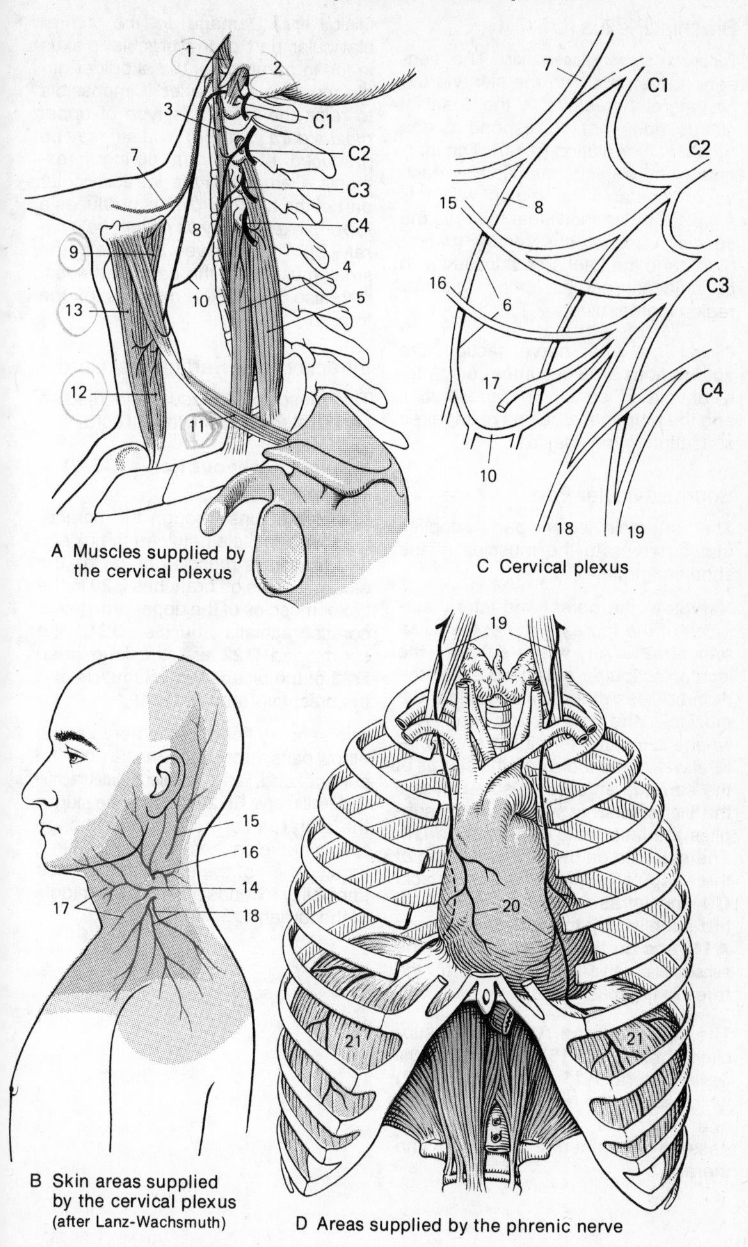

A Muscles supplied by the cervical plexus

B Skin areas supplied by the cervical plexus (after Lanz-Wachsmuth)

C Cervical plexus

D Areas supplied by the phrenic nerve

Brachial Plexus (C5–T1)

Peripheral sensory innervation. The sensory nerve supply to the skin via the peripheral nerves from the brachial plexus does not correspond to the radicular innervation (p. 60). The margins of the regions supplied by each nerve overlap. The region which is supplied by an individual nerve is the specific *autonomous* region (darker blue), and the total area supplied also by adjacent nerves is the *maximal* region (lighter blue).

Clinical Tips. After nerve section, the autonomous area becomes completely devoid of sensation *(anesthetic)*, and the marginal areas have impaired sensation *(hypesthesia)*.

Supraclavicular Part

The supraclavicular part supplies motor nerves to the muscles of the shoulder girdle.

Nerves to the dorsal and lateral surfaces of the thorax include: the **dorsal scapular nerve A1,** which supplies the levator scapulae muscle **C2** and the rhomboideus major **C4** and minor **C3** muscles, the **long thoracic nerve A5,** whose branches terminate on the lateral wall of the thorax on the slopes of the serratus anterior muscle **B6,** and the **thoracodorsal nerve A7,** which supplies the latissimus dorsi muscle **C8.** The muscles on the dorsal surface of the scapula, supraspinatous muscle **C9** and infraspinatous muscle **C10,** are supplied by the **suprascapular nerve A11,** and on the ventral surface, the **subscapular nerve A12** passes to the teres major muscle **C13.**

The **subclavius nerve A14** (to the subclavius muscle **B15**) and the **pectoral nerves** (lateral **A16** and medial **A17**) which supply the pectoralis major **B18** and pectoralis minor muscles **B19** pass to the anterior surface of the thorax.

Clinical tips. Damage to the supraclavicular part of the brachial plexus leads to paralysis of the shoulder girdle muscles and makes it impossible to raise the arm. This type of *upper plexus paralysis (Erb's palsy)* may be produced of the arm during anesthesia. Damage to the infraclavicular part of the brachial plexus produces a *lower plexus paralysis (Klumpke's paralysis)*. This involves principally the small muscles of the hand and possibly also the flexor muscles of the forearm.

Infraclavicular Part Lateral Cord

The musculocutaneous and median nerves stem from the lateral cord.

Musculocutaneous Nerve (C5–7) D–F

The nerve runs through the coracobrachialis muscle between the biceps and brachialis muscles as far as the elbow. It gives off branches **E20** to the flexor muscles of the upper arm: to the coracobrachialis muscle **D21,** the short head **D22** and the long head **D23** of the biceps brachii muscle and the brachialis muscle **D24.**

The sensory fibers of the nerve at the elbow pass through the fascia onto the surface and, as the **lateral antebrachial cutaneous nerve EF25,** supply the skin in the lateral part of the forearm. Damage to this nerve causes sensory loss in a small region of the elbow; diminished sensibility extends as far as the middle of the forearm.

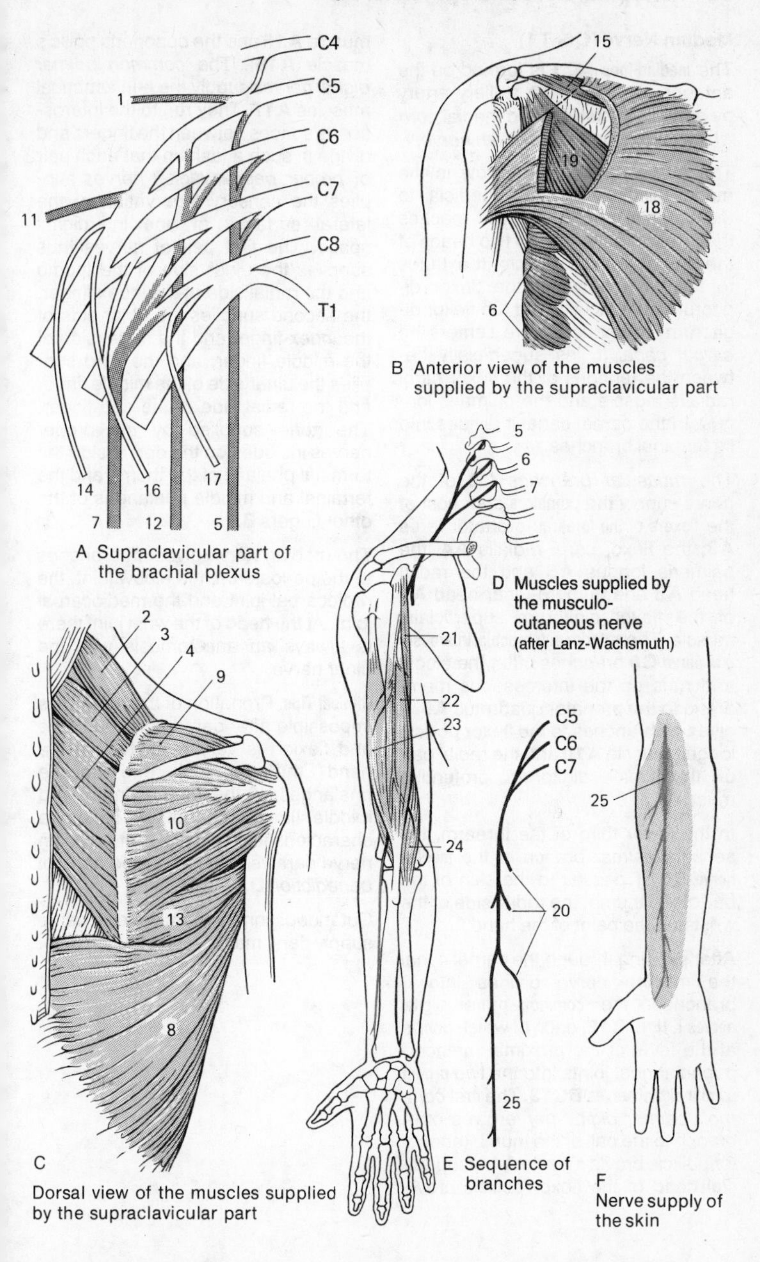

A Supraclavicular part of the brachial plexus

B Anterior view of the muscles supplied by the supraclavicular part

C Dorsal view of the muscles supplied by the supraclavicular part

D Muscles supplied by the musculo-cutaneous nerve (after Lanz-Wachsmuth)

E Sequence of branches

F Nerve supply of the skin

Median Nerve (C6–T1)

The **median loop AC1** is formed on the anterior surface of the axillary artery by parts of the lateral and medial cord which join to form the *median nerve.*

The nerve runs to the elbow in the medial bicipital sulcus, superficial to the brachial artery, where it reaches the forearm between the two heads of the pronator teres muscle. It extends to the wrist between the flexor digitorum superficialis and the flexor digitorum profundus. Before it enters the *carpal canal* it lies superficially between the tendons of the flexor carpi radialis muscle and the palmaris longus. In the carpal canal it divides into its terminal branches.

The *muscular branches* **C2** of the nerve supply the **pronators** and most of the **flexors of the forearm:** pronator teres **A3,** the flexor carpi radialis **A4,** the palmaris longus **A5** and the radial head **A6** and humero-ulnar head **A7** of the flexor digitorum superficialis muscle. The **anterior antebrachial interosseous nerve C8** branches off at the elbow and runs on the interosseous membrane to the pronator quadratus **A9.** It gives off branches to the flexor pollicis longus muscle **A10** and the radial part of the flexor digitorum profundus muscle.

In the lower third of the forearm, the sensory palmar branch of the **median nerve BC11** passes to the skin of the ball of the thumb, the radial side of the wrist and the palm of the hand.

After passing through the carpal canal the median nerve divides into 3 branches: the **common palmar digital nerves** I, II, III **C12,** each of which divide at the level of the proximal metacarpophalangeal joints into the two **proper palmar digital nerves BC13.** The first *common palmar digital nerve* gives off a branch to the ball of the thumb (abductor pollicis brevis muscle **A14,** superficial head of the flexor pollicis brevis muscle **A15** and the opponens pollicis muscle **A16**). The *common palmar digital nerves* supply the I–III lumbrical muscles **A17.** They run to the interosseous spaces between the fingers and divide in such a fashion that each pair of *proper palmar digital nerves* supplies the sensory innervation of the lateral surfaces of one interdigital space. The first pair of nerves thus supplies the ulnar side of the thumb and the radial side of the index finger, the second supplies the ulnar side of the index finger and the radial side of the middle finger, and the third supplies the ulnar side of the middle finger and the radial side of the ring finger. The zone supplied by the proper nerves includes on the dorsal side the terminal phalanx of the thumb and the terminal and middle phalanges of the other fingers **B.**

The median nerve gives off branches to the periosteum, the elbow joint, the radiocarpal joint and the mediocarpal joint. At the head of the wrist joint there is always an anastomosis with the ulnar nerve.

Clinical Tips. Pronation of the forearm is impossible after damage to the nerve and flexion is very restricted. In the hand the terminal and middle phalanges of the thumb, index and middle fingers cannot be flexed, a characteristic indication of median nerve paralysis, the so-called *hand of benediction* **D.**

Cutaneous innervation: specific nerve supply dark, maximal distribution pale.

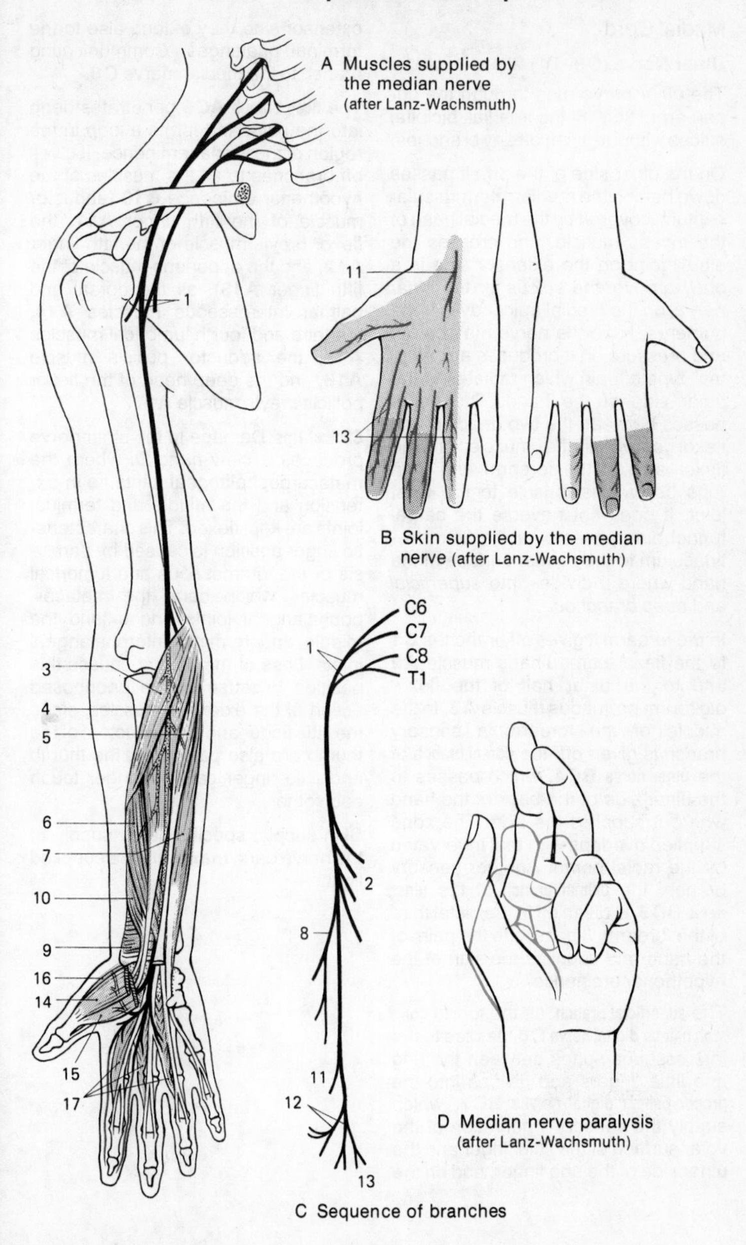

A Muscles supplied by the median nerve (after Lanz-Wachsmuth)

B Skin supplied by the median nerve (after Lanz-Wachsmuth)

C Sequence of branches

D Median nerve paralysis (after Lanz-Wachsmuth)

Medial Cord

Ulnar Nerve (C8–T1)

The ulnar nerve runs through the upper arm, first in the medial bicipital sulcus without giving off any branches.

On the ulnar side of the arm it passes down behind the medial intermuscular septum, covered by the medial head of the triceps muscle, and crosses the elbow joint on the extensor side in a bony groove, the *sulcus* for the *ulnar nerve,* on the medial epicondyle of the humerus. Here the nerve may be felt and pressure on it produces an "electric" type of pain which radiates to the ulnar side of the hand. The nerve passes between the two heads of the flexor carpi ulnaris muscle on the flexor side of the forearm and then runs below this muscle to the wrist joint. It does not traverse the carpal tunnel but passes over the flexor retinaculum to the palmar surface of the hand where it divides into *superficial* and *deep branches.*

In the forearm it gives off branches **C1** to the flexor carpi ulnaris muscle **A2** and to the ulnar half of the flexor digitorum profundus muscle **A3.** In the middle of the forearm a sensory branch is given off, the **dorsal branch** of the **ulnar nerve BC4,** which passes to the ulnar side of the back of the hand where it supplies the skin. The zone supplied overlaps with that innervated by the radial nerve. Another sensory branch, the **palmar branch** of the **ulnar nerve BC5,** is given off in the distal third of the forearm. It passes to the palm of the hand and supplies the skin of the hypothenar eminence.

The **superficial branch,** as the fourth **common palmar digital nerve C6,** passes to the interosseous space between the ring and little fingers and divides into the **proper palmar digital nerves BC7,** which supply the sensory innervation of the volar surface of the little finger and the ulnar side of the ring finger, and on the extensor side they extend also to the terminal phalanges. Communicating branch to the median nerve **C8.**

The **deep branch AC9** penetrates deep into the palm and forms a loop in the region of the thenar eminence. It gives off branches to all the muscles of the hypothenar eminence **C10** (abductor muscle of the fifth finger **A11,** the flexor brevis muscle of the fifth finger **A12,** and the opponens muscle of the fifth finger **A13**), all the dorsal and palmar interosseous muscles **A14,** the third and fourth lumbrical muscles **A15,** the adductor pollicis muscle **A16,** and the deep head of the flexor pollicis brevis muscle **A17.**

Clinical Tips. Damage to the ulnar nerve produces a *claw hand* **D,** where the metacarpophalangeal joints lie in extension and the middle and terminal joints are kept flexed. This characteristic finger position is caused by paralysis of the interosseous and lumbrical muscles which bend the metacarpophalangeal joints and extend the middle and terminal interphalangeal joints. Loss of the flexors causes this position because of the unopposed action of the extensor muscles, since the little finger and the adductors of the thumb are also paralyzed, the thumb and little finger can no longer touch each other.

Skin supply: specific nerve supply of the nerve dark, maximal area supplied light.

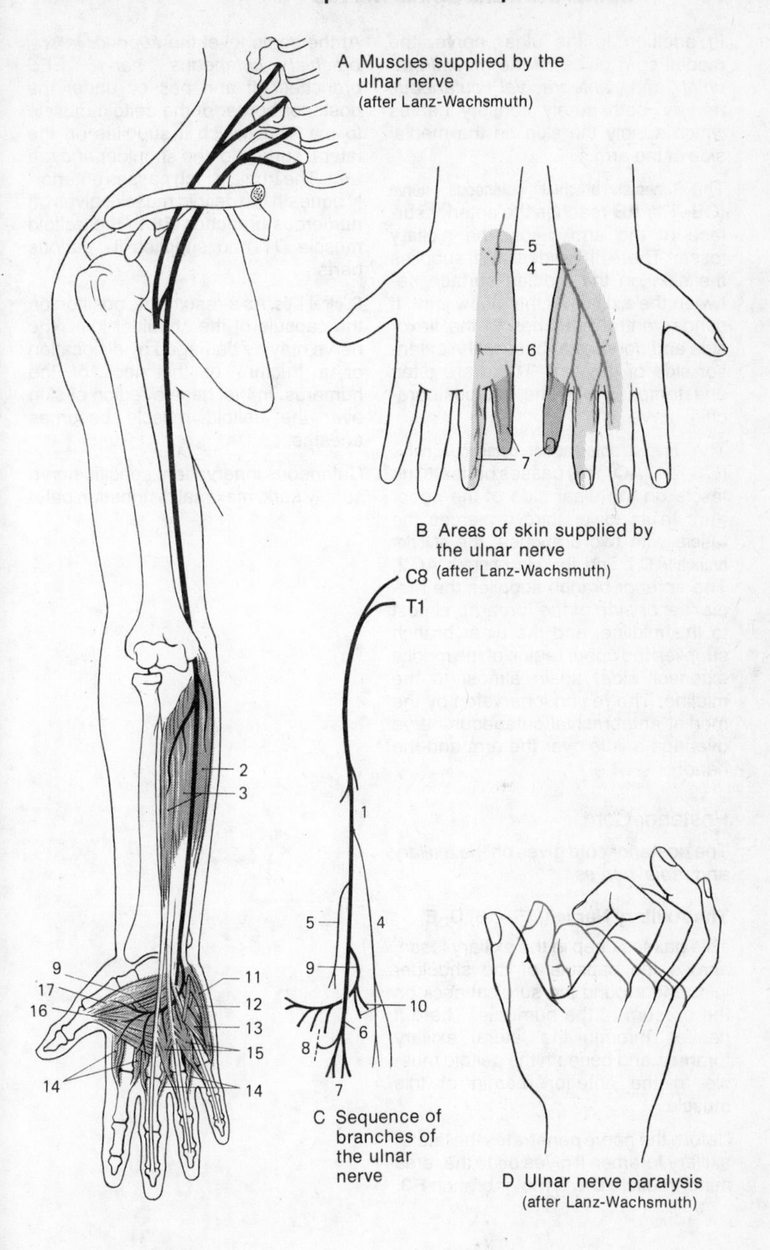

A Muscles supplied by the ulnar nerve (after Lanz-Wachsmuth)

B Areas of skin supplied by the ulnar nerve (after Lanz-Wachsmuth)

C Sequence of branches of the ulnar nerve

D Ulnar nerve paralysis (after Lanz-Wachsmuth)

In addition to the ulnar nerve, the medial cord gives off the *medial brachial* and *antebrachial cutaneous nerves,* both purely sensory nerves which supply the skin on the medial side of the arm.

The **medial brachial cutaneous nerve** (C8–T1) **AB** reaches the anterior surface of the arm below the axillary fossa. There it divides and supplies the skin on the median surface between the axilla and the elbow joint. It sends ventral branches to the flexor side and dorsal branches to the extensor side of the arm. There are often anastomoses with the intercostobrachial nerve.

The **medial antebrachial cutaneous nerve** (C8–T1) **AC.** This passes beneath the fascia on the ulnar side of the upper arm. In its lower third it pierces the fascia with two branches, the **anterior branch AC1** and the **ulnar branch AC2.** The anterior branch supplies the medial flexor side of the forearm, almost to the midline, and the ulnar branch supplies the upper region of the medial extensor side, again almost to the midline. The region innervated by the medial antebrachial cutaneous nerve overlaps a little over the arm and the hand.

Posterior Cord

The posterior cord gives off the *axillary* and *radial nerves.*

The Axillary Nerve (C5–C6) D–E

This passes deep in the axillary fossa, across the capsule of the shoulder joint and around the surgical neck on the dorsum of the humerus. There it passes through the lateral axillary foramen and beneath the deltoid muscle to the anterior margin of this muscle.

Before the nerve penetrates the lateral axillary foramen it gives off to the teres minor muscle **D4,** a motor branch **F3.**

At the same level the *superior lateral brachial cutaneous nerve* **EF5** branches off and passes under the posterior border of the deltoid muscle to the skin, which it supplies on the lateral surface of the shoulder and the arm. The trunk, which passes anteriorly beneath the deltoid muscle, gives off numerous branches **F6** to the deltoid muscle **D7** and supplies its various parts.

Clinical Tips. As a result of its position on the capsule of the shoulder joint, the nerve may be damaged by dislocation or a fracture of the neck of the humerus. In that case a region of skin over the deltoid muscle becomes anesthetic.

Cutaneous innervation: specific nerve supply dark, maximal distribution pale.

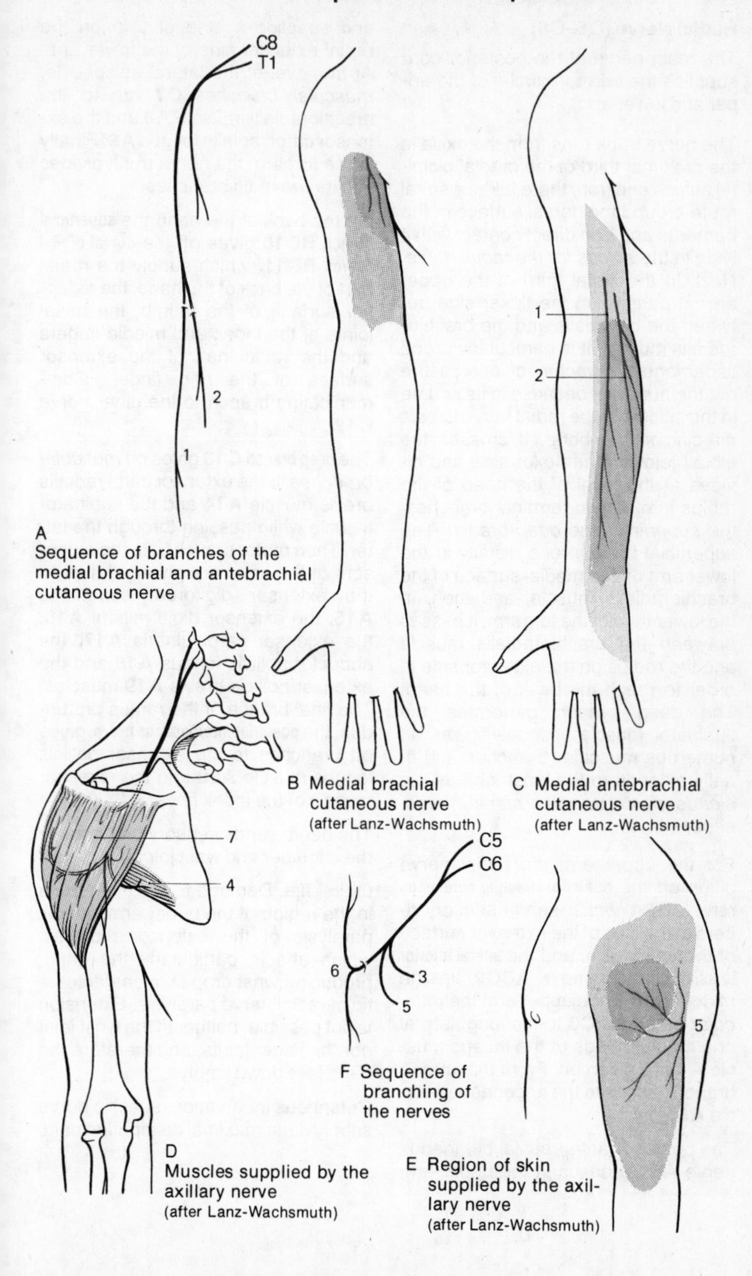

A
Sequence of branches of the medial brachial and antebrachial cutaneous nerve

B Medial brachial cutaneous nerve
(after Lanz-Wachsmuth)

C Medial antebrachial cutaneous nerve
(after Lanz-Wachsmuth)

D
Muscles supplied by the axillary nerve
(after Lanz-Wachsmuth)

E Region of skin supplied by the axillary nerve
(after Lanz-Wachsmuth)

F Sequence of branching of the nerves

Radial Nerve (C5–C8)

The main nerve of the posterior cord supplies the **extensor muscles** of the upper and lower arm.

The nerve trunk runs from the axilla in the proximal third of the medial bicipital sulcus and from there takes a spiral route around the dorsal surface of the humerus and is in direct contact with it lying in the *sulcus for the radial nerve.* Next, in the distal third of the upper arm, it passes on the flexor side between the brachialis and the brachioradialis muscle. It is particularly prone to damage by a fracture of, or pressure on, the humerus because in its course in the sulcus for the radial nerve it rests directly on the bone. It crosses the elbow joint on the flexor side and divides at the level of the head of the radius into its two terminal branches, the *superficial* and *deep branch.* The superficial branch runs distally in the lower arm on the medial surface of the brachioradialis muscle, and then, in the lower third of the forearm, it passes between the brachioradialis muscle and the radius on the extensor side in order to reach the back of the hand. The deep branch perforates the supinator muscle obliquely, gives off numerous muscular branches and finally extends to the wrist joint as the *slender posterior antebrachial interosseous nerve.*

For the upper arm the radial nerve gives off the **posterior brachial cutaneous nerve ABC1** which sends sensory fibers to the skin of the extensor surface of the upper arm, and the **lateral inferior brachial cutaneous nerve ABC2.** In the middle third of the upper arm the *muscular branches* **C3** to the long, lateral and medial heads of the triceps muscle **A4** are given off. From the latter a branch passes to the anconaeus muscle **A5.**

The posterior antebrachial cutaneous nerve **ABC6** arises in the upper arm and supplies a strip of skin on the radial extensor side of the lower arm. At the level of the lateral epicondyle, *muscular branches* **C7** run to the brachioradialis muscle **A8** and the extensor carpi radialis longus **A9.** Finally in the forearm the nerve trunk divides into its two main branches.

On the back of the hand the **superficial branch BC10** gives off the **dorsal digital nerves BC11,** which supply the radial part of the back of the hand, the extensor surface of the thumb, the basal joints of the index and middle fingers and the radial half of the extensor surface of the ring finger. Communicating branch to the ulnar nerve **C12.**

The **deep branch C13** gives off *muscular branches* to the extensor carpi radialis brevis muscle **A14** and the supinator muscle while passing through the latter. Then motor branches to the extensors of the hand are given off, i. e. to the extensor digitorum communis **A15,** the extensor digiti minimi **A16,** the extensor carpi ulnaris **A17,** the abductor pollicis longus **A18** and the extensor pollicis brevis **A19** muscles. The final branch of the ramus profundus, the **posterior interosseous nerve,** gives off branches to the extensor pollicis longus muscle **A20** and the extensor muscle of the index finger **A21.**

The nerve sends sensory branches to the shoulder and wrist joints.

Clinical Tips. Damage to the main trunk in the region of the upper arm causes paralysis of the extensor muscles, which affects particularly the hand, producing wrist drop **D,** a characteristic of radial nerve paralysis. Extension is not possible, neither in the wrist joint nor the finger joints, and therefore the hand falls down limply.

Cutaneous innervation: specific nerve supply dark, maximal distribution pale.

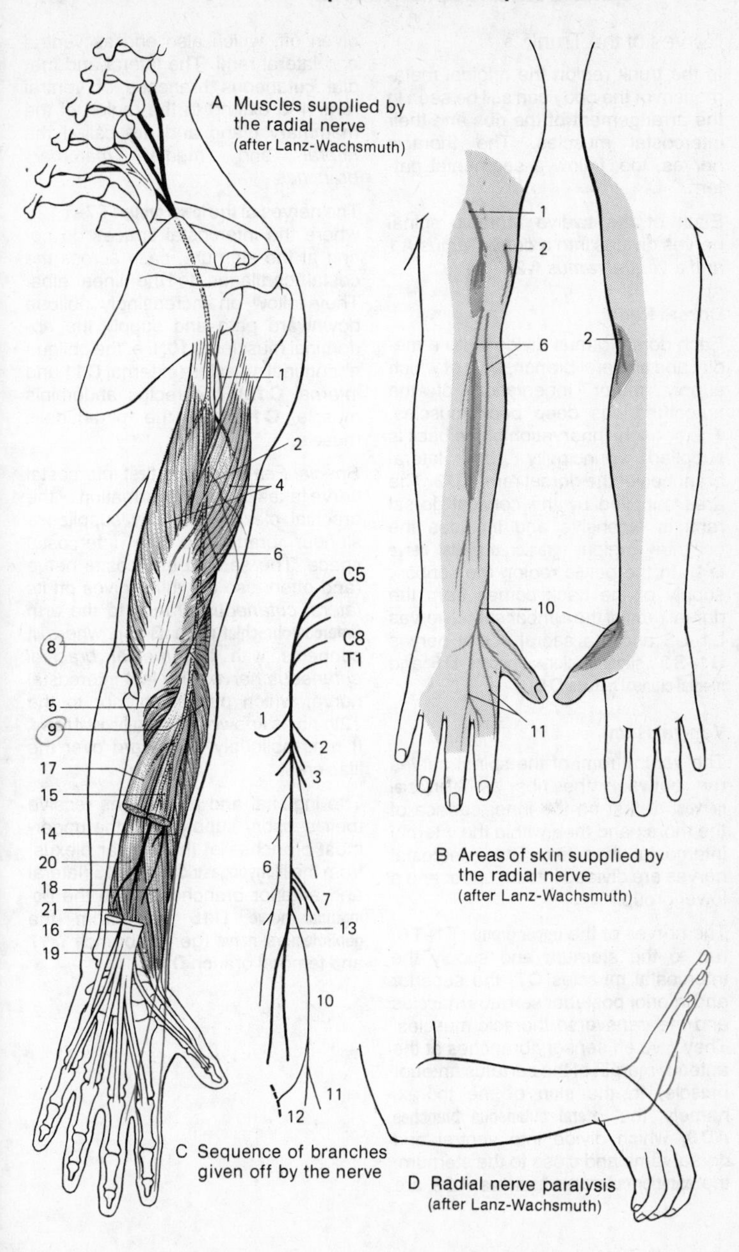

A Muscles supplied by the radial nerve (after Lanz-Wachsmuth)

B Areas of skin supplied by the radial nerve (after Lanz-Wachsmuth)

C Sequence of branches given off by the nerve

D Radial nerve paralysis (after Lanz-Wachsmuth)

Nerves of the Trunk

In the trunk region the original metamerism of the body can still be seen in the arrangement of the ribs and their intercostal muscles. The thoracic nerves, too, follow a segmental pattern.

Each of the twelve thoracic spinal nerves divides into a *dorsal ramus* **A 1** and a *ventral ramus* **A 2.**

Dorsal Rami

Each dorsal ramus divides into a medial and a lateral branch, both of which supply motor innervation of the autochthonous deep back muscles. The sensory innervation of the back is supplied principally by lateral branches of the dorsal rami **AD 3.** The area supplied by the cervical dorsal rami is extensive and includes the occipital region (**greater occipital nerve D 4**). In the pelvic region the sensory supply of the back comes from the dorsal rami of the lumbar spinal nerves L 1–L 3 and the sacral spinal nerves S 1–S 3 (**superior cluneal nerves D 5** and **medial cluneal nerves D 6**).

Ventral Rami

The ventral rami of the spinal nerves run between the ribs as **intercostal nerves,** at first on the inner surface of the thorax and then within the internal intercostal muscles. The intercostal nerves are divided into an upper and a lower group.

The nerves of the **upper group** (T 1–T 6) run to the sternum and supply the intercostal muscles **C 7,** the superior and inferior posterior serratus muscles and the transverse thoracic muscles. They give off sensory branches at the anterior margin of the serratus anterior muscle, to the skin of the thorax, namely the **lateral cutaneous branches AD 8,** which divide into ventral and dorsal rami, and close to the sternum, the **anterior cutaneous branches AD 9** are given off, which also end as ventral and lateral rami. The lateral and medial cutaneous branches of ventral rami 4–6 extend to the region of the mammary gland and are called the *lateral* and *medial mammary branches.*

The nerves of the **lower group** (T 7–T 12), where the intercostal spaces do not end at the sternum, pass across the costal cartilages to the linea alba. They follow an increasingly oblique downward path and supply the abdominal muscles **C 10,** i. e. the oblique abdominal muscles (external **C 11** and internal **C 12**), the rectus abdominis muscle **C 13** and the pyramidalis muscle.

Special Features: The first intercostal nerve takes part in the formation of the brachial plexus and only supplies a slender branch to the intercostal space. The second intercostal nerve (and often also the third) gives off its *lateral cutaneous branch* to the arm (**intercostobrachial nerve B 14**), where it connects with the *medial brachial cutaneous nerve.* The last intercostal nerve, which passes inferior to the 12th rib, is known as the **subcostal nerve.** It runs obliquely downward over the iliac crest.

The inguinal and hip regions receive their sensory supply from the uppermost branches of the lumbar plexus; from the **iliohypogastric nerve D 15** (lateral and anterior branches), from the **ilioinguinal nerve D 16** and from the **genitofemoral nerve** (genital branch **D 17** and femoral branch **D 18**).

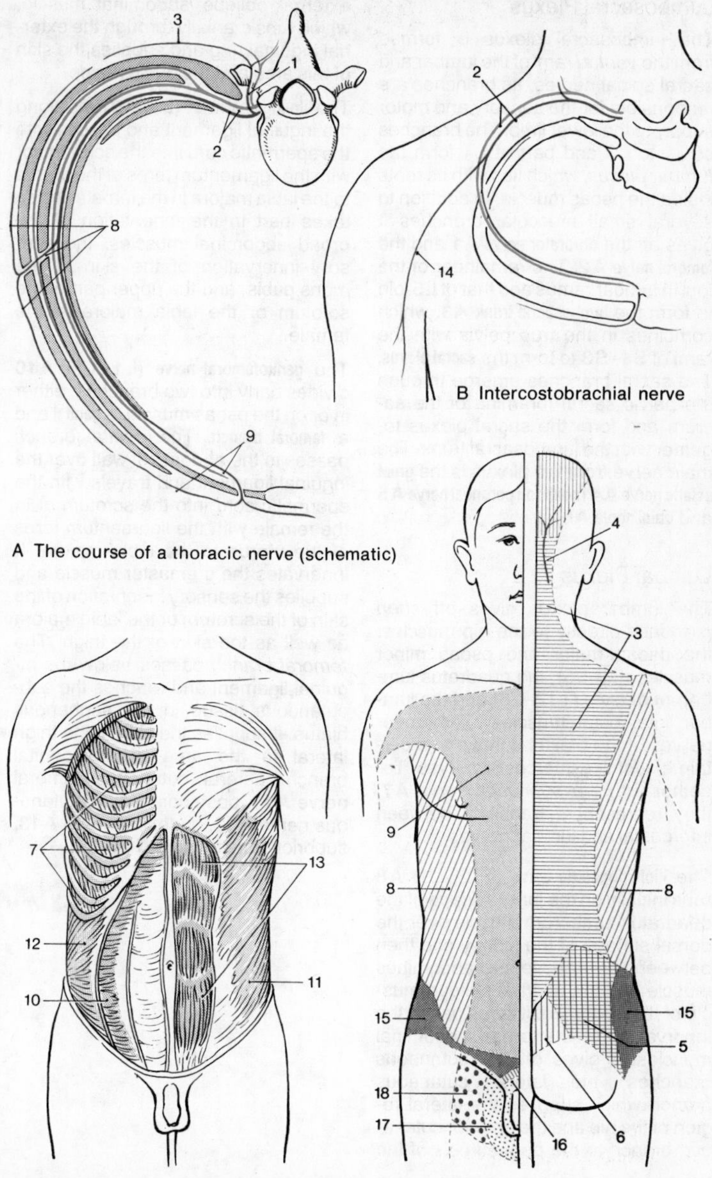

A The course of a thoracic nerve (schematic)

B Intercostobrachial nerve

C Muscles supplied by intercostal nerves

D Cutaneous supply of the trunk

Lumbosacral Plexus

The lumbosacral plexus is formed from the *ventral rami* of the lumbar and sacral spinal nerves. Its branches are responsible for the sensory and motor supply to the lower limb. The branches of L1 to L3 and part of L4 form the *lumbar plexus,* which lies with its roots inside the psoas muscle. In addition to several small muscular branches it gives off the **obturator nerve A1** and the **femoral nerve A2.** The remainder of the fourth lumbar ramus and that of L5 join to form the **lumbosacral trunk A3,** which combines in the true pelvis with the rami of S1–S3 to form the **sacral plexus.** The sacral branches emerge through the pelvic sacral foramina of the sacrum and form the sacral plexus together with the lumbosacral trunk. The main nerve from the plexus is the **great sciatic nerve A4 (common peroneal nerve A5** and **tibial nerve A6**).

Lumbar Plexus

The lumbar plexus gives off *short branches* directly to the hip muscles: the psoas major and psoas minor muscles (L1–L5), the quadratus lumborum muscle (T12–L3) and the lumbar intercostal muscles. The upper nerves of the plexus behave to a certain extent like intercostal nerves. Together with the *subcostal nerve* **A7,** they represent a transition between intercostal and lumbar nerves.

The **iliohypogastric nerve** (T12, L1) **A8** runs initially on the inner surface of the quadratus lumborum muscle over the dorsal surface of the kidney and then between the transversus abdominus muscle and the internal oblique muscle of the abdomen. It takes part in the innervation of the broad abdominal muscles. It gives off two cutaneous branches, the lateral cutaneous branch which supplies the lateral region of the hip and the anterior cutaneous branch to the aponeurosis of the external oblique abdominal muscle, which runs cranially through the external inguinal ring and supplies the skin of this area and the pubic region.

The **ilio-inguinal nerve** (L1) **A9** runs along the inguinal ligament and passes with the spermatic cord into the scrotum, or with the ligamentum teres of the uterus to the labia majora in the female. It also takes part in the innervation of the broad abdominal muscles, the sensory innervation of the skin of the mons pubis, and the upper part of the scrotum or the labia majora in the female.

The **genitofemoral nerve** (L1, L2) **A10** divides early into two branches, either in or on the psoas muscle, a **genital** and a **femoral branch.** The *genital branch* passes in the abdominal wall over the inguinal ligament and travels with the spermatic cord into the scrotum or in the female with the ligamentum teres of the uterus to the labia majora. It innervates the cremaster muscle and supplies the sensory innervation of the skin of the scrotum or the labia majora as well as the skin of the thigh. The *femoral branch* passes below the inguinal ligament and reaches the subcutaneous tissue in the saphenous hiatus. It supplies the skin of the thigh lateral to the area of the genital branch. Lateral cutaneous femoral nerve **A11,** posterior femoral cutaneous nerve **A12,** pudendal nerve **A13,** superior gluteal nerve **A14.**

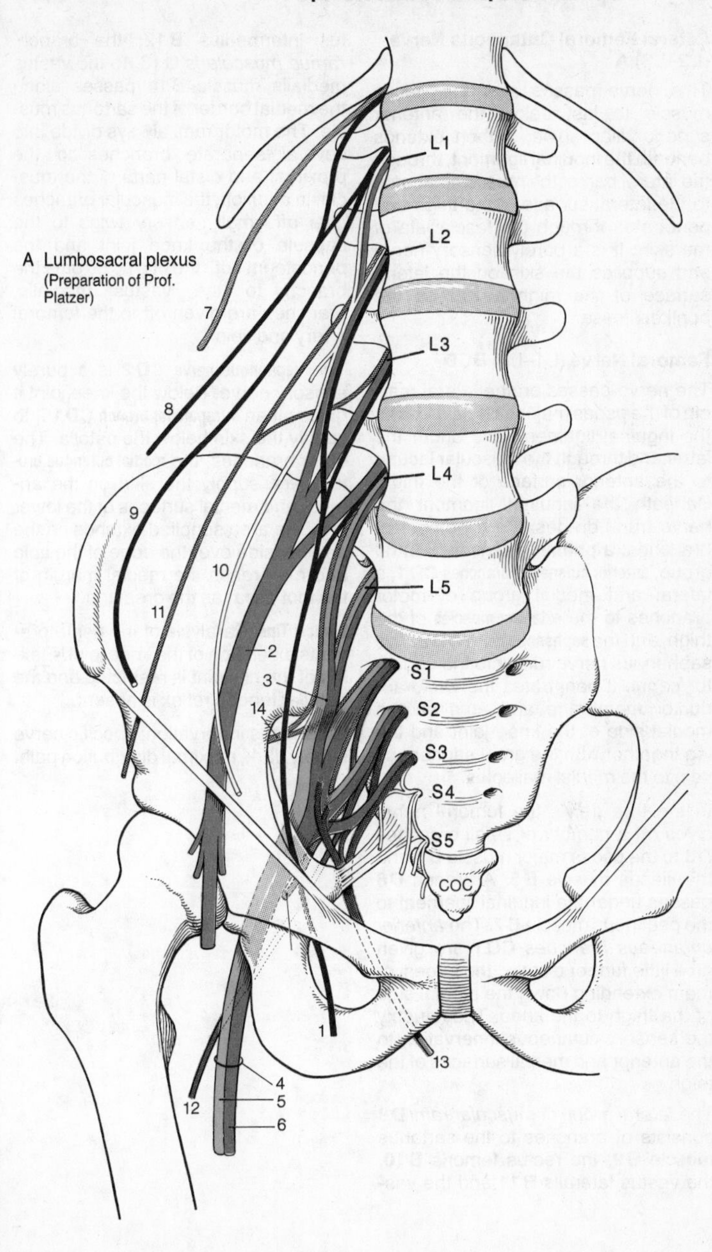

A Lumbosacral plexus
(Preparation of Prof. Platzer)

Lateral Femoral Cutaneous Nerve (L2–L3) **A**

This nerve passes over the iliacus muscle to just below the anterior superior iliac spine. It then extends beneath the inguinal ligament, through the lateral part of the muscular lacuna, to the lateral surface of the thigh and penetrates through the fascia lata of the skin. It is a purely sensory nerve and supplies the skin on the lateral surface of the thigh as far as the popliteal fossa.

Femoral Nerve (L1–L4) **BCD**

The nerve passes on the lateral margin of the psoas major muscle as far as the inguinal ligament, and under the latter and through the muscular lacuna to the anterior surface of the thigh. Beneath the inguinal ligament the nerve trunk divides into a number of branches: a primarily sensory ventral group, **anterior cutaneous branches CD1,** a lateral and medial group of motor branches to the **extensor muscles** of the thigh, and the **saphenous nerve CD2.** The saphenous nerve runs into the adductor canal. It penetrates the vasto-adductor membrane and runs on the medial side of the knee joint and the leg together with the great saphenous vein to the medial malleolus.

In the true pelvis the femoral nerve gives off a number of small branches **D3** to the psoas major muscle **B4** and the iliacus muscle **B5.** A branch **D6** passes under the inguinal ligament to the pectineus muscle **B7.** The *anterior cutaneous branches* **CD1** are given off a little further distally, the largest of them extending down the medial side of the thigh to the knee. They supply the sensory cutaneous innervation to the anterior and medial surfaces of the thigh.

The *lateral group* of *muscular rami* **D8** consists of branches to the sartorius muscle **B9,** the rectus femoris **B10,** the vastus lateralis **B11** and the vastus intermedius **B12;** the branch, *ramus muscularis* **D13,** to the vastus medialis muscle **B14** passes along the medial border of the sartorius muscle. The motor rami always divide into several separate branches to the proximal and distal parts of the muscle. In addition, the muscular branches give off small sensory twigs to the capsule of the knee joint and the periosteum of the tibia. From the branch to the vastus medialis, branches are given off to the femoral artery and vein.

The **saphenous nerve CD2** is a purely sensory nerve. Below the knee joint it gives off an **intrapatellar branch CD15,** to supply the skin below the patella. The other branches, the **medial cutanous crural nerves,** supply the skin on the anterior and medial surfaces of the lower leg. The area supplied extends on the anterior side over the edge of the tibia and may reach the medial margin of the foot as far as the great toe.

Clinical Tips. Paralysis of the nerve prevents extension of the knee joint. Flexion of the hip joint is restricted and the patellar tendon reflex is absent.

Cutaneous innervation: specific nerve supply dark, maximal distribution pale.

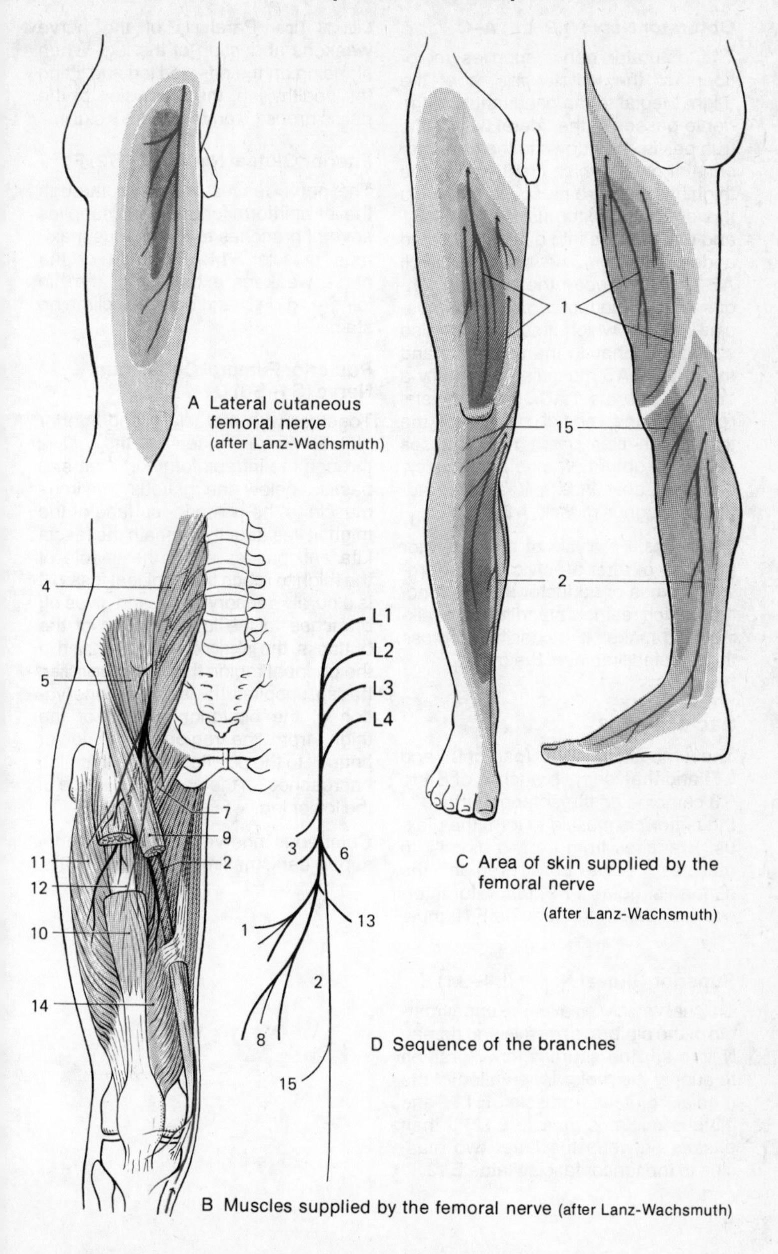

A Lateral cutaneous femoral nerve (after Lanz-Wachsmuth)

C Area of skin supplied by the femoral nerve (after Lanz-Wachsmuth)

D Sequence of the branches

B Muscles supplied by the femoral nerve (after Lanz-Wachsmuth)

Obturator Nerve (L2–L4) A–C

The obturator nerve supplies motor fibers to the **adductor muscles** of the thigh. Medial to the psoas muscle the nerve passes to the lateral wall of the true pelvis, then down to the obturator canal through which it extends to the thigh. It supplies a muscular branch to the obturator externus muscle **AB1** and then divides into a superficial and a deep branch. The **superficial branch AB2** runs between the adductor longus **A3** and adductor brevis **A4** muscles, both of which it supplies. It also sends branches to the pectineus and the gracilis **A5** muscles, and finally a *cutaneous branch* **ABC6** to the distal region of the medial surface of the thigh. The **deep branch AB7** passes over the obturator externus muscle and then descends to reach the adductor magnus muscle **A8.**

Clincial Tips. Paralysis of the obturator nerve, e.g. after a pelvic fracture, results in loss of adductor muscle function, which restricts standing and walking, and makes it impossible to cross the affected limb over the other.

Sacral Plexus

The *lumbosacral trunk* (part of L4 and L5) and the ventral branches of S1 to S3 combine on the anterior surface of the piriformis muscle to form the plexus. Branches from it pass directly to muscles in the pelvic region: the piriformis, gemelli **F9,** obturator internus and quadratus femoris **F10** muscles.

Superior Gluteal Nerve (L4–S1) E

This nerve passes over the upper margin of the piriformis muscle and dorsally through the suprapiriform foramen to supply the motor innervation of the medial gluteus muscle **E11** and gluteus minimus muscle **E12.** It then passes between the latter two muscles to the tensor fasciae latae **E13.**

Clinical Tips. Paralysis of the nerve weakens abduction of the leg. When standing on the affected leg and lifting the healthy leg, the other side of the pelvis drops (Trendelenburg's sign).

Inferior Gluteal Nerve (L5–S2) F

The nerve leaves the pelvis through the infrapiriform foramen and supplies several branches to the gluteus maximus muscle **F14.** Paralysis of the nerve weakens extension of the hip joint, e. g. in standing up or climbing stairs.

Posterior Femoral Cutaneous Nerve (S1–S3) D

Together with the sciatic and inferior gluteal nerves it leaves the pelvis through the infrapiriform foramen and passes below the gluteus maximus muscle to the posterior surface of the thigh. It lies directly beneath the fascia lata and passes along the middle of the thigh to reach the popliteal fossa. It is a purely sensory nerve and gives off branches to the lower margin of the buttocks, the **inferior cluneal nerves,** and in the perineal region the **perineal branches.** It also supplies the sensory innervation of the posterior surface of the thigh, from the region of the lower buttock to the popliteal fossa, and also encroaches on the proximal surface of the lower leg.

Cutaneous innervation: specific nerve supply dark, maximal distribution pale.

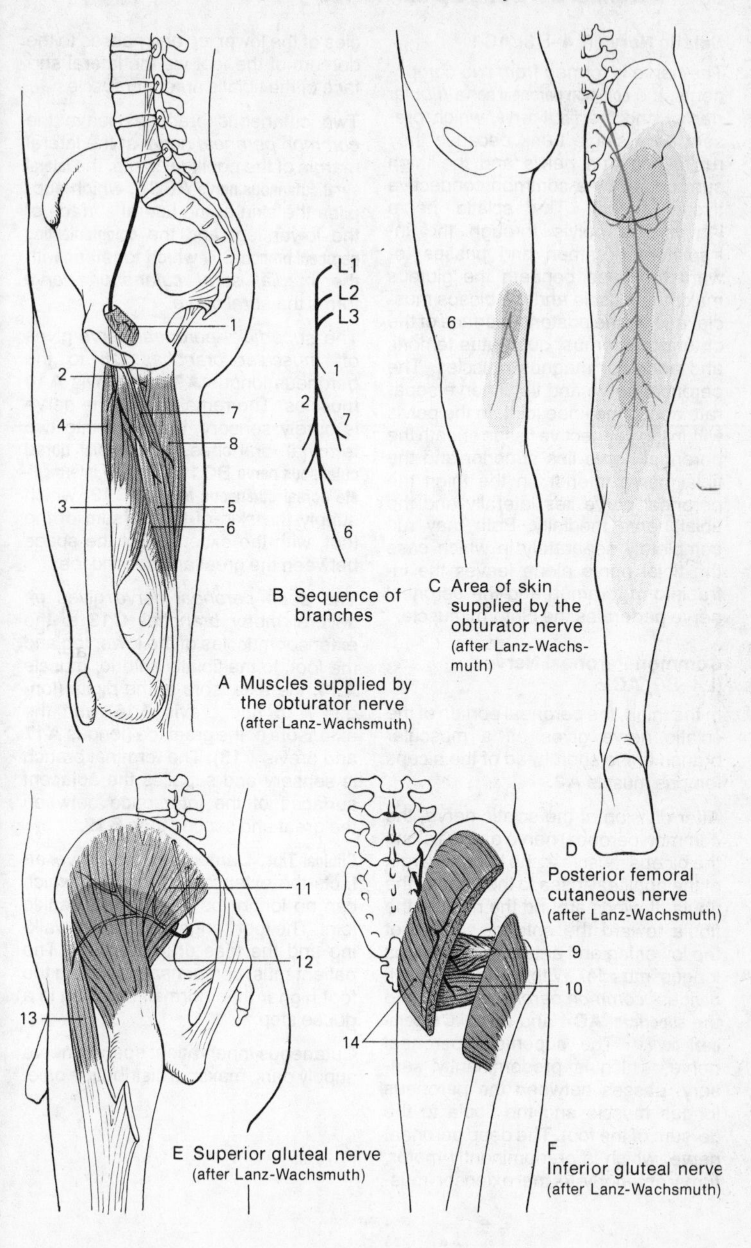

A Muscles supplied by the obturator nerve (after Lanz-Wachsmuth)

B Sequence of branches

C Area of skin supplied by the obturator nerve (after Lanz-Wachsmuth)

D Posterior femoral cutaneous nerve (after Lanz-Wachsmuth)

E Superior gluteal nerve (after Lanz-Wachsmuth)

F Inferior gluteal nerve (after Lanz-Wachsmuth)

Sciatic Nerve (L4–S3) **AC1**

The nerve is formed from two components, the **common peroneal nerve** *(fibular nerve)* and the **tibial nerve,** which present as a single trunk because they run in the true pelvis and the thigh surrounded by a common connective tissue sheath. The sciatic nerve leaves the pelvis through the infrapiriform foramen and passes toward the knee beneath the gluteus maximus muscle and the biceps muscle, and on the posterior surface of the obturator internus, quadratus femoris and adductor magnus muscles. The peroneal nerve and tibial nerve separate above the knee joint. In the pelvis within the connective tissue sheath the peroneal nerve lies superior and the tibial nerve inferior; in the thigh the peroneal nerve lies laterally and the tibial nerve medially. Both may run completely separately in which case the tibial nerve alone leaves the infrapiriform foramen and the peroneal nerve perforates the piriform muscle.

Common Peroneal Nerve (L4–S2) **AC2**

In the thigh, the peroneal portion of the sciatic nerve gives off a muscular branch to the short head of the biceps femoris muscle **A3.**

After division of the sciatic nerve, the common peroneal nerve passes along the biceps muscle at the lateral margin of the popliteal fossa to the head of the fibula. It winds around the neck of the fibula toward the anterior surface of the lower leg and enters the peroneus longus muscle. Within the muscle it divides *(common peroneal nerve)* into the **superficial AC4** and **deep AC5 peroneal nerves.** The superficial peroneal nerve, which is predominantly sensory, passes between the peroneus longus muscle and the fibula to the dorsum of the foot. The deep peroneal nerve, which is predominantly motor, turns anteriorly to the extensor muscles of the lower leg and passes to the dorsum of the foot via the lateral surface of the tibialis anterior muscle.

Two cutaneous branches leave the *common peroneal nerve* at the lateral margin of the popliteal fossa, the **lateral sural cutaneous nerve ABC6,** which supplies the skin of the lateral surface of the lower leg, and the **communicating peroneal branch C7,** which together with the *medial sural cutaneous nerve* forms the *sural nerve.*

The *superficial peroneal nerve* gives off *muscular branches* **C8** to the peroneus longus **A9** and brevis **A10** muscles. The remainder of the nerve is purely sensory; it divides into two terminal branches, the **medial dorsal cutaneous nerve BC11** and the **intermediate dorsal cutaneous nerve BC12,** which supply the skin of the dorsum of the foot, with the exception of the space between the great and second toes.

The *deep peroneal nerve* gives off several *motor branches* **C13** to the extensor muscles of the lower leg and the foot, to the tibialis anterior muscle **A14,** the extensors of the digits (longus **A15** and brevis **A16**) and the extensors of the great toe (longus **A17** and brevis **A18**). The terminal branch is sensory and supplies the adjacent surfaces of the interspace between the great and second toes **B19.**

Clinical Tips. Damage to the nerve affects the extensors of the foot, which can no longer be raised at the ankle joint. The foot hangs down whilst walking and the toes drag the floor. The patient must compensate by lifting the foot higher than normal, resulting in a goose step.

Cutaneous innervation: specific nerve supply dark, maximal distribution pale.

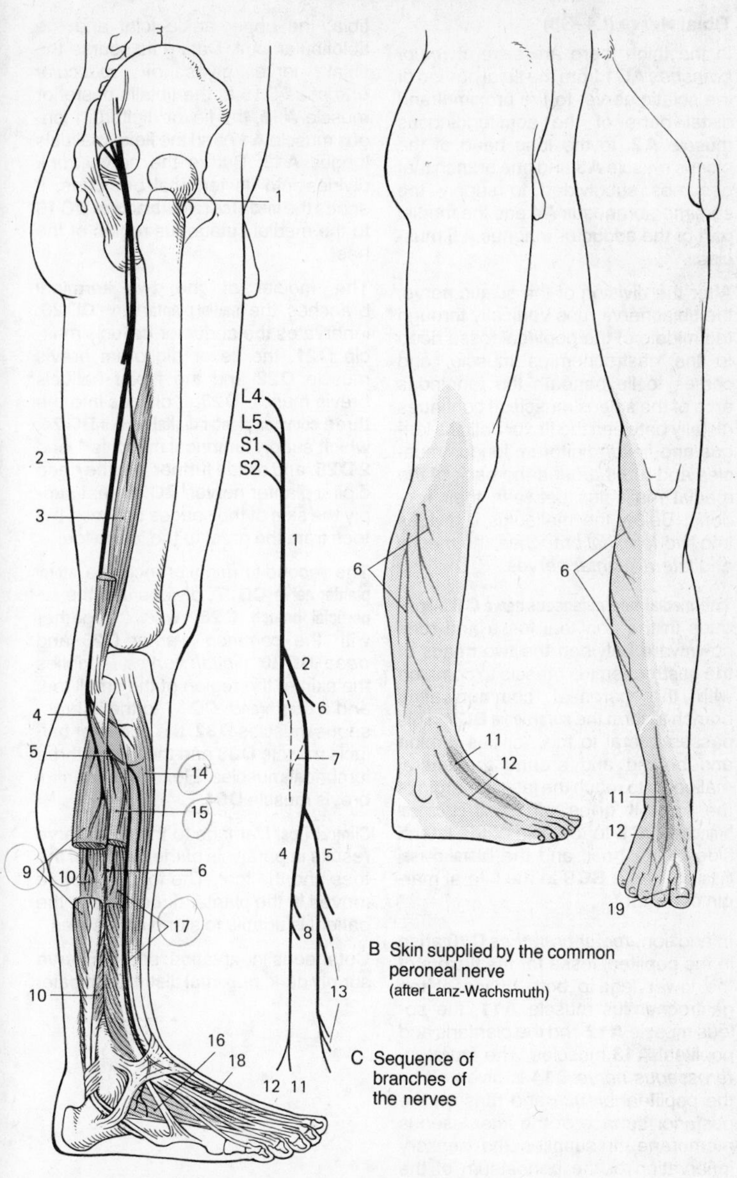

B Skin supplied by the common peroneal nerve (after Lanz-Wachsmuth)

C Sequence of branches of the nerves

A Muscles supplied by the common peroneal nerve (after Lanz-Wachsmuth)

Tibial Nerve (L4–S3)

In the thigh there are several *motor branches* **AC1** from the tibial portion of the sciatic nerve: to the proximal and distal parts of the semitendinosus muscle **A2**, to the long head of the biceps muscle **A3** and one branch that becomes subdivided to supply the semimembranosus **A4** and the medial part of the adductor magnus **A5** muscles.

After the division of the sciatic nerve, the tibial nerve runs vertically through the middle of the popliteal fossa deep to the gastrocnemius muscle, and comes to lie beneath the tendinous arch of the soleus muscle. It continues distally between the flexor hallucis longus and flexor digitorum longus muscles and turns around the back of the medial malleolus between their tendons. Below the malleolus it divides into two *terminal branches*, the *medial* and *lateral plantar nerves*.

The **medial sural cutaneous nerve C6** separates in the popliteal fossa and runs downward between the two heads of the gastrocnemius muscle to combine with the peroneal communicating branch to form the **sural nerve BC7**. This passes lateral to the Achilles tendon and behind and around the lateral malleolus to reach the lateral margin of the foot. It gives off **lateral calcaneal branches BC8** to the skin of the lateral side of the heel, and the **lateral dorsal cutaneous nerve BC9** to the lateral margin of the foot.

In addition, *motor branches* **C10** arise in the popliteal fossa for the flexors of the lower leg: to both heads of the gastrocnemius muscle **A11**, the soleus muscle **A12** and the plantaris and popliteal **A13** muscles. The crural *interosseous nerve* **C14** is given off by the popliteal branch and runs on the posterior surface of the interosseous membrane. It supplies the sensory innervation of the periosteum of the tibia, the upper ankle joint and the tibiofibular joint. During its course the tibial nerve gives off *muscular branches* **C15** to the tibialis posterior muscle **A16**, the flexor digitorum longus muscle **A17** and the flexor hallucis longus **A18**. Before the nerve trunk divides into its terminal branches, it sends the **medial calcaneal branches BC19** to the medial cutaneous region of the heel.

The medial of the two terminal branches, the **medial plantar nerve CD20**, innervates the abductor hallucis muscle **D21**, the flexor digitorum brevis muscle **D22** and the flexor hallucis brevis muscle **D23**. It divides into the three **common plantar digital nerves BC24**, which supply lumbrical muscles 1 and 2 **D25**, and divide further into the deep digital plantar nerves **BC26**, that supply the skin of the spaces between the toes from the great to the fourth toe.

The second terminal branch, the **lateral plantar nerve CD27**, divides into the **superficial branch C28**, which, together with the *common plantar* **C29** and *deep* **BC30** *digital nerves*, supplies the skin of the region of the small toe, and a **deep branch CD31** to the interosseous muscles **D32**, the adductor hallucis muscle **D33** and the lateral three lumbrical muscles. Flexor digiti minimi brevis muscle **D34**.

Clinical Tips. Damage to the tibial nerve results in paralysis of the flexors of the toes and the foot. The foot cannot be moved in the plantar direction and the patient is unable to stand on his toes.

Cutaneous innervation: specific nerve supply dark, maximal distribution pale.

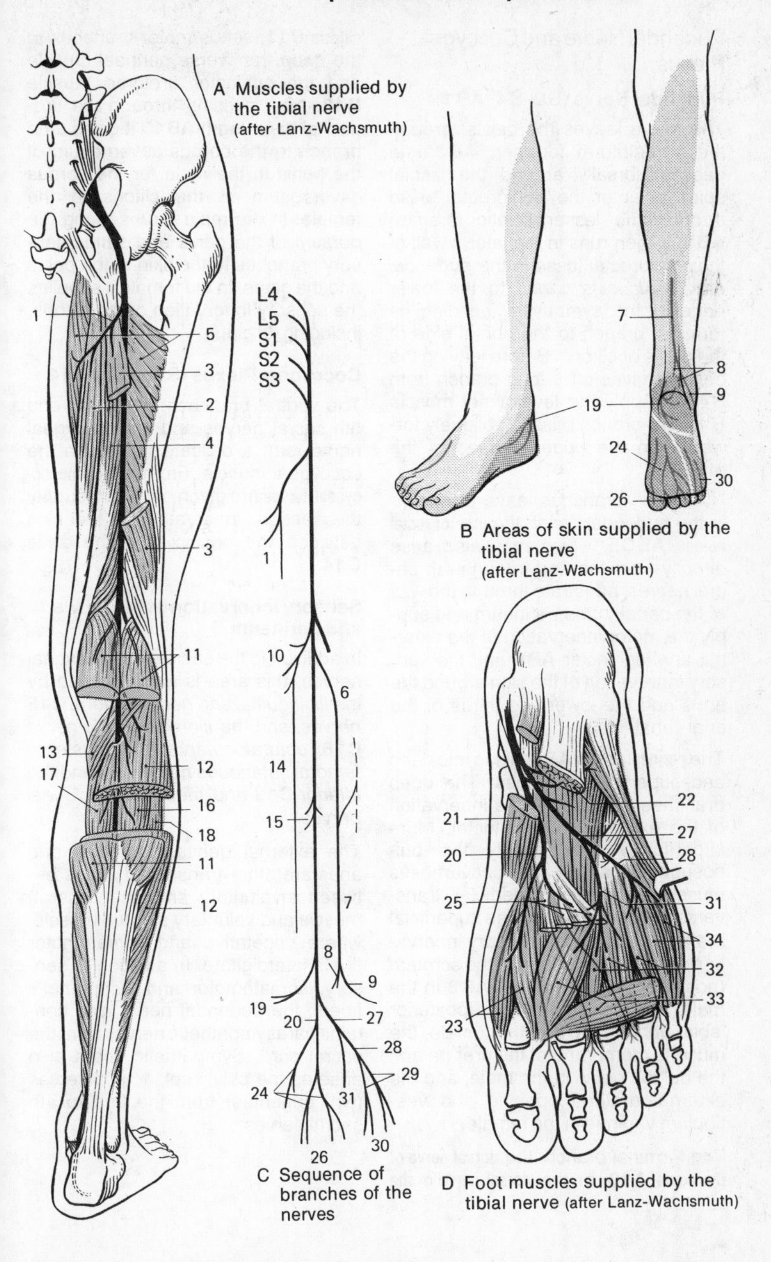

A Muscles supplied by the tibial nerve (after Lanz-Wachsmuth)

B Areas of skin supplied by the tibial nerve (after Lanz-Wachsmuth)

C Sequence of branches of the nerves

D Foot muscles supplied by the tibial nerve (after Lanz-Wachsmuth)

Pudendal Nerve and Coccygeal Plexus

Pudendal Nerve (S2–S4) AB1

The nerve leaves the pelvis through the infrapiriform foramen **AB2** and passes dorsally around the ischial spine to enter the ischiorectal fossa through the lesser sciatic foramen **AB3.** It then runs in the lateral wall of the ischiorectal fossa in the *pudendal canal (Alcock's canal)* to the lower edge of the symphysis, sending its terminal branch to the dorsal side of the penis or clitoris. Before leaving the pelvis it gives off a long branch from S4 to supply the levator ani muscle **B4**. This branch passes obliquely forwards on the upper surface of the muscle.

Numerous branches leave the nerve in the pudendal canal: the **inferior rectal nerves ABC5,** which may also arise directly from the 2nd through 4th sacral nerves, penetrate through the wall of the canal to the perineum and supply the motor innervation of the external anal sphincter **AB6** and the sensory innervation of the skin around the anus and the lower two thirds of the anal canal.

The **perineal nerves AB7** divide into deep and superficial branches. The deep branches take part in the innervation of the external anal sphincter. More superficially they supply the bulbospongiosus and ischiocavernosus muscles and the superficial transverse perineal muscle. The superficial branches supply the sensory innervation of the posterior part of the scrotum (posterior scrotal nerves) **AC8** in the male, and the labia majora (posterior labial nerves) **BC9** in the female, the mucous membrane of the urethra and the bulbus penis in the male, and the external urethral opening and vestibulum vaginae in the female.

The terminal branch, the **dorsal nerve of the penis A10,** or the **dorsal nerve of the clitoris B11,** sends motor branches to the deep transverse perineal muscle and the sphincter urethrae muscle **B12.** After passing through the urogenital diaphragm **AB13** it gives off a branch to the corpus cavernosum of the penis in the male, or the corpus cavernosum of the clitoris in the female. In the male it runs along the dorsum of the penis and sends sensory branches to the skin of the penis and the glans. In the female it supplies the sensory innervation of the clitoris, including its glans.

Coccygeal Plexus (S4–C0) AB14

The ventral branches of the 4th and 5th sacral nerves and the coccygeal nerve form a delicate plexus on the coccygeal muscle. From it the **anococcygeal nerves** are given off, which supply the sensory innervation to the skin between the coccyx and the anus **C14.**

Sensory Innervation of the Pelvis and Perineum C

In addition to the sacral and coccygeal nerves, this area is also supplied by the *ilioinguinal* and *genitofemoral* **C15** *nerves,* and the *iliohypogastric nerve* **C16,** *obturator nerve* **C17,** *posterior femoral cutaneous nerve* **C18** and the *inferior* **C19** and *medial* **C20** *cluneal nerves.*

The external genitalia, urethra and anus are other transitional zones between involuntary smooth (visceral) muscle and voluntary striated muscle, where vegetative and somatomotor fibers interdigitate. In addition to sensory, somatomotor and sympathetic fibers, the pudendal nerve also contains parasympathetic nerves from the sacral cord. Sympathetic fibers also arise as the pelvic splanchnic nerves (Nn. erigentes) from the 2nd to 4th sacral nerves.

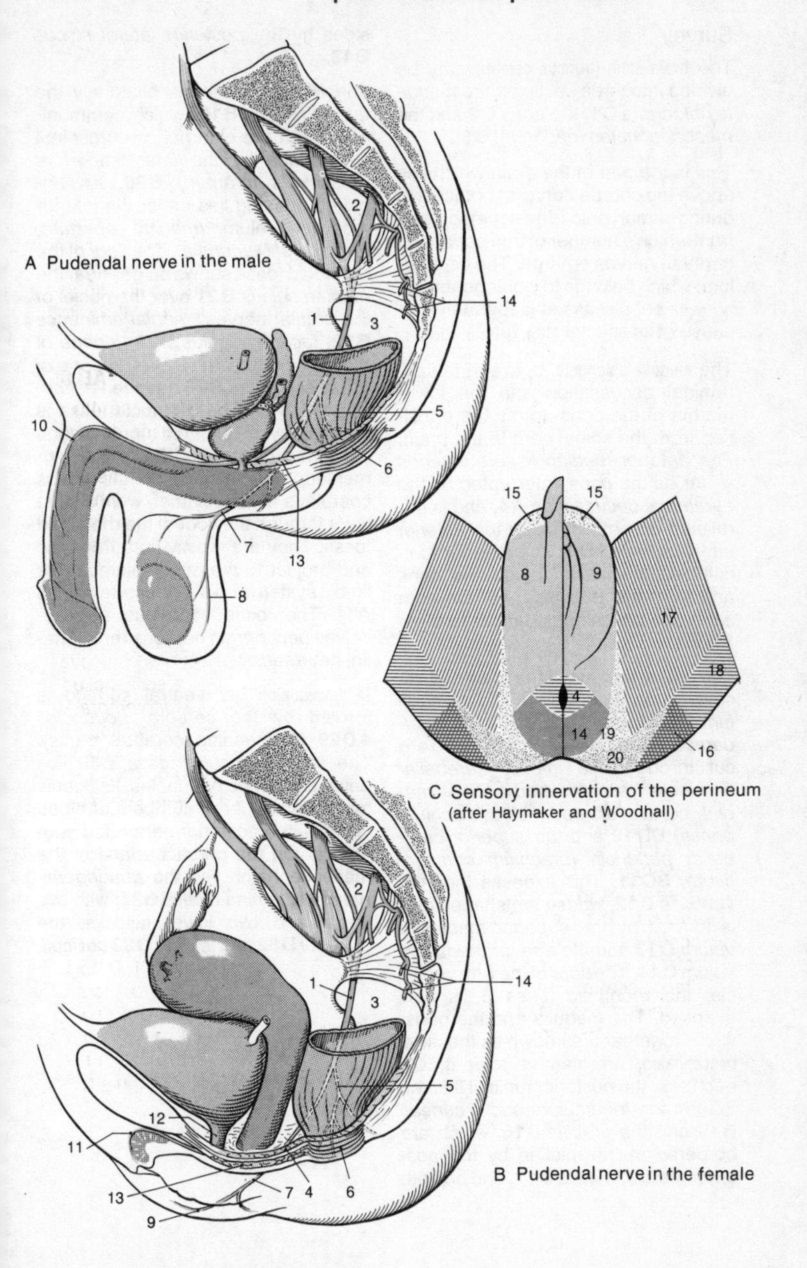

A Pudendal nerve in the male

C Sensory innervation of the perineum (after Haymaker and Woodhall)

B Pudendal nerve in the female

Survey

The **brain stem (truncus cerebri)** may be divided into three sections: the *medulla oblongata* **C1,** the *pons* **C2** and the *mesencephalon (midbrain)* **C3.**

This is the part of the brain which lies above the chorda dorsalis (notochord) during embryonic development and from which a number of true peripheral (cranial) nerves emerge. The cerebellum, which belongs to it ontogenetically, will be discussed separately because of its special structure.

The **medulla oblongata,** between the pyramidal decussation and the lower margin of the pons, forms the transition from the spinal cord to the brain. The *anterior median fissure* extends as far as the pons, interrupted by the *pyramidal decussation* **A4,** and is paralleled on both sides by the *anterior lateral sulcus* **AD5.** The anterior funiculi become thicker below the pons and form the *pyramids* **A6.** On both sides the *olives* bulge lateral to them **AD7.**

The dorsal surface of the brain stem is covered on each side by the *cerebellum* **C8.** Upon removal of it the three *cerebellar peduncles* on each side are cut through: the *inferior cerebellar peduncle (restiform body)* **BD9,** *medial cerebellar peduncle (brachium pontis)* **BD10** and the *superior cerebellar peduncle (brachium conjunctivum)* **BD11.** This exposes the *IVth ventricle* **C12,** whose tentshaped roof is formed by the *superior medullary velum* **C13** and the *inferior medullary velum* **C14.** The floor of the IVth ventricle, the *rhomboid fossa* B, is now exposed. The medulla and the pons, which together are known as the **rhombencephalon,** are named after it. On each side the posterior funiculi thicken to form the *tuberculum nuclei cuneati* **B15** and the *gracilis* **B16,** which are bordered in the midline by the *posterior median sulcus* **B17,** and on both sides by the *posterior lateral sulcus* **B18.**

The **IVth ventricle** forms bilaterally the *lateral recess* **B19,** which communicates with the external cerebrospinal fluid space by the *lateral aperture (Luschka's foramen)* **B20.** An unpaired opening lies under the inferior medullar velum *(median aperture, foramen of Magendie).* The floor of the rhomboid fossa shows bulges near the *median sulcus* **B21** over the nuclei of the cranial nerves: *medial eminence* **B22,** *facial colliculus* **B23,** *trigone of the hypoglossal nerve* **B24,** *trigone of the vagus nerve* **B25** and the *vestibular area* **B26.** The rhomboid fossa is crossed by myelinated nerve fibers, the *striae medullares* **B27.** The pigmented nerve cells of the locus coeruleus **B28** glimmer with a blue color through the floor of the rhomboid fossa. They are mostly noradrenergic and project to the hypothalamus, the limbic system and the neocortex (p. 28 A1). The locus coeruleus also receives peptidergic neurons (encephalin, neurotensin).

Mesencephalon. Its ventral surface is formed by the *cerebral peduncles* **AD29** (descending cerebral tracts). The *interpeduncular fossa* **A30** lies between the two peduncles. Its floor is perforated by large numbers of blood vessels, the posterior perforated substance. On the dorsal surface of the mesencephalon lies the *quadrigeminal plate, lamina tecti,* **BD31,** with two upper and two lower hillocks, the *superior* **D32** and *inferior* **D33** *colliculi.*

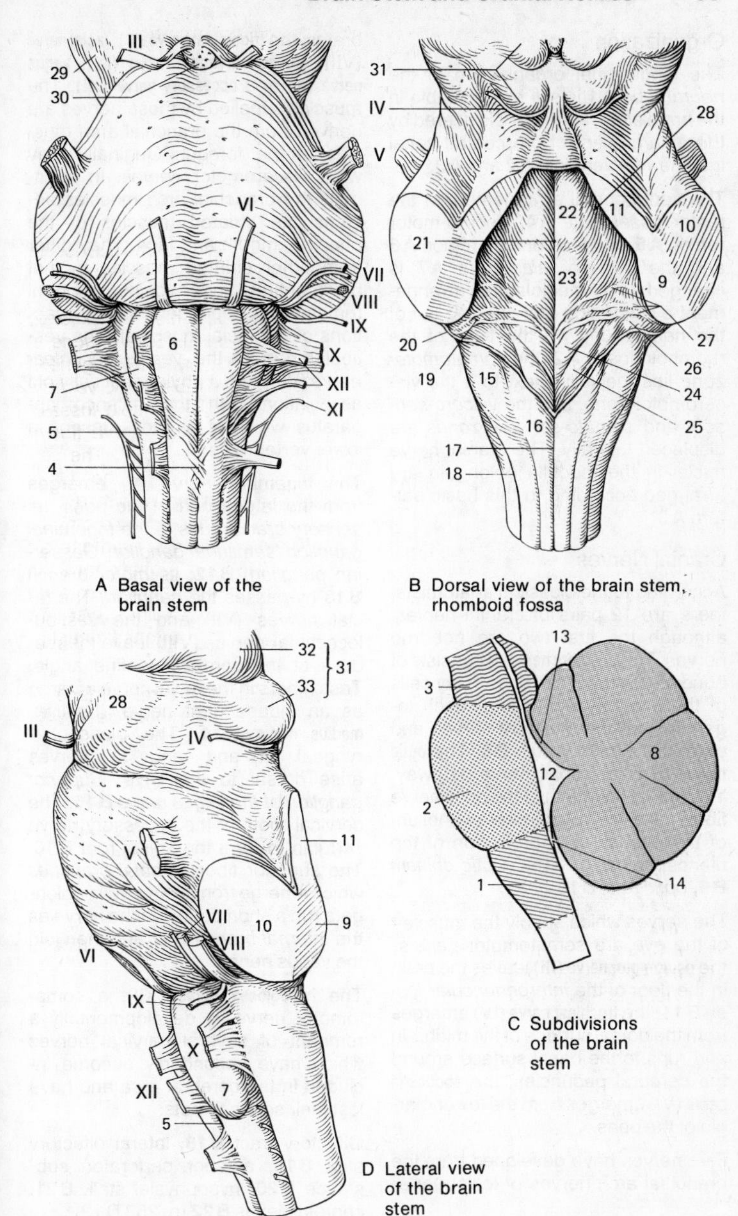

A Basal view of the brain stem

B Dorsal view of the brain stem, rhomboid fossa

C Subdivisions of the brain stem

D Lateral view of the brain stem

Organization

The longitudinal organization of the *neural tube* **A1** is still recognizable in the brain stem, although it is altered by the enlargement of the central canal into the IVth ventricle **A2** and **A3.**

The ventrodorsal arrangement of the motor *basal plate* **A4,** viscero-motor region **A5,** viscero-sensory region **A6** and the *sensory alar plate* **A7** is changed to a mediolateral arrangement by flattening and opening out of the neural tube on the floor of the rhomboid fossa **A2:** the *somatomotor zone* lies medially, next to it the visceromotor zone, and the *viscero-sensory* and *somato-sensory zones* are displaced laterally. The cranial nerve nuclei in the medulla oblongata are arranged according to this basic pattern **A3.**

Cranial Nerves

According to the classical anatomists, there are 12 pairs of cranial nerves, although the first two are not true nerves. The **olfactory nerve** (I) consists of bundles of processes of sensory cells of the olfactory epithelium which together form the olfactory nerves and enter the olfactory bulb **B8.** The **optic nerve** (II) is a cranial nerve pathway. The retina, the origin of the optic nerve fibers, like the pigmentary epithelium of the eyeball, is a projection of the diencephalon (p. 316). *Optic chiasm* **B9,** *optic tract* **B10.**

The nerves which supply the muscles of the eye are somatomotor nerves: the **oculomotor nerve** (III) leaves the brain in the floor of the *interpeduncular fossa* **B11;** the **trochlear nerve** (IV) emerges from the dorsal surface of the midbrain and runs to the basal surface around the cerebral peduncles; the **abducens nerve** (VI) emerges from the lower margin of the pons.

Five nerves have developed from the branchial arch nerves of lower vertebrates, the **trigeminal nerve** (V), **facial nerve** (VII), **glossopharyngeal nerve** (IX), **vagus nerve** (X) and **accessory nerve** (XI). The muscles supplied by these nerves are derived from the branchial arch muscles of the foregut. Originally they were visceromotor nerves. In mammals the branchial arch muscles became the striated muscles of the pharynx, mouth and face. They differ from other striated muscles by not being entirely under voluntary control (emotionally dependent, mimic reactions of the facial muscles). The vestibular part of the *vestibulocochlear nerve* (VIII) has a phylogenetically old association with the balance apparatus which is already present in lower vertebrates.

The trigeminal nerve (V) emerges from the lateral part of the pons. Its *sensory branch* runs to the *trigeminal ganglion (semilunar ganglion, Gasserian ganglion)* **B12;** its *motor branch* **B13** by-passes the ganglion. The facial nerves (VII) and the vestibulocochlear nerves (VIII) leave the medulla at the cerebellopontine angle. Taste fibers in the facial nerve emerge as an independent nerve, the **intermedius nerve B14.** The glossopharyngeal (IX) and vagus (X) nerves arise dorsal to the olive. *Superior ganglion of the vagus nerve* **B15.** The cervical roots of the accessory nerve (XI) join to form the spinal root **B16.** The superior fibers *cranial branches* which emerge from the medulla oblongata run a short course in the nerve as the *internal branch* **B17** and then join the vagus nerve.

The **hypoglossal nerve** (XII), a somatomotor nerve, is developmentally a remnant of several cervical nerves which have secondarily become included in the cerebral area and have lost their sensory roots.

Olfactory tract **B18,** lateral olfactory stria **B19,** anterior perforated substance **B20,** hypophysial stalk **B21,** choroid plexus **B22** (p. 262 D12).

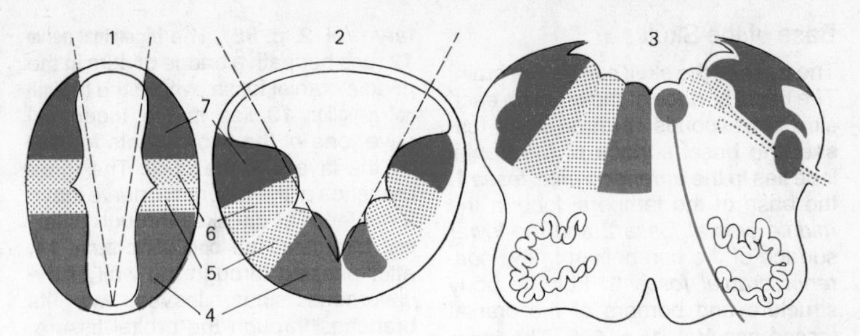

A Longitudinal organization of the medulla oblongata (after Herrick)

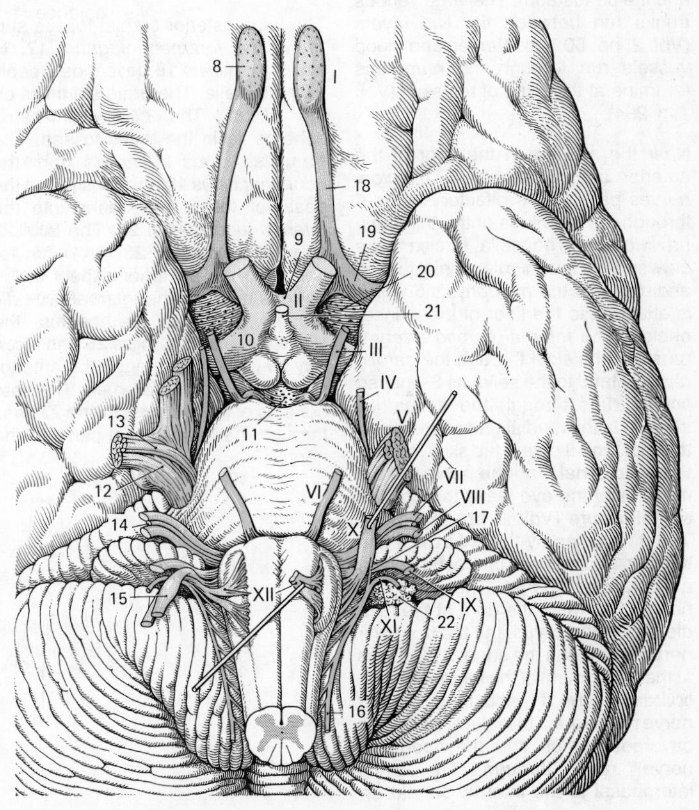

B Cranial nerves, base of the brain

Base of the Skull

The base of the skull carries the brain. The basal surface of the brain on each side corresponds to three bony fossae: the basal surface of the frontal lobe lies in the *anterior cranial fossa* **1,** the base of the temporal lobe in the *middle cranial fossa* **2** and the lower surface of the cerebellum in the *posterior cranial fossa* **3.** For the bony structure and borders of the cranial fossae see Vol. 1, p. 290. The inner surface of the skull is covered by the *dura mater.* It consists of two layers which form the covering of the brain and the periosteum. The large venous trunks run between the two layers (Vol. 2, pp. 60, 62). Nerves and blood vessels run through the numerous foramina at the base of the skull (Vol. 1, p. 294).

Near the midline, in the floor of the anterior cranial fossa, the *olfactory nerves* pass to the olfactory bulbs **4** through the openings of the thin *lamina cribrosa.* The sella turcica rises between the two middle cranial fossae and contains the *hypophysis* **5** which is attached to the floor of the diencephalon. The *internal carotid artery* **6** runs into the skull through the *carotid canal,* lateral to the sella. Its S-shaped course runs through the *cavernous sinus* **7.** In the medial part of the fossa, the **optic nerve 8** enters the skull through the optic canal, and the nerves to the muscles of the eye leave through the orbital fissure (Vol. 1. pp. 286, 330). The path taken by the **abducens 9** and **trochlear 10 nerves** are characterised by an intradural course. The abducens nerve penetrates the dura at the middle part of the clivus and the trochlear nerve runs along the edge of the clivus at the attachment of the tentorium. The **oculomotor nerve 11** and the trochlear nerve run through the lateral wall of the cavernous sinus and the abducens nerve runs through the sinus laterobasal to the internal carotid artery (Vol. 2, p. 82). The **trigeminal nerve 12** runs beneath a bridge of dura in the middle cranial fossa, where the **trigeminal ganglion 13** lies in the trigeminal cave, one of the two pockets formed by the layers of the dura. The three branches of the trigeminal nerve leave the interior of the skull through different foramina: the **ophthalmic nerve 14,** after passing through the wall of the cavernous sinus, leaves with its branches through the orbital fissure. The **maxillary nerve 15** leaves through the foramen rotundum, and the **mandibular nerve 16** through the foramen ovale.

The two posterior cranial fossae surround the *foramen magnum* **17,** to which the *clivus* **18** descends steeply from the sella. The brain stem lies on the clivus. The cerebellar hemispheres lie in the basal fossae. The *tranverse sinus* **20** arises from the *confluent sinus* **19** and runs round the posterior fossa and opens into the *internal jugular vein* **21.** The **facial 22** and **vestibulocochlear 23 nerves** run together to the posterior aspect of the petrous bone, to the *internal acoustic meatus.* Basal to the meatus, the **glossopharyngeal 34, vagus 25** and **accessory 26 nerves** run through the anterior part of the *jugular foramen.* The fiber bundles of the **hypoglossal nerve 27** pass through the hypoglossal canal to the nerve.

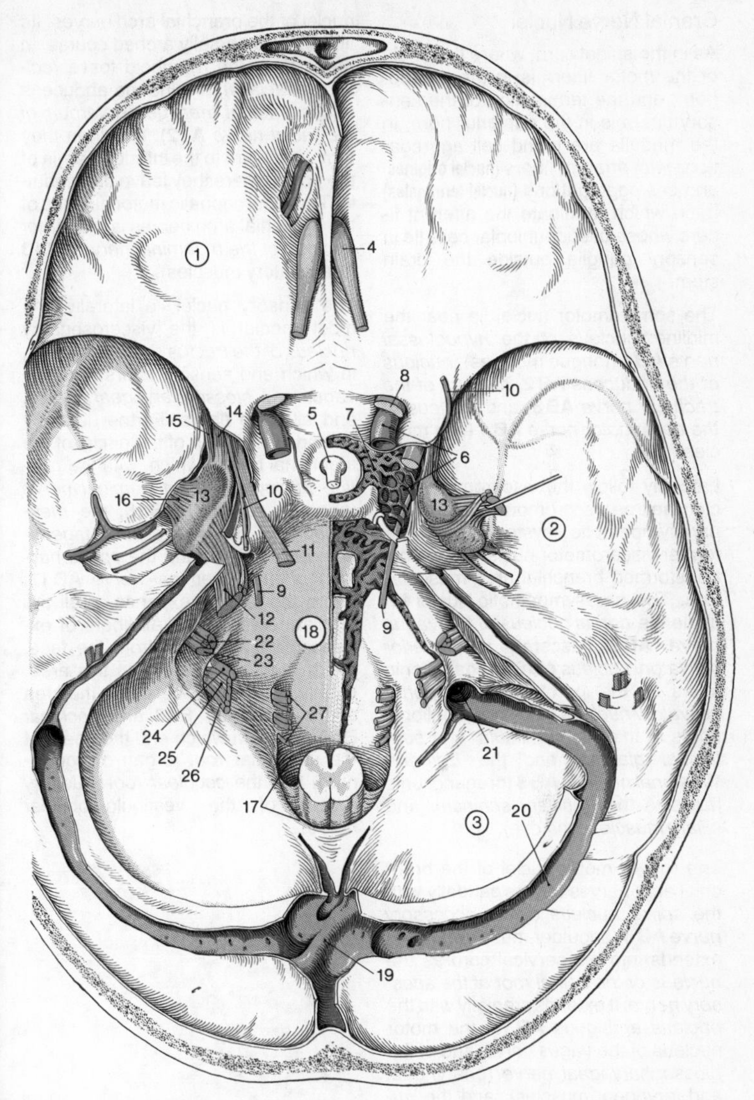

View of the base of the skull from above (Preparation of Prof. Platzer)

Cranial Nerve Nuclei

As in the spinal cord, where the origin of the motor fibers is in the anterior horn, and the termination of the sensory fibers is in the posterior horn, in the medulla are found cell aggregations with efferent fibers **(nuclei origines)** and cell aggregations **(nuclei terminales)** upon which terminate the afferent fibers whose pseudounipolar cells lie in sensory ganglia outside the brain stem.

The somatomotor nuclei lie near the midline: nucleus of the *hypoglossal nerve* **AB 1** (tongue muscles), *nucleus of the abducens* **AB 2,** *nucleus of the trochlear nerve* **AB 3,** and *nucleus of the oculomotor nerve* **AB 4** (eye muscles).

Laterally follow the visceromotor nuclei: the true visceromotor nuclei of the parasympathetic system, and the former visceromotor nuclei of the now transformed branchial arch musculature. The parasympathetic nuclei include the *dorsal nucleus of the vagus nerve* **AB 5** (viscera), the *inferior salivatory nucleus* **AB 6** (preganglionic fibers to the parotid), the *superior salivatory nucleus* **AB 7** (preganglionic fibers to the *submandibular* and *sublingual glands*), and the *Edinger-Westphal nucleus* **AB 8** (preganglionic fibers to the *pupillary sphincter* and *ciliary muscle of the eye*).

The row of motor nuclei of the branchial arch nerves begins caudally with the *spinal nucleus of the accessory nerve* **AB 9** (shoulder muscles), which extends into the cervical cord as the *nucleus of the spinal root of the acessory nerve.* It extends cranially with the *nucleus ambiguus* **AB 10,** the motor nucleus of the *vagus nerve,* and of the *glossopharyngeal nerve* (pharyngeal and laryngeal muscles), and the *nucleus of the facial nerve* **AB 11** (facial muscles). The nucleus of the facial nerve lies deeply, as do all the motor nuclei of the branchial arch nerves. Its fibers run a dorsally arched course, in the floor of the rhomboid fossa *(colliculus facialis),* around the abducens nucleus *(facial knee, genu internum of the facial nerve* **A 12**), and then they turn downward to the inferior margin of the pons, where they leave the medulla. The most cephalic motor nucleus of the branchial arch nerves is the *motor nucleus of the trigeminal nerve* **AB 13** (masticatory muscles).

The sensory nuclei lie laterally: the most medial is the viscerosensory *nucleus of the tractus solitarius* **AB 14,** in which end sensory fibers from the *vagus* and *glossopharyngeal nerves,* and all taste fibers. Further laterally extends the region of the nuclei of the trigeminal nerve, which, with the *pontine nucleus of the trigeminal nerve (principal nucleus)* **AB 15,** the *mesencephalic nucleus of the trigeminal nerve* **AB 16,** and the *spinal nucleus of the trigeminal nerve* **AB 17,** forms the largest expanse of all the cranial nerve nuclei. All fibers of exteroceptive sensation from the face, mouth and sinuses end in this area. Finally, lying most laterally, is the area of the *vestibular* **B 18** and *cochlear* **B 19** *nuclei,* in which end the fibers of the vestibular *root* (organ of equilibrium) and the *cochlear root* (auditory organ) of the vestibulo-cochlear nerve.

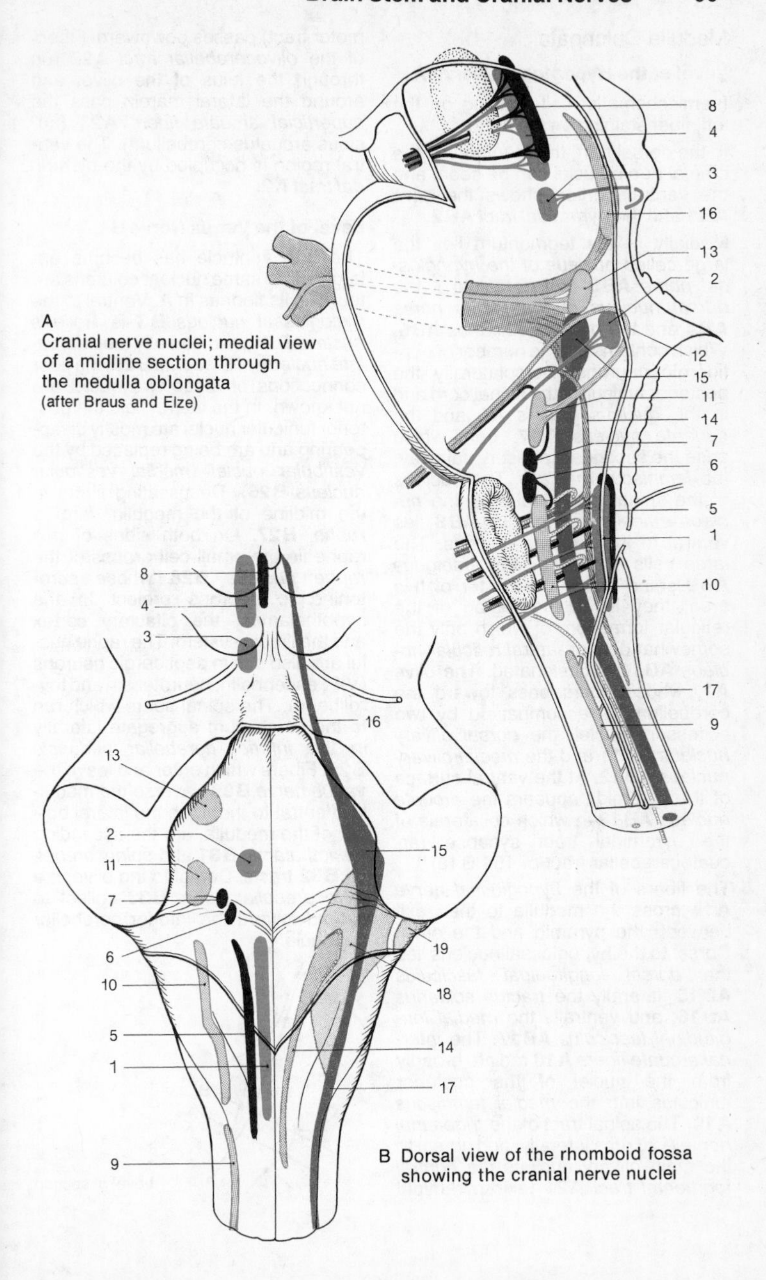

A
Cranial nerve nuclei; medial view of a midline section through the medulla oblongata
(after Braus and Elze)

B Dorsal view of the rhomboid fossa showing the cranial nerve nuclei

Medulla Oblongata

Level of the Hypoglossal Nerve A

(semischematic; cell staining on the left, fiber staining on the right)

In the dorsal part, the *tegmentum,* the cranial nerve nuclei can be seen, and the ventral portion shows the *olive* **AB1** and the *pyramidal tract* **AB2.**

Medially in the tegmentum lies the large-celled *nucleus of the hypoglossal nerve* **AB3** and dorsal to it the *dorsal nucleus of the vagus nerve* **AB4** and the *nucleus solitarius* **AB5.** Which contains a large number of peptidergic neurons. Dorsolaterally the posterior funiculi of the spinal cord end in the *nucleus gracilis* **A6** and the *cuneate nucleus* **AB7,** from which arise the secondary sensory pathway, the *lemniscus medialis.* The nucleus of the spinal trigeminal root, the *nucleus spinalis n. trigemini* **AB8** lies ventral to the cuneate nucleus. The large cells of the *nucleus ambiguus* **AB9** stand out in the center of this area, they lie in the region of the reticular formation, of which only the somewhat denser *lateral reticular nucleus* **AB10** is designated. The *olive* **A1,** whose fibers pass toward the cerebellum, is accompanied by two accessory nuclei, the *dorsal olivary nucleus* **AB11** and the *medial olivary nucleus* **AB12.** At the ventral surface of the pyramids appears the *arcuate nucleus* **AB13** in which collaterals of the pyramidal tract synapse (arcuatocerebellar tract p. 154 B18).

The fibers of the *hypoglossal nerve* **A14** cross the medulla to their exit between the pyramid and the olive. Dorsal to the hypoglossal nucleus lies the *dorsal longitudinal fasciculus* **AB15,** laterally the *tractus solitarius* **AB16,** and ventrally the *medial longitudinal fasciculus* **AB17.** The *internal arcuate fibers* **A18** radiate broadly from the nuclei of the posterior funiculus into the *medial lemniscus* **A19.** The spinal tract of the *trigeminal nerve* **A20** runs laterally, and dorsal to the main olivary nucleus the *central tegmental tract* **A21** (extrapyramidal motor tract) passes downward. Fibers of the *olivocerebellar tract* **A22** run through the hilus of the olive, and around the lateral margin pass the *superficial arcuate fibers* **A23** (nucleus arcuatus-cerebellum). The ventral region is occupied by the *pyramidal tract* **A2.**

Level of the Vagus Nerve B

The IVth ventricle has become enlarged. The same nuclear columns are found in its floor as in **A.** Ventral to the *hypoglossal nucleus* **B3** is *Roller's nucleus* **B24,** and dorsally the *intercalate nucleus of Staderini* **B25;** the fiber connections of these two nuclei are not known. In the lateral field the posterior funicular nuclei are mostly disappearing and are being replaced by the *vestibular nuclei* (*medial vestibular nucleus* **B26**). Decussating fibers in the midline of the medulla, form a *raphe,* **B27.** On both sides of this raphe lie the small cell groups of the *raphe nucleus* **B28** whose serotoninergic neurons project to the hypothalamus, the olfactory cortex and the limbic system. The raphe nuclei are also rich in peptidergic neurons (VIP, encephalin, neurotensin and thyroliberin). The spinal fibers which run to the cerebellum aggregate laterally in the *inferior cerebellar peduncle* **B29.** Fibers which enter and leave the *vagus nerve* **B30** traverse the medulla. Ventral to them, on the lateral border of the medulla, are the ascending *spinothalamic* **B31** and *spinocerebellar* **B32** *tracts.* Dorsal to the olive, the *olivocerebellar fibers* **B33** collect to extend toward the inferior cerebellar peduncle.

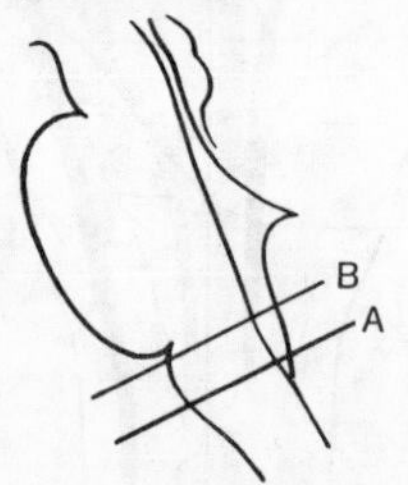

Level of section

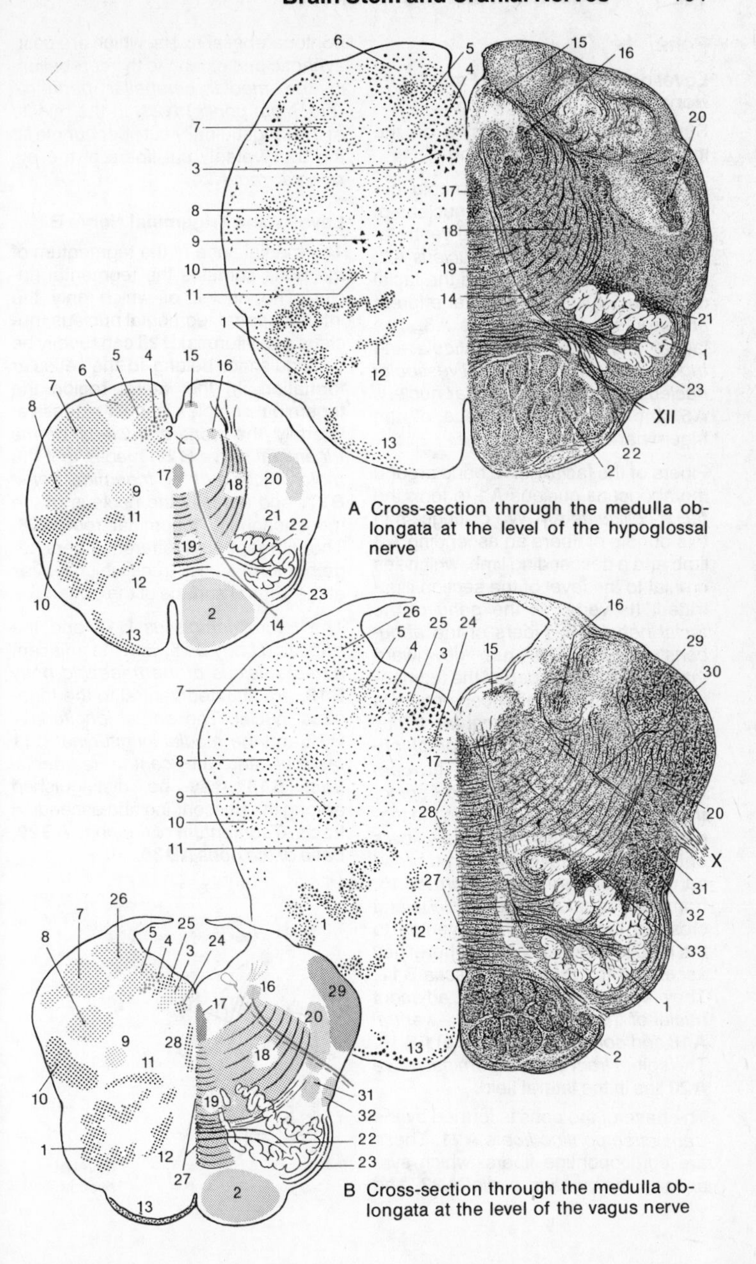

A Cross-section through the medulla oblongata at the level of the hypoglossal nerve

B Cross-section through the medulla oblongata at the level of the vagus nerve

Pons

Level of the Genu of the Facial Nerve A

Semischematic; cell staining on the left, fiber staining on the right)

Beneath the floor of the rhomboid fossa lie the large cells of the nucleus of the *abducens nerve* **A1** and ventrolateral to it the *facial nerve nucleus* **A2.** Between the abducens and the facial nerve nuclei lies the visceral efferent *superior salivatory nucleus* **A3.** The lateral field contains the *vestibular* and *trigeminal nuclei;* the *medial vestibular nucleus* **A4,** *lateral vestibular nucleus* **A5** and the spinal nucleus of the trigeminal root **A6.**

Fibers of the facial nerve bend around the abducens nucleus **A1** to form the *facial colliculus* **A7.** We distinguish in this bundle of fibers an ascending **A8** limb and a descending limb, which lies cranial to the level of the section illustrated. Its vertex is the *genu of the facial nerve* **A9.** Fibers of the *abducens nerve* **A10** run downward through the medial part of the tegmentum. Medial to the abducens nucleus lies the *medial longitudinal fasciculus* **AB11** and dorsal to it the *dorsal longitudinal fasciculus* **AB12.** The *central tegmental* **AB13** and *spinothalamic* **A14** *tracts* lie deep in the tegmentum of the pons. Secondary fibers of the auditory tract collect in a broad band, the *corpus trapezoideum* **AB15,** from the ventral cochlear nucleus and cross to the opposite side ventral to the *medial lemniscus* **A16,** where they ascend in the *lateral lemniscus* **B17.** They synapse partly in the adjacent *nuclei of the trapezoid body – ventral* **A18** and *dorsal* (superior olive) **AB19.** The spinal *tract of the trigeminal nerve* **A20** lies in the lateral field.

The base of the pons is formed by the *transverse pontine fibers* **A21.** These are corticopontine fibers, which synapse in the *pontine nuclei* **A22,** and pontocerebellar fibers, which are postsynaptic and extend to the cerebellum in the *medial cerebellar peduncle (brachium pontis)* **A23.** In the middle of the longitudinally cut fiber bundle lie the transversely cut fibers of the *pyramidal tract* **A24.**

Level of the Trigeminal Nerve B

The medial zone of the tegmentum of the pons contains the tegmental nuclei. The nuclei, of which only the inferior central tegmental nucleus (nucleus papilliformis) **B25** can readily be distinguished, belong to the reticular formation. In the lateral region the trigeminal complex is most extensive: laterally the *pontire nucleus of the trigeminal nerve* **B26,** medial to it the *motor nucleus of the trigeminal nerve* **B27,** and dorsally the nucleus of the mesencephalic trigeminal root **A28.** The afferent and efferent fibers together form a thick trunk which leaves at the ventral surface of the pons.

The *lateral lemniscus* **B17** and the *trapezoid body* **B15,** with the adjacent *dorsal nucleus of the trapezoid body* **B19,** are situated ventral to the trigeminal nuclei. The *dorsal longitudinal* **B12** and the *medial longitudinal* **B11** *fasciculi* and the *central tegmental tract* **B13** may be distinguished among the descending and ascending tracts. *Tegmentum of the pons* **AB29,** *base of the pons* **AB30.**

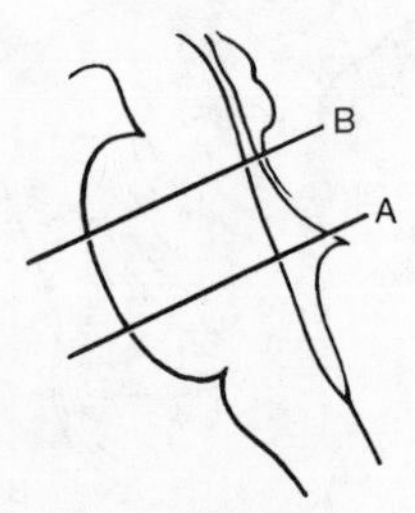

Level of section

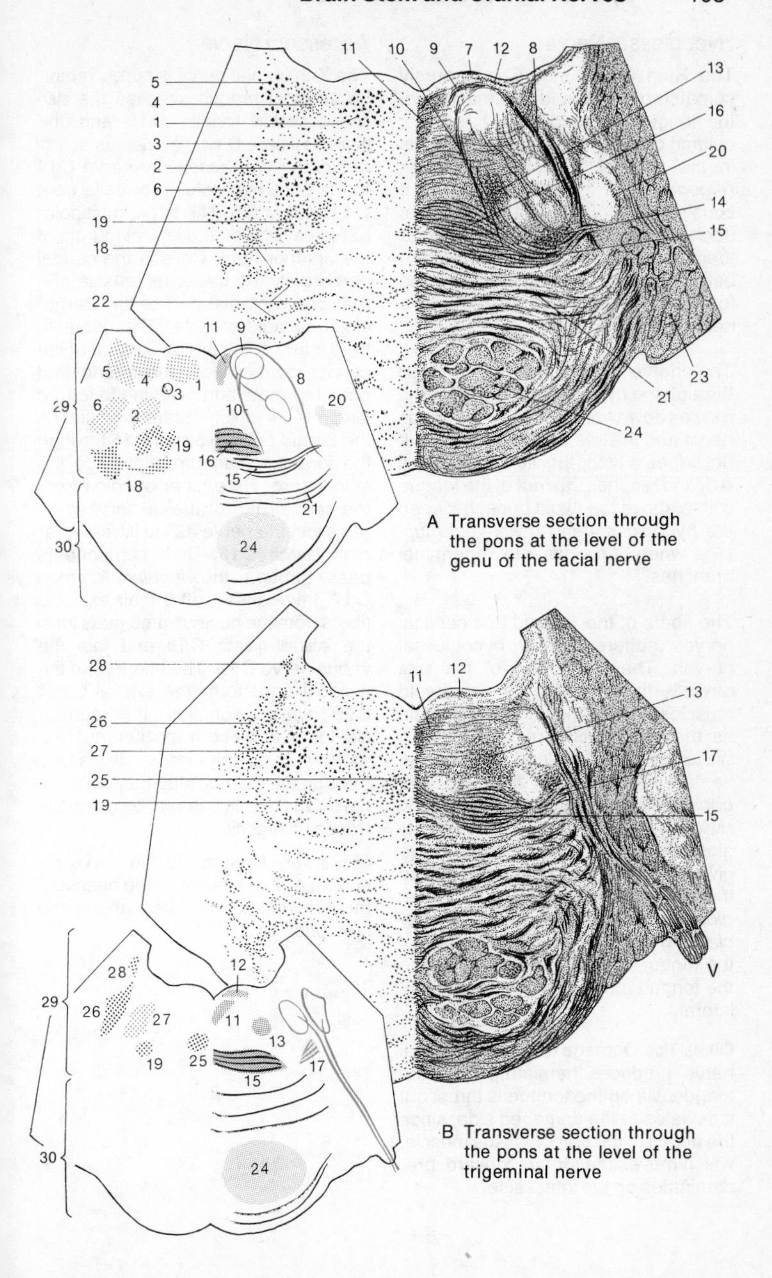

A Transverse section through the pons at the level of the genu of the facial nerve

B Transverse section through the pons at the level of the trigeminal nerve

Hypoglossal Nerve

The XIIth cranial nerve is a purely somatomotor nerve for the muscles of the tongue. Its nucleus **B1** forms a column of large, multipolar nerve cells in the floor of the rhomboid fossa *(trigone of the hypoglossal nerve).* It consists of a number of cell groups, each of which supplies a particular tongue muscle. The nerve fibers leave between the pyramid and the olive and form two bundles which combine into a nerve trunk.

The nerve then leaves the skull through the *hypoglossal canal* **B2** and passes downward, lateral to the vagus nerve and the internal carotid artery. It describes a loop, the **arcus n. hypoglossi A3,** and reaches the root of the tongue a little above the hyoid bone, between the *hypoglossal* and *mylohyoid muscles,* where it divides into its terminal branches.

The fibers of the 1st and 2nd cervical nerves adhere to the hypoglossal nerves. They form part of the **ansa cervicalis** (branches to the infrahyoid muscles), where they branch off again as the **superior root A4** and join the **inferior root A5** (2nd and 3rd cervical nerves). The cervical fibers for the *geniohyoid muscle* **A6** and the **thyroid muscle A7** extend further in the hypoglossal nerve. The hypoglossal nerve gives off branches, the **rami linguales,** to the *hypoglossal muscle* **A8,** *genioglossal muscle* **A9,** *styloglossal muscle* **A10** and the intrinsic muscles of the tongue **A11.** The nerve supply to the tongue musculature is strictly unilateral.

Clinical Tips. Damage to the hypoglossal nerve produces hemiatrophy of the tongue. When the tongue is thrust out it deviates to the damaged side, since the action of the genioglossus muscle, which moves the tongue forward, predominates on the intact side.

Accessory Nerve

The XIth cranial nerve is purely motor. Its external branch supplies the **sternocleidomastoid muscles D12** and the **trapezius muscle D13.** Its nucleus *spinal nucleus of the accessory nerve* **C14** forms a narrow column of cells from C1 to C5 or C6. The large, multipolar nerve cells lie on the lateral margin of the anterior horn. Cells in the caudal part supply the trapezius muscle and that of the cranial part of the sternocleidomastoid muscle. The nerve fibers leave at the lateral surface of the cervical cord between the anterior and posterior roots and combine to form a single trunk which enters the skull as the **external** (spinal) **branch C15** through the foramen magnum, alongside the spinal cord. Here, fiber bundles from the caudal part of the *nucleus ambiguus* join the nerve as the **internal** (cranial) **branch C16.** Both components pass through the *jugular foramen* **C17.** Immediately after their exit, the fibers from the nucleus ambiguus form the **internal branch C18** and join the vagus nerve **C19.** The fibers from the cervical cord form the **external branch C20** which supplies the sternocleidomastoid and trapezius muscles as the *accessory nerve.* It passes through the sternocleidomastoid muscle and its terminal branches reach the trapezius muscle.

Clinical Tips. Damage to the nerve results in a tilted position of the head and the arm cannot be lifted above the horizontal.

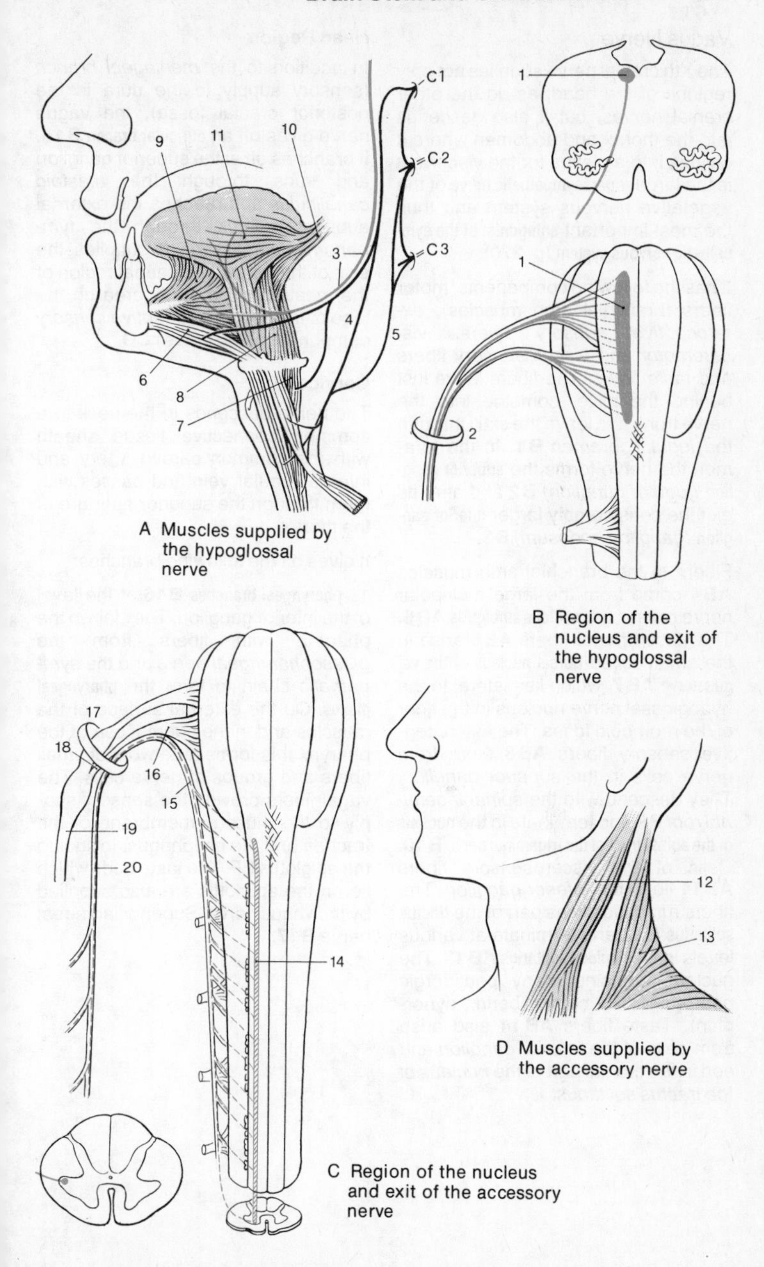

A Muscles supplied by the hypoglossal nerve

B Region of the nucleus and exit of the hypoglossal nerve

C Region of the nucleus and exit of the accessory nerve

D Muscles supplied by the accessory nerve

Vagus Nerve

The Xth cranial nerve supplies not only regions of the head, as do the other cranial nerves, but it also descends into the thorax and abdomen where it divides into a plexus for the viscera. It is the largest **parasympathetic nerve** of the vegetative nervous system and thus the most important **antagonist of the sympathetic nervous system** (p. 270).

It has the following components: motor fibers (branchial arch muscles), *exteroceptive sensory* fibers, *visceromotor* and *viscerosensory* fibers and *taste fibers*. The fibers leave just behind the olive, combine into the nerve trunk and leave the skull through the *jugular foramen* **B1**. In the foramen the nerve forms the **superior ganglion** *(jugular ganglion)* **B2** and after its exit the considerably larger **inferior ganglion** *(ganglion nodosum)* **B3.**

Fibers to the branchial arch muscles **AB4** come from the large multipolar nerve cells of the **nucleus ambiguus AB5.** The visceromotor fibers **AB6** arise in the small-celled **dorsal nucleus of the vagus nerve AB7,** which lies lateral to the hypoglossal nerve nucleus in the floor of the rhomboid fossa. The exteroceptive sensory fibers **AB8** stem from nerve cells in the *superior ganglion.* They descend with the *spinal trigeminal root* **B9** and terminate in the **nucleus of the spinal tract of the trigeminal nerve B10.** Cells of the viscerosensory fibers **AB11** lie in the *inferior ganglion.* The fibers run caudally as part of the **tractus solitarius B12** and terminate at various levels in the **nucleus solitarius AB13.** The nucleus contains many peptidergic neurons (VIP, corticoliberin, dynorphin). Taste fibers **AB14** also arise from cells of the *inferior ganglion* and end in the cranial part of the *nucleus of the tractus solitarius.*

Head Region

In addition to the *meningeal branch* (sensory supply to the dura in the posterior cranial fossa), the vagus nerve gives off an **auricular branch B15.** It branches off at the superior ganglion and runs through the *mastoid canaliculus* and reaches the external auditory meatus through the *tympanomastoid fissure.* It supplies the skin of the dorsal and caudal region of the meatus and a small area on the lobe of the ear *(exteroceptive sensory* component of the nerve) **CD.**

Cervical Part

The nerve descends in the neck in a common connective tissue sheath with the common carotid artery and internal jugular vein and passes with them through the superior aperture of the thorax.

It gives off the following branches:

1. **pharyngeal branches B16** at the level of the inferior ganglion. They join in the pharynx with fibers from the *glossopharyngeal nerve* and the *sympathetic* chain to form the **pharyngeal plexus.** On the external surface of the muscles and in the submucosa of the pharynx this forms a network of small fibers and groups of nerve cells. The vagal fibers provide the sensory supply to the mucous membrane of the trachea and the esophagus, including the epiglottis EF. The taste buds which lie on the epiglottis are also supplied by the vagus nerve. Superior laryngeal nerve **B17.**

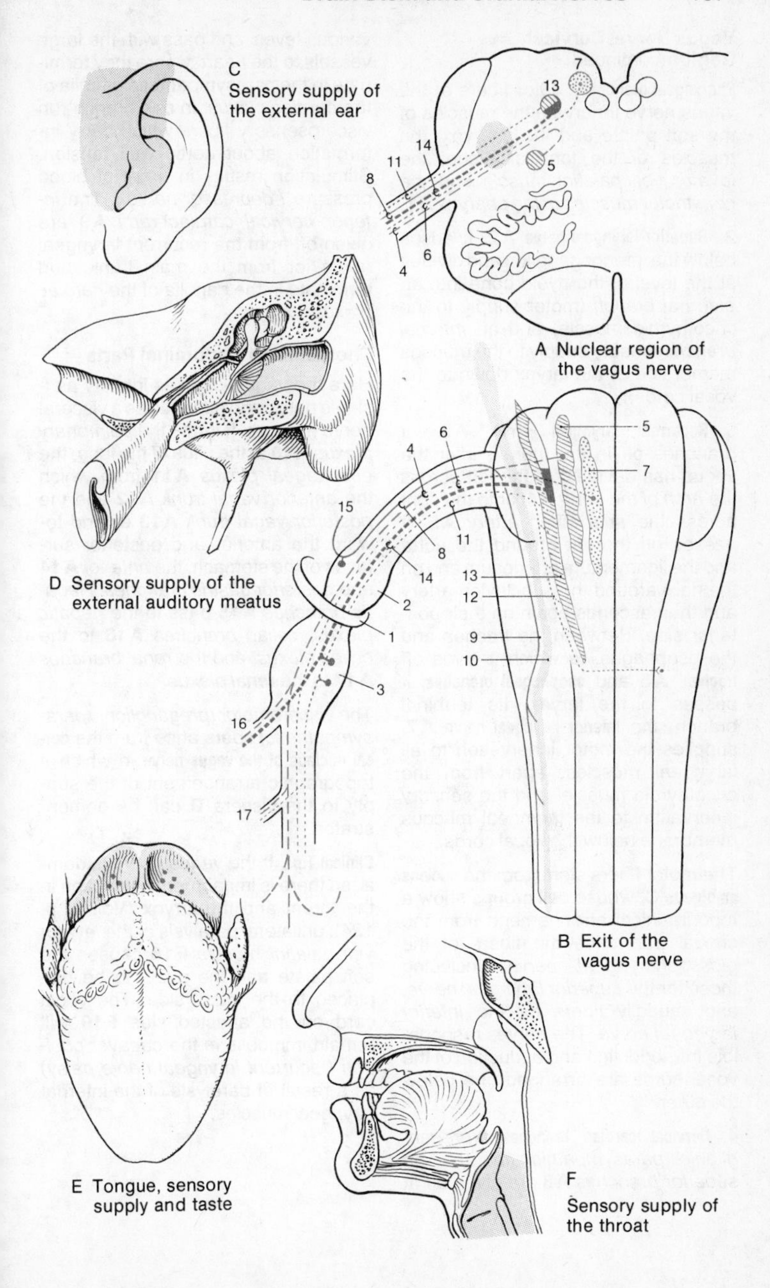

A Nuclear region of the vagus nerve

B Exit of the vagus nerve

C Sensory supply of the external ear

D Sensory supply of the external auditory meatus

E Tongue, sensory supply and taste

F Sensory supply of the throat

Vagus Nerve, Cervical Region (continued)

Pharyngeal branches. Motor fibers of the vagus nerve innervate the muscles of the soft palate and the pharynx: the muscles of the tonsillar niche, the *levator veli palatini muscle* and the *constrictor muscles of the pharynx* **B1.**

2. **Superior laryngeal nerve A2.** It arises below the inferior ganglion and divides at the level of the hyoid bone into an *external branch* (motor supply to the cricothyroid muscle) and an *internal branch* (sensory supply to the mucous membraine of the larynx down to the vocal cords).

3. **Recurrent laryngeal nerve A3.** It branches off in the thorax, after the vagus has deviated to the left across the *arch of the aorta* **A4,** or to the right across the *subclavian artery* **A5.** It passes on the left around the aorta and the ligamentum arteriosum and on the right around the subclavian artery and then ascends again on their posterior side. Between the trachea and the esophagus, to which it gives off **tracheal A6** and **esophageal branches,** it passes to the larynx. Its terminal branch, the **inferior laryngeal nerve A7,** supplies the motor innervation to all laryngeal muscles, apart from the cricothyroid muscle, and the sensory innervation to the laryngeal mucous membrane below the vocal cords.

The motor fibers stem from the **nucleus ambiguus C,** whose cell groups show a topographical arrangement: from the cranial part arise the fibers of the *glossopharyngeal nerve,* including those for the *superior laryngeal nerve,* and caudally fibers for the *inferior laryngeal nerve.* The nerves responsible for abduction and adduction of the vocal cords are arranged one below the other.

4. **Cervical cardiac branches.** *(preganglionic parasympathic fibers).* The *superior branches* **A8** are given off at various levels and pass with the large vessels to the heart, where they terminate in the parasympathetic ganglia of the *cardiac plexus.* In one branch run viscerosensory fibers which carry information about aortic wall tension. Stimulation results in a fall of blood pressure *('depressor nerve').* The *inferior cervical cardiac rami* **A9** are given off from the recurrent laryngeal nerve, or from the main trunk, and terminate in the ganglia of the *cardiac plexus.*

Thoracic and Abdominal Parts

Here the vagus loses its identity as a single nerve and expands as a visceral nerve network. It forms the *pulmonary plexus* **A10** at the hilus of the lung, the *esophageal plexus* **A11** from which the *anterior vagal trunk* **A12** and the *posterior vagal trunk* **A13** extend toward the anterior and posterior surfaces of the stomach, the *anterior* **A14** and *posterior gastric branches. Hepatic branches* **A15** pass to the *hepatic plexus, celiac branches* **A16** to the *celiac plexus,* and the *renal branches* **A17** to the *renal plexus.*

The *visceromotor (preganglionic parasympathetic)* fibers arise from the **dorsal nucleus of the vagus nerve** in which a topographic arrangement of the supply to the viscera **D** can be demonstrated.

Clinical Tips. If the vagus nerve is damaged there is impairment of function in the throat and the larynx (Vol. 2, p. 124): unilateral paralysis of the *levator veli palatini muscle* **F18** causes the soft palate and the uvula to be displaced to the intact side. The vocal cord on the affected side **F19** will remain immobile in the *cadaver position (recurrent laryngeal nerve palsy)* as a result of paralysis of the internal laryngeal muscles.

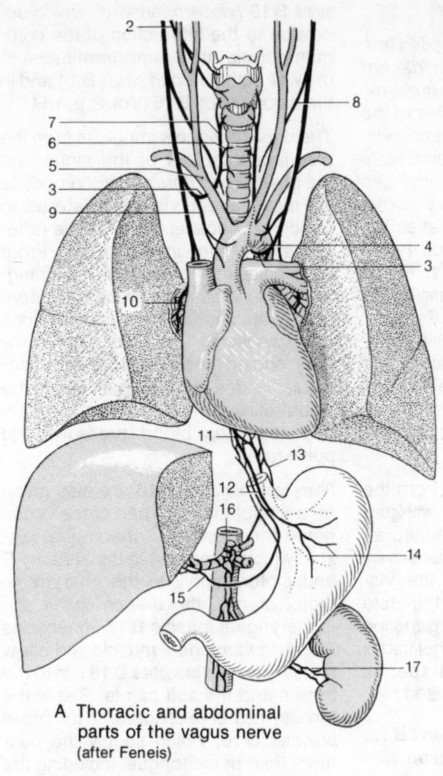

A Thoracic and abdominal parts of the vagus nerve (after Feneis)

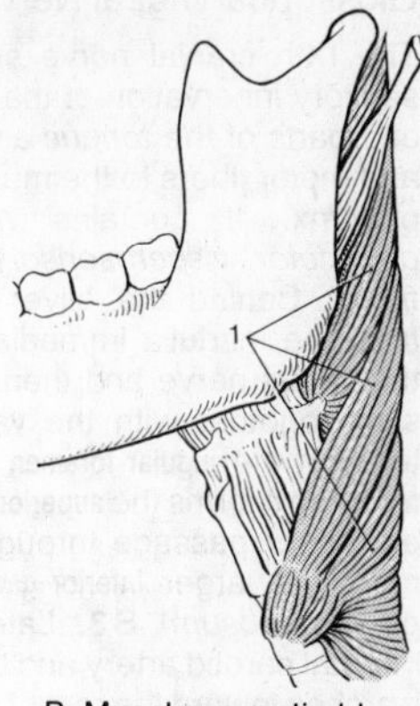

B Muscles supplied by the vagus nerve

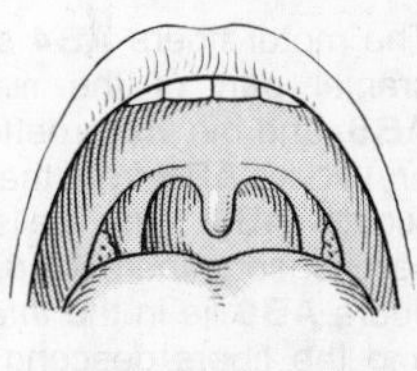

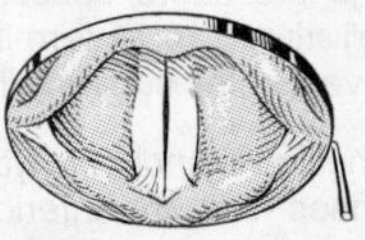

E Normal position of the soft palate and vocal cords

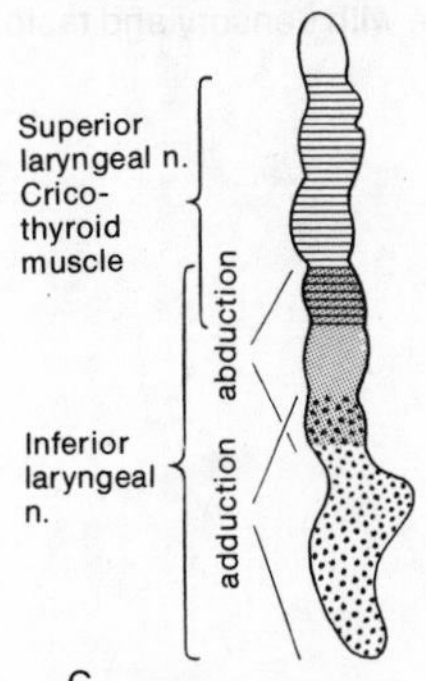

C Somatotopic organization of the nucleus ambiguus (after Crosby, Humphrey, Lauer)

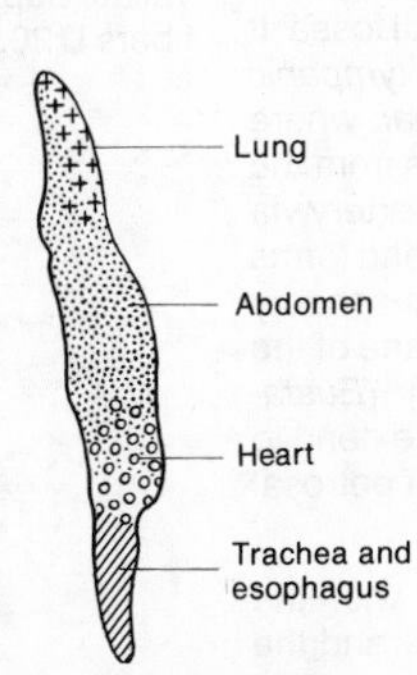

D Somatotopic organization of the dorsal nucleus of the vagus nerve (after Getz and Sienes)

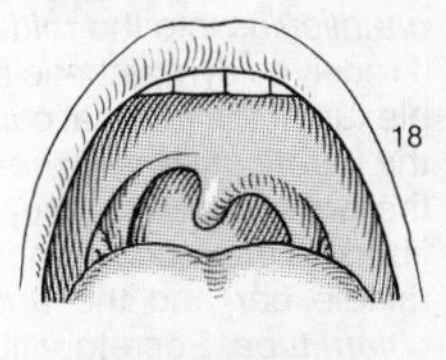

F Position of the soft palate and vocal cords after paralysis of the left vagus nerve

Glossopharyngeal Nerve

The IXth cranial nerve supplies the sensory innervation of the *middle ear* and parts of the *tongue* and *pharynx,* and motor fibers to the muscles of the pharynx. It contains *motor, visceromotor, viscerosensory* and *taste fibers.* Behind the olive it emerges from the medulla immediately above the vagus nerve and then leaves the skull together with the vagus nerve through the **jugular foramen B1.** In the foramen it forms the **superior ganglion B2** and, after passage through the foramen, the larger **inferior ganglion** *(ganglion petrosum)* **B3.** Lateral to the internal carotid artery and the pharynx it arches toward the root of the tongue, where it divides into a number of terminal branches.

The motor fibers **AB4** stem from the cranial part of the **nucleus ambiguus AB5,** and the visceroefferent (secretory) fibers **AB6** from the **inferior salivary nucleus AB7.** The cells of the viscerosensory fibers **AB8** and the taste fibers **AB9** lie in the *inferior ganglion* and the fibers descend in the **tractus solitarius B10** to terminate at specific levels of the **nucleus solitarius AB11.**

The first branch, the **tympanic nerve B12,** arises from the inferior ganglion with viscerosensory and preganglionic secretory fibers in the petrosal fossa. It passes through the *inferior tympanic canaliculus* into the *middle ear,* where it receives sympathetic fibers from the plexus of the internal carotid artery via the *caroticotympanic nerve,* and forms the **tympanic plexus.** It supplies sensory fibers to the mucous membrane of the middle ear and the *auditory (Eustachian) tube.* Secretory fibers extend to the otic ganglion as the lesser petrosal nerve (p. 122).

In addition to connections with the vagus nerve, the facial nerve and the sympathetic system, the inferior ganglion gives off the **branch to the carotid sinus B13** *(viscerosensory),* which descends to the bifuraction of the common carotid artery and terminates in the wall of the *carotid sinus* **B14** and in the *carotid body* **B15** (Vol. 2, p. 164C).

The nerve conducts impulses from the *mechanoreceptors* of the sinus and the *chemoreceptors* of carotid body to the medulla, and via the collaterals to the dorsal nucleus of the vagus (afferent limb of the sinus reflexes). From the nucleus of the vagus, preganglionic fibers pass to groups of nerve cells in the cardiac atria, whose axons (postganglionic parasympathetic fibers) end on the sinotrial and atrioventricular nodes (efferent pathway of the sinus reflexes). This system registers and regulates blood pressure and pulse rate.

Pharyngeal branches B16 are also given off and together with part of the vagus nerve they form the pharyngeal plexus, which takes part in the sensory E and motor supply to the pharynx. A motor branch, the *branch to the stylopharyngeal muscle* **B17**, innervates the *stylopharyngeal muscle* and a few sensory **tonsillar branches D18** run to the tonsils and the soft palate. Below the tonsils, the nerve divides into the **lingual branches D19,** which supply the posterior third of the tongue, including the vallate papillae with sensory and taste fibers **D20.**

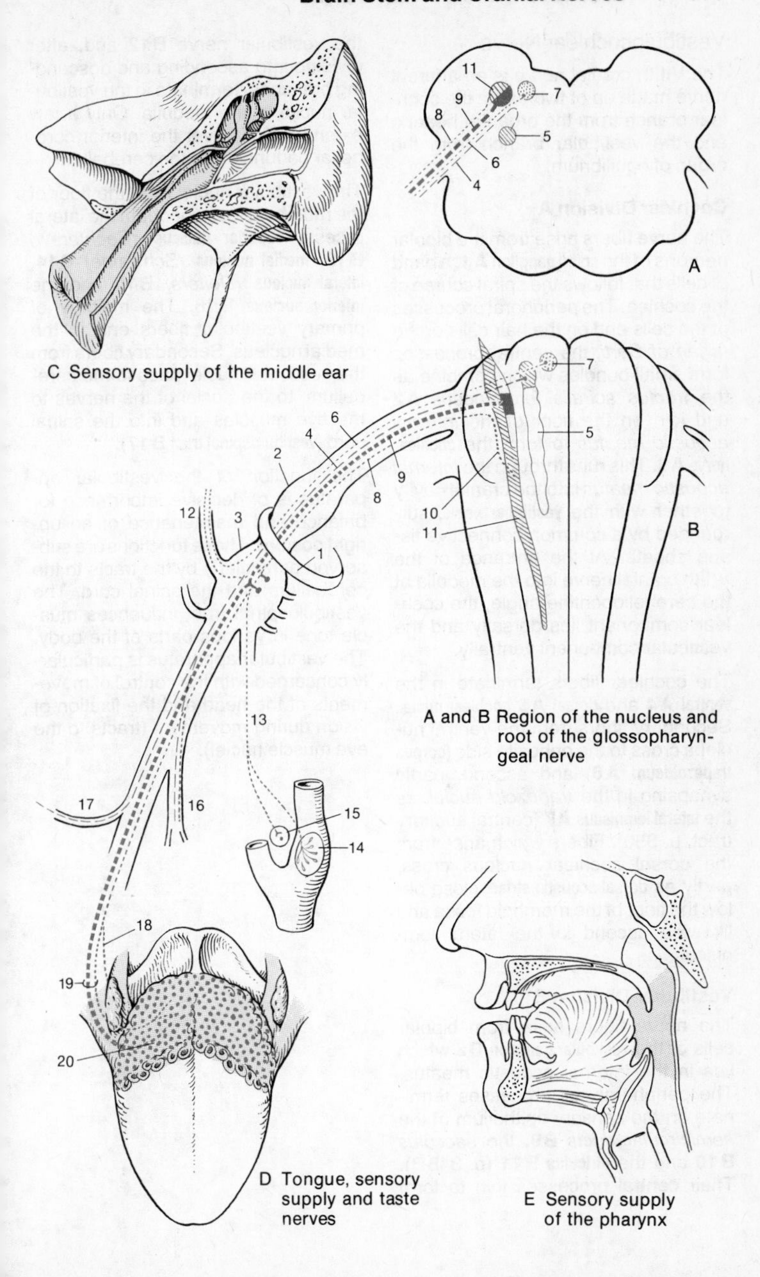

C Sensory supply of the middle ear

A and B Region of the nucleus and root of the glossopharyngeal nerve

D Tongue, sensory supply and taste nerves

E Sensory supply of the pharynx

Vestibulocochlear Nerve

The VIIIth cranial nerve is an afferent nerve made up of two parts: the cochlear branch from the organ of hearing and the vestibular branch from the organ of equilibrium.

Cochlear Division A

The nerve fibers arise from the bipolar neurons of the **spiral ganglion A1**, a band of cells that follows the spiral course of the cochlea. The peripheral processes of the cells end on the hair cells or the *organ of Corti;* the central processes form small bundles which combine as the *tractus spiralis foraminosus* **A2** and join on the floor of the *internal acoustic meatus* to form the **cochlear nerve A3.** This runs through the *internal acoustic meatus* into the cranial cavity together with the **vestibular nerve,** surrounded by a common connective tissue sheath. At the entrance of the VIIIth cranial nerve into the medulla at the cerebellopontine angle, the cochlear component lies dorsally and the vestibular component ventrally.

The cochlear fibers terminate in the **ventral A4** and **dorsal A5 cochlear nuclei.** Secondary fibers from the ventral nucleus cross to the opposite side (**corpus trapezoideum A6**) and ascend, partly synapsing in the *trapezoid nuclei,* as the **lateral lemniscus A7** (central auditory tract, p. 350). Fibers which arise from the dorsal cochlear nucleus cross, partly as **dorsal acoustic striae,** close below the floor of the rhomboid fossa and likewise ascend in the lateral lemniscus.

Vestibular Division B

The nerve fibers arise from bipolar cells of the **vestibular ganglion B8** which lies in the internal acoustic meatus. The peripheral cell processes terminate on the sensory epithelium of the *semicircular ducts* **B9,** the *sacculus* **B10** and the **utriculus B11** (p. 348D). Their central processes join to form the vestibular nerve **B12** and, after dividing into ascending and descending branches, terminate in the vestibular nuclei of the medulla. Only a few extend directly over the inferior cerebellar peduncle into the cerebellum.

The vestibular nuclei lie in the floor of the rhomboid fossa beneath the lateral recess: **superior nucleus** *(Bechterew)* **B13, medial nucleus** *(Schwalbe)* **B14, lateral nucleus** *(Deiters)* **B15** and the **inferior nucleus B16.** The majority of primary vestibular fibers end in the medial nucleus. Secondary fibers from the vestibular nuclei pass to the cerebellum, to the nuclei of the nerves to the eye muscles and into the spinal cord (**vestibulospinal tract B17).**

The function of the vestibular apparatus is of decisive importance for balance and maintenance of an upright posture. These functions are subserved particularly by the tracts to the cerebellum and the spinal cord. The vestibulospinal tract influences muscle tone in various parts of the body. The vestibular apparatus is particularly concerned with the control of movements of the head and the fixation of vision during movement (tracts to the eye muscle nuclei).

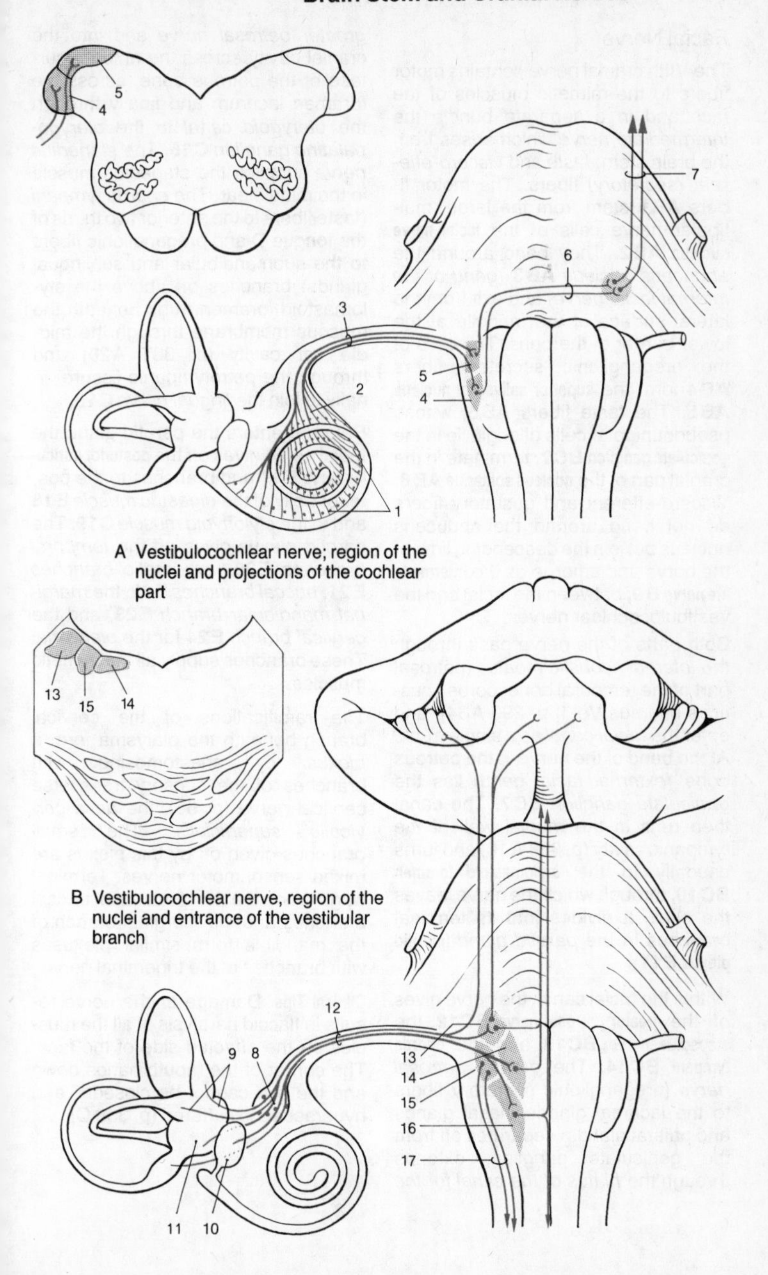

A Vestibulocochlear nerve; region of the nuclei and projections of the cochlear part

B Vestibulocochlear nerve, region of the nuclei and entrance of the vestibular branch

Facial Nerve

The VIIth cranial nerve contains motor fibers to the mimetic muscles of the face and in a separate bundle the *intermediate nerve* which arises from the brain stem, *taste* and *viscero-efferent (secretory)* fibers. The motor fibers **AB1** stem from the large, multipolar nerve cells of the **facial nerve nucleus AB2.** They bend around the *abducens nucleus* **AB3** *(genu of the internal facial nerve)* and exit from the lateral surface of the medulla at the lower border of the pons. The cells of the preganglionic secretory fibers **AB4** form the **superior salivatory nucleus AB5.** The taste fibers **AB6,** whose pseudounipolar cells of origin lie in the **geniculate ganglion BC7,** terminate in the cranial part of the **nucleus solitarius AB8.** Viscero-efferent and gustatory fibers do not bend around the abducens nucleus but join the descending limb of the nerve and emerge as the **intermediate nerve B9,** between the facial and the vestibulocochlear nerves.

Both parts of the nerve pass through the *internal acoustic meatus* (petrosal part of the temporal bone, porus acusticus internus Vol I, p. 290 A34), and enter the *facial canal* as a single trunk. At the bend of the nerve in the petrous bone *(external facial genu)* lies the *geniculate ganglion* **BC7.** The canal then runs in the medial wall of the tympanic cavity (p. 338 A10) and turns caudally to the **stylomastoid foramen BC10,** through which the nerve leaves the skull. It divides into its terminal branches in the *parotid gland* **(parotid plexus E11).**

Within the facial canal, the nerve gives off the **greater petrosal nerve BC12,** the **stapedius nerve BC13** and the **chorda tympani BC14.** The *greater petrosal nerve* (preganglionic secretory fibers to the lacrimal glands, nasal glands and palatal glands) separates off from the geniculate ganglion, extends through the *hiatus of the canal for the greater petrosal nerve* and into the cranial cavity across the anterior surface of the petrous bone, across the foramen lacerum and finally through the *pterygoid canal* to the *pterygopalatine ganglion* **C15.** The *stapedius nerve* supplies the stapedius muscle in the middle ear. The *chorda tympani* (taste fibers to the anterior two thirds of the tongue D and preganglionic fibers to the submandibular and sublingual glands) branches off above the stylomastoid foramen, runs beneath the mucous membrane through the middle ear cavity (p. 337 A20) and through the petrotympanic fissure, finally to join the *lingual nerve* **C16.**

Before it enters the parotid gland the facial nerve gives off the **posterior auricular nerve E17** and branches to the posterior belly of the *digastric muscle* **E18** and to the *stylohyoid muscle* **C19.** The parotid plexus gives off the *temporal branches* **E20,** *zygomatic branches* **E21,** *buccal branches* **E22,** the *marginal mandibular branch* **E23,** and the *cervical branch* **E24** for the *platysma.* These branches supply all the mimetic muscles.

The ramifications of the cervical branch beneath the platysma form a *plexus* by anastomosing with branches of the sensory transverse cervical nerve (p. 66), the *ansa cervicalis superficialis.* The small branches given off by this plexus are mixed sensorimotor nerves. Terminal twigs from the temporal and buccal branches and the marginal branch of the mandible form similar plexuses with branches of the trigeminal nerve.

Clinical Tips. Damage to the nerve results in flaccid paralysis of all the muscles on the affected side of the face. The corner of the mouth hangs down and the eye cannot be closed **F** and hyperacusis is suffered (p. 338 C).

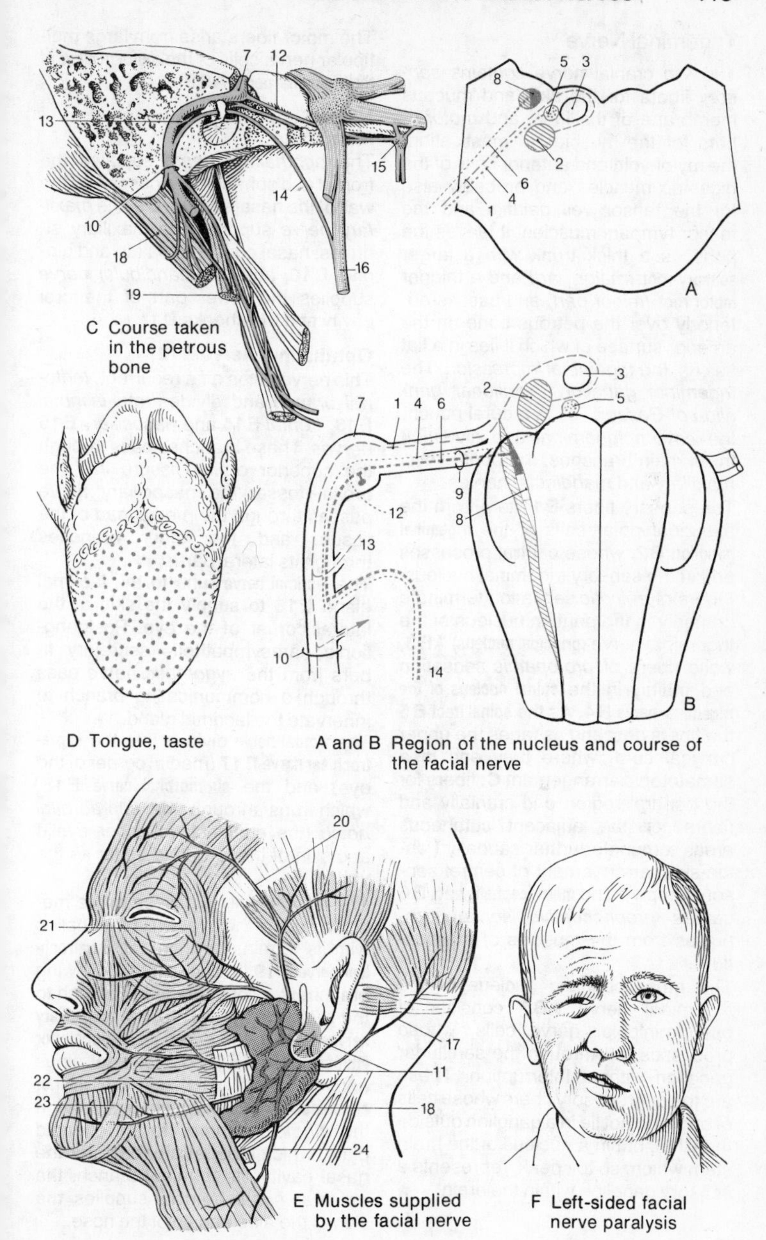

C Course taken in the petrous bone

D Tongue, taste

A and B Region of the nucleus and course of the facial nerve

E Muscles supplied by the facial nerve

F Left-sided facial nerve paralysis

Trigeminal Nerve

The Vth cranial nerve contains *sensory* fibers for the skin and mucous membrane of the face, and *motor* fibers for the muscles of mastication, the mylohyoid and anterior belly of the digastric muscles and probably also for the tensor veli palatini and the tensor tympani muscles. It leaves the pons as a thick trunk with a larger **sensory root** *(major part)* and a thinner **motor root** *(minor part)* and passes anteriorly over the petrous bone on the anterior surface of which it lies in a flat sulcus, the *trigeminal impression*. The *trigeminal ganglion (semilunar ganglion of Gasser)* lies in a dural pouch, the *cavum trigeminal* and gives off three main branches: the *ophthalmic, maxillary* and *mandibular* nerves.

The sensory fibers **B1** stem from the pseudounipolar cells of the **trigeminal ganglion B2,** whose central processes end in the sensory trigeminal nucleus. Fibers of *epicritic* sensation terminate primarily in the pontine nucleus of the trigeminal nerve **(principal nucleus) AB3,** while fibers of *protopathic* sensation end mainly in the **spinal nucleus of the trigeminal nerve B4.** As the **spinal tract B5** the fibers descend as far as the upper cervical cord, where they end in a somatotopic arrangement **C:** fibers for the perioral region end cranially and fibers for the adjacent cutaneous areas terminate further caudally ('onion-skin' arrangement of central sensory supply). The **mesencephalic tract B6** carries *proprioceptive sensory* impulses from the muscles of mastication.

The mesencephalic nucleus of the trigeminal nerve **AB7** consists of pseudounipolar nerve cells, whose processes run through the semilunar ganglion without interruption. These are the only sensory fibers whose cells of origin do not lie in a ganglion outside the CNS, but in a nucleus of the brain stem which, so to speak, represents a sensory ganglion within the brain.

The motor fibers arise from large multipolar nerve cells of the **motor nucleus of the trigeminal nerve AB8.**

Nerve Supply of the Mucous Membrane

The *ophthalmic nerve* supplies the frontal and sphenoidal sinuses and the wall of the nasal septum **D9,** the *maxillary nerve* supplies the maxillary sinuses, nasal conchae, palate and gingiva **D10,** and the *mandibular nerve* supplies the lower part of the oral cavity and the cheeks **D11.**

Ophthalmic Nerve E12

This nerve gives off a recurrent, *tentorial branch* and divides into *lacrimal* **E13,** *frontal* **E14** and *nasociliary* **E15** *nerves.* These branches pass through the superior orbital fissure into the orbital fossa; the nasociliary nerve passes through the medial part of the fissure and the other branches through its lateral part.

The **lacrimal nerve** runs to the lacrimal gland **E16** to supply the skin at the lateral corner of the eye. Postganglionic parasympathetic secretory fibers from the *zygomatic nerve* pass through a communicating branch to innervate the lacrimal gland.

The **frontal nerve** divides into the **supratrochlear nerve E17** (medial corner of the eye) and the **supraorbital nerve E18,** which runs through the *supraorbital notch* (the conjunctiva, upper eyelid and skin of the forehead as far as the vertex).

The **nasociliary nerve** passes to the medial corner of the eye, which it supplies with its terminal branch, the **infratrochlear nerve E19.** It gives off the following branches: a communicating branch to the *ciliary ganglion* **E20,** the **long ciliary nerves E21** to the eye ball, the **posterior ethmoidal nerve E22** (sphenoidal sinus and ethmoidal cells) and the **anterior ethmoidal nerve E23,** which runs through the *ethmoid foramen* to the ethmoid plate, which it penetrates to enter the nasal cavity. Its terminal branch, the *external nasal branch,* supplies the skin of the alae and tip of the nose.

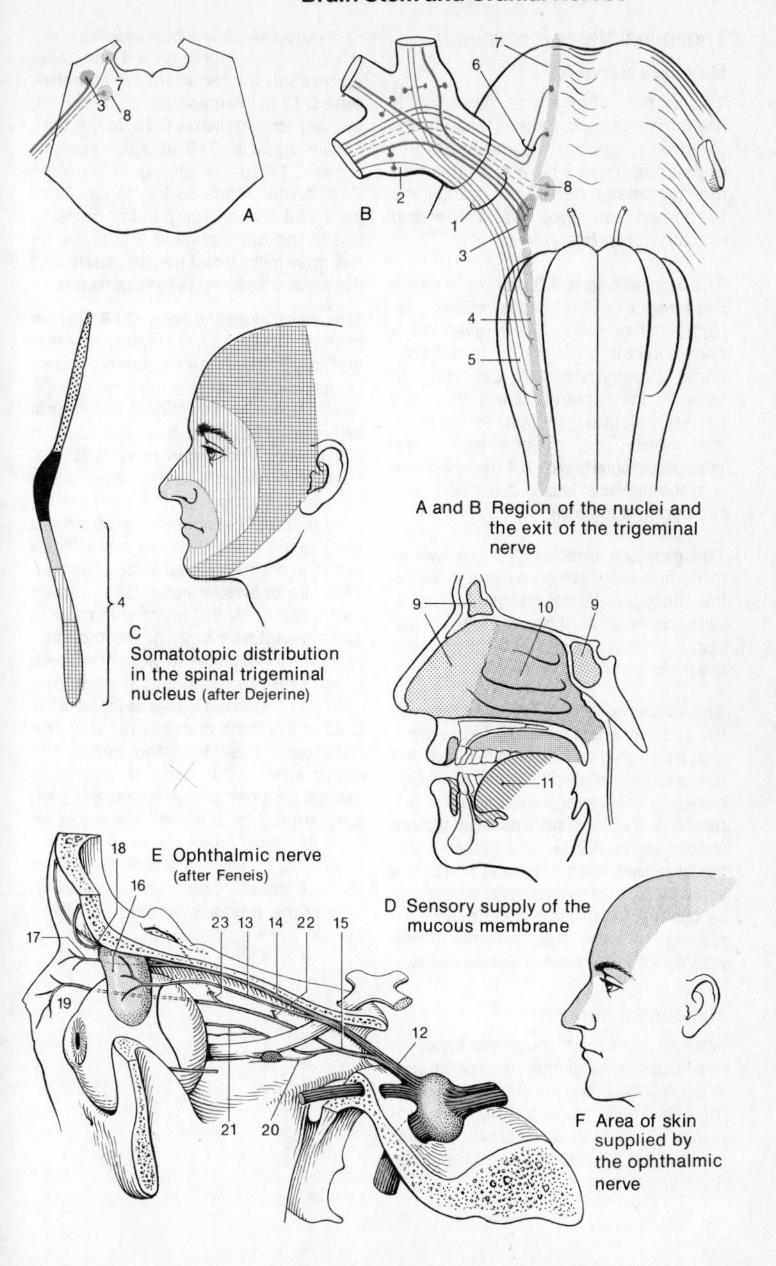

A and B Region of the nuclei and the exit of the trigeminal nerve

C Somatotopic distribution in the spinal trigeminal nucleus (after Dejerine)

D Sensory supply of the mucous membrane

E Ophthalmic nerve (after Feneis)

F Area of skin supplied by the ophthalmic nerve

Trigeminal Nerve (continued)

Maxillary Nerve A1

This nerve, after it has given off a *meningeal branch,* passes through the *foramen rotundum* **A2** into the pterygopalatine fossa where it divides into the *zygomatic nerve,* the *ganglionic branches (pterygopalatine nerves)* and the *infraorbital nerve.*

The **zygomatic nerve A3** passes through the inferior orbital fissure onto the lateral wall of the orbit. It gives off a communicating branch (postganglionic, parasympathetic secretory fibers to the lacrimal gland from the pterygopalatine ganglion) to the lacrimal nerve and divides into a **zygomaticotemporal branch A4** (temple) and a **zygomaticofacial branch A5** (skin over the zygomatic arch).

The **ganglionic branches A6** are two to three fine twigs which extend down to the pterygopalatine ganglion. The fibers carry a sensory supply to the upper pharynx, the nasal cavity and the hard and soft palates.

The **infraorbital nerve A7** passes through the inferior orbital fissure into the orbit, and through the infraorbital canal **A8** to the cheek, where it supplies the skin between the lower eyelid and the upper lip **B**. It gives off the **posterior superior alveolar nerves A9** (to the molars), the **medial superior alveolar nerve A10** (to the premolars) and the **anterior superior alveolar nerve A11** (to the canine and incisors). Above the alveoli, these nerves form the **superior alveolar plexus.**

Mandibular Nerve C

After its passage through the *foramen ovale,* and after giving off a *meningeal branch* **C12,** in the infratemporal fossa the nerve divides into the *auriculotemporal, lingual* and *inferior alveolar nerves,* the *buccal nerve* and the purely *motor branches.*

The pure **motor branches** leave the mandibular nerve shortly after it has passed through the foramen: the **masseteric nerve C13** to the masseter muscle **F14,** the **deep temporal nerves C15,** to the temporalis muscle **F16** and the **pterygoid nerves C17** to the pterygoid muscles **F18.** Motor fibers to the tensor tympani and tensor veli palatini muscles run to the otic ganglion and leave as the **nerve to the tensor tympani muscle** and the **nerve to the tensor veli palatini muscle.**

The **auriculotemporal nerve C19** (to the skin of the temporal region, external auditory meatus and tympanic membrane) usually arises from two roots, which surround the middle meningeal artery (p. 123), and then join to form the nerve. The **lingual nerve C20** descends in an arch to the base of the tongue and supplies the sensory innervation to the anterior two thirds of the tongue **D**. It receives taste fibers from the chorda tympani (facial nerve). The **inferior alveolar nerve C21** carries motor fibers for the mylohyoid muscle and the anterior belly of the digastric muscle, in addition to sensory fibers which enter the mandibular canal and give off numerous **inferior dental branches C22** to the teeth of the lower jaw. The cutaneous branch of the nerve, the **mental nerve C23**, exits through the mental foramen and provides the sensory supply for the chin, the lower lip and the skin over the body of the mandible **E**. The **buccal nerve C24** runs through the buccinator muscle **C25** to supply the mucous membrane of the cheek.

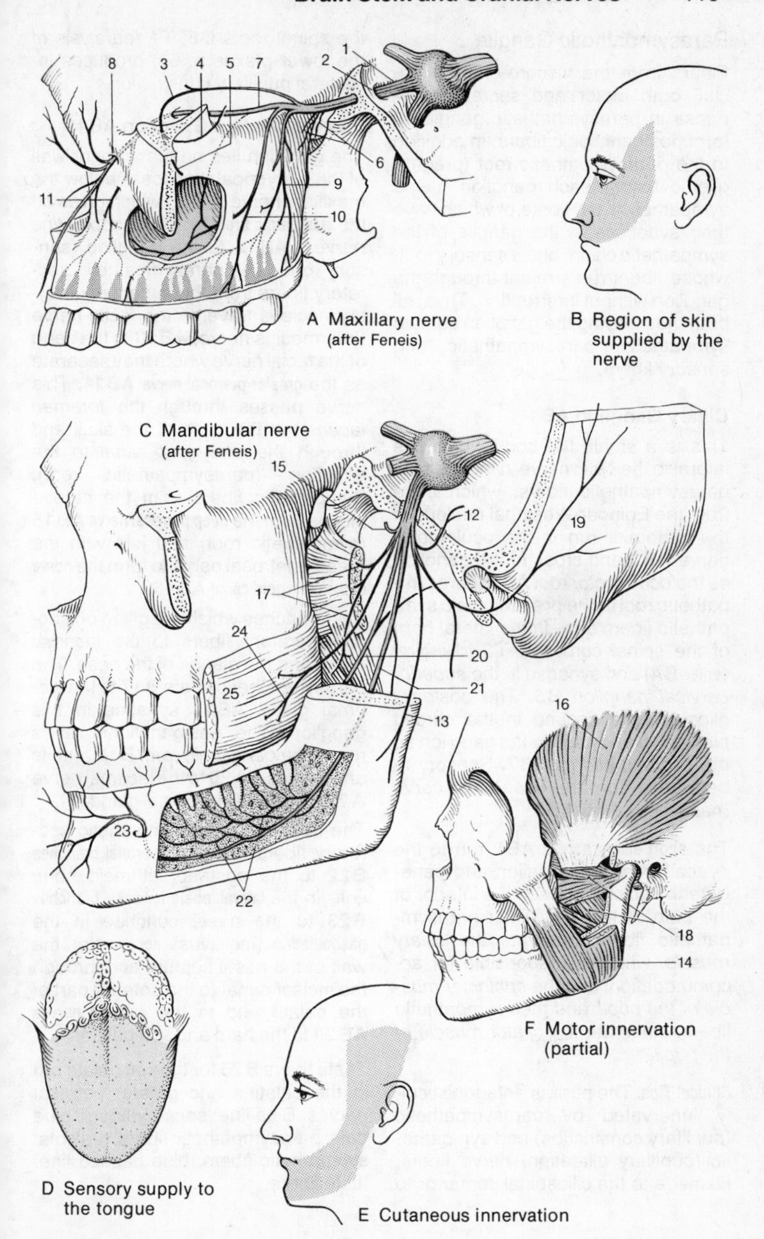

A Maxillary nerve (after Feneis)

B Region of skin supplied by the nerve

C Mandibular nerve (after Feneis)

D Sensory supply to the tongue

E Cutaneous innervation

F Motor innervation (partial)

Parasympathetic Ganglia

Fibers from the viscero-efferent nuclei, both motor and secretory, synapse in parasympathetic ganglia to form postganglionic fibers. In addition to the parasympathetic root (preganglionic fibers) each ganglion has a sympathetic root, fibers of which have their synapses in the ganglia of the sympathetic chain, and a sensory root, whose fibers run straight through the ganglion without interruption. Thus, all branches leaving the ganglion contain sympathetic, parasympathetic and sensory fibers.

Ciliary Ganglion AB 1

This is a small, flat body which lies lateral to the optic nerve in the orbit. Its parasympathetic fibers, which arise from the Edinger-Westphal nucleus in the midbrain, run in the oculomotor nerve **AB 2** and cross to the ganglion as the *oculomotor root* **AB 3** (parasympathetic root). The preganglionic sympathetic fibers arise in the lateral horn of the spinal cord, C8–T2 (**ciliospinal center B 4)** and synapse in the *superior cervical ganglion* **B 5.** The postganglionic fibers ascend in the carotid plexus **B 6** and pass to the ganglion as the *sympathetic root* **B 7.** Sensory fibers arise from the *nasociliary nerve* (nasociliary root) **AB 8.**

The **short ciliary nerves AB 9** run to the eyeball from the ganglion and penetrate the sclera to reach the interior of the bulb of the eye. Their parasympathetic fibers innervate the *ciliary muscle,* which is responsible for accommodation, and the *sphincter muscle of the pupil,* and their sympathetic fibers, innervate the *dilator muscle of the pupil.*

Clinical Tips. The pupil is antagonistically innervated by parasympathetic (pupillary constriction) and sympathetic (pupillary dilatation) nerve fibers. Damage to the ciliospinal center or to the spinal roots C8, T1 (paralysis of the lower plexus p. 68) produces ipsilateral pupillary constriction.

Pterygopalatine Ganglion AB 10

The ganglion lies on the anterior wall of the pterygopalatine fossa below the maxillary nerve **AB 11,** which gives off the **ganglionic branches** *(pterygopalatine nerves)* **AB 12** to the ganglion (sensory root). The parasympathetic secretory fibers from the superior salivatory nucleus travel in the facial nerve (intermedius nerve) **AB 13** to the genu of the facial nerve where they separate as the **greater petrosal nerve AB 14.** This nerve passes through the *foramen lacerum* in the base of the skull and through the *pterygoid canal* to the ganglion (parasympathetic root). Sympathetic fibers from the carotid plexus form the **deep petrosal nerve AB 15** (sympathetic root) and join with the greater petrosal nerve to form the **nerve of the pterygoid canal AB 16.**

The branches which are given off supply secretory fibers to the lacrimal gland and the glands of the nose. The parasympathetic fibers **B 17** to the lacrimal gland **AB 18** synapse in this ganglion. The postganglionic fibers run in the ganglionic rami **B 12** and its anastomosis **A 20** to the lacrimal nerve **A 21** to reach the lacrimal gland.

The remaining parasympathetic secretory fibers run in the **orbital branches B 22** to the posterior ethmoid bone cells in the **lateral, posterior nasal branches B 23** to the nasal conchae in the **nasopalatine** (incisivus) **nerve** over the wall of the nasal septum and through the incisor canal to the anterior part of the palate and in the **palatine nerves AB 24** to the hard and soft palates.

Taste fibers **B 25** for the soft palate run in the palatine and greater petrosal nerves. Blue line: sensory fibers, blue dots: parasympathetic fibers, red dots: sympathetic fibers, blue dashed line: taste fibers.

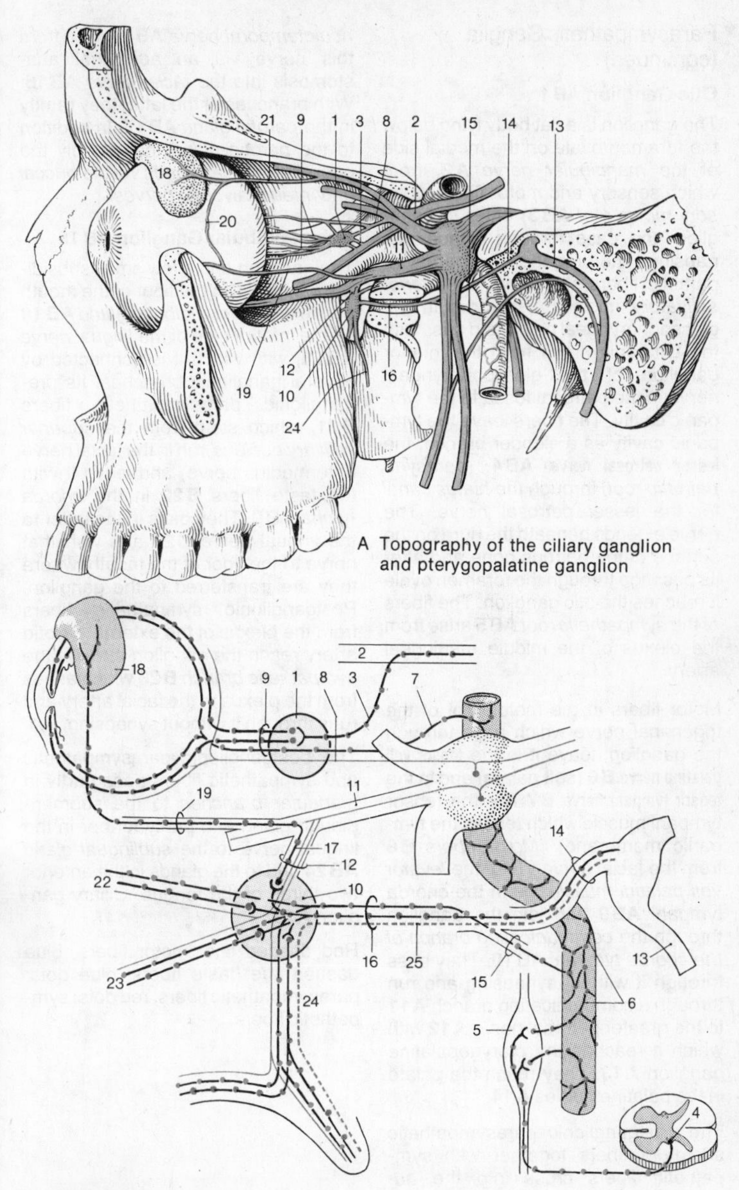

A Topography of the ciliary ganglion and pterygopalatine ganglion

B Conduction pathways of the ciliary ganglion and the pterygopalatine ganglion

Parasympathetic Ganglia (continued)

Otic Ganglion AB 1

The ganglion is a flat body lying below the foramen ovale on the medial side of the *mandibular nerve* **A2,** from which sensory and motor fibers *(sensory-motor root* **AB3)** enter the ganglion and pass through without synapsing. The preganglionic parasympathetic fibers stem from the *inferior salivatory nucleus.* They run in the glossopharyngeal nerve and pass with the tympanic nerve from the inferior ganglion of the glossopharyngeal nerve in the petrosal fossa to the tympanic cavity. The fibers leave the tympanic cavity as a slender branch, the **lesser petrosal nerve AB4** *(parasympathetic root)* through the *hiatus canal* for the lesser petrosal nerve. The nerve extends beneath the dura on the surface of the petrous bone and after its passage through the foramen ovale it reaches the otic ganglion. The fibers of the *sympathetic root* **AB5** arise from the plexus of the middle meningeal artery.

Motor fibers in the motor root of the trigeminal nerve which pass through the ganglion, leave it in the **tensor veli palatine nerve B6** (soft palate) and in the **tensor tympani nerve B7** (for the tensor tympani muscle which tenses the tympanic membrane). *Motor fibers* **B8** from the facial nerve VII for the *levator veli palatini muscle* run in the chorda tympani **AB9** and join the ganglion through the *communicating branch of the chorda tympani* **AB10.** They pass through it without synapsing and run through a communicating branch **A11** to the greater petrosal nerve **A12** with which it reaches the pterygopalatine ganglion **A13.** They reach the palate in the palatine nerves **A14.**

The postganglionic parasympathetic secretory fibers together with sympathetic fibers cross into the *auriculotemporal nerve* **AB15,** and from this nerve via an additional anastomosis into the *facial nerve* **AB16.** With branches of the latter they ramify in the *parotid gland* **AB17.** In addition to the parotid gland they supply the buccal and labial glands via the *buccal* and *inferior alveolar nerves.*

Submandibular Ganglion AB 18

The ganglion and a few small subsidiary ganglia lie in the floor of the mouth above the *submandibular gland* **AB19** on the inferior side of the *lingual nerve* **AB20**, with which it is connected by several ganglionic branches. Its preganglionic parasympathetic fibers **B21,** which stem from the *superior salivary nucleus,* run in the facial nerve (intermedius nerve) and leave it with the taste fibers **B22** in the *chorda tympani* **B9.** They pass in the latter to the lingual nerve **B20** and with that nerve to the floor of the mouth, where they are transferred to the ganglion. Postganglionic sympathetic fibers from the plexus of the external carotid artery reach the ganglion through the *sympathetic branch* **B23** which arises from the plexus of the facial artery and runs through it without synapsing.

The postganglionic parasympathetic and sympathetic fibers pass partly in *glandular branches* to the submandibular gland and partly further in the lingual nerve to the *sublingual gland* **AB24** and to the glands in the anterior two thirds of the tongue. Ciliary ganglion **A25.**

Red dashed line: motor fibers, blue dashed line: taste fibers, blue dots: parasympathetic fibers, red dots: sympathetic fibers.

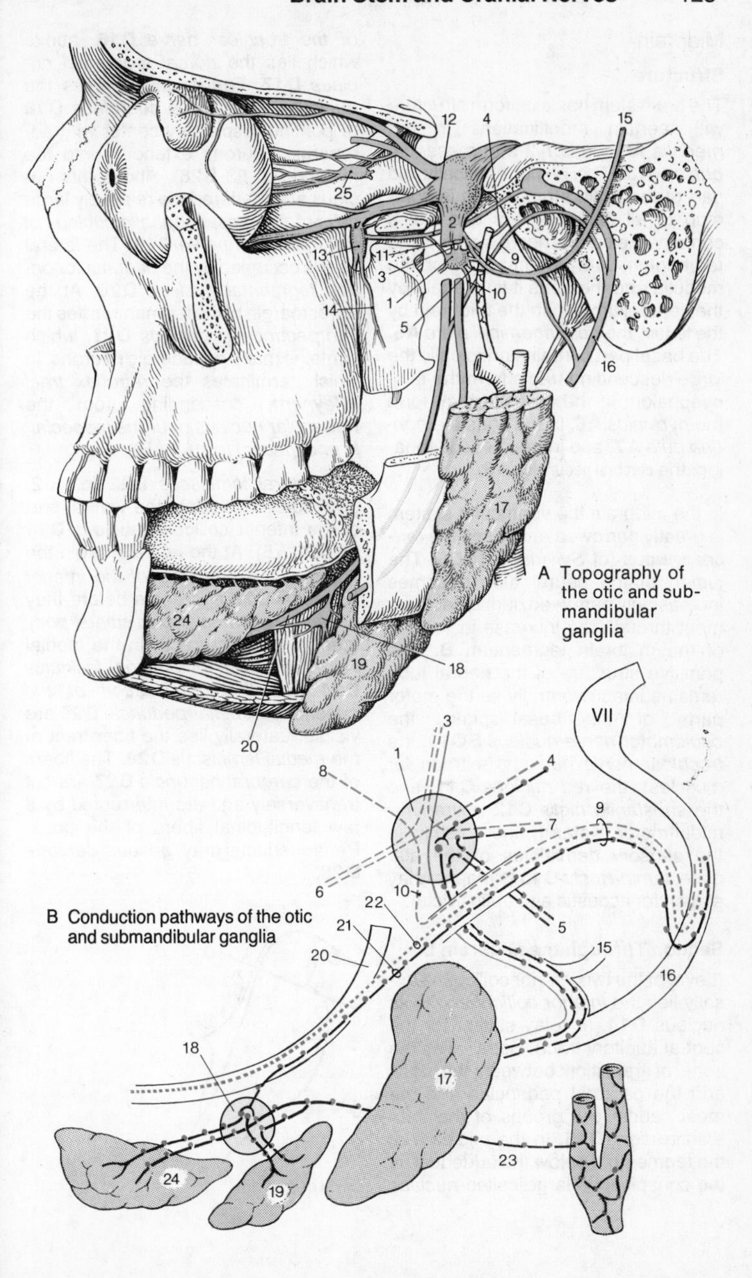

A Topography of the otic and submandibular ganglia

B Conduction pathways of the otic and submandibular ganglia

Midbrain

Structure

The brain stem has a uniform structure with certain modifications in the *medulla* **A1**, *pons* **A2** and *mesencephalon* **A3**. The phylogenetically old part of the brain stem, which is common to all three regions and which contains the cranial nerve nuclei, is the *tegmentum* **A4.** At the level of the medulla and the pons it is overlaid by the cerebellum and in the midbrain by the **tectum,** the *quadrigeminal plate* **A5.** The basal part contains principally the large descending tracts from the telencephalon: in the medulla they form the *pyramids* **A6,** in the pons the *pontine bulb* **A7,** and in the mesencephalon, the **cerebral peduncles A8.**

In the midbrain the ventricular system is greatly narrowed and forms the **cerebral aqueduct** (of *Sylvius)* **ABCD9.** The lumen of the neural tube becomes increasingly narrowed during development through the increase in volume of the midbrain tegmentum **B.** The primitive structure of the neural tube remains intact: ventrally lie the motor parts of the basal plate, the *oculomotor nerve nucleus* **BC10,** the *trochlear nerve nucleus* (extraocular muscles), the *red nucleus* **C11,** and the *substantia nigra* **C12** (extrapyramidal-motor system); and dorsally lie the sensory derivatives of the alar plate, *lamina tecta* **C13** (synaptic relay station for acoustic and optic tracts).

Section Through the Midbrain D

(Level of the two inferior colliculi.) Dorsally lies the *inferior colliculus* with its nucleus **D14,** (a relay station for the central auditory tract). Basally lies the zone of transition between the pons and the cerebral peduncles and the most caudal cell groups of the *substantia nigra* **D15.** In the midzone of the tegmentum below the aqueduct, is the conspicuous large-celled *nucleus of the trochlear nerve* **D16,** above which lies the *dorsal tegmental nucleus* **D17.** Further laterally lies the *nucleus of the locus coeruleus* **D18** (a pontine respiratory center with adrenergic neurons extending into the midbrain p. 93 B28). Above this nucleus are scattered the relatively large cells of the *mesencephalic nucleus of the trigeminal nerve* **D19.** The lateral field is occupied by the *pedunculopontine tegmental nucleus* **D20.** At the lower margin of the tegmentum lies the *interpeduncular nucleus* **D21,** which contains many peptidergic neurons, in which terminates the *retroflex tract (Meynert),* descending from the *habenular nucleus neurons (especially encephalin)* (p. 164 AB).

The *lateral lemniscus* **D22** (p. 112, 350) spreads out into the ventral part of the inferior collicular nucleus **D14** (p. 351 A5). At the lateral margin the fibers of the *peduncle of the inferior colliculus* **D23** aggregate before they pass to the *medial geniculate body* (central auditory tract). In the medial field the *medial longitudinal fasiculus* **D24** and the *decussation of the superior cerebellar peduncle* **D25** are visible. Laterally lies the fiber tract of the *medial lemniscus* **D26.** The fibers of the *cerebral peduncle* **D27** are cut transversely and are interrupted by a few longitudinal fibers of the pons. Periaqueductal gray, *griseum centrale* **D28.**

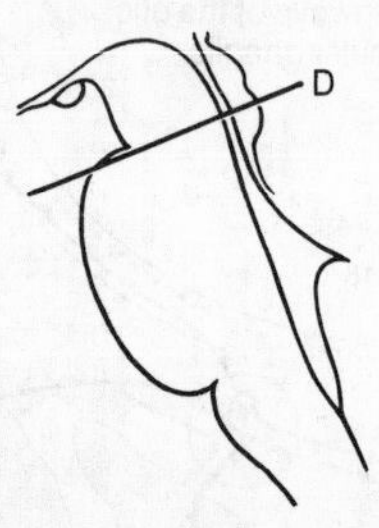

Level of section

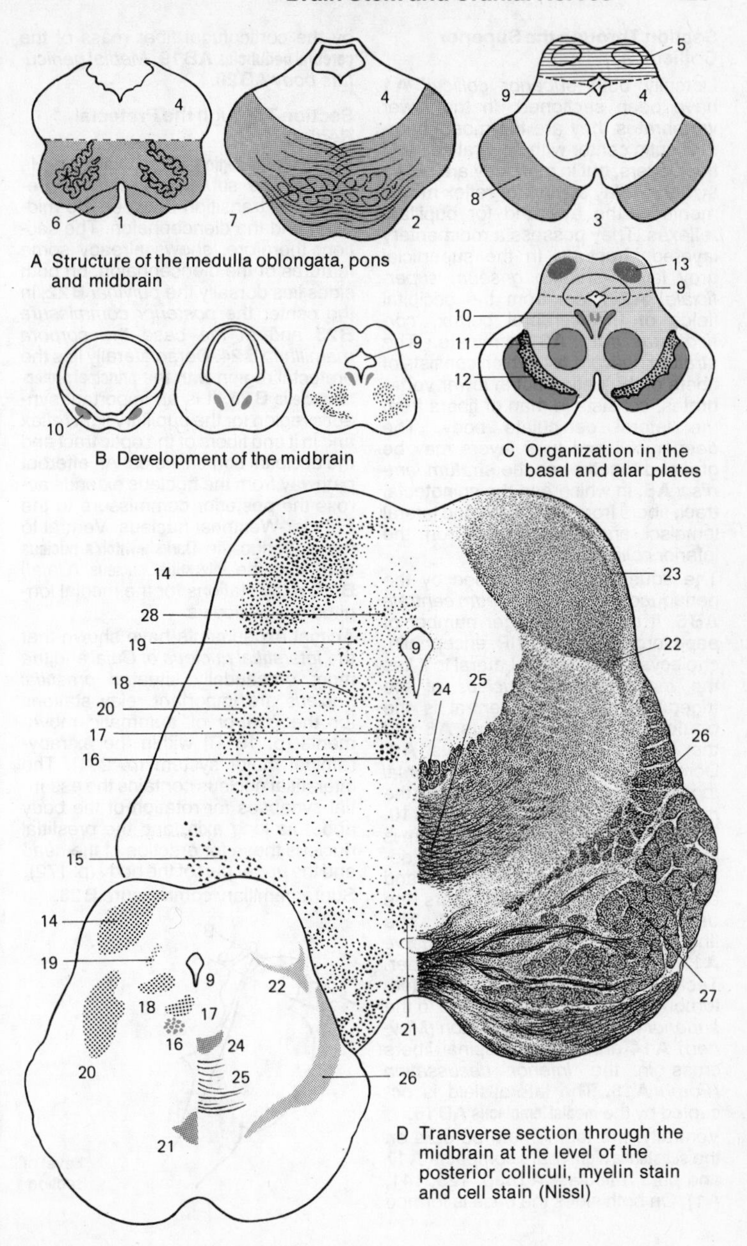

A Structure of the medulla oblongata, pons and midbrain

B Development of the midbrain

C Organization in the basal and alar plates

D Transverse section through the midbrain at the level of the posterior colliculi, myelin stain and cell stain (Nissl)

Section Through the Superior Colliculi A

Dorsally both *superior colliculi* **A1** have been sectioned. In the lower vertebrates they are the most important optic center with several cell and fiber layers, but in man they are only a synaptic relay station for reflex movements of the eye and for pupillary reflexes. They possess a rudimentary layered structure. In the superficial grey layer *(stratum griseum superficiale)* **A2** fibers from the occipital fields of the cerebral cortex, *corticotectal tract* **A3** terminate. The *stratum opticum* **A4,** which consists of fibers of the optic tract in lower vertebrates, consists in man of fibers from the lateral geniculate body. The deeper cell and fiber layers may be grouped together as the *stratum lemnisci* **A5,** in which end the spinotectal tract, fibers from the medial and lateral lemnisci, and fiber bundles from the inferior colliculi.

The aqueduct is surrounded by the periaqueductal gray, *griseum centrale* **AB6**. It contains a larger number of peptidergic neurons (VIP, encephalin, cholecystokinin et al.). Lateral to it lies the *mesencephalic nucleus* of the trigeminal nerve **A7,** ventral is the **oculomotor nerve nucleus A8** and the Edinger-Westphal nucleus **A9**. Dorsal to both nuclei passes the *dorsal longitudinal fasciculus* and ventral the *medial longitudinal fasciculus* **A10.** The chief nucleus in the tegmentum is the **red nucleus AB11,** which is delimited by its capsule of afferent and efferent fibers, amongst others the *dentatorubral fasciculus* **A12.** The fiber bundles of the *oculomotor nerve* **A13** run ventrally at its medial border. Tectospinal (pupillary reflex) and tectorubral fibers cross the midline in the *superior tegmental decussation (Meynert)* **A14** and tegmentospinal fibers cross in the *inferior decussation (Forel)* **A15.** The lateral field is occupied by the **medial lemniscus AB16.**

Ventrally, the tegmentum borders on the **substantia nigra** *(pars compacta* **A17** and *pars reticulata* **A18,** p. 128, 141, A1). On both sides the base is formed by the corticofugal fiber mass of the **cerebral peduncles AB19.** *Medial geniculate body* **AB20.**

Section Through the Pretectal Region

The **pretectal region B21,** which lies in front of the superior colliculi, represents the transition between the midbrain and the diencephalon. The section, therefore, shows already some features of the diencephalon: on both sides lies dorsally the *pulvinar* **B22,** in the center the *posterior commissure* **B23** and at the base the *corpora mamillaria* **B24.** Dorsolaterally lies the pretectal region with the **principal pretectal nucleus B25.** It is an important synaptic region for the pupillary light reflex and in it end fibers of the optic tract and the occipital cortical fields. An efferent pathway from the nucleus extends across the posterior commissure to the Edinger-Westphal nucleus. Ventral to the aqueduct lie **Darkshewitch's nucleus B26** and the **interstitial nucleus** *(Cajal)* **B27,** relay stations for the medial longitudinal fasciculus.

Animal experiments have shown that the *interstitial nucleus of Cajal* and the more orocaudally situated *prestitial nucleus* are important relay stations for the control of automatic movements (p. 181 B) within the extrapyramidal motor system (p. 284). The interstitial nucleus contains the essential synapses for rotation of the body about its long axis, and the prestitial nucleus those for erection of the head and the upper part of the body (p. 172). Supramamillary commissure **B28.**

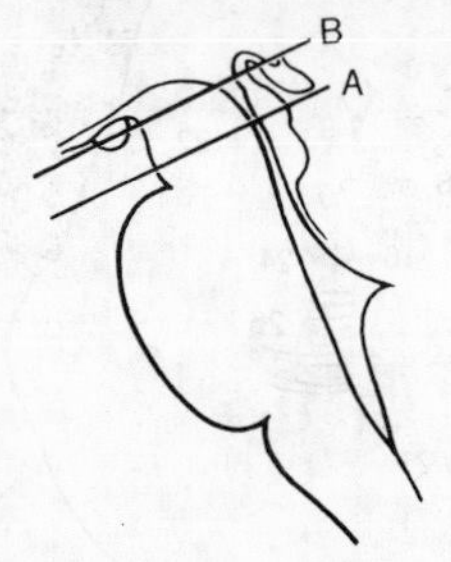

Level of section

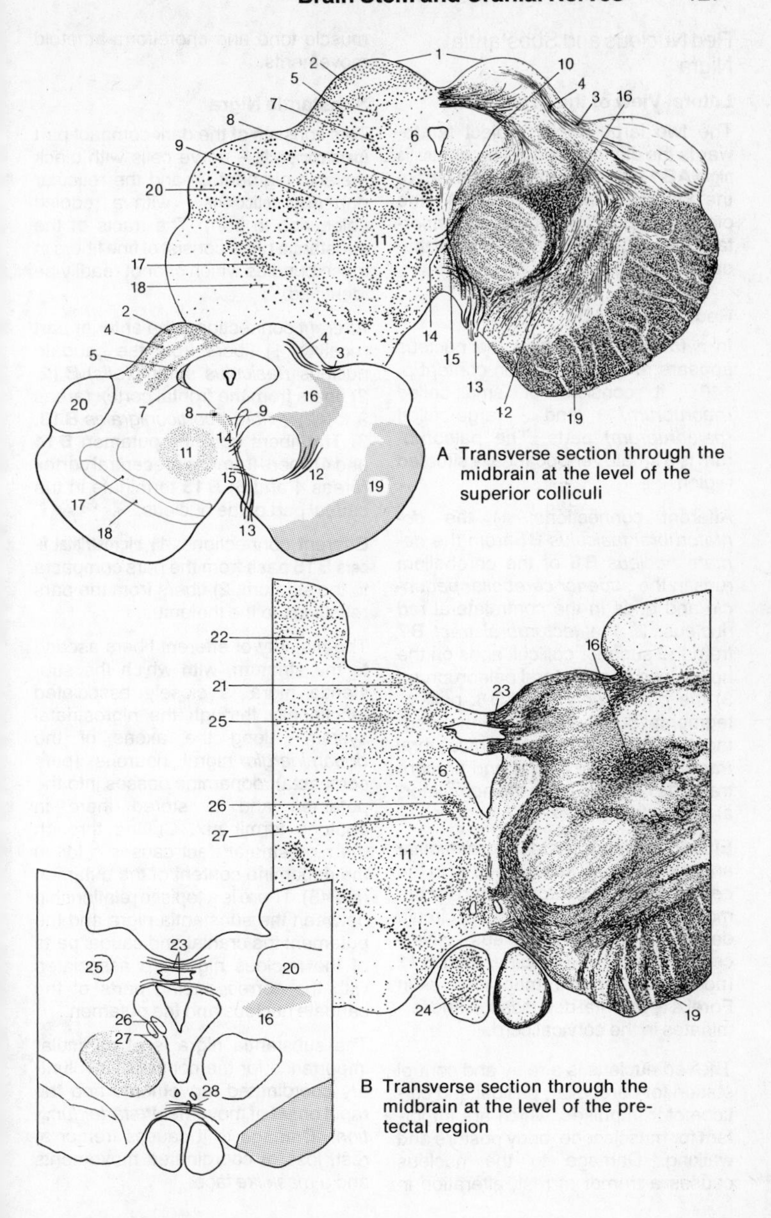

A Transverse section through the midbrain at the level of the superior colliculi

B Transverse section through the midbrain at the level of the pretectal region

Red Nucleus and Substantia Nigra

Lateral View of the Brain Stem

The two large nuclei project far towards the diencephalon. The **substantia nigra AB 1** extends from the oral part of the *pons* **A 2** to the *pallidum* **AB 3** in the diencephalon. Both nuclei are important relay stations of the extrapyramidal system (p. 284).

Red Nucleus AB 4

In a fresh brain section the nucleus appears reddish (high iron content p. 140). It consists of small-celled *(neorubrum)* and large-celled *(paleorubrum)* parts. The paleorubrum is a small, ventrocaudally situated region.

Afferent connections: 1) the *dentatorubral fasiculus* **B 5** from the *dentate nucleus* **B 6** of the cerebellum runs in the *superior cerebellar peduncle* and ends in the contralateral red nucleus. 2) the *tectorubral tract* **B 7** from the superior colliculi ends on the homo- and contralateral paleorubrum. 3) the *pallidorubral tract* **B 8,** pallidotegmental bundle, from the inner segment of the pallidum. 4) *corticorubral tract* **B 9** from the frontal and precentral cortex terminates in the homolateral red nucleus.

Efferent connections: 1) *rubroreticular* and *rubro-olivary fibers* **B 10** run in the *central tegmental tract* and end for the most part in the olive (neuronal circuit: dentate nucleus – red nucleus – olive – cerebellum). 2) *rubrospinal tract* **B 11** (poorly developed in man) crosses in Forel's tegmental decussation and terminates in the cervical cord.

The red nucleus is a relay and control station for cerebellar, pallidal and corticomotor impulses, which are important for muscle tone, body posture and walking. Damage to the nucleus causes a tremor at rest, alteration in muscle tone and choreiform-athetoid movements.

Substantia Nigra

This consists of the dark compact part (**pars compacta** – nerve cells with black melanin pigment **C**) and the reticular part (**pars reticularis** – with a reddish color, rich in iron). The tracts of the substantia nigra consist of fine fibers in loose bundles which cannot readily be identified.

Afferent connections: the anterior part receives 1) fibers from the caudate nucleus *(fasiculus strionigralis)* **B 12,** 2) fibers from the frontal cortex (areas 9 to 12) – *fibrae corticonigrales* **B 13.** 3) The fibers from the putamen **B 14** and 4) fibers from the precentral cortex (areas 4 and 6) **B 15** terminate in the caudal part of the nucleus.

Efferent connections: 1) **nigrostriatal fibers B 16** pass from the pars compacta to the striatum, 2) fibers from the pars reticularis to the thalamus.

The majority of efferent fibers ascend to the striatum, with which the substantia nigra is closely associated functionally through the nigrostriatal system. Along the axons of the *dopaminergic* nigral neurons (pars compacta), dopamine passes into the putamen and is stored there in boutons terminaux. Cutting through the nigrostriatal tract causes a fall in the dopamine content of the putamen (p. 248). There is a topistic relationship between the substantia nigra and the putamen: the cranial and caudal parts of the nucleus niger are associated with the corresponding parts of the caudate nucleus and the putamen.

The substantia nigra is of particular importance for the control of involuntary coordinated movements and the rapid onset of movement *(starter function).* Damage to it causes tremor at rest, loss of coordinated movements and a *masklike* face.

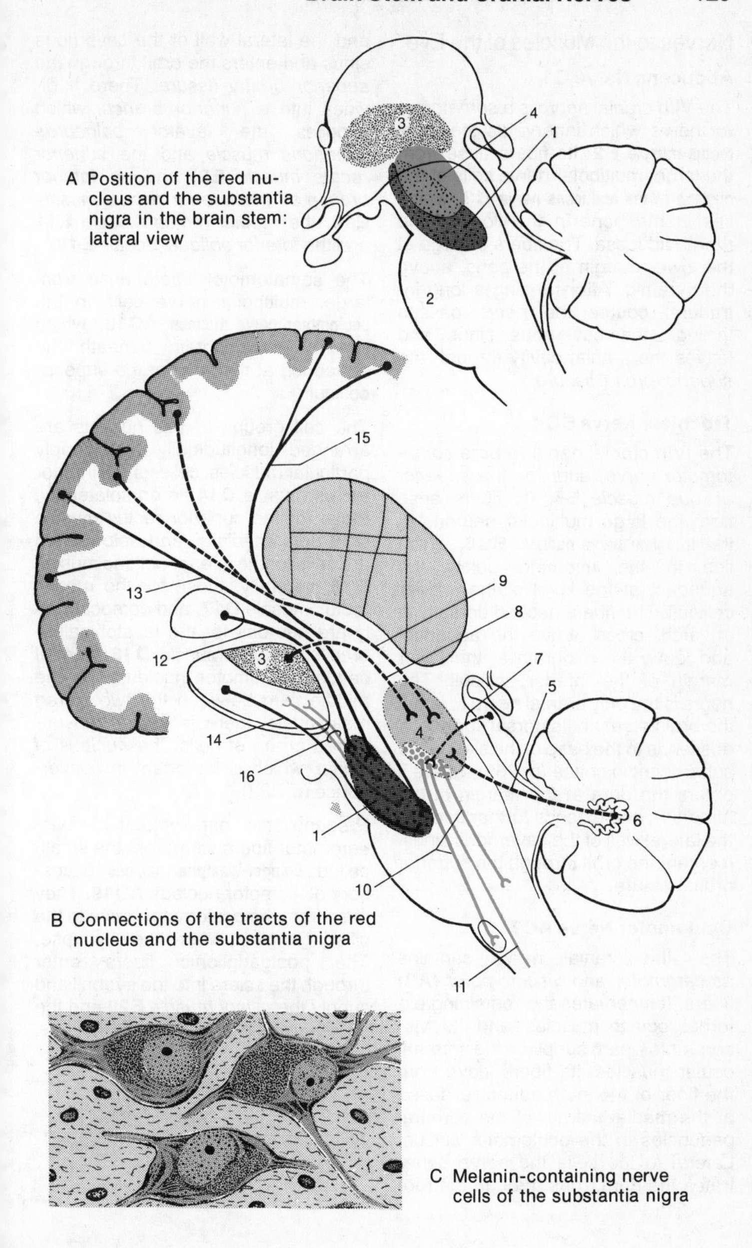

A Position of the red nucleus and the substantia nigra in the brain stem: lateral view

B Connections of the tracts of the red nucleus and the substantia nigra

C Melanin-containing nerve cells of the substantia nigra

Nerves to the Muscles of the Eye

Abducens Nerve C1

The VIth cranial nerve is a *somatomotor* nerve which innervates the **lateral rectus muscle E2.** Its fibers arise from the large, multipolar nerve cells of the **nucleus of the abducens nerve C3,** which lies in the pons in the floor of the rhomboid fossa. The fibers emerge at the lower margin of the pons, above the pyramid. After running a long intradural course the nerve passes through the cavernous sinus and leaves the cranial cavity through the *superior orbital fissure.*

Trochlear Nerve BC4

The IVth cranial nerve (a pure *somatomotor* nerve) supplies the *superior oblique muscle* **E5.** Its fibers arise from the large multipolar neurons of the **trochlear nerve nucleus BC6,** which lies in the midbrain below the aqueduct at the level of the inferior colliculi. The fibers ascend dorsally in an arch, cross above the aqueduct and leave the midbrain at the lower margin of the inferior colliculi. The nerve is the only cranial nerve to leave the brain stem on its dorsal surface. It descends to the base of the skull in the subarachnoid space (p. 268), where it enters the dura at the margin of the tentorium and extends further through the lateral wall of the cavernous sinus. It enters the orbit through the superior orbital fissure.

Oculomotor Nerve AC7

The IIIrd cranial nerve contains *somatomotor* and *visceromotor* **(A8)** fibers. It innervates the remaining external ocular muscles and its visceromotor part supplies the internal ocular muscles. Its fibers leave from the floor of the interpeduncular fossa at the medial margin of the cerebral peduncles in the oculomotor sulcus. Lateral to the sella the nerve penetrates the dura, runs through the roof and the lateral wall of the cavernous sinus and enters the orbit through the *superior orbital fissure.* There it divides into a *superior branch,* which supplies the *levator palpebrae superioris muscle* and the *superior rectus muscle* **E9,** and an *inferior branch* to the *inferior rectus muscle* **E10,** the *medial rectus muscle* **E11** and the *inferior oblique muscle* **E12.**

The somatomotor fibers arise from large, multipolar nerve cells in the **oculomotor nerve nucleus AC13,** which lies in the midbrain beneath the aqueduct, at the level of the superior colliculi.

The cell groups of this nucleus are arranged longitudinally and supply particular muscles: cells for the inferior rectus muscle **D14** lie dorsolaterally, those for the superior rectus muscle **D15** dorsomedially, and below them are cells for the inferior oblique muscle **D16,** ventrally, those for the medial rectus muscle **D17,** and dorsocaudally are neurons for the levator palpebrae superioris muscle **D18** (central caudal oculomotor nucleus). In the middle third between the two paired main nuclei there is usually an unpaired group of cells, the *nucleus of Perlia,* which is important in convergence (p. 332).

Preganglionic parasympathetic visceromotor fibers stem from the small-celled **Edinger-Westphal nucleus** accessory oculomotor nucleus **AD19.** They pass from the oculomotor nerve to the ciliary ganglion, where they synapse. The postganglionic fibers enter through the sclera into the eyeball and supply the *ciliary muscle* **F20** and the *pupillary sphincter muscle* **F21.**

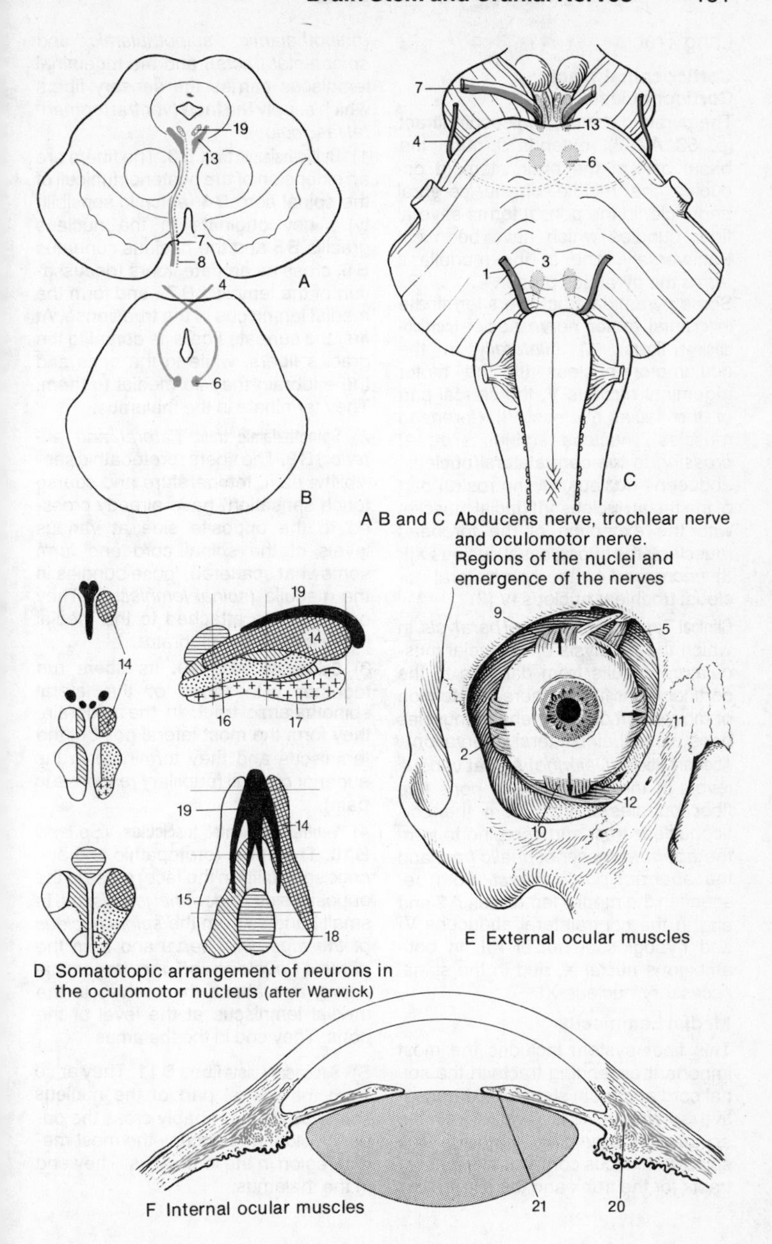

A B and C Abducens nerve, trochlear nerve and oculomotor nerve. Regions of the nuclei and emergence of the nerves

D Somatotopic arrangement of neurons in the oculomotor nucleus (after Warwick)

E External ocular muscles

F Internal ocular muscles

Long Tracts

Corticospinal Tract, Corticonuclear Fibers

The pyramidal tract, corticospinaltract (p. 53 A282), extends through the basal part of the brain stem. It occupies the middle of the cerebral peduncle, in the pons it forms several fiber bundles which have been cut transversely, and in the medulla it forms the pyramids proper.

Some pyramidal tract fibers terminate in cranial motor nerve nuclei (**corticonuclear fibers**): 1) *bilaterally* in the oculomotor nucleus III, the motor trigeminal nucleus V, the caudal part of the facial nucleus VII (forehead muscles), nucleus ambiguus X; 2) *crossing* to the contralateral nucleus: abducens nucleus VI, the rostral part of the facial nucleus VII (facial muscles with the exception of the forehead muscles), the hypoglossal nucleus XII; 3) *uncrossed* on the homolateral nucleus: trochlear nucleus IV (?).

Clinical Tips. In *central facial paralysis,* in which the paralysis of the facial musculature results from damage to the corticobulbar fibers, there is retention of movement of the forehead muscles because of their bilateral innervation.

Aberrant fibers *(Déjérine)* **A1:** at various levels in the midbrain and pons fine fiber bundles branch from the corticonuclear tract and combine to form the *aberrant mesencephalic tract* and the *aberrant pontine tract.* Both descend in the medial lemniscus **A2** and end in the contralateral abducens VI and hypoglossal nuclei XII, in both ambiguus nuclei X, and in the spinal accessory nucleus XI.

Medial Lemniscus

This fiber system includes the most important ascending tracts in the spinal cord and brain stem for exteroceptive sensation. It is divided into the *spinal* and *trigeminal lemnisci.* The spinal lemniscus contains the sensory tracts for the trunk and the extremities *(bulbothalamic, spinothalamic* and *spinotectal tracts),* and the trigeminal lemniscus carries the sensory fibers which supply the face *(ventral tegmental fasciculus).*

1) **Bulbothalamic tract B3.** The fibers are an extension of the posterior funiculi of the spinal cord **B4** (epicritic sensibility). They originate in the nucleus gracilis **B5** and the nucleus cuneatus **B6,** cross as arcuate fibers (decussation of the lemnisci **B7**). and form the medial lemniscus in the true sense. At first the cuneate fibers lie dorsal to the gracilis fibers, while in the pons and the midbrain they lie medial to them. They terminate in the thalamus.

2) **Spinothalamic tract** *(lateral and anterior)* **B8.** The fibers (protopathic sensibility: pain, temperature and coarse touch sensation) have already crossed to the opposite side at various levels of the spinal cord and form somewhat scattered, loose bundles in the medulla *(spinal lemniscus).* They only become attached to the medial lemniscus in the midbrain.

3) **Spinotectal tract B9.** Its fibers run together with those of the lateral spinothalamic tract. In the midbrain, they form the most lateral point of the lemniscus and they terminate in the superior colliculi (pupillary response to pain).

4) **Ventral tegmental fasciculus** *(Spitzer)* **B10.** The fibers (protopathic and epicritic sensibility in the face) cross to the opposite side *(trigeminal lemniscus)* in small bundles from the *spinal nucleus of the trigeminal nerve* and from the *pontine nucleus of the trigeminal nerve,* and attach themselves to the medial lemniscus at the level of the pons. They end in the thalamus.

5) **Secondary taste fibers B11.** They arise from the rostral part of the nucleus solitarius **B12,** probably cross the opposite side, and occupy the most medial region in the lemniscus. They end in the thalamus.

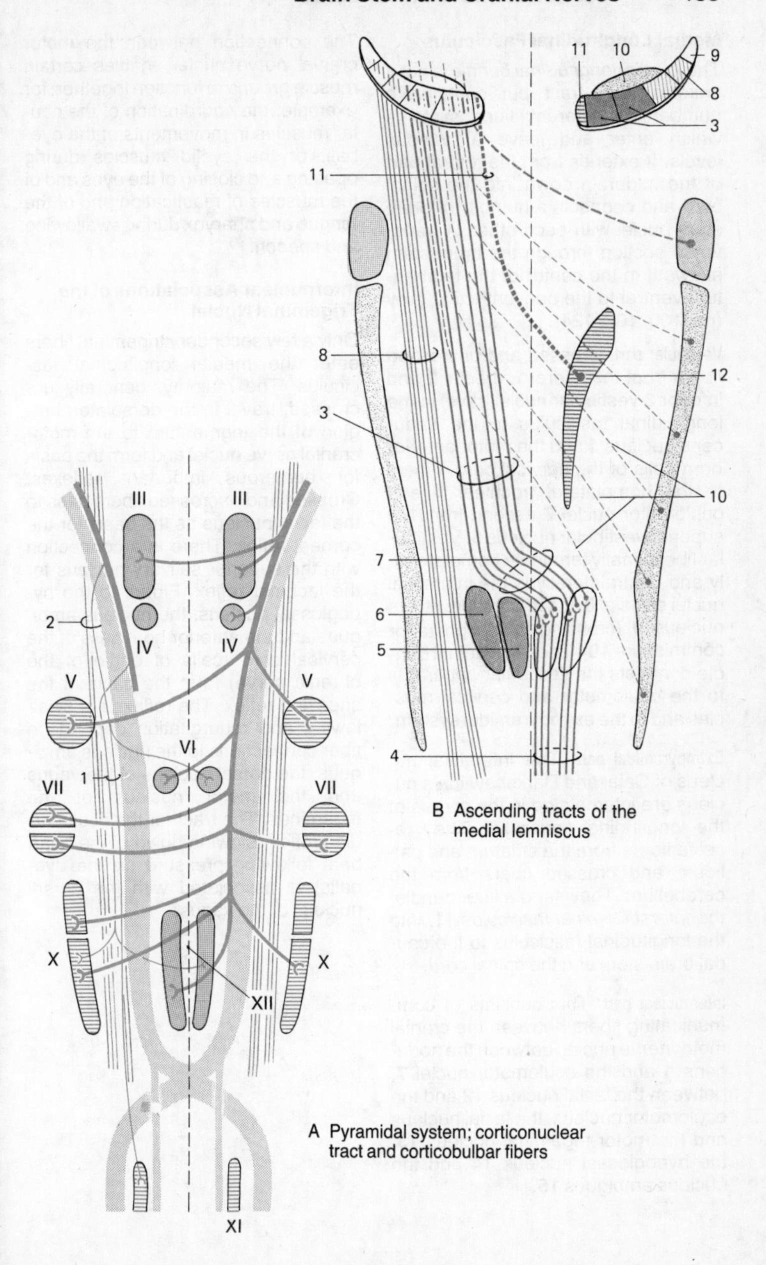

B Ascending tracts of the medial lemniscus

A Pyramidal system; corticonuclear tract and corticobulbar fibers

Medial Longitudinal Fasciculus

The *medial longitudinal bundle* is not a uniform fiber tract but contains a number of different fiber systems which enter and leave at various levels. It extends from the rostral part of the midbrain down into the spinal cord and connects a number of brain stem nuclei with each other. A transverse section through the brain stem shows it in the center of the tegmentum ventral to the periventricular grey (pp. 100, 102, 124).

Vestibular part. Crossed and uncrossed fibers from the lateral **1,** medial **2** and inferior **3** vestibular nuclei travel in the longitudinal fasciculus to the abducens nucleus **4** and the motor anterior horn cells of the cervical cord. Fibers to the homolateral trochlear **6** and oculomotor nuclei **7** ascend from the superior vestibular nucleus **5.** Vestibular fibers finally terminate homolaterally and contralaterally in the interstitial nuclei of Cajal **8** and in Darkshewitch's nucleus **9** (crossing in the posterior commissure **10**). The longitudinal bundle connects the vestibular apparatus to the oculomotor and cervical muscles and to the extrapyramidal system.

Extrapyramidal part. The interstitial nucleus of Cajal and Darkshewitch's nucleus are intercalated in the course of the longitudinal fasciculus. They receive fibers from the striatum and pallidum and crossed fibers from the cerebellum. They send a fiber bundle, the *interstitiospinal fasciculus* **11,** into the longitudinal fasciculus to the caudal brain stem and the spinal cord.

Internuclear part. This consists of communicating fibers between the cranial motor nerve nuclei: between the abducens **4** and the oculomotor nuclei **7,** between the facial nucleus **12** and the oculomotor nucleus, the facial nucleus and the motor trigeminal nucleus **13,** the hypoglossal nucleus **14** and the nucleus ambiguus **15.** The connection between the motor cranial nerve nuclei enables certain muscle groups to function together; for example, the coordination of the ocular muscles in movements of the eyeball, of the eyelid muscles during opening and closing of the eyes and of the muscles of mastication and of the tongue and pharynx during swallowing and speech.

Internuclear Associations of the Trigeminal Nuclei

Only a few secondary trigeminal fibers enter the medial longitudinal fasciculus. The majority, generally uncrossed, travel in the dorsolateral region of the tegmentum to the motor cranial nerve nuclei and form the basis for numerous important reflexes. Crossed and uncrossed fibers pass to the facial nucleus as the basis for the corneal reflex. There is a connection with the superior salivary nucleus for the lacrimal reflex. Fibers to the hypoglossal nucleus, the nucleus ambiguus and the anterior horn cells of the cervical cord (cells of origin of the phrenic nerve) form the basis of the sneezing reflex. The reflexes of swallowing and regurgitation depend on fiber connections to the nucleus ambiguus, the dorsal nucleus of the vagus and the motor nucleus of the trigeminus. The tract for the oculocardiac reflex (slowing down of the heart beat following pressure on the eyeballs) is associated with the dorsal nucleus of the vagus.

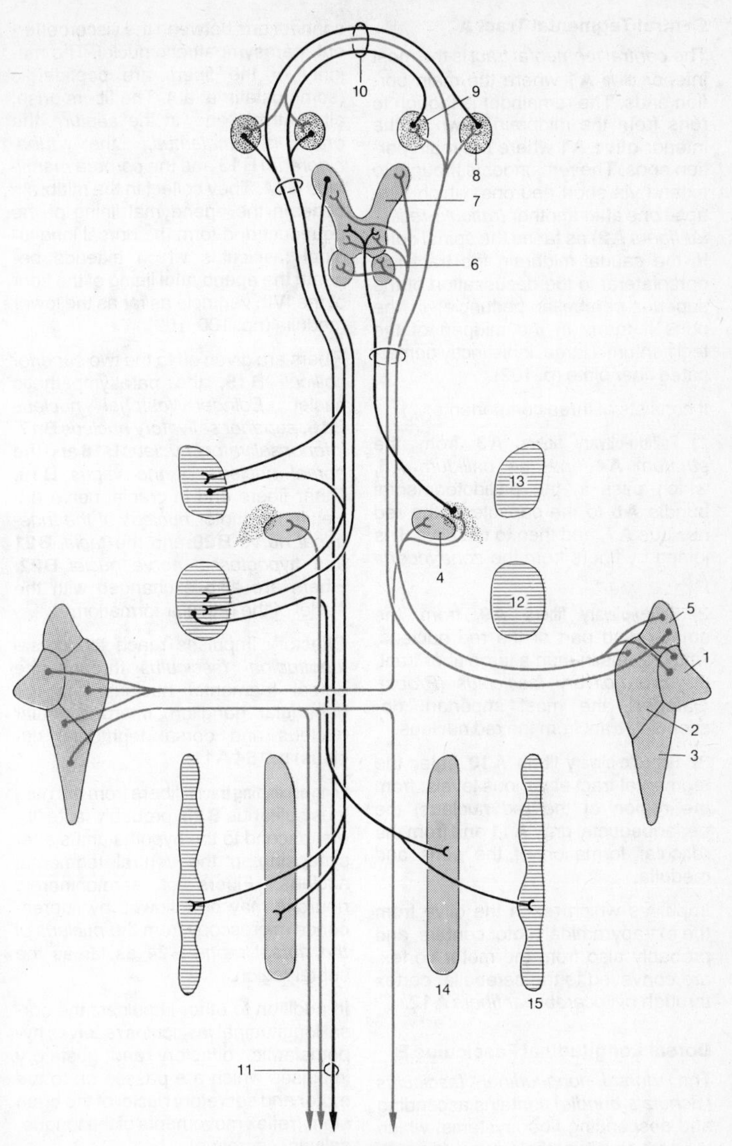

Medial longitudinal fasciculus
(after Crosby, Humphrey and Lauer)

Central Tegmental Tract A

The *central tegmental tract* is the most inferior **olive A1** where the main portion ends. The remainder is though to runs from the midbrain down to the inferior **olive A1** where the main portion ends. The remainder is thought to extend via short neurons, which synapse one after another *(reticulo-reticular fibers* **A2)** as far as the spinal cord. In the caudal midbrain the tract lies dorsolateral to the decussation of the superior cerebellar peduncle; in the pons it forms in the midpart of the tegmentum a large, indistinctly demarcated fiber plate (p. 102).

It consists of three components:

1) **Pallido-olivary fibers A3** from the *striatum* **A4** and the *pallidum* **A5,** which pass in the pallidotegmental bundle **A6** to the capsule of the red nucleus **A7,** and then to the olive. It is joined by fibers from the *zona incerta* **A8.**

2) **Rubro-olivary fibers A9** from the small-celled part of the red nucleus, which forms in man a large fiber tract, the *rubro-olivary fasciculus (Probst-Gamper),* the most important descending tract from the red nucleus.

3) **Reticulo-olivary fibers A10** enter the tegmental tract at various levels: from the region of the red nucleus, the periaqueductal gray **A11** and from the reticular formation of the pons and medulla.

Impulses which reach the olive from the extrapyramidal motor centers, and probably also from the motor cortex, are conveyed to the cerebellar cortex through *olivocerebellar fibers* **A12.**

Dorsal Longitudinal Fasciculus B

The *dorsal longitudinal fasciculus (Schütz's bundle)* contains ascending and descending fiber systems, which connect the hypothalamus with various brain stem nuclei and provide connections between the visceroefferent, parasympathetic nuclei. The majority of the fibers are peptidergic (somatostatin et al.). The fibers arise, alternatively end, in the *septum,* the *oral hypothalamus,* the *tuber cinereum* **B13** and the *corpora mamillaria* **B14.** They collect in the midbrain beneath the ependymal lining of the aqueduct and form the dorsal longitudinal fasciculus which extends beneath the ependymal lining of the floor of the IVth ventricle as far as the lower medulla (pp. 100, 102).

Fibers are given off to the two *superior colliculi* **B15,** the parasympathetic nuclei: *Edinger-Westphal nucleus* **B16,** *superior salivatory nucleus* **B17,** *inferior salivatory nucleus* **B18** and the *dorsal nucleus of the vagus* **B19.** Other fibers end in cranial nerve nuclei: in the motor *nucleus of the trigeminal nerve* **B20** and the *facial* **B21** and *hypoglossal nerve nuclei* **B22.** Fibers are also exchanged with the nuclei of the reticular formation.

Olfactory impulses reach the *dorsal longitudinal fasciculus* through the dorsal tegmental nucleus (via the habenular ganglion, interpeduncular nucleus and dorsal tegmental nucleus) p. 164 A11.

Long ascending tracts: fibers from the nucleus solitarius **B23,** probably taste fibers, ascend to the hypothalamus after synapsing in the ventral tegmental nucleus. Fibers of serotoninergic neurons may be followed by fluorescence microscopy from the *nucleus of the dorsal raphe* **B24** as far as the septal region.

In addition to other impulses, the dorsal longitudinal fasciculus receives hypothalamic, olfactory and gustatory impulses which are passed on to the motor and secretory nuclei of the brain stem (reflex movements of the tongue, salivary secretion).

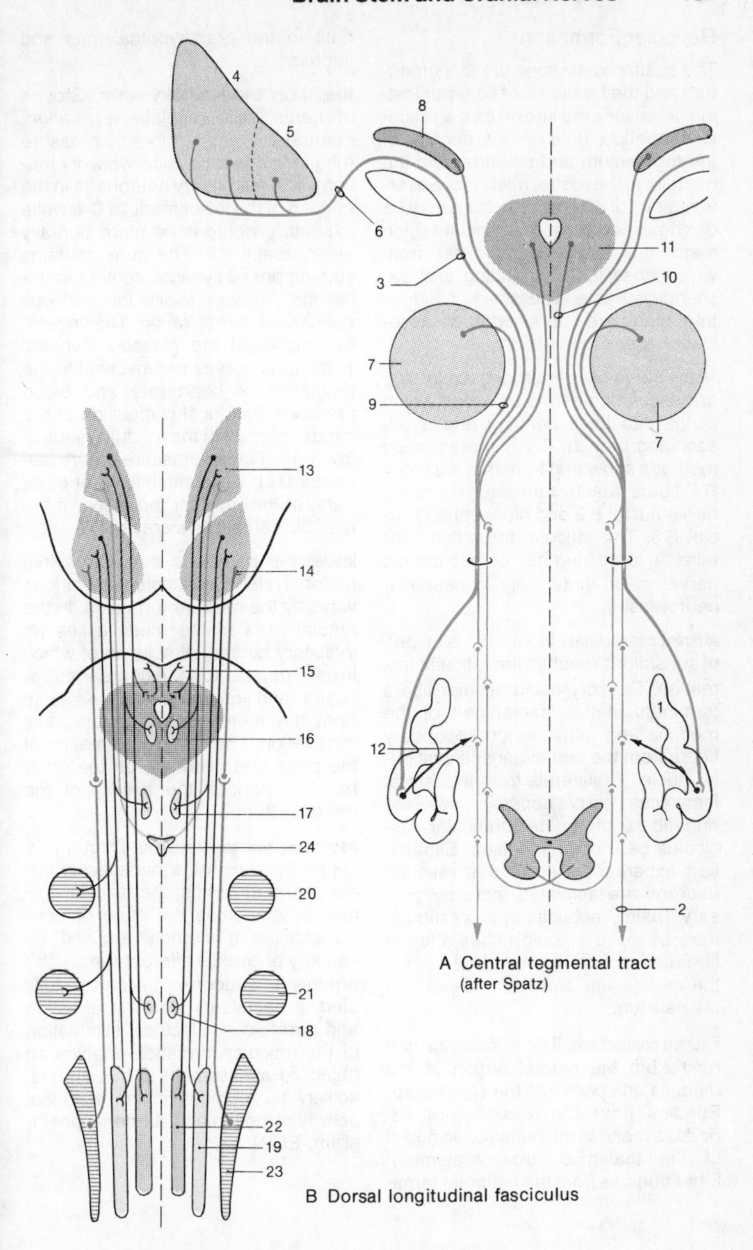

A Central tegmental tract
(after Spatz)

B Dorsal longitudinal fasciculus

Reticular Formation

The scattered neurons of the tegmentum and their network of communicating processes are known as the reticular formation. It lies in the midpart of the tegmentum and extends from the medulla to the rostral midbrain. Several regions of different structure may be distinguished **A.** In the medial region there are large-celled nuclei from which arise long ascending and descending tracts. The small-celled lateral strips are supposed to be association areas.

Many nerve cells have long ascending or descending axons or axons which divide into an ascending and a descending branch. Golgi impregnation methods show that from one such cell **B1** fibers reach both caudal cranial nerve nuclei **B2** and diencephalic nuclei **B3.** The reticular formation contains a large number of peptidergic nerve cells (inter al. encephalin, neurotensin).

Afferent connections. Input from all types of sensations reaches the reticular formation. Sensory spinoreticular fibers terminate in the medial part of the medulla and pons, as do secondary fibers from the trigeminal and vestibular nuclei. Collaterals from the lateral lemniscus carry acoustic impulses and fibers of the tectoreticular fasciculus bear optic impulses. Stimulation experiments show that reticular neurons are activated more by sensory (pain), acoustic and vestibular than by optic stimuli. Other afferent fibers arise from the cerebral cortex, the cerebellum, the red nucleus and the pallidum.

Efferent connections. The **reticulospinal tract** runs from the medial region of the medulla and pons into the spinal cord. Bundles from the **reticulothalamic fasciculus** ascend to the nonspecific nuclei of the thalamus (truncothalamus). Fiber bundles from the midbrain terminate in the oral hypothalamus and septum.

Respiratory and circulatory center. Groups of nerve cells regulate respiration, heartbeat and blood pressure (changes due to physical work or emotion). The inspiratory neurons lie in the midpart of the lower medulla **C4,** while expiratory neurons lie more dorsally and laterally **C5.** The pons contains superimposed synaptic centers for inhibition and facilitation *(locus coeruleus)* of respiration. The vegetative nuclei of the glossopharyngeal and vagus nerves are involved in the regulation of heart rate and blood pressure. Electrical stimulation of the caudal midpart of the medulla leads to lowering of blood pressure *(depressor center* **D6),** while stimulation of other parts of the reticular formation of the medulla causes an increase **D7.**

Influence on the motor system. The spinal motor system is affected in various ways by the reticular formation. In the medial area of the medulla lies an inhibitory center, stimulation of which lowers muscle tone, extinguishes reflexes and prevents any response from the motor cortex on electrical stimulation. The reticular formation of the pons and midbrain, on the other hand, enhances the activity of the motor system.

Ascending activation system. Through its connections with the nonspecific nuclei of the thalamus, the reticular formation influences the state of consciousness. If strongly aroused by sensory of cortical afferent stimuli, the organism suddenly becomes fully alert, a necessary state for attention and perception. Electrical stimulation of the reticular formation enables an objective assessment of this arousal activity by changes in the electrical activity of the brain (electroencephalogram, EEG).

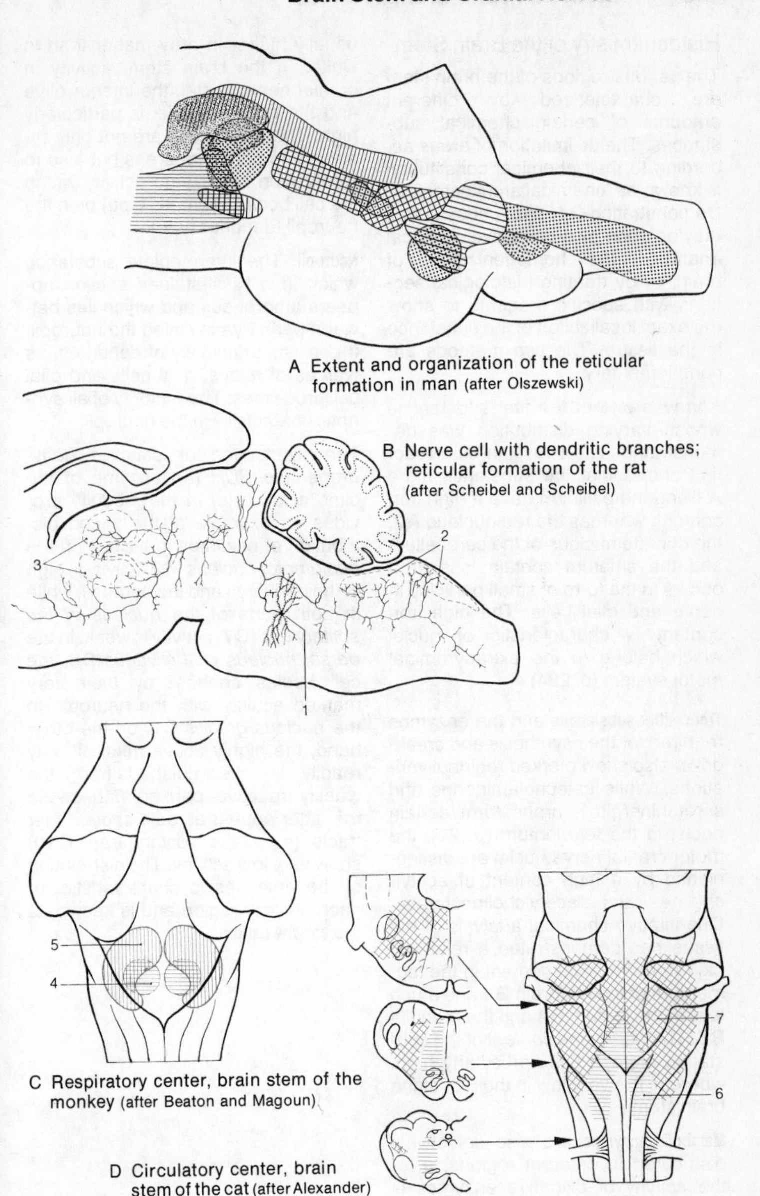

A Extent and organization of the reticular formation in man (after Olszewski)

B Nerve cell with dendritic branches; reticular formation of the rat (after Scheibel and Scheibel)

C Respiratory center, brain stem of the monkey (after Beaton and Magoun)

D Circulatory center, brain stem of the cat (after Alexander)

Histochemistry of the Brain Stem

The various regions of the brain stem are characterized by different amounts of certain chemical substances. The delimitation of areas according to their chemical constitution is known as 'chemical architectonics'. Demonstration of such substances may be done by quantitative chemical analysis after homogenization of brain, or by treating histological sections with specific reagents to show the exact localization of the substance in the tissue. The two methods are complementary.

Iron was one of the first substances whose varying distribution was demonstrated. The Prussian-blue reaction shows that the substantia nigra **A1** and the pallidum have a high iron content, whereas the red nucleus **A2,** the dentate nucleus of the cerebellum and the striatum contain less. Iron occurs in the form of small particles in nerve and glial cells. The high iron content is characteristic of nuclei which belong to the extrapyramidal motor system (p. 284).

Transmitter substances and the enzymes required for their synthesis and breakdown also show marked regional variations. While catecholaminergic and serotoninergic neurons form certain nuclei in the tegmentum (p. 28), the motor cranial nerve nuclei are distinguished by a high content of acetylcholine and acetylcholinesterase. Quantitative chemical analysis of the tissue has demonstrated a relatively high *noradrenaline* content in the tegmentum of the midbrain **B3** and much less in the tectum **B4** and the medulla **B5.** The *dopamine content* of the substantia nigra **B1** is particularly high, whereas it is very low in the rest of the brain stem.

Metabolic enzymes are also selectively distributed in different regions, e. g. the activity of oxidative enzymes is usually higher in grey matter than in white. In the brain stem, activity in cranial nerve nuclei, the inferior olive and the pontine nuclei is particularly high. The differences are not only restricted to particular areas but also to localization of enzyme activity within the cell bodies (somatic type) or in the neuropil (dendritic type).

Neuropil. The intermediate substance which in a Nissl-stained section appears amorphous and which lies between perikarya is called the neuropil. It consists principally of dendrites, as well as of axons, glial cells and glial cell processes. The majority of all synaptic contacts lie in the neuropil.

The distribution of *succinic dehydrogenase SDH* (an enzyme of the citric acid cycle) in the medulla provides an example of the varied distribution of enzymes in tissues: in the *oculomotor nucleus* **C6** activity is high in the perikarya and the neuropil, while in both parts of the *nucleus of the solitary tract* **C7,** activity is weak. In the *dorsal nucleus of the vagus* **C8,** the cell bodies contrast by their very marked activity with the neuropil. In the *nucleus gracilis* **C9,** on the other hand, the highly active neuropil may readily be distinguished from the weakly reactive perikarya, because the latter appear as pale spots. Fiber tracts (e. g. the *solitary tract* **C10)** show very low activity. The distribution of the enzymes is characteristic for each nuclear region and is known as the **enzyme pattern.**

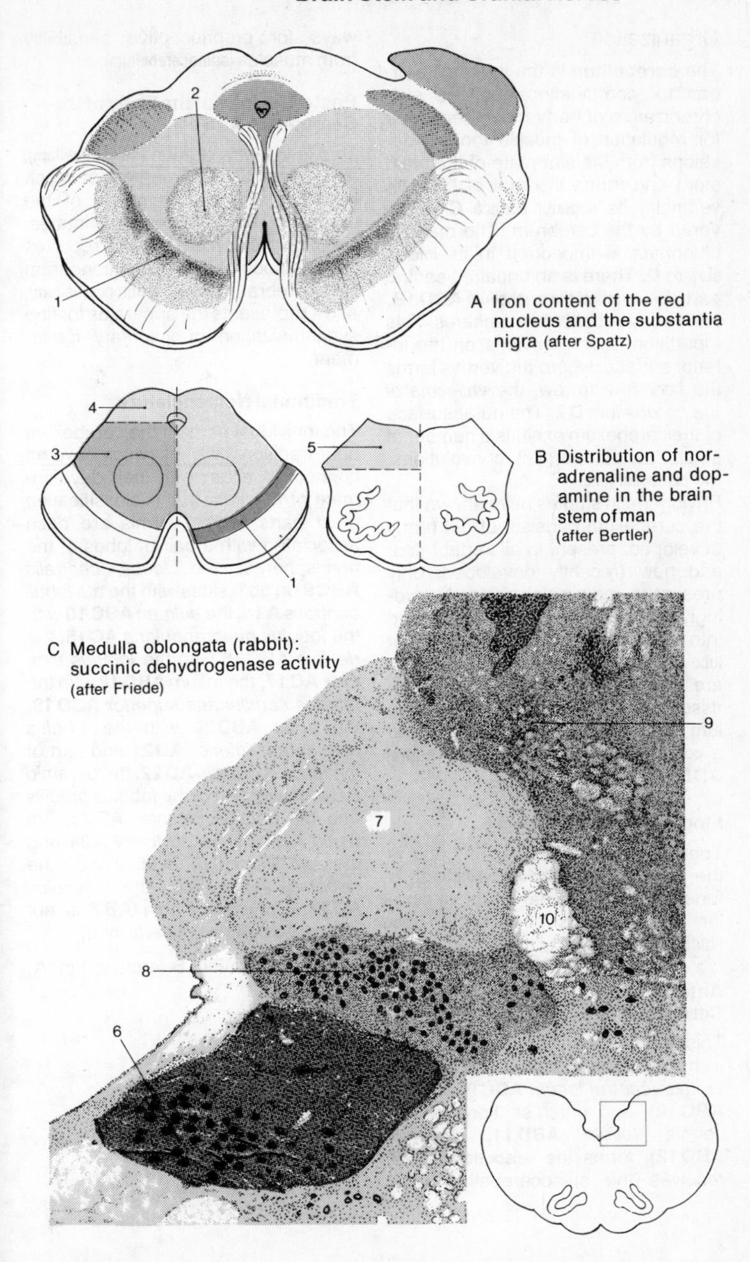

A Iron content of the red nucleus and the substantia nigra (after Spatz)

B Distribution of noradrenaline and dopamine in the brain stem of man (after Bertler)

C Medulla oblongata (rabbit): succinic dehydrogenase activity (after Friede)

Organization

The **cerebellum** is the integrative organ for coordination and fine synchronization of body movements and for regulation of muscle tone. It develops from the alar plate of the brain stem and forms the roof of the IVth ventricle. Its **superior surface C** is covered by the cerebrum. The medulla oblongata is imbedded in its **inferior surface D.** There is an unpaired central part, the **vermis of the cerebellum ACD 1 B,** and two *cerebellar hemispheres.* This tripartition is only obvious on the inferior surface, where the vermis forms the floor of a furrow, the *vallecula of the cerebellum* **D 2.** The outer surface of the cerebellum exhibits a number of small, almost parallel convolutions, the **cerebellar folia.**

Phylogenetic studies have shown that the cerebellum consists of old (early developed, present in all vertebrates) and new (recently developed, only present in mammals) parts. Accordingly, the cerebellum can be divided into two components: the **flocculonodular lobe** and the **cerebellar body A 3.** The two are separated by the *posterolateral fissure* **A 4.** The body of the cerebellum is again divided into an **anterior** and a **posterior lobe** by the *primary fissure* **AC 5.**

Flocculonodular Lobe A 6

Together with the lingula **AB 7,** this is the oldest part **(archicerebellum).** It is functionally connected to the vestibular nuclei by its fiber tracts **(vestibulocerebellum).**

Anterior Lobe of the Body of the Cerebellum A 8

This is an old portion which, together with its middle section belonging to the vermis *(central lobule* **ABC 9,** *culmen* **ABC 10),** and other sections of the vermis *(uvula* **ABD 11,** *pyramid* **ABD 12),** forms the **paleocerebellum.** It receives the spinocerebellar pathways for proprioceptive sensibility from muscles **(spinocerebellum).**

Posterior Lobe of the Body of the Cerebellum A 13

This is the new portion **(neocerebellum),** the substantial enlargement of which contributes to the formation of the cerebellar hemispheres in primates. Through the pontine nuclei it receives the corticocerebellar projections from the cerebral cortex **(pontocerebellum),** and it represents the apparatus for fine synchronization of voluntary movement.

Traditional Nomenclature

The individual parts of the cerebellum bear traditional names which have no connection either with their development or function. In this classification most parts of the vermis are each associated with a pair of lobes of the hemispheres: the *lobus centralis* **ABC 9** on both sides with the *ala lobuli centralis* **A 14,** the *culmen* **ABC 10** with the *lobulus quadrangularis* **AC 15,** the *declive* **ABC 16** with the *lobulus simplex* **AC 17,** the *folium* **ABC 18** with the *lobulus semilunaris superior* **ACD 19,** the *tuber* **ABD 20** with the *lobulus semilunaris inferior* **AD 21** and part of the *lobulus gracilis* **AD 22,** the *pyramid* **ABD 12** with part of the *lobulus gracilis* and the *lobulus biventer* **AD 23,** the *uvula* **AB 11** with the *tonsil* **A 24** and the *paraflocculus* **A 25,** and the *nodulus* **AB 26** with the *flocculus* **AD 27.** Only the *lingula* **AB 7** is not associated with any lateral lobe.

The arrow A in Fig. **B** refers to Fig. **A,** p. 145.

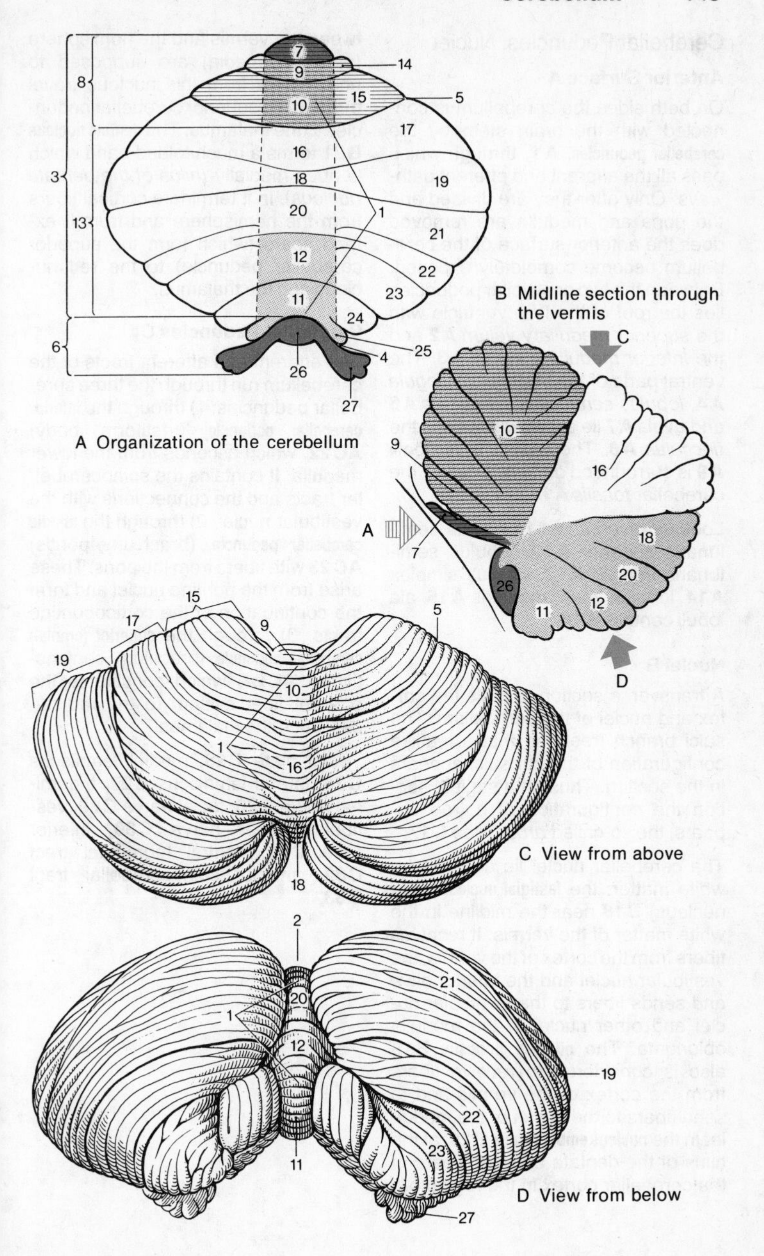

A Organization of the cerebellum

B Midline section through the vermis

C View from above

D View from below

Cerebellar Peduncles, Nuclei

Anterior Surface A

On both sides the cerebellum is connected with the brain stem by the **cerebellar peduncles, A1,** through which pass all the afferent and efferent pathways. Only after they are divided and the pons and medulla are removed does the anterior surface of the cerebellum become completely exposed. Between the two cerebellar peduncles lies the roof of the IVth ventricle with the *superior medullary velum* **A2** and the *inferior medullary velum* **A3.** The ventral parts of the vermis, the *lingula* **A4,** *lobulus centralis* **A5,** *nodulus* **A6** and *uvula* **A7** lie exposed, as does the *flocculus* **A8.** The vallecula cerebelli **A9** is surrounded on both sides by the cerebellar *tonsils* **A10.**

Lobulus biventer **A11,** lobulus semilunaris superior **A12,** lobulus semilunaris inferior **A13,** lobulus simplex **A14,** lobulus quadrangularis **A15,** ala lobuli centalis **A16.**

Nuclei B

A transverse section reveals the cortex and nuclei of the cerebellum. The sulci branch freely so that a leaflike configuration of the cut sulci appears in the section. Thus, in a sagittal section the configuration of a tree appears, the so-called *arbor vitae* **C17.**

The cerebellar nuclei lie deep in the white matter: the **fastigial nucleus** (roof nucleus) **B18** near the midline in the white matter of the vermis. It receives fibers from the cortex of the vermis, the vestibular nuclei and the inferior olive and sends fibers to the vestibular nuclei and other nuclei of the medulla oblongata. The **nucleus globosus B19** also is considered to receive fibers from the cortex of the vermis and to send fibers to the nuclei of the medulla. In the **nucleus emboliformis B20,** at the hilus of the dentate nucleus, fibers of the cerebellar cortex in the region between the vermis and the hemisphere (pars intermedia) are supposed to end. Fibers from this nucleus travel through the superior cerebellar peduncles to the thalamus. The **dentate nucleus B21** forms a much folded band which is open medially *(hilus of the dentate nucleus).* In it terminate cortical fibers from the hemisphere and from it extend fibers (which form the superior cerebellar peduncle) to the red nucleus and the thalamus.

Cerebellar Peduncles C

The efferent and afferent tracts of the cerebellum run through the three cerebellar peduncles: 1) through the **inferior cerebellar peduncle** (restiform body) **AC22,** which ascends from the lower medulla. It contains the spinocerebellar tracts and the connections with the vestibular nuclei. 2) through the **middle cerebellar peduncle** (brachium pontis) **AC23** with fibers from the pons. These arise from the pontine nuclei and form the continuation of the corticopontine tracts. 3) through the **superior (cranial) cerebellar peduncle** (brachium conjunctivum) **AC24,** which constitutes the efferent fiber system to the red nucleus and the thalamus.

Tectal plate **C25,** medial lemniscus **C26,** lateral lemniscus **C27,** trigeminal nerve **C28,** facial nerve **C29,** vestibulocochlear nerve **C30,** inferior olive **C31,** central tegmental tract **C32,** anterior spinocerebellar tract **C33.**

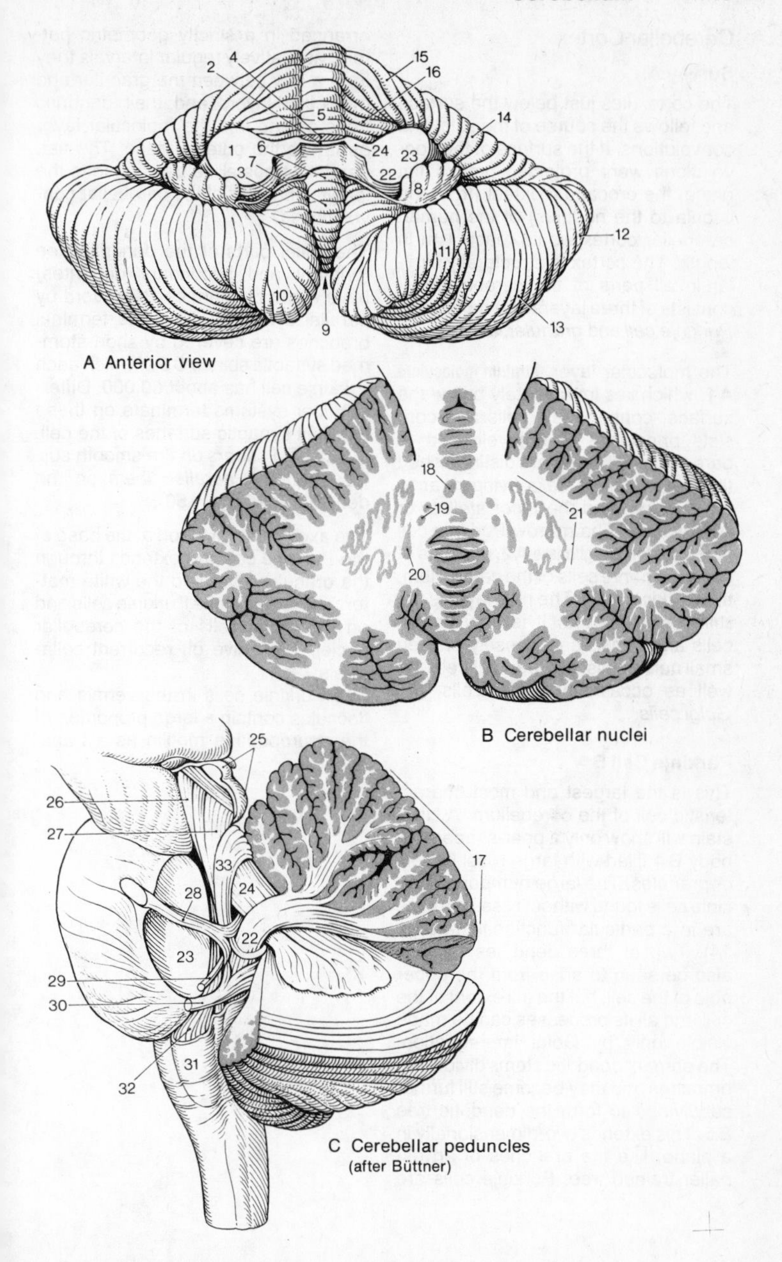

A Anterior view

B Cerebellar nuclei

C Cerebellar peduncles
(after Büttner)

Cerebellar Cortex

Survey A

The cortex lies just below the surface and follows the course of the sulci and convolutions. If the surface of the convolutions were projected onto a flat plane, the orocaudal extent (from the lingula to the nodulus) of the human cerebellar cortex would reach 1 m in length. The cortex is structurally similar in all parts of the cerebellum. It consists of three layers: the *molecular, Purkinje cell* and *granular layers.*

The molecular layer, **stratum moleculare A1,** which lies immediately below the surface, contains few cells and consists principally of unmyelinated fibers. Among its cells are distinguished the *outer stellate cells* (lying near to the surface) and the *inner stellate* or *basket cells.* The narrow Purkinje cell layer, **stratum ganglionare A2,** consists of the large nerve cells of the cerebellum, the *Purkinje cells.* The next layer is the **stratum granulare A3.** It is very rich in cells and consists of densely packed small nerve cells, the granular cells, as well as occasional larger cells, the *Golgi cells.*

Purkinje Cell B

This is the largest and most characteristic cell of the cerebellum. A Nissl stain will show only a pear-shaped cell body **B4** filled with large Nissl bodies or granules. The large number of Purkinje cells found without Nissl granules are in a particular functional state (p. 14). Two or three dendrites **B5** can also be seen to arise from the upper pole of the cell, but the full extent of the cell and all its processes can be made visible only by Golgi impregnation. The primary dendritic stems divide into branches and they become still further subdivided to form the dendritic tree **B6**. This extends two-dimensionally in a plane, like the branches of an espalier trained tree. Purkinje cells are arranged in a strictly geometric pattern: at relatively regular intervals they form a row between the granular and molecular layers and their dendritic trees extend into the molecular layer as far as the outer surface. The flat, dendritic trees all lie transverse to the longitudinal axis of the cerebellar convolutions **D.**

The initial parts of the dendritic tree (primary and secondary dendrites) have a smooth surface **C7** coverd by parallel synapses. The fine terminal branches are covered by short-stemmed synaptic spines **C8** of which each Purkinje cell has about 60 000. Different fiber systems terminate on these different synaptic surfaces of the cell: the *climbing fibers* on the smooth surface and the *parallel fibers* on the dendritic spines (p. 150).

The axon **B9** is given off at the base of the Purkinje cell and extends through the granular layer into the white matter. The axons of the Purkinje cells end on the nerve cells of the cerebellar nuclei. They give off recurrent collaterals.

The Purkinje cells in the vermis and flocculus contain a large proportion of the neuropeptide motilin as a transmitter.

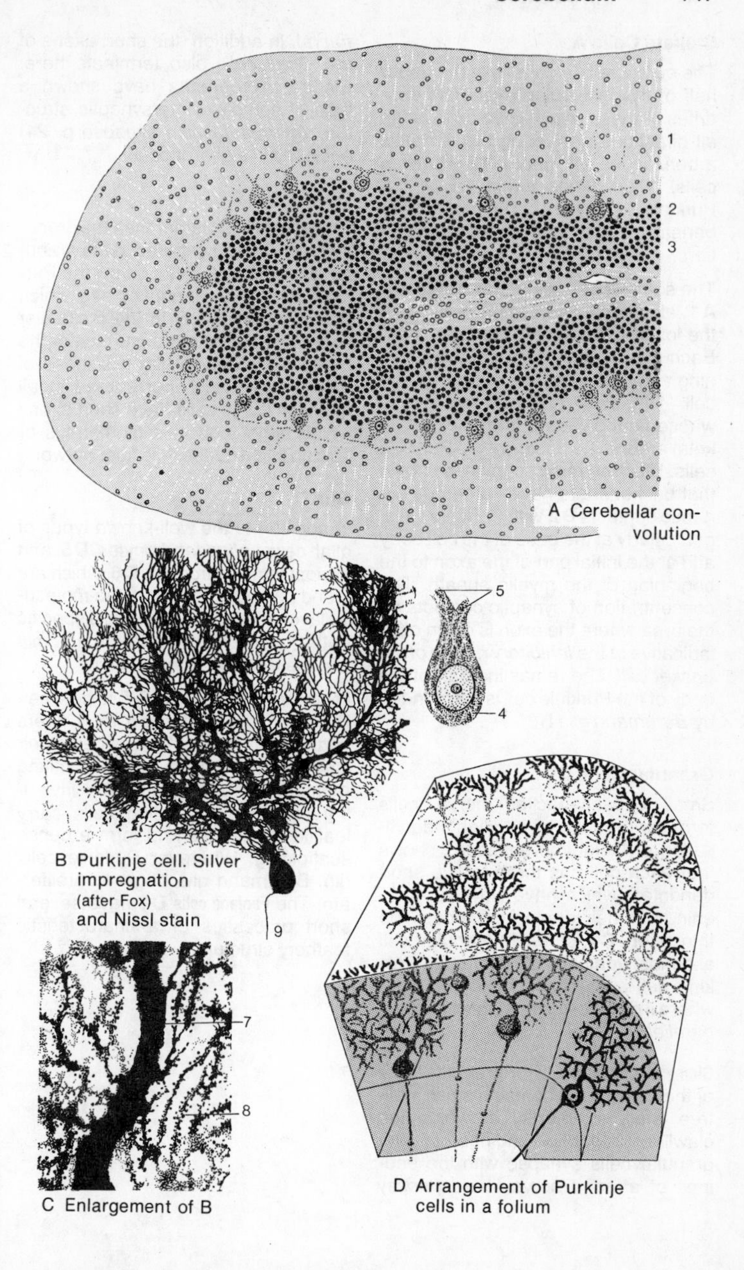

A Cerebellar convolution

B Purkinje cell. Silver impregnation (after Fox) and Nissl stain

C Enlargement of B

D Arrangement of Purkinje cells in a folium

Stellate Cells A

The outer stellate cells lie in the upper half of the molecular layer. The dendrites of these small nerve cells run in all directions and reach the dendritic arborizations of about 12 Purkinje cells. Their axons terminate on the Purkinje cell bodies or run horizontally beneath the surface of the convolution.

The somewhat larger inner **stellate cells A1,** also known as basket cells, lie in the lower third of the molecular layer. Each has a long horizontal axon running above the bodies of the Purkinje cell. The axon gives off collaterals whose final divisions form plexi (baskets) around the bodies of the Purkinje cells. Electron microscopy has shown that basket fibers have numerous synaptic contacts **B2** with Purkinje cells, particularly at the base of the cell body and at the initial part of the axon to the beginning of the myelin sheath. The concentration of synaptic contacts on the area where the axon is given off is indicative of the *inhibitory nature of the basket cell.* The remaining part of the body of the Purkinje cell is surrounded by *Bergmann glia* **B3.**

Granular Cells C

Small densely-packed nerve cells form the granular layer. High magnification of Golgi impregnations shows that each cell has three to five short dendrites, which have clawlike thickenings on their terminal branches. The thin axon **C4** of each granular cell ascends vertically through the Purkinje cell layer into the molecular layer where it divides at right angles into two *parallel fibers.*

Glomeruli cerebellosi. The granular layer of the cerebellum contains small, cell-free islets (glomeruli) in which the clawlike dendritic endings of the granular cells synapse with the endings of afferent nerve fibers *(mossy fibers).* In addition, the short axons of the *Golgi cells* also terminate here. Electronmicrographs have shown a type of complex, intersynaptic structure *(glomerulus-like synapse* p. 24) enclosed in a glial capsule.

Golgi Cells E

These are much larger than the granular cells and lie scattered in the granular layer, mainly a little below the Purkinje cells. Their dendritic trees, which arborize particularly in the molecular layer and extend to the surface of the cerebellum, are not flat like those of the Purkinje cells, but spread in all directions. The cells have short axons which either end in a glomerulus or split up into a dense, fine fiber network.

Glia D

In addition to the well-known types of glial cells, *oligodendrocytes* **D5** and protoplasmic *astrocytes* **D6,** which are found particularly often in the molecular layer, there are types of glia specific to the cerebellum: the *Bergmann glia* and the feathery *Fanjana cells.*

Bergmann supporting cells D7 lie between Purkinje cells and send long fibers vertically to the surface, where their small end-feet form a glial limiting membrane abutting onto the investing pia. The supporting cells carry leaflike processes and form a dense scaffolding. Wherever Purkinje cells die, Bergmann glia begin to proliferate. The **Fanjana cells D8** have several short processes of a characteristic feathery structure.

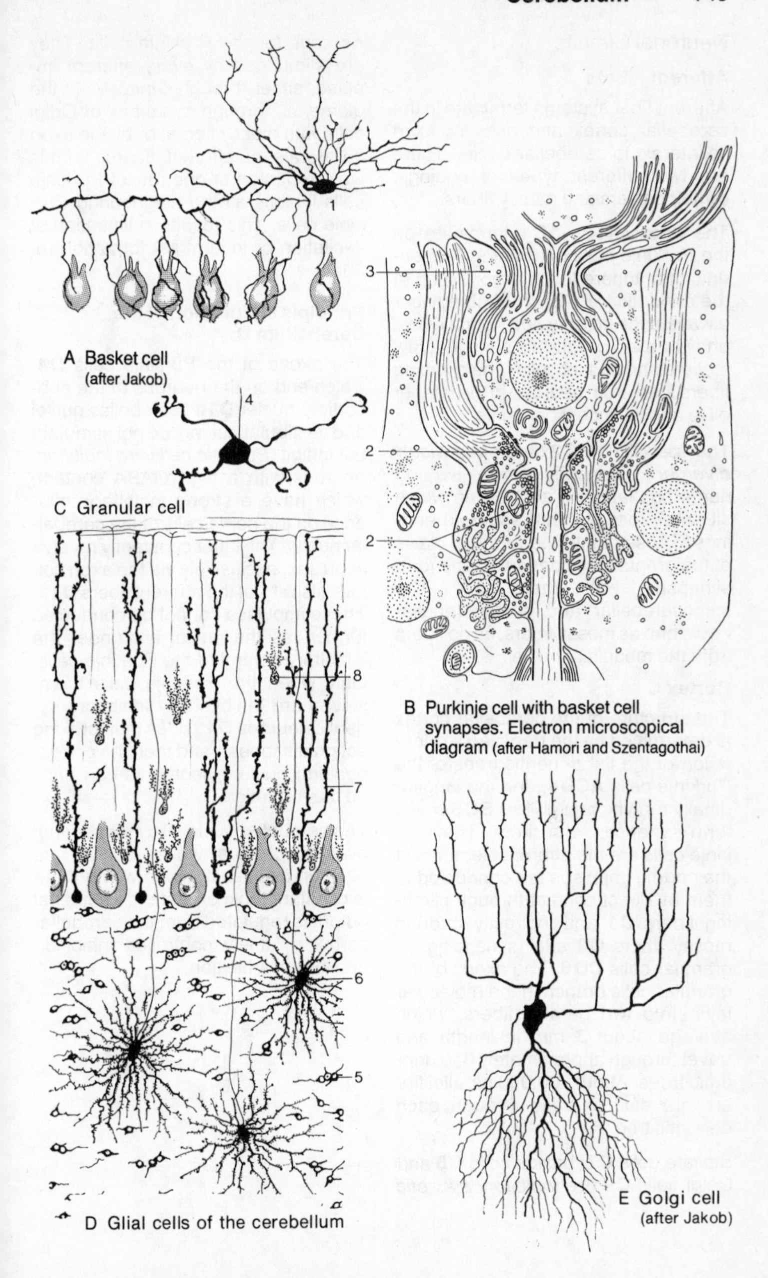

A Basket cell (after Jakob)

B Purkinje cell with basket cell synapses. Electron microscopical diagram (after Hamori and Szentagothai)

C Granular cell

D Glial cells of the cerebellum

E Golgi cell (after Jakob)

Neuronal Circuits

Afferent Fibers

Afferent fiber systems terminate in the cerebellar cortex and give off axon collaterals to cerebellar nuclei. There are two different types of endings: climbing fibers and mossy fibers.

The **climbing fibers AC1** terminate on the Purkinje cells by dividing like tendrils that adhere to the arborization of the dendritic tree. Each climbing fiber always ends on *only one* Purkinje cell, and via axon collaterals also on a few stellate and basket cells. The climbing fibers stem from neurons in the inferior olive and its accessory nuclei.

The **mossy fibers BC2** divide into *widely divergent branches* and give off numerous lateral branches, on which sit small rosettes with spheroid endings. These fit into the terminal claws of the granular cell dendrites and form synapses with them **B3.** The spinocerebellar and pontocerebellar tracts end as mossy fibers, as do fibers from the medullary nuclei.

Cortex C

The structure of the cerebellar cortex is determined by the transverse orientation of the flat dendritic trees of the Purkinje cells **ACD4,** and the longitudinally running **parallel fibers BC5** which form synapses with them. The Purkinje cells are the efferent elements of the cortex. Impulses are conducted to them by direct contact through climbing fibers **C1** and indirectly through mossy fibers **C2** after synapsing on granular cells **CD6.** The axons of the granular cells branch in the molecular layer into two parallel fibers, which average about 3 mm in length and travel through approximately 350 dendritic trees. About 200,000 parallel fibers are said to pass through each dendritic tree.

Stellate cells **C7,** basket cells **C8** and Golgi cells **C9** are *inhibitory synaptic neurons* for the Purkinje cells. They are stimulated by every afferent impulse, either through synapses in the glomeruli, through synapses of Golgi cells with mossy fibers, or by the axon collaterals of afferent fibers. In this way excitation of one row of Purkinje cells inhibits all the neighboring Purkinje cells. This results in limitation of excitation or in contrast formation (p. 30D).

Principle of Function of the Cerebellum D

The axons of the Purkinje cells **D4,** which end on the neurons of the subcortical nuclei **D10** (cerebellar nuclei and vestibular nuclei), do not stimulate but inhibit. *Purkinje cells are inhibitory neurons* with a high GABA content, which have a strong inhibitory influence on the nerve cells of the cerebellar nuclei. The latter constantly receive impulses, exclusively via the axon collaterals **D11** of the afferent fibers **D12.** These impulses cannot be conducted further, as the nuclei are under the inhibitory control of the Purkinje cells. Only when the Purkinje cells themselves are inhibited by inhibitory synaptic neurons **D13** does their braking action disappear, and then the corresponding nuclear segments can pass on the excitation.

The cerebellar nuclei are *independent synaptic centers,* which receive and pass on impulses, and in which there is constant tonic excitation. Their final output is regulated from the cerebellar cortex by finely controlled inhibition and loss of inhibition.

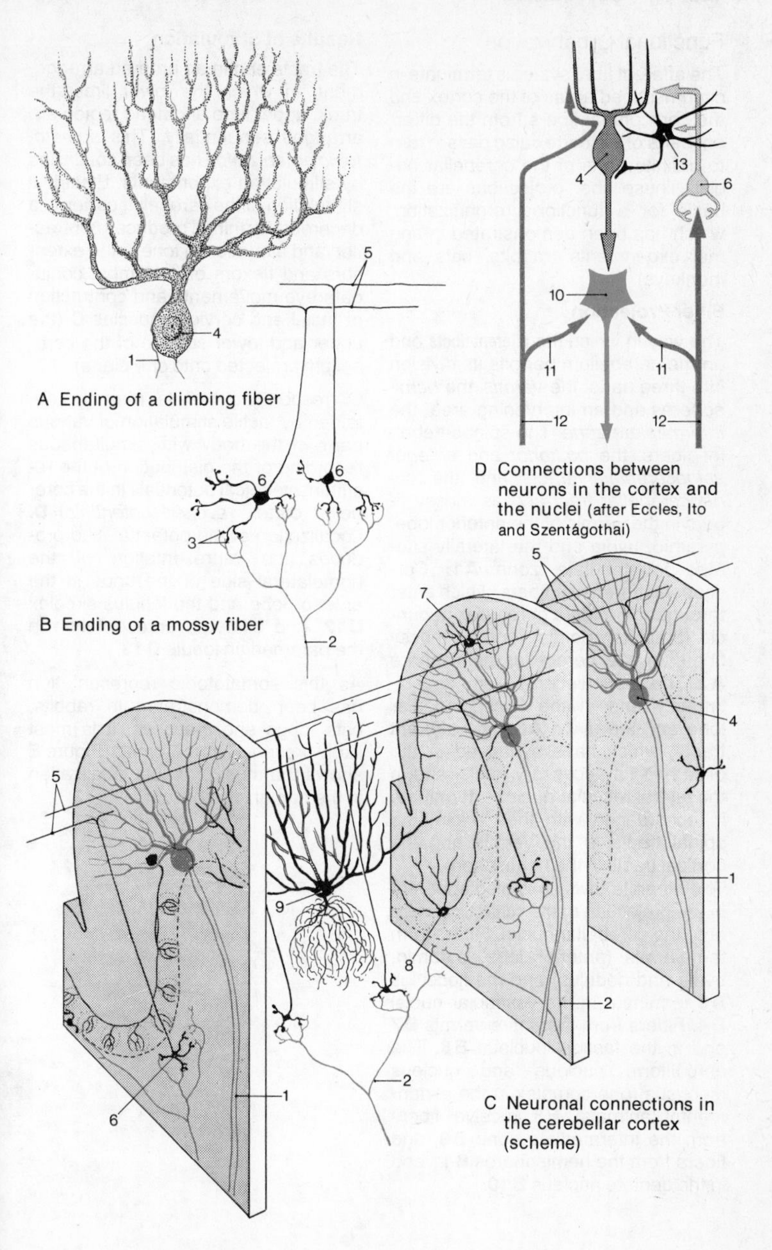

A Ending of a climbing fiber

B Ending of a mossy fiber

D Connections between neurons in the cortex and the nuclei (after Eccles, Ito and Szentágothai)

C Neuronal connections in the cerebellar cortex (scheme)

Functional Organization

The afferent fiber systems terminate in circumscribed areas of the cortex and the corticofugal fibers from the different parts of the cortex also pass in turn to definite parts of the cerebellar nuclei. These fiber projections are the basis for a functional organization, which has been demonstrated in animal experiments (rabbits, cats and monkeys).

Fiber Projection

The way in which the **afferent fibers** end on the cerebellum permits its division into three parts: the *vermis,* the *hemispheres* and an intervening area, the *intermediate zone.* The spinocerebellar fibers, the *posterior* and *anterior spinocerebellar tracts* and the *cuneocerebellar tract* end as mossy fibers in the vermis of the anterior lobe, pyramid, uvula and the laterally situated intermediate zone **A1.** Corticopontocerebellar fibers, which enter through the middle cerebellar peduncle (brachium pontis), end as mossy fibers in the cerebellar hemisphere **A2.** The vestibulocerebellar axons ultimately end in the flocculonodular lobe and the uvula **A3.** The afferent tracts, which have synapsed in the olive and its accessory nuclei and in the lateral reticular nucleus, terminate in accordance with their origin: the spinal tracts in the vermis and the cortical tracts in the hemisphere.

The tripartite division is also apparent in the projection of the **corticofugal axons** onto the cerebellar nuclei. Fibers from the vermis (anterior lobe, pyramid, uvula and nodulus) and the flocculus **B5** terminate in the vestibular nuclei **B4.** Fibers from the entire vermis **B7** end in the fastigal nucleus **B6.** The emboliform nucleus and nucleus globosus (one complex in the experimental animals) **B8** receive fibers from the intermediate zone **B9,** and fibers from the hemispheres **B11** end in the dentate nucleus **B10.**

Results of Stimulation

The tracts end in an ordered arrangement, in which the lower limb, the trunk, upper limb and head region are arranged sequentially. This *somatotopic organization* has been confirmed by stimulation experiments. Electrical stimulation of the cerebellar cortex in a decerebrate animal produces contraction and alteration in tone in the extensors and flexors of the limbs, conjugate eye movements and contraction of facial and cervical muscles **C** (the upper and lower surface of the cerebellum projected onto one plane).

Corresponding results may be obtained by tactile stimulation of various parts of the body with simultaneous recording of the distribution of the resultant electrical potentials in the cerebellar cortex *(evoked potentials)* **D.** Localization of the potential also produces the representation of the homolateral side of the body in the anterior lobe and the lobulus simplex **D12,** and bilateral representation in the paramedian lobule **D13.**

As this somatotopic representation has been demonstrated in rabbits, cats, dogs and monkeys, it is most likely present in all mammals. Figure **E** shows the most probable localization in the human cerebellum.

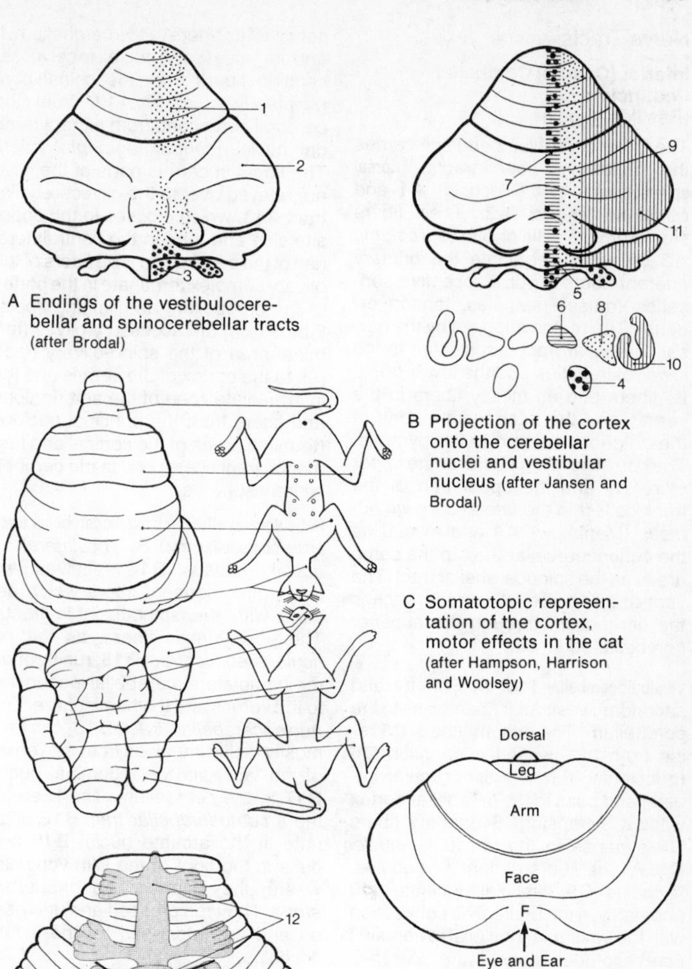

A Endings of the vestibulocerebellar and spinocerebellar tracts (after Brodal)

B Projection of the cortex onto the cerebellar nuclei and vestibular nucleus (after Jansen and Brodal)

C Somatotopic representation of the cortex, motor effects in the cat (after Hampson, Harrison and Woolsey)

D Somatotopic representation of the cortex, "evoked potentials" due to sensory stimulation (after Snider)

E Hypothetical somatotopic representation in man (after Hampson, Harrison and Woolsey)

Nerve Tracts

Inferior (Caudal) Cerebellar Peduncle

(Restiform body)

The lower cerebellar peduncle carries the following fiber tracts: **dorsal spinocerebellar tract** (Flechsig) **A1** and **cuneocerebellar tract A2.** Their fibers stem from the cells of *Clarke's column* **A3** on which terminate the primary afferent fibers for proprioceptive sensation (muscle spindles, tendon organs). The region supplied by the dorsal cerebellar tract is restricted to the lower extremities and the lower trunk. Its fibers end as mossy fibers in the vermis and the intermediate zone of the anterior lobe and in the pyramid. The corresponding fibers for the upper extremity and the upper part of the trunk collect in the *lateral cuneate nucleus* (Monakow) **A4** and extend as the cuneocerebellar tract to the same areas as the spinocerebellar tract. The ventral spinocerebellar tract reaches the cerebellum through the superior cerebellar peduncle.

Vestibulocerebellar Tract C. Primary and secondary vestibular fibers pass to the cerebellum. The primary fibers **C5** arise from the *vestibular ganglion* **C6** (principally from the semicircular canals) and pass to the cerebellar cortex without synapsing. Secondary fibers **C7** synapse in the *vestibular nuclei* **C8.** All the fibers end in the nodulus, flocculus **C9,** fastigial nucleus **C10** and partly in the uvula. The connection with the vestibular nuclei also contains cerebellofugal fibers, the *cerebellovestibular tract,* which stems from the terminal regions mentioned and from the vermis of the anterior lobe. Part synapses in the lateral vestibular nucleus and passes to the spinal cord in the *vestibulospinal tract.*

Olivocerebellar Tract A. The inferior olive **A11,** which may be regarded as a ventrally displaced cerebellar nucleus, sends all its fibers to the cerebellum. It and its accessory nuclei receive ascending fibers from the spinal cord *(spino-olivary tract)* **A12,** from the cerebral cortex and from extrapyramidal nuclei *(central tegmental tract).* The fibers in certain parts of the olive are relayed over to the olivocerebellar tract **A13,** which crosses to the opposite side and runs to the contralateral half of the cerebellum. The fibers of the olivary complex terminate in the cortex of the cerebellum as climbing fibers. Fibers from the accessory olives (terminal area of the spino-olivary tract) run to the cortex of the vermis and the intermediate zone of the anterior lobe and fibers from the principal nucleus (terminal areas of the cortical and tegmental tract fibers) end in the cerebellar hemispheres.

Reticulocerebellar tract, nucleocerebellar tract, arcuatocerebellar tract B. The *lateral reticular nucleus* **B14** receives exteroceptive sensory fibers which ascend with the spinothalamic tracts. The postsynaptic fibers, as the *reticulocerebellar tract* **B15,** run through the homolateral cerebellar peduncles to the vermis and the hemisphere. The *nucleocerebellar tract* **B16** conveys mostly tactile impulses from the region of the face, via the trigeminal nuclei **B17** to the cerebellum. The fibers of the *arcuatocerebellar tract* **B18** originate in the arcuate nuclei **B19** and pass to the floor of the IVth ventricle, where they appear as *medullary striae.* They run crossed and uncrossed and are supposed to end in the flocculus.

The existence of an *uncinate fasciculus,* a cerebellospinal tract, which has its origin in the contralateral fastigial nucleus has not been clearly demonstrated in man.

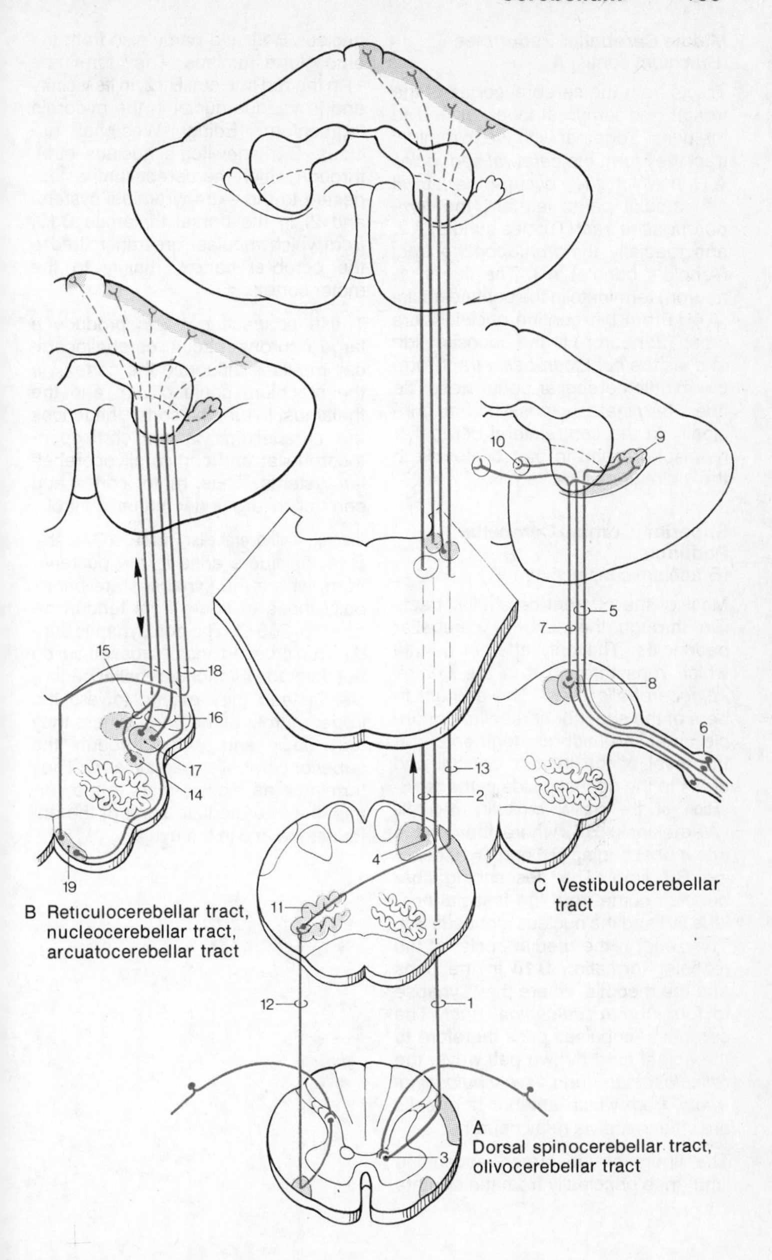

A Dorsal spinocerebellar tract, olivocerebellar tract

B Reticulocerebellar tract, nucleocerebellar tract, arcuatocerebellar tract

C Vestibulocerebellar tract

Middle Cerebellar Peduncles
(Brachium pontis) **A**

Tracts from the cerebral cortex of the frontal and temporal lobes extend to the pons. Together with the pyramidal tract they form the *cerebral peduncles* **A1,** in which they occupy the lateral and medial parts: laterally the *temporopontine tract* (Türck's bundle) **A2,** and medially the *frontopontine tract* (Arnold's bundle) **A3.** The fibers (1. neuron) terminate in the pontine nuclei (**A4**). From the pontine nuclei, fibers cross (2. neuron) to the opposite side and, as the *pontocerebellar tract,* form the middle cerebellar peduncles. The fibers terminate as mossy fibers, principally in the contralateral cerebellar hemisphere and in part bilaterally in the midregion of the vermis.

Superior (Cranial) Cerebellar Peduncle
(Brachium conjunctivum) **B**

Most of the efferent cerebellar tracts run through the superior cerebellar peduncles. The only afferent bundle which runs through it is the *ventral spinocerebellar tract.* The efferent fibers of the superior cerebellar peduncle enter the midbrain tegmentum at the level of the inferior colliculi and cross to the opposite side in the **decussation of the cranial cerebellar peduncle** (Wernekinck) **B5,** where they divide into a descending **B6** and an ascending **B7** limb. The descending fiber bundles come from the fastigial nucleus **B8** and the nucleus globosus **B9.** They end in the medial nuclei of the reticular formation **B10** in the pons and the medulla, where they synapse to form the *reticulospinal tract.* The cerebellar impulses pass therefore to the spinal cord by two pathways, the *reticulospinal* and *vestibulospinal tracts,* from which anterior horn cells are influenced via relay neurons.

The fibers of the larger, ascending limb arise principally from the dentate nucleus **B11** and partly also from the emboliform nucleus. They terminate 1) in the red nucleus **B12,** in its vicinity and in various nuclei in the midbrain tegmentum (Edinger-Westphal nucleus, Darkshewitch's nucleus etc.), through which the cerebellum is connected to the extrapyramidal system, and 2) in the dorsal thalamus **B13,** from which impulses are transmitted to the cerebral cortex, mainly to the motor cortex.

These connecting tracts produce a large neuronal circuit: cerebellocortical impulses influence the cortex via the brachium conjunctivum and the thalamus. In turn the cortex influences the cerebellum via the corticopontocerebellar and cortico-olivocerebellar systems. Thus, motor cortex and cerebellum are under mutual control.

Ventral spinocerebellar tract (Gowers) **B14.** Its fibers arise in the posterior horn, where the synapses are principally those of fibers from tendon organs (p. 286 E). The postsynaptic bundles run crossed and uncrossed but do not enter the inferior cerebellar peduncle, instead they extend toward the upper margin of the pons, where they turn back and enter through the superior cerebellar peduncles **C.** They terminate as mossy fibers in the vermis, the intermediate zone of the anterior lobe and in the uvula.

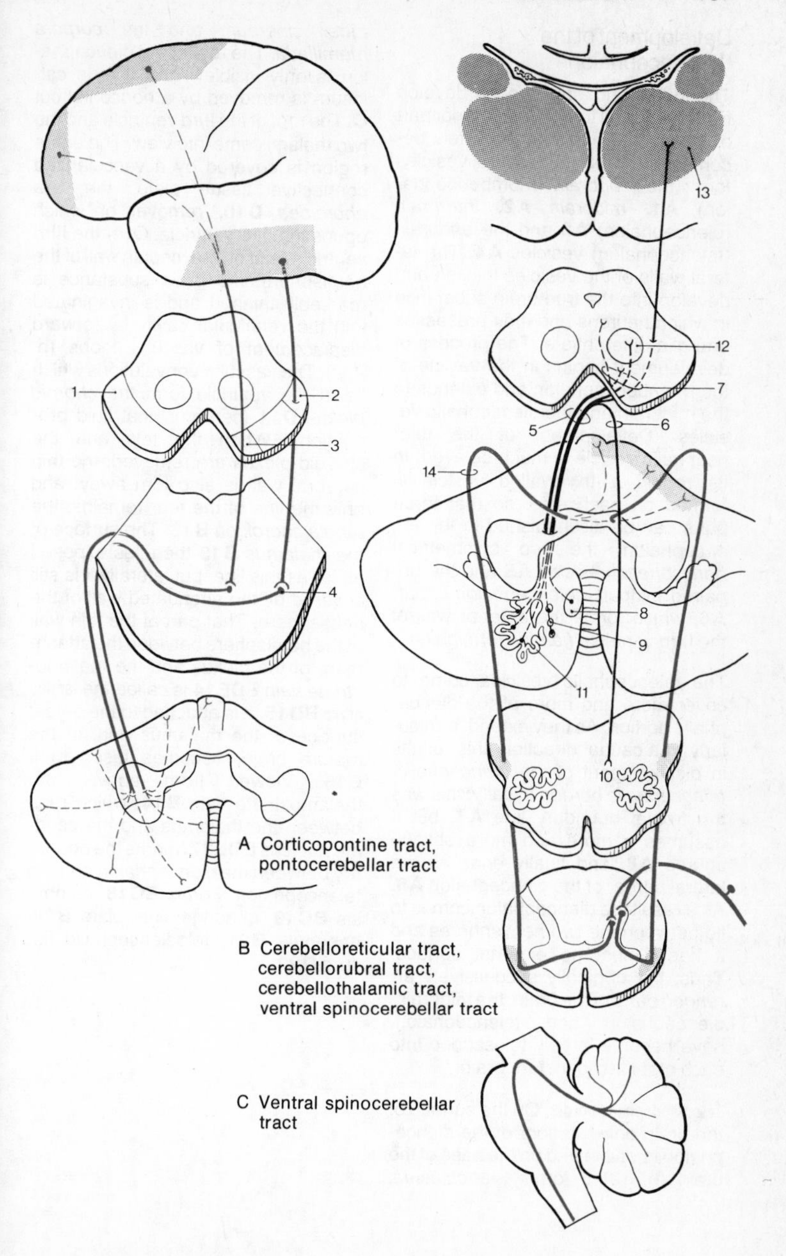

A Corticopontine tract, pontocerebellar tract

B Cerebelloreticular tract, cerebellorubral tract, cerebellothalamic tract, ventral spinocerebellar tract

C Ventral spinocerebellar tract

Development of the Prosencephalon

The brain and spinal cord develop from the neural tube in the anterior part of which (the part that ultimately becomes the cerebrum) several vesicles form: the *hindbrain* (rhombencephalon) **A1,** *midbrain* **A2,** *interbrain* (diencephalon) **A3** and the *endbrain* (telencephalon) vesicles **A4.** The lateral walls of the vesicles thicken and develop into the true brain substance in which neurons and their processes and glia differentiate. The process of development begins in the vesicle of the rhombencephalon and extends to the midbrain and the diencephalic vesicles. Development of the telencephalic vesicle is much delayed. In its region a thin-walled vesicle is formed on each side, so that three parts can be distinguished in the telencephalon: the two symmetrical *hemispheric vesicles* **A5** and the unpaired midpart *(telencephalon impar)* **A6,** which forms the anterior wall of the IIIrd ventricle *(lamina terminalis).*

The telencephalic vesicles come to cover more and more of the diencephalic portion. As they extend, particularly in a caudal direction, this results in displacement of the *diencephalic-telencephalic border.* Initially, this was the frontal boundary line **A7,** but it assumes a more and more oblique course **A8,** and finally ends as the lateral border of the diencephalon **A9.** As a result the diencephalon comes to lie between the two hemispheres and it has almost no external surface. Thus, the originally successively arranged parts of the brain, the midbrain, diencephalon and telencephalon, have become largely telescoped into each other in the mature brain.

Telodiencephalic border. On the surface of the brain only the floor of the diencephalon is visible and on the base of the brain (p. 12) it forms the *chiasma, tuber cinereum* and the *corpora mamillaria.* The roof of the diencephalon is only visible if the corpus callosum is removed by a horizontal cut **C.** The roof of the IIIrd ventricle and the two thalami come into view. The entire region is covered by a vascularized connective tissue plate, the *tela choroidea* **D10,** removal of which opens the IIIrd ventricle. Over the IIIrd ventricle, and on the median wall of the hemisphere, the brain substance is markedly thinned and is invaginated into the ventricular cavity by forward displacement of vascular loops (p. 262). The vascular convolutions which lie in the ventricle form the *choroid plexus* **D11** (cerebrospinal fluid production). When the tela and the choroid plexus are removed, the thin cerebral wall is also torn away, and only the line of the tear remains, the *taenia choroidea* **B12.** The surface of the *thalamus* **B13** then lies exposed as far as this line, but laterally it is still covered by the attenuated wall of the hemisphere. That part of the thin wall of the hemisphere between the attachment of the plexus and the *thalamostriate vein* **BDE14** is called the **lamina affixa BD15.** It is attached to the dorsal surface of the thalamus and, in the mature brain, becomes fused to it **E16.** Viewed from above, the thalamostriate vein **B14,** which runs between the thalamus and the caudate nucleus **BDE17,** marks the boundary between the diencephalon and the telencephalon. Fornix **BC18,** epiphysis **BC19,** quadrigeminal plate **B20,** habenula **B21,** telodiencephalic fissure **D22.**

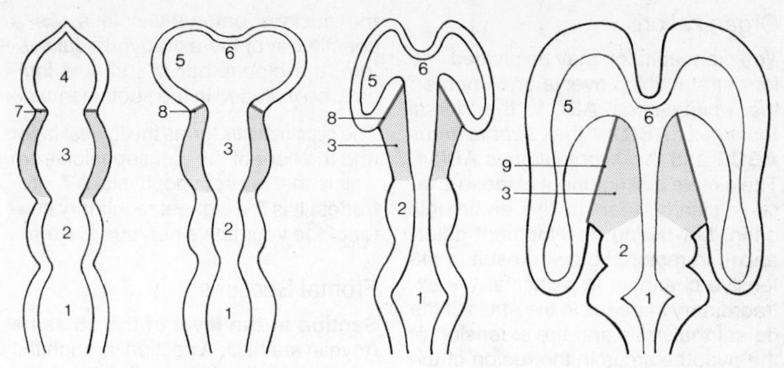

A Development of the diencephalon (after Schwalbe)

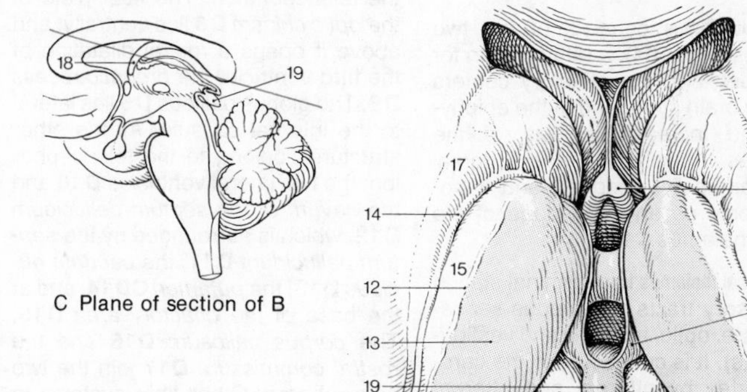

C Plane of section of B

B Diencephalon from above, horizontal section after removal of the corpus callosum, fornix and choroid plexus

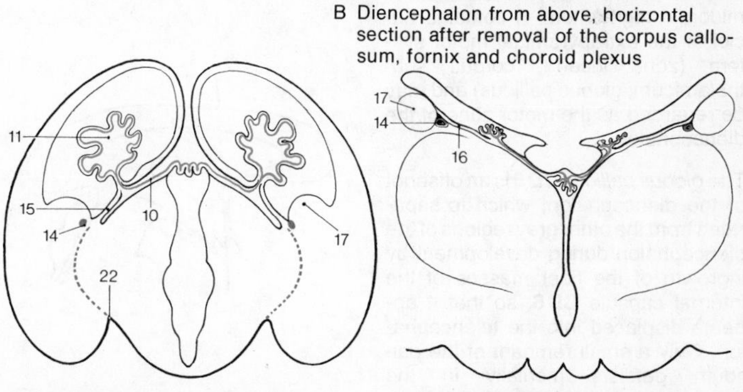

D Lamina affixa, embryonic brain, frontal section

E Lamina affixa, in the brain of an adult, frontal section

Organization

The diencephalon may be divided into four strata which overlie one another: the *epithalamus* **ABC1,** the *dorsal thalamus* **ABC2,** the *subthalamus* **ABC3** and the *hypothalamus* **ABC4.** The simple arrangement of these strata is clearly seen in the embryonic brain, but during development differences in regional growth result in extensive changes. In particular the extraordinary increase in the mass of the dorsal thalamus and the extension of the hypothalamus in the region of the tuber cinereum determine the structure of the mature diencephalon.

The **epithalamus** consists of the two habenulae (p. 164), a relay station for tracts between the olfactory centers and the brain stem, and of the epiphysis. The increase in size of the thalamus displaces the dorsal epithalamus **B1** medially and it appears only as an appendage of the dorsal thalamus **C1.**

The **dorsal thalamus** is a terminal station for sensory tracts (cutaneous sensation, taste, optic, acoustic and vestibular tracts). It is connected to the cerebral cortex by afferent and efferent systems of fibers.

The **subthalamus** is a continuation of the midbrain tegmentum. It contains nuclei of the extrapyramidal motor system (zona incerta, corpus subthalamicum, globus pallidus) and may be regarded as the motor zone of the diencephalon.

The *globus pallidus* **CD5** is an offshoot of the diencephalon, which is separated from the other grey regions of the diencephalon during development by ingrowth of the fiber masses of the internal capsule **CD6,** so that it appears displaced into the telencephalon. Only a small remnant of the pallidum persists medially in the framework of the diencephalon; it is the nucleus entopeduncularis. As a constituent of the extrapyramidal system, the globus pallidus should logically be included in the subthalamus.

The **hypothalamus** forms the lowest layer and the floor of the diencephalon from which the neurohypophysis **A7** protrudes. It is the highest regulatory center of the vegetative nervous system.

Frontal Sections

Section at the level of the chiasma (myelin stain) **D.** A section through the anterior wall of the IIIrd ventricle shows parts of the diencephalon and the telencephalon. The fiber plate of the *optic chiasm* **D8** lies ventrally, and above it opens a rostral dilatation of the IIIrd ventricle, the *preoptic recess* **D9.** The globus pallidus **D5** lies lateral to the internal capsule. All the other structures belong to the telencephalon: the two lateral ventricles **D10** and the *cavum of the septum pellucidum* **D12,** which is surrounded by the *septum pellucidum* **D11,** the *caudate nucleus* **D13,** the *putamen* **CD14,** and at the base of the *olfactory area* **D15.** The *corpus callosum* **D16** and the *rostral commissure* **D17** join the two hemispheres. Other fiber systems in the section are the *fornix* **D18** and the *lateral olfactory stria* **D19.**

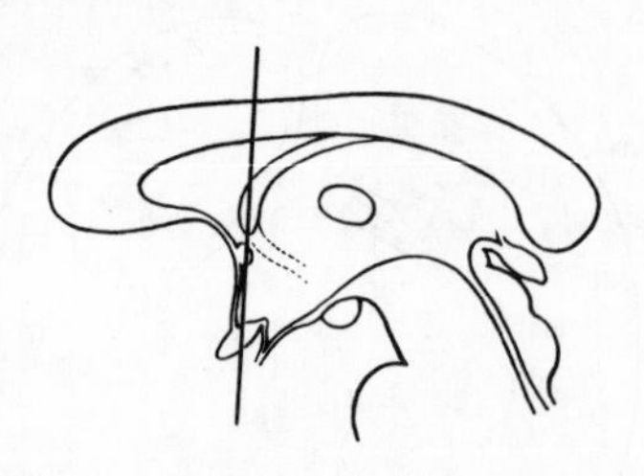

Level of section

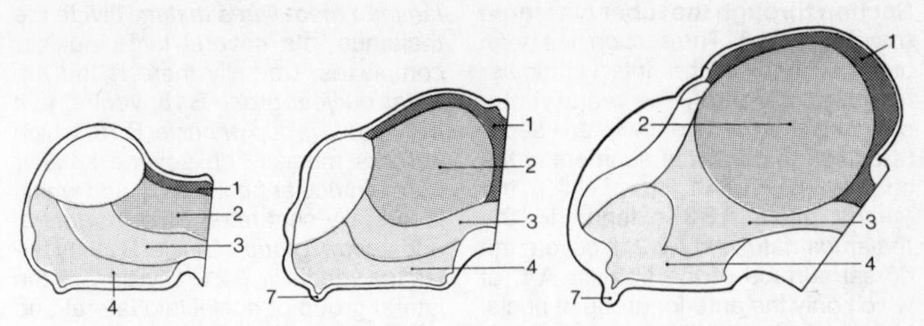

A Development of the strata of the diencephalon

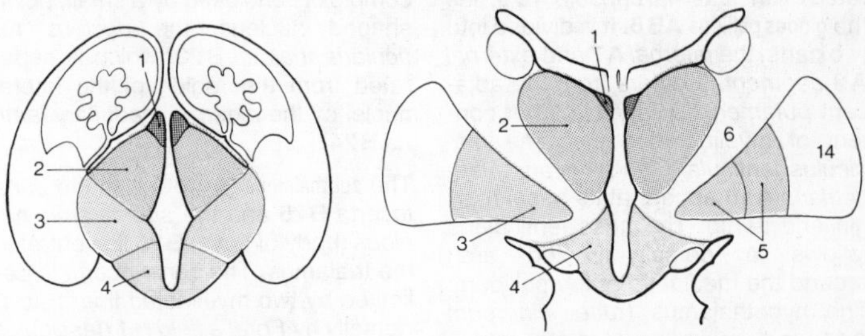

B Structure of the diencephalon in an embryonic brain

C Structure of the diencephalon in an adult

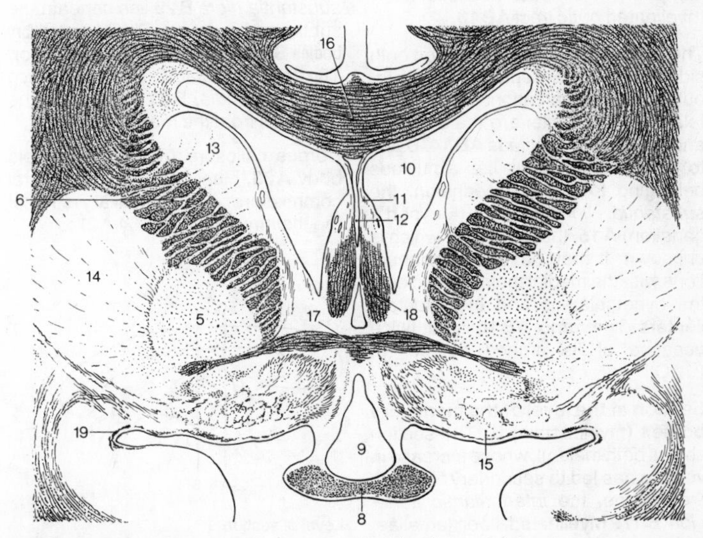

D Frontal section through the rostral wall of the IIIrd ventricle (after Villiger and Ludwig)

Section through the tuber cinereum (myelin stain) **A.** The section has been cut just behind the interventricular foramen (of Monro). The lateral ventricles and the third ventricle are separated by the narrow attachment of the choroid plexus **A1,** from which the *lamina affixa* **AB3** extends to the *thalamostriate vein* **AB2.** It covers the dorsal surface of the **thalamus A4,** of which only the anterior group of nuclei is visible. Ventrolateral to this, separated by the *internal capsule* **AB5,** lies the **globus pallidus AB6.** It is divided into two parts, the *internal* **A7** and *external* **A8** *segment.* It differs from the adjacent *putamen* **AB9** by its higher content of myelinated fibers. The *fasciculus lenticularis* and the *ansa lenticularis* **A10** appear at its basal margin and its tip. The ansa lenticularis follows a dorsally-directed arch around the medial tip of the pallidum. The hypothalamus (*tuber cinereum* **A11** and *infundibulum* **A12**), which forms the ventral part of the diencephalon, here appears markedly myelin-poor, in contrast to the densely myelinated *optic tract* **AB13.**

The diencephalon is enclosed on both sides by the telencephalon, but without a clear demarcation. The nearest telencephalic nuclei are the putamen and the *caudate nucleus* **AB14.** Basal to the globus pallidus lies a nucleus belonging to the telencephalon, the *substantia innominata* (Meynert's ganglion) **A15,** the function of which is unknown. It may have fiber connections with the medial thalamic nucleus, the hypothalamus and the olfactory centers. The *fornix* **AB16** is cut twice because of its arched course (p. 219C).

Section at the level of the mamillary bodies (myelin stain) **B.** The section shows both thalami, whose increase in volume has led to secondary fusion in the midline, the *interthalamic adhesion* **B17.** Myelinated fiber lamellae, *laminae medullares thalami* divide the thalamus into several large nuclear complexes. Dorsally there is the *anterior nuclear group* **B18,** ventral to it the *medial group of nuclei* **B19,** which borders medially on several smaller paraventricular nuclei **B20,** and which is laterally separated from the *lateral* and *ventral group of nuclei* **B22** by the lamina medialis **B21.** Division of the lateral group of nuclei into lateral and ventral areas is less clear. The entire complex is enclosed by a small, bowl-shaped nucleus, the *nucleus reticularis thalami* **B23,** which is separated from the region of the lateral nuclei by the *external medullary lamina* **B24.**

The **subthalamus** (p. 180) with the *zona incerta* **B25** and the *subthalamic nucleus* (body of Luys) **B26** lie ventral to the thalamus. The zona incerta is delimited by two myelinated fiber tracts, dorsally by *Forel's field H1 (fasciculus subthalamicus)* **B27** and ventrally by *Forel's field H2 (fasciculus lenticularis)* **B28.** The rostral pole of the *substantia nigra* **B29** lies beneath the subthalamic nucleus. The two **mamillary bodies B30** form the diencephalic floor. The *mamillothalamic fasciculus (Vicq d'Azyr's bundle)* **B31** ascends to the thalamus from the mamillary body.

Corpus callosum **AB32,** amygdaloid body **A33,** hippocampus **B34,** rostral commissure **A35.** Medullary stria of the thalamus **B36** (p. 164 A2).

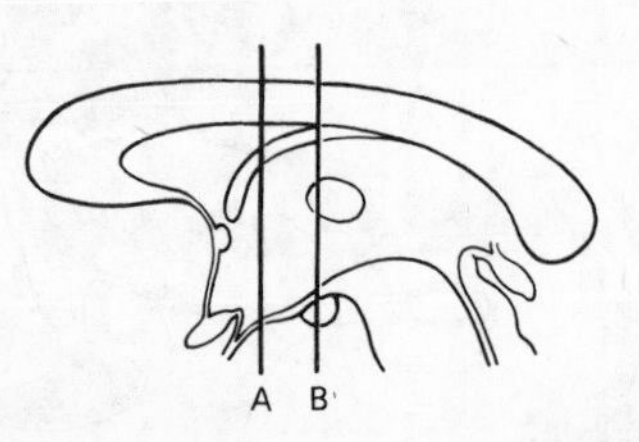

Level of section

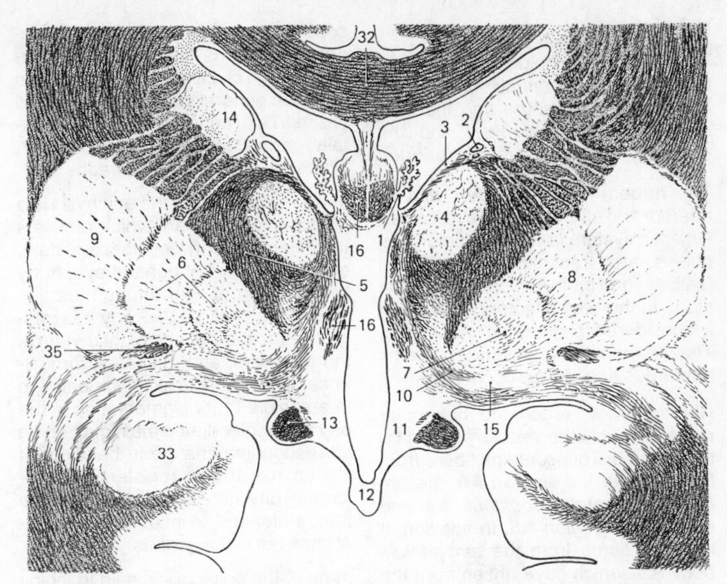

A Section through the diencephalon at the level of the tuber cinereum
(after Villiger and Ludwig)

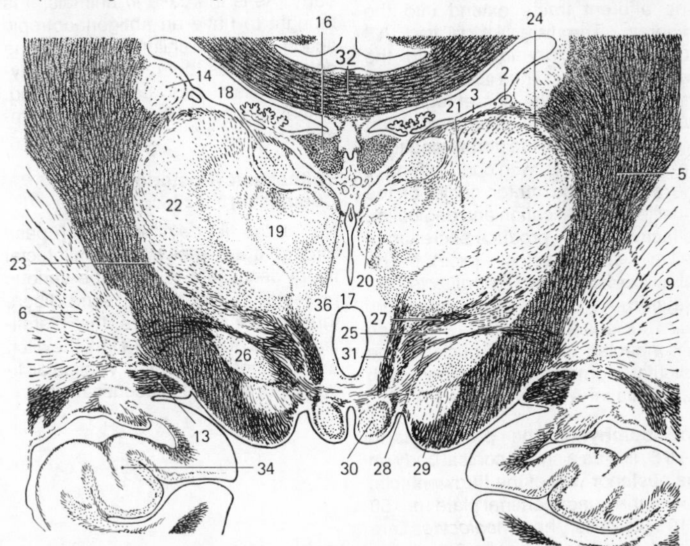

B Section through the diencephalon at the level of the mamillary bodies
(after Villiger and Ludwig)

Epithalamus

The epithalamus includes the *habenula* with the *habenular nuclei,* the *habenular commissure* and the *stria medullaris.* The *epiphysis* and the *epithalamic commissure (posterior).*

The **habenula A1** (p. 158) with its afferent and efferent pathways forms a synaptic system in which olfactory impulses are transmitted to efferent (salivary and motor) nuclei of the brain stem. In this way olfactory stimulation is considered to influence food intake. The habenular nucleus contains numerous peptidergic nerve cells.

Afferent tracts reach the habenular nuclei through the **medullary stria of the thalamus A2.** This contains fibers from the *nuclei of the septum* **A3,** the *anterior perforated substance* **A4** and the *preoptic region* **A5.** In addition, it receives fibers from the *amygdaloid body* **A6,** which have arisen from the *stria terminalis* **A7.**

The efferent tracts extend into the midbrain. The **habenulotectal tract A8** conducts olfactory impulses to the superior colliculi. The **habenulotegmental tract A9** ends in the *dorsal tegmental nucleus* **A10,** where a communication is established with the dorsal longitudinal tract (p. 136B) with connections to the salivary and motor nuclei of the muscles of mastication and deglutition (olfactory stimuli produce secretion of saliva and gastric juice). The **habenulopeduncular tract** *(retroflex tract of Meynert)* **A11** ends in the *interpeduncular nucleus* **A12** (p. 124 D21), which is connected to various nuclei of the reticular formation. Olfactory bulb **A13,** chiasm **A14,** hypophysis **A15.**

The **epiphysis,** the **pineal gland A16B,** lies as a small, conical body on the posterior wall of the IIIrd ventricle, above the quadrigeminal plate (p. 158 B19). Its cells, the *pinealocytes,* are joined together by connective tissue septa into small lobes. In silver impregnated sections they have long processes with clublike terminal swellings **C,** which generally terminate on blood vessels **D.** In adults the epiphysis contains large foci of calcification **B17,** which are visible on radiographs.

In lower vertebrates the epiphysis is a photosensitive organ, which registers the difference between light and dark, either by a special parietal eye, or by transmission of light through the thin roof of the skull. In this way it is integrated into the day and night rhythm of the organism, and regulates, for instance, the change in skin pigmentation in amphibia (dark pigmentation in the day and lighter during the night), and a corresponding change in behavior of the animal. It also registers the prolonged daylight of summer and the dark winter and so influences seasonal changes in the gonads.

In man, the epiphysis is said to inhibit maturation of the genitals before puberty. As is the case in animals, it is thought to have an antigonadotropic action. In certain children, in whom the epiphysis has been destroyed, hypergenitalism has been observed. Pineal recess **B18,** habenular commissure **B19.**

Epithalamic (Posterior) Commissure B20

Not all the fiber systems which pass through it are known. Habenulotectal fibers decussate in it. Of the various pretectal nuclei which send fibers through the commissure, the interstitial nuclei of Cajal and Darkshewitch are the most important. Vestibular fibers, too, are said to decussate in it.

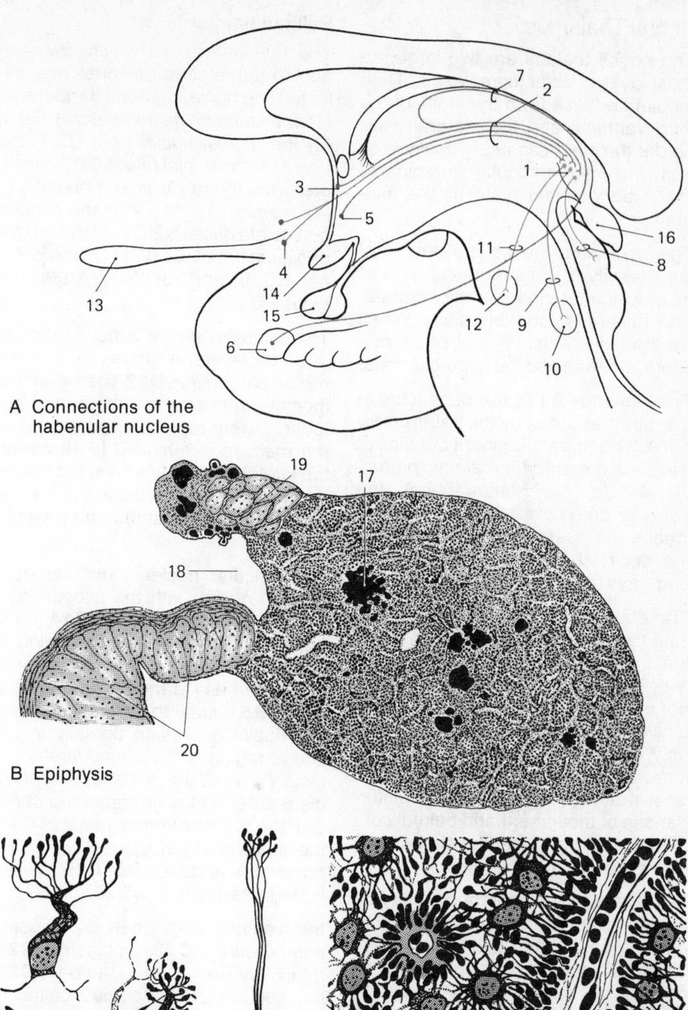

A Connections of the habenular nucleus

B Epiphysis

C Pinealocytes, silver impregnation (after Hortega)

D Histological appearance of the epiphysis. Silver impregnation (after Hortega)

Dorsal Thalamus

The dorsal thalami are two large, almost oval nuclear complexes. Their medial surfaces form the walls of the third ventricle and their lateral walls border the inner capsule. They stretch from the interventricular foramina to the quadrigeminal plate of the midbrain.

The two thalami are the end stations of the majority of sensory tracts, almost all of which terminate in the contralateral thalamus. Fiber bundles connect the thalamus with the cerebellum, pallidum, striatum and the hypothalamus.

The thalamus **A1** is also connected to the cerebral cortex by the **thalamic radiation A.** The more prominent bundles of it include the *anterior thalamic peduncle* **A2** (to the frontal lobes), the *superior peduncle* **A3** (to the parietal lobes), the *posterior peduncle* **A4** (to the occipital lobes) and the *inferior peduncle* (to the temporal lobes).

The abundant fiber connections indicate the central control function of the thalamus, which is intercalated directly or indirectly into most systems. It is not a uniform structure but consists of a diversely organized complex of variably constructed groups of nuclei. A knowledge of its elementary structure is of practical importance, as disturbances of movement and painful conditions may be treated by stereotactic operations on the thalamus.

Two types of thalamic nuclei may be distinguished on the basis of their fiber connection: 1. the nuclei which have fiber connections with the cerebral cortex are called *palliothalamic* or *specific thalamic nuclei;* 2. the nuclei which have no relation with the cortex but are connected to the brain stem form the *truncothalamic* or *nonspecific thalamic nuclei.*

Palliothalamus

The thalamic nuclei which are connected to the cerebral cortex are collected into nuclear groups (territories): (1) the *anterior nuclei* (yellow) **BD5,** (2) the *medial nuclei* (red) **BD6,** the *ventrolateral nuclei* (blue) **CD7,** which are divided into (3) *lateral nuclei* and (4) *ventral nuclei,* (5) the *lateral geniculate nucleus* **BC8** (6) the *medial geniculate nucleus* **BC9,** (7) the *pulvinar* **BC10,** and (8) the *reticular nucleus* **BC11.**

The different nuclear zones are separated by layers of fibers: the *internal medullary lamina* **D12** (between the medial group and the lateral and anterior nuclear regions), and the *external medullary lamina* **D13** (between the lateral nuclear area and the reticular nucleus of the thalamus, which surrounds the thalamus on its lateral surface).

The reticular nucleus and the nonspecific nuclei, with the exception of the *centromedian nucleus* **B14,** have been omitted in the reconstruction of the nuclear zones. The most anterior nuclear area is formed by the anterior nuclei **B5** caudal to which lie the medial nuclei **B6.** Lying dorsally in the lateral region is the *dorsal lateral nucleus* **C15** and the *posterior lateral nucleus* **C16,** and a ventral group of nuclei, the *anterior ventral nucleus* **C17,** the *lateral ventral nucleus* **C18** and the *posterior ventral nucleus* **C19.** Superficial dorsal nucleus **BC20.**

Interventricular foramen **B21,** rostral commissure **C22,** chiasm **C23,** mamillary body **C24,** optic tract **C25,** cut surface of the corpus callosum **A26.**

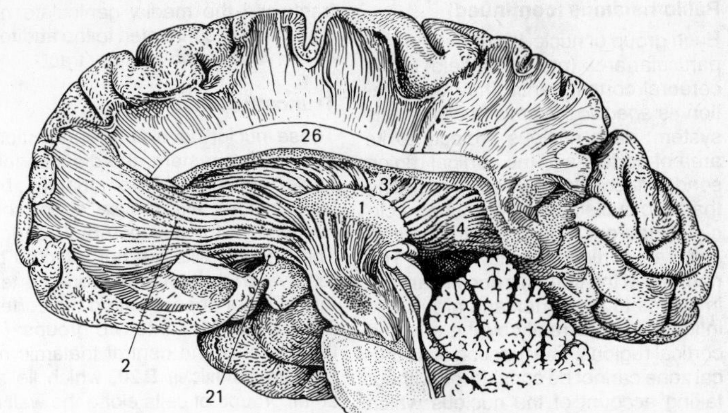

A Thalamic radiations, fiber preparation (after Ludwig and Klingler)

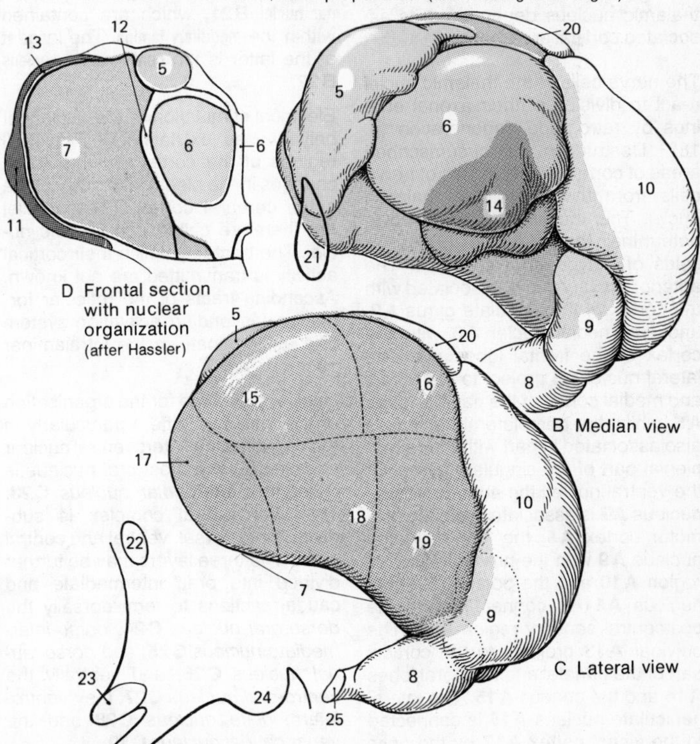

D Frontal section with nuclear organization (after Hassler)

B Median view

C Lateral view

BC Diagrammatic representation of the thalamic regions

Palliothalamus (continued)

Each group of nuclei is connected to a particular area (projection field) of the cerebral cortex, hence their designation as *specific thalamic nuclei.* In this system, each nucleus projects to its area of cortex, and the cortical region sends fibers to the corresponding thalamic nucleus. This produces a circle of neurons with a thalamocortical and a corticothalamic limb. The neurons of the thalamic nuclei transmit impulses to the cortex and are in turn influenced recurrently by the affected cortical regions. The function of a cortical zone cannot be considered without taking account of the nucleus which belongs to it, and the function of the thalamic nucleus depends on its associated cortical area.

The nerve cells of the thalamic nuclei react to division of their axonal endings by retrograde degeneration (p. 18). Destruction of circumscribed areas of cortex results in loss of nerve cells from the associated thalamic nuclei. This method may be used to determine the thalamic projection fields of the cerebral cortex **A.** The anterior nuclei **A1** are associated with the cortex of the cingulate gyrus **A2,** and the medial nuclei **A3** with the cortex of the frontal lobes **A4.** The lateral nuclei **A5** project to the dorsal and medial cortex of the parietal lobes **A6,** whilst the dorsolateral nucleus is also associated in part with the retrosplenial part of the cingulate gyrus. Of the ventral nuclei, the anterior ventral nucleus **A7** is associated with the premotor cortex **A8,** the lateral ventral nucleus **A9** with the precentral motor region **A10** and the posterior ventral nucleus **A11** is connected with the postcentral sensory region **A12.** The pulvinar **A13** projects to the cortical part of the parietal and temporal lobes **A14** and the cuneus **A15.** The lateral geniculate nucleus **A16** is connected to the visual cortex **A17** by the optic tract, and the medial geniculate nucleus **A18** is connected to the auditory cortex **A19** by the auditory tract.

Truncothalamus

These nuclei have fiber connections with the brain stem, the diencephalic nuclei and the corpus striatum, but lack anatomically proven direct connections with the cerebral cortex. Their nerve cells are not affected by removal of the entire cerebral cortex; they are not dependent on the cortex. They are divided into two groups: (1) the nuclei of the central thalamic region **(median nuclei) B20,** which lie as small groups of cells along the wall of the third ventricle, and (2) the **intralaminar nuclei B21,** which are contained within the **medullary lamina.** The largest of the latter is the **centromedian nucleus B22.**

Electrical stimulation of the nuclei not only causes excitation of individual regions of the cortex, but results in changes in the electrical activity of the entire cerebral cortex. These nuclei are therefore called *non-specific nuclei.* The tracts by which their cortical activity is transmitted are not known. Ascending tracts of the reticular formation (ascending activation system p. 138) terminate in the intralaminar nuclei.

Hassler's scheme for the organization of the thalamus differs particularly in the division of the lateroventral nuclear complexes. The most oral nucleus is called the *lateropolar nucleus* **C23.** The lateroventral complex is subdivided into dorsal, ventral and central parts. The three layers may be further divided into oral, intermediate and caudal sections to form dorsally the *dorso-oral nucleus* **C24,** *dorso-intermediate nucleus* **C25,** and *dorsocaudal nucleus* **C26,** and ventrally the *ventro-oral nucleus* **C27,** they *ventro-intermediate nucleus* **C28** and the *ventrocaudal nucleus* **C29.**

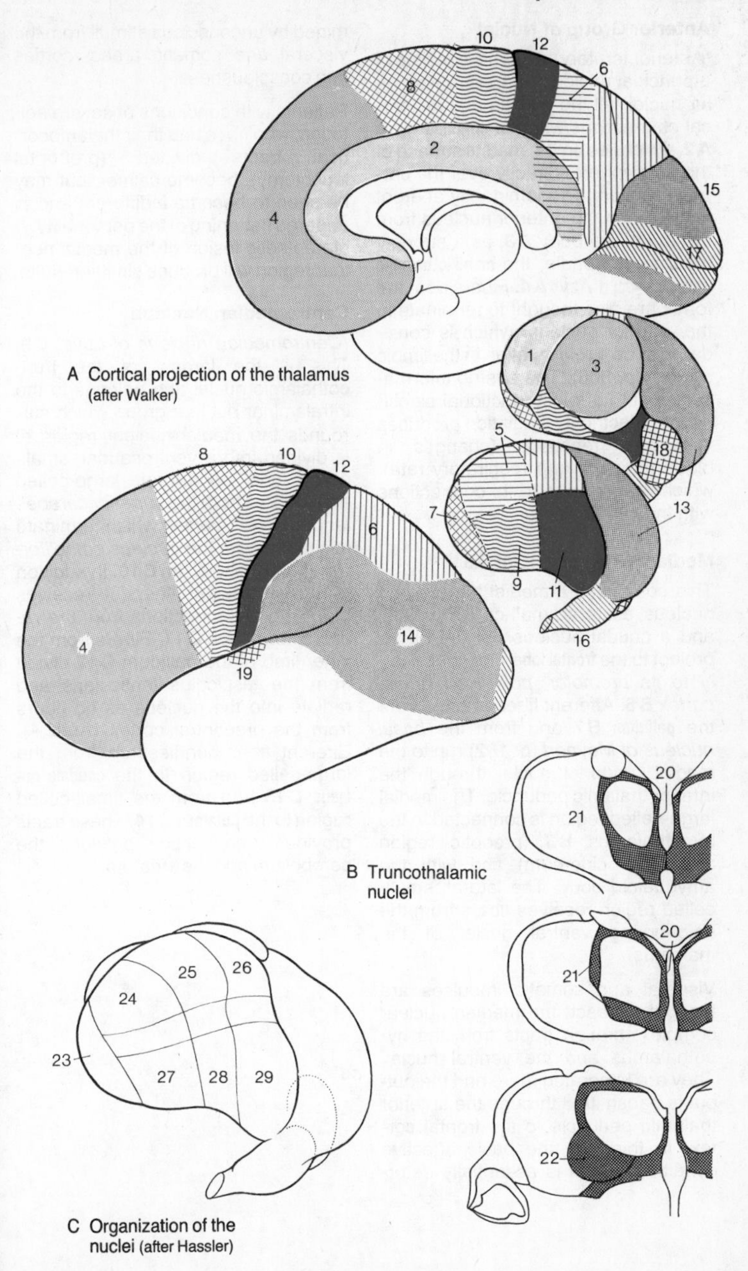

A Cortical projection of the thalamus (after Walker)

B Truncothalamic nuclei

C Organization of the nuclei (after Hassler)

Anterior Group of Nuclei

(Anterior territory) **A1.** This consists of a principal nucleus and several smaller nuclei. All the nuclei have reciprocal connections with the **cingulate gyrus A2,** which lies on the medial surface of the hemisphere, directly over the corpus callosum. Predominantly afferent fibers reach the anterior nucleus from the *mamillary body* **A3,** as a strongly myelinated bundle, the **mamillothalamic tract** of Vicq d'Azyr **A4.** Fibers from the fornix are also thought to terminate in the anterior nucleus, which is considered to be a relay station in the limbic system (p. 306). There is no information about its exact functional significance. Electrical stimulation produces vegetative reactions (changes in blood pressure and respiratory rate), which may be due to its connections with the hypothalamus.

Medial Group of Nuclei B5

This comprises a medial large-celled nucleus, a lateral small-celled nucleus and a caudal nucleus. All the nuclei project to the **frontal lobe,** and specifically to its *premotor, polar* and *orbital cortex* **B6.** Afferent fiber bundles from the **pallidum B7** and from the *basal nucleus of Meynert* (p. 162) run to the medial group of nuclei through the inferior thalamic peduncle. The medial large-celled region is connected to the hypothalamus **B8** (preoptic region and tuber cinereum) and with the amygdaloid body. The lateral, small-celled region receives fibers from the neighboring ventral nuclei of the thalamus.

Visceral and somatic impulses are thought to reach the median nuclear complex through tracts from the hypothalamus and the ventral nuclei. They are integrated there and the output is transmitted through the anterior thalamic peduncle to the frontal cortex. In this way the basic affective condition, which is essentially determined by unconscious stimuli from the visceral and somatic areas, comes into consciousness.

Patients with conditions of severe agitation, who have had their thalamocortical tracts divided (prefrontal leucotomy), become calmer, but may be seen to become indifferent and to undergo flattening of the personality. A stereotactic lesion of the medial nuclear region will produce similar results.

Centromedian Nucleus

(Centromedian nucleus of Luys) **C9.** This is the largest of the truncothalamic nuclei and belongs to the intralaminar nuclear group which surrounds the medial nuclear region. It is divided into a ventrocaudal, small-celled and dorso-oral, large-celled area. Fibers from the *superior cerebellar peduncle (cranial)* which terminate in it, arise from the *nucleus emboliformis of the cerebellum* **C10.** In addition to these crossed fibers it receives homolateral connections from the *reticular formation* **C11.** Fibers from the inner limb of the pallidum **C12** divide from the fasciculus lenticularis and radiate into the nucleus as do fibers from the precentral cortex (Area 4). Efferent fiber bundles run from the large-celled region to the **caudale nucleus C13** and from the small-celled region to the **putamen C14.** These tracts provide a connection between the cerebellum and the striatum.

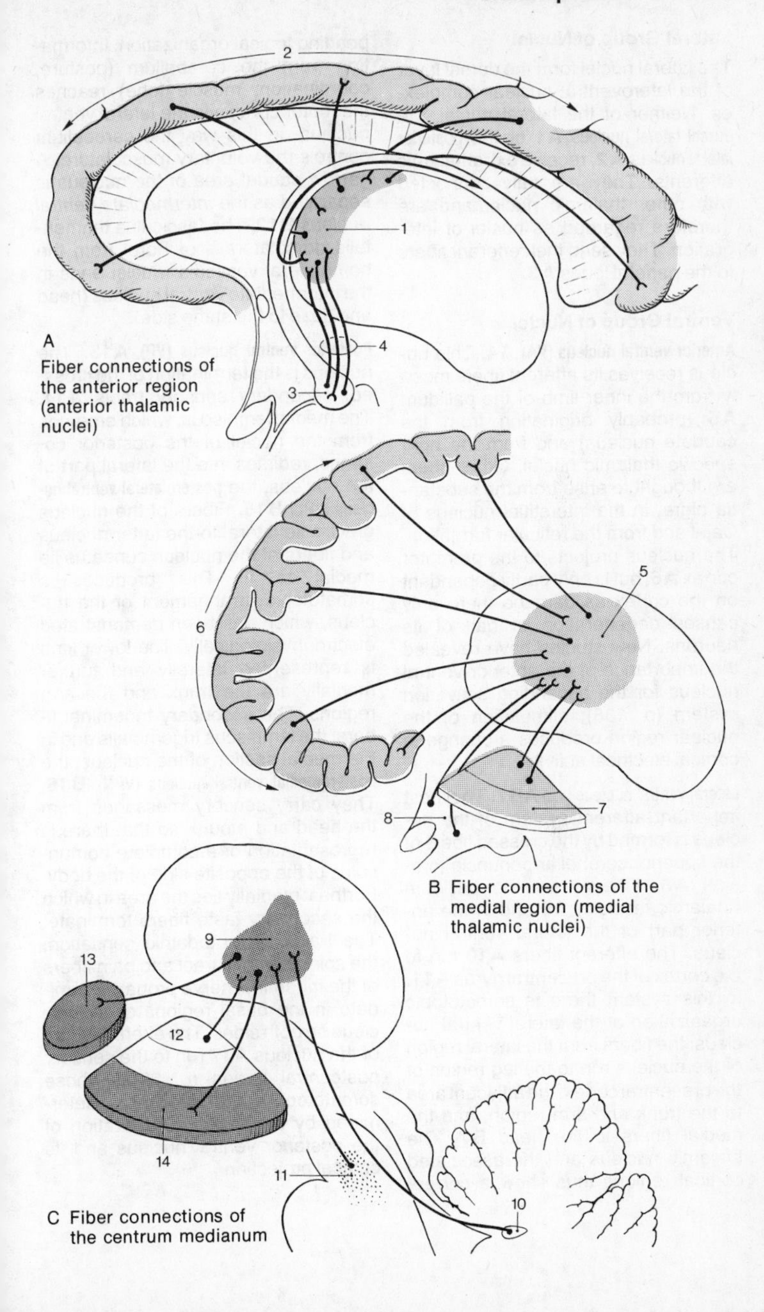

A
Fiber connections of the anterior region (anterior thalamic nuclei)

B Fiber connections of the medial region (medial thalamic nuclei)

C Fiber connections of the centrum medianum

Lateral Group of Nuclei

The lateral nuclei form the dorsal layer of the lateroventral nuclear complexes. Neither of the lateral nuclei, the **dorsal lateral nucleus A1** or the **posterior lateral nucleus A2**, receive extrathalamic afferents. They are only connected with other thalamic nuclei and are therefore regarded as nuclei of integration. They send their efferent fibers to the parietal lobes **A3.**

Ventral Group of Nuclei

Anterior ventral nucleus (VA) A4. This nucleus receives its afferent fibers mostly from the inner limb of the pallidum **A5** (probably originating from the caudate nucleus) and from the nonspecific thalamic nuclei. Other fibers are thought to arise from the substantia nigra, in the interstitial nucleus of Cajal and from the reticular formation. The nucleus projects to the premotor cortex **A6,** but is only partly dependent on the cortex as damage there only causes degeneration of half of its neurons. New studies have revealed the importance of the anterior ventral nucleus for the ascending activation system (p. 138); stimulation of the nuclear region produces a change in cortical electrical activity.

Lateral ventral nucleus (VL) AB7. The most important afferent system of the nucleus is formed by the crossed fibers of the superior cerebellar peduncle (cranial) **A8.** Fibers from the pallidum (thalamic fascicle) **A9** end in the anterior part of the lateral ventral nucleus. The efferent fibers **A10** run to the cortex of the precentral gyrus **A11.** In this system there is somatotopic organization of the lateral ventral nucleus: the fibers from the lateral region of the nucleus run to the leg region of the precentral cortex, the adjacent area to the trunk and arm region, and the medial fibers to the head **B7.** The thalamic nucleus and the associated cortical regions thus show a corresponding topical organization. Information from the cerebellum (posture, coordination, muscle tone) reaches the motor cortex via the lateral ventral nucleus. In this way the cerebellum controls the voluntary musculature. A narrow caudal area of the nucleus is separated as the *intermediate ventral nucleus* **B12.** The fasciculus tegmentalis dorsolateralis of Forel from the homolateral vestibular nuclei ends in the intermediate ventral nucleus (head and gaze to the same side).

Posterior ventral nucleus (VP) A13. The nucleus is the termination of the crossed, secondary sensory tracts **A14.** The medial lemniscus, which emerges from the nuclei of the posterior columns, radiates into the lateral part of the nucleus, the **posterolateral ventral nucleus (VPL) B15.** Fibers of the nucleus gracilis lie lateral to the latter nucleus and fibers of the nucleus cuneatus lie medial to it. This produces a somatotopic arrangement of the nucleus, which has been demonstrated electrophysiologically. The lower limb is represented laterally and further medially are the trunk and the arm regions. The secondary trigeminal fibers, the lemniscus trigeminalis end in the medial section of the nucleus, the **posteromedial ventral nucleus (VPM) B16.** They carry sensory messages from the head and mouth, so that there is representation of a complete homunculus of the opposite side of the body. Furthest medially lies the area in which the secondary taste fibers terminate. The tracts of protopathic sensation, the spinothalamic tract and pain fibers of the trigeminal nerve probably terminate in the basal regions of the nucleus on both sides. The efferent fibers of the nucleus **A17** run to the sensory postcentral region (p. 234), whose somatotopic organization is determined by the topical organization of the posterior ventral nucleus and its cortical projection.

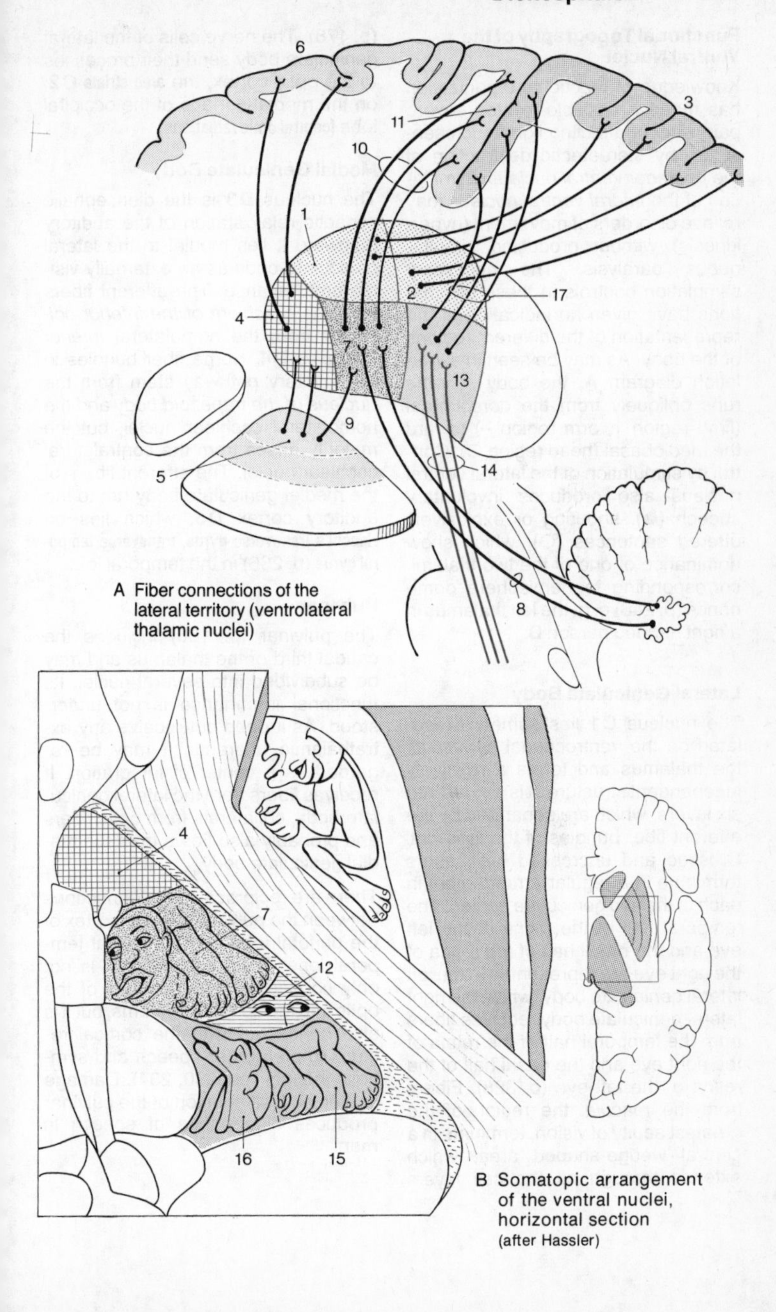

A Fiber connections of the lateral territory (ventrolateral thalamic nuclei)

B Somatopic arrangement of the ventral nuclei, horizontal section (after Hassler)

Functional Topography of the Ventral Nuclei

Knowledge of functional organization has made it possible to stop severe pain without affecting light touch sensation by stereotactic destruction of the *posterior ventral nucleus.* Elimination of the *lateral ventral nucleus* may relieve disorders of movement (hyperkinesis) without producing simultaneous paralysis. The necessary stimulation controls in these eliminations have given an indication of the representation of the different regions of the body. As may be seen in stimulation diagram **A,** the body scheme runs obliquely from the dorsolateral (limb region |, arm region –) toward the mediobasal (head region ○). Control by stimulation of the lateral ventral nucleus also produces involuntary speech (●), shouting or explosively uttered sentences (○), which show dominance of one of the two thalami, corresponding to hemisphere dominance (p. 246) e. g. the left thalamus in a right handed person **B.**

Lateral Geniculate Body

The nucleus **C1** lies somewhat isolated on the ventrocaudal surface of the thalamus and forms a relatively independent structure. It is divided into six layers, which are separated by the afferent fiber bundles of the **optic tract.** Crossed and uncrossed optic fibers terminate in a regular arrangement in each of the two geniculate bodies. The temporal half of the retina of the left eye and the nasal half of the retina of the right eye are represented in the left lateral geniculate body, whilst the right lateral geniculate body receives fibers from the temporal half of the retina of the right eye and the nasal half of the retina of the left eye (p. 328). Fibers from the *macula,* the region of the greatest acuity of vision, terminate in a central wedge-shaped area, which extends through all the cell layers (p. 178). The nerve cells of the lateral geniculate body send their processes to the optic cortex, the **area striata C2,** on the medial surface of the occipital lobe **(central optic radiations).**

Medial Geniculate Body

The nucleus **D3** is the diencephalic synaptic relay station of the auditory pathway. It lies medial to the lateral geniculate body as an externally visible protuberance. The afferent fibers form the *brachium of the inferior colliculus* from the homolateral *inferior colliculus* **D4.** Some fiber bundles in the auditory pathway stem from the *nucleus of the trapezoid body* and the homolateral cochlear nuclei, but the majority arises from the contralateral cochlear nuclei. The efferent fibers of the medial geniculate body run to the auditory cortex **D5,** which lies on **Heschl's transverse gyrus, transverse temporal gyrus** (p. 236) in the temporal lobe.

Pulvinar

The pulvinar (p. 166) includes the caudal third of the thalamus and may be subdivided into several nuclei. Its functional importance is not understood. As it does not receive any extrathalamic afferents, it may be regarded as a center of integration. It receives fibers from the lateral geniculate body, collaterals from optic fibers and probably also fibers from the medial geniculate body.

There are reciprocal fiber connections between the pulvinar and the cortex of the parietal lobe and the dorsal temporal lobe. Thus, the pulvinar is not only connected in the circuits of the optic and the acoustic systems, but it is also connected with the cortical regions important for speech and symbolic thought (pp. 230, 234). Damage or electrical stimulation of the pulvinar produces disturbance of speech in man.

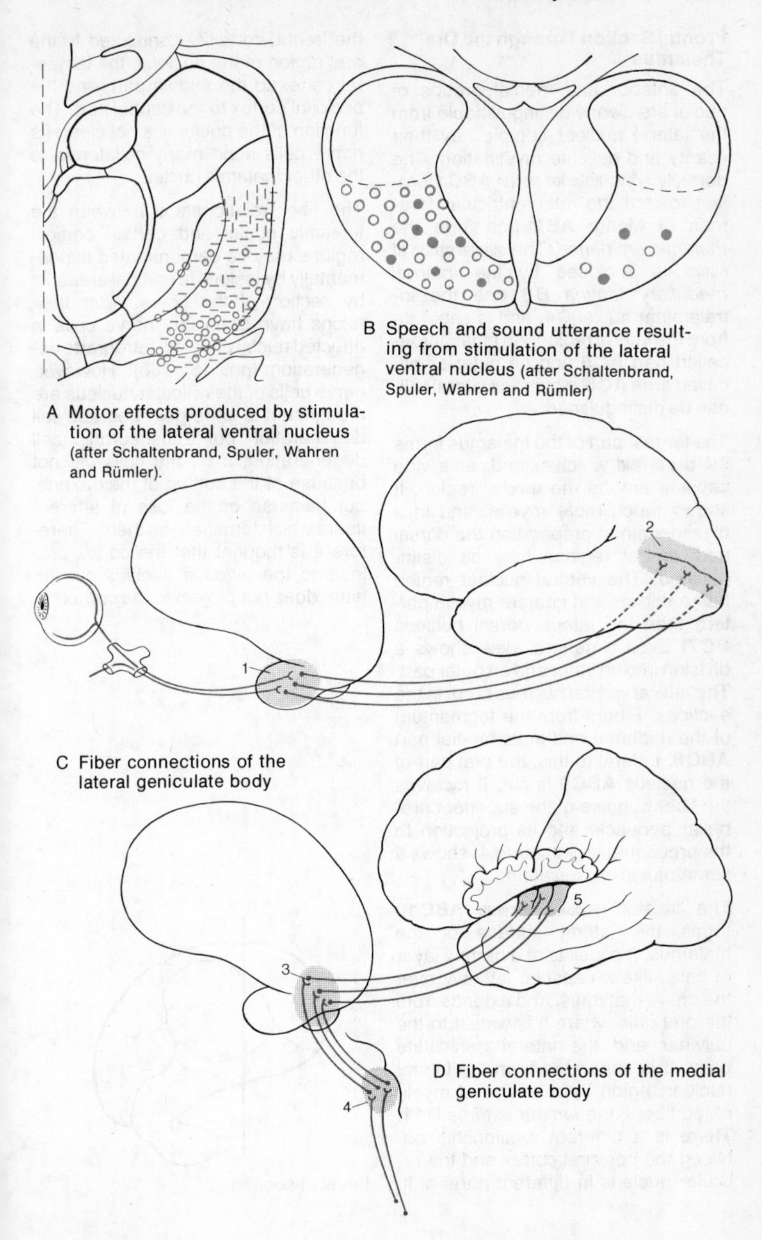

A Motor effects produced by stimulation of the lateral ventral nucleus (after Schaltenbrand, Spuler, Wahren and Rümler).

B Speech and sound utterance resulting from stimulation of the lateral ventral nucleus (after Schaltenbrand, Spuler, Wahren and Rümler)

C Fiber connections of the lateral geniculate body

D Fiber connections of the medial geniculate body

Frontal Section Through the Oral Thalamus

The anterior and medial groups of nuclei are clearly distinguishable from the lateral nuclear complex by their scanty and delicate myelination. The dorsally lying **anterior nuclei** **ABC1** project toward the interventricular foramen of Monro **AB2** and form the *thalamic eminence.* The **medial group of nuclei** is enclosed by the *internal medullary lamina* **B3** and the intralaminar nuclei **C4,** and is separate from the lateral region. An inner, large-celled part **AC5** and an outer small-celled area **AC6,** which lies lateral to it, can be distinguished.

The largest part of the thalamus forms the **lateral field,** which extends as a wide capsule around the medial region. It shows much more myelin, and in a myelin-stained preparation the dorsal and ventral regions may be distinguished. The ventral nuclear region has a thicker and coarser myelin pattern than the lateral dorsal nucleus **BC7.** Even a general view shows a division into an inner and an outer part. The *lateral ventral nucleus* is cut in the sections. Fibers from the tegmentum of the midbrain end in its medial part **ABC8.** Lateral to this, the oral part of the nucleus **ABC9** is cut. It receives the fiber bundles of the superior cerebellar peduncle and its projection to the precentral region (area 4) shows a somatotopic organization.

The **thalamic reticular nucleus** **ABC10** forms the lateral surface of the thalamus. It consists of a narrow layer of cells, like a capsule, laterally over the entire thalamus, and extends from the oral pole where it is widest to the pulvinar and the lateral geniculate body. It is separated from the lateral nuclear region by a lamina of myelinated fibers, the *lamina externa* **B11.** There is a different relationship between the cerebral cortex and the reticular nucleus in different parts of it; the frontal cortex is connected to the oral region of the nucleus, the temporal cortex to the middle part, and the occipital cortex to the caudal part. The function of the nucleus is not clear. Its nerve cells send many collaterals to the other thalamic nuclei.

The fiber relationships between the thalamic nuclei and certain cortical regions may be demonstrated experimentally by lesions in cortical areas, or by section of the fibers. After their axons have been cut, nerve cells in affected nuclei undergo retrograde degeneration (pp. 18, 168). However, nerve cells of the reticular nucleus are thought not to undergo retrograde cell degeneration but transneuronal cell degeneration, i. e., the cells die not because of the cutting of their axons, but because of the loss of afferent fibers which terminate on them. Therefore it is thought that the cortex projects to the reticular nucleus but the latter does not project to the cortex.

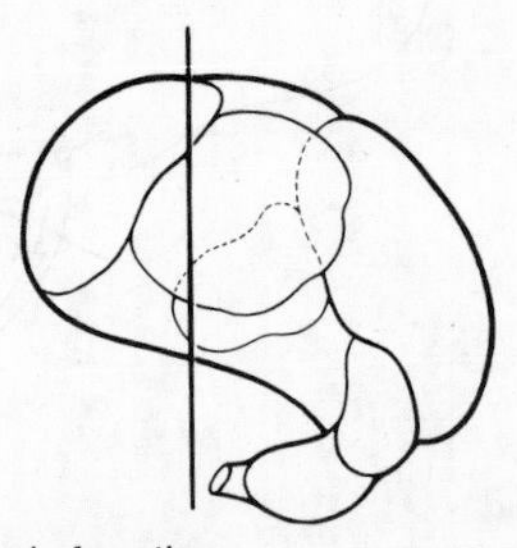

Level of section

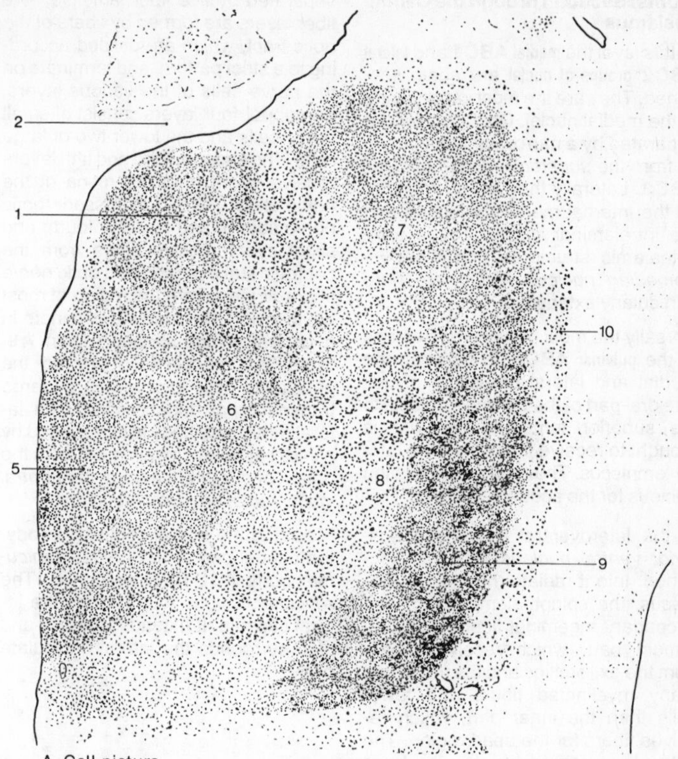

A Cell picture

Frontal section through the rostral thalamus (semischematic)

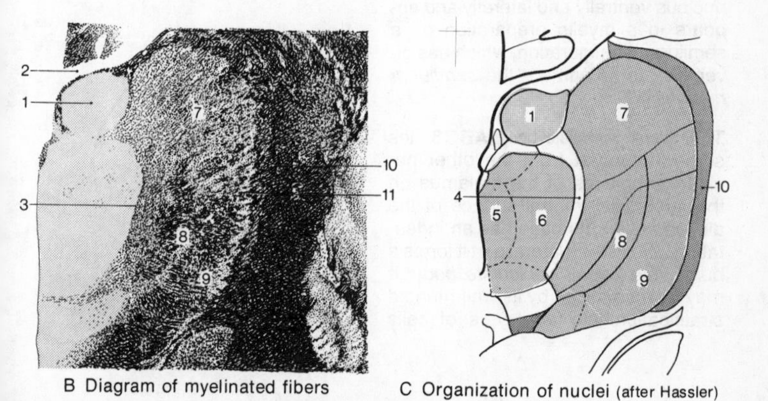

B Diagram of myelinated fibers

C Organization of nuclei (after Hassler)

Frontal Section Through the Caudal Thalamus

At this level the **medial ABC 1** and **lateral ABC 2 groups of nuclei** are again sectioned. They are the most caudal parts of the medial nuclei. Dorsally they are separated by a narrow lamina of myelin from the *superficial dorsal nucleus* **ABC 3.** Laterally they are surrounded by the internal medullary lamina and the intralaminar nuclei. These truncothalamic structures and the *centromedian nucleus* **ABC 4** are here particularly extensive.

Dorsally the most oral nuclear regions of the **pulvinar ABC 5** lie between the medial and lateral groups of nuclei. The oral part of the pulvinar projects to the superior temporal gyri and is thought to receive fibers from the lateral lemniscus. Thus it is an integration nucleus for the acoustic system.

In the lateroventral region the *posterior ventral nucleus* **ABC 6** is sectioned. Into it radiate the *medial lemniscus,* the spinothalamic tracts and secondary trigeminal fibers. The outermost part, which receives fibers from the extremities and the trunk, has many myelinated fibers and fewer cells than the inner part, which receives fibers for the head region. The latter is recognizable by its large number of cells and more thinly myelinated fibers. It surrounds the central nucleus ventrally and laterally and appears in a myelin preparation as a semilunar configuration, which has given rise to its name – the *semilunar nucleus* **B 7.**

The **lateral geniculate body ABC 8** lies somewhat apart from the other nuclear complexes of the thalamus on the superficial ventral surface of the diencephalon. Its base has an indentation, and with its lateral part it forms a bulge, the *lateral geniculate body.* It may by recognized by its well-marked stratification into six layers of cells separated by five fiber lamellae. The fiber layers are formed by fibers of the optic tract, which are divided according to a strict pattern, and terminate on the nerve cells of the various layers. The upper four layers consist of small nerve cells and the lower two of large cells. The second, third and fifth layers receive fibers from the retina of the homolateral eye (uncrossed optic nerve fibers), and the first, fourth and sixth layers receive fibers from the contralateral eye (crossed optic nerve fibers). Fibers from the region of most acute vision, the *macula,* terminate in a central wedge-shaped region **A 9.** When the macula is destroyed the geniculate cells in this area undergo transneuronal degeneration. The lateral geniculate body is surrounded by a thick capsule of myelinated fibers the dorsal and lateral fibers of the *central visual radiation (optic radiation).*

Medial to the lateral geniculate body, the caudal part of the *medial geniculate body* **ABC 10** is sectioned. The *reticular nucleus* **AC 11** forms the lateral capsule. It extends ventrally and also surrounds the lateral geniculate body.

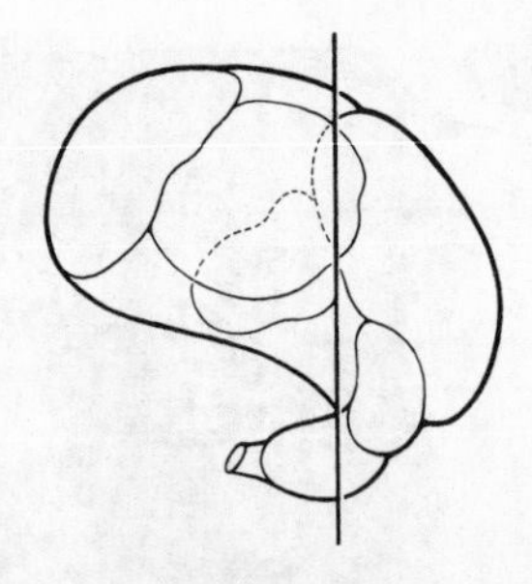

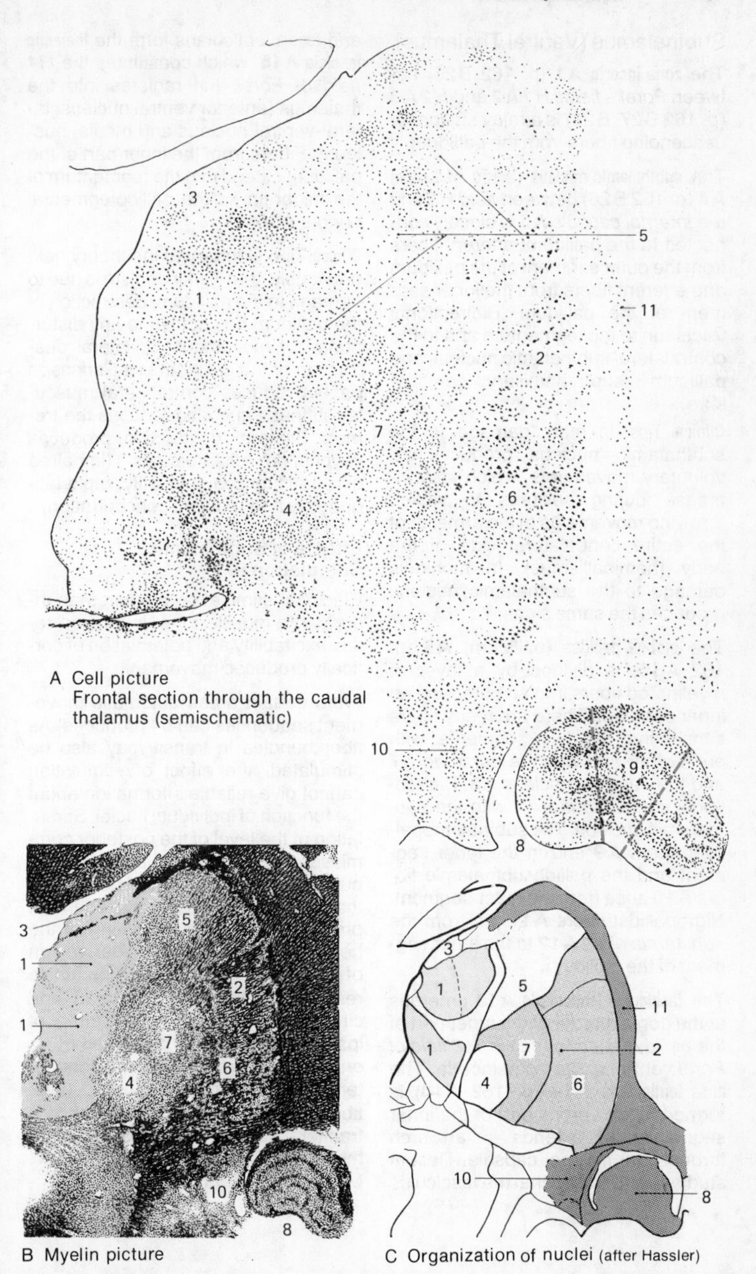

A Cell picture
Frontal section through the caudal thalamus (semischematic)

B Myelin picture

C Organization of nuclei (after Hassler)

Subthalamus (Ventral Thalamus)

The **zona incerta A1** (p. 162 B25) between *Forel's fields H1* **A2** and *H2* **A3** (p. 162 B27, B28) is a relay station for descending fibers from the pallidum.

The **subthalamic nucleus** (body of Luys) **A4** (p. 162 B26) between *field H2* and the *internal capsule* **A5** is closely connected to the pallidum: afferent fibers from the outer segment of the pallidum and efferent fibers from the inner segment of the pallidum. Bidirectional tracts run to the tegmentum and to the contralateral subthalamic nucleus and pallidum (supramamillary commissure).

Clinical Tips. In man, damage to the subthalamic nucleus produces involuntary movements, which may increase during seizures to violent throwing movements of the arms, or of the entire contralateral side of the body (hemiballismus). In monkeys damage to the subthalamic nucleus produces the same signs.

The **globus pallidus** (pallidum) **A6** (p. 162 AB6) is divided by a layer of myelinated fibers into an outer and an inner segment. There are many fibers connecting the two segments with each other and with the *putamen* **A7** and the *caudate nucleus* **A8.** Bidirectional connections exist with the *subthalamic nucleus*. The subthalamopallidal fibers **A9** end in the inner segment and the pallidosubthalamic fibers **A10** arise from the outer segment. Nigropallidal fibers **A11** run from the *substantia nigra* **A12** to the inner segment of the pallidum.

The **fasciculus lenticularis A13** emerges at the dorsal border of the inner part of the pallidum and forms the *H2 field of Forel* ventral to the zona incerta. The **ansa lenticularis A14** (p. 162 A10) is formed in the ventral part of the inner segment and extends in an arch through the internal capsule. Recent studies have shown that the fasciculus and ansa lenticularis form the **thalamic fascicle A15,** which constitutes the *H1 field of Forel* and radiates into the thalamus (anterior ventral nucleus, lateral ventral nucleus and medial nucleus). Fibers from the inner part of the pallidum extend into the tegmentum of the midbrain as the pallidotegmental bundle **A16.**

Clinical Tips. Contrary to the theory held previously, that Parkinsonism is due to damage to the pallidum, destruction of the pallidum has not led to any disturbance of movement. Unilateral destruction of the pallidum in a Parkinson patient reduces contralateral muscular stiffness and also reduces the tremor. Bilateral destruction produces psychiatric disturbances (impaired cerebral function, irritability, rapid fatiguability and reduced concentration).

Effects of Subthalamic Stimulation B

Electrical stimulation produces an increase in muscle tone, increased reflex excitability and potentiation of cortically produced movement.

From certain areas automatic movement sequences can be produced. As fiber bundles in transit may also be stimulated, the effect of stimulation cannot give reliable information about the function of individual nuclei. Stimulation at the level of the posterior commissure (fiber region of the interstitial nucleus) produces forward bending of the head (↓), medial stimulation in the prerubral field produces lifting of the head (↑). Stimulation of fiber region of the superior cerebellar peduncle results in rotation of the head and circling movements ↻. The area of ipsilateral turning of the head (←) is encountered in the field of the dorsolateral tegmental fasciculus (vestibulothalamic tract). The area of contralateral turning (→) corresponds to the zona incerta. Fornix **B17.** Mamillothalamic tract **B18.**

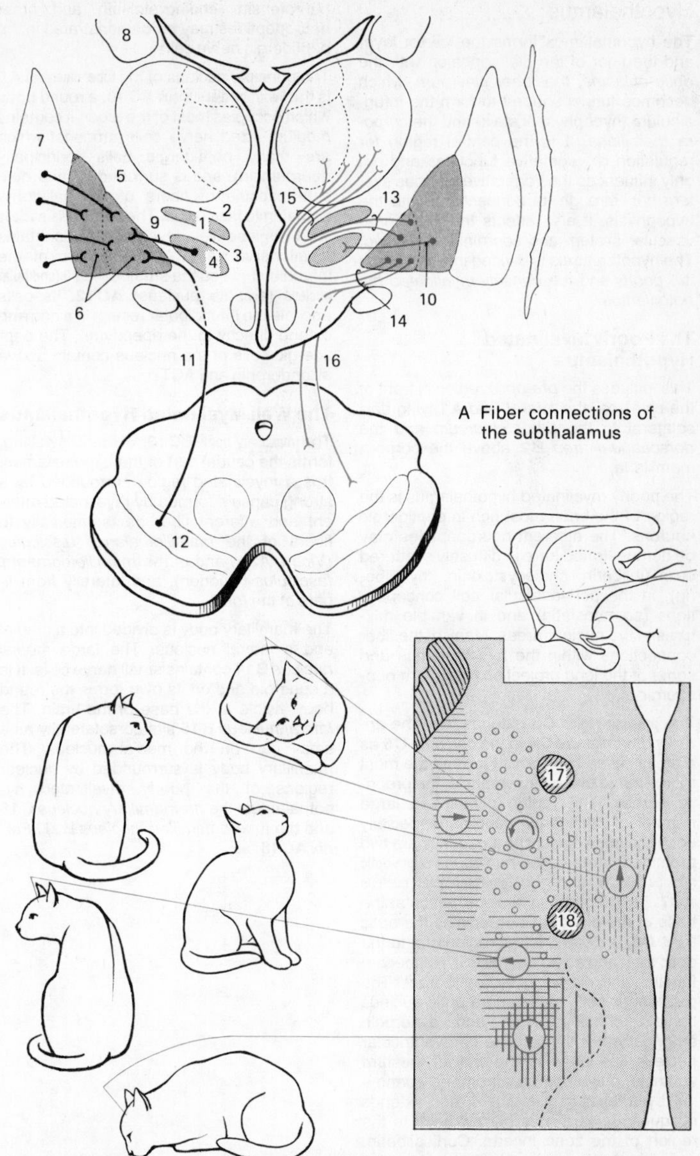

A Fiber connections of the subthalamus

B Motor responses to stimulation of the subthalamus and the tegmentum, horizontal section through the diencephalon of the cat (after Hess)

Hypothalamus

The hypothalamus forms the lowest layer and the floor of the diencephalon with the *optic chiasma,* the *tuber cinereum,* which becomes funnel shaped to form the infundibulum (hypophyseal stalk) and the *corpora mamillaria.* It is the central region for regulation of vegetative functions and not only influences the vegetative nervous system but, through its connections with the hypophysis, it also affects the endocrine-vascular system and coordinates the two. The hypothalamus is divided into two areas the *poorly* and the *densely myelinated hypothalamus.*

The Poorly Myelinated Hypothalamus

This includes the *pre-optic region* in front of the chiasma, the *lateral field* **A1** lying dorsolateral to the *tuber cinereum* and the *dorsocaudal area* **B2** above the corpora mamillaria.

The poorly myelinated hypothalamus is the region of the brain most rich in peptidergic neurons. The different neuropeptides may be demonstrated here in diffusely scattered cells (luliberin, cholecystokinin, thyroliberin), in the periventricular cell concentrations (somatostatin) and in variable mixtures in the nuclear areas. Many of the fiber connections within the hypothalamus and some of the long projection tracks are peptidergic.

The **pre-optic region C3** extends from the *anterior commissure* **C4** to the *chiasma* **C5** as a small-celled field, which frames the most rostral bulge of the IIIrd ventricle, the preoptic recess. The region contains a large number of peptidergic neurons (especially encephalin). Adjacent to it caudally are two prominent large-celled nuclei: the **supra-optic nucleus AC6** and the **paraventricular nucleus AC7.** The supraoptic nucleus, which at the base of the diencephalon adjoins the optic tract **A8,** is divided into a region oral to the optic tract and a part caudal to it. Neither are functionally related to the optic tract. The peptidergic cells of the supra-optic nucleus contain cholecystokinin and dynorphin amongst other things. The paraventricular nucleus lies close to the wall of the IIIrd ventricle, only separated from the ependyma by a layer of glial fibers, and extends obliquely upward as a narrow band to the region of the zona incerta. Corticoliberin, neurotensin, cholecystokinin and other neuropeptides may be demonstrated in the peptidergic nerve cells.

The principal nucleus of the **tuber cinereum A9** is the **ventromedial nucleus AC10,** a round body which occupies most of the tuber. It contains medium-sized nerve cells amongst which are many peptidergic cells (principally neurotensin), and is surrounded by a delicate capsule of fibers of the pallidohypothalamic fasciculus. The **dorsomedial nucleus AC11** is less well developed and contains small nerve cells. At the bottom of the infundibulum lies the small-celled **infundibular nucleus** (arcuate nucleus) **AC12.** Its cells encircle the infundibular recess in a ring and extend directly to the ependyma. The peptidergic cells of the nucleus contain above all endorphin and ACTH.

The Well Myelinated Hypothalamus

The **mamillary body BC13,** a round swelling, forms the caudal part of the hypothalamus. It is a myelinated region surrounded by a strong capsule formed by myelinated afferent and efferent fiber tracts: medially to fibers of the *mamillothalamic fasciculus (Vicq d'Azyr)* and of the *mamillotegmental fasciculus* (Gudden), and laterally from fibers of the *fornix.*

The mamillary body is divided into a medial and a lateral nucleus. The large *medial nucleus* **B14** contains small nerve cells. It is a spheroid and on its own forms the round body visible at the base of the brain. The *lateral nucleus* **B15** sits dorsolaterally as a small cap on the medial nucleus. The mamillary body is surrounded by nuclear regions of the poorly myelinated hypothalamus: the *premamillary nucleus* **C16** and the *tuberomamillary nucleus* **B17.** *Fornix* **AC18.**

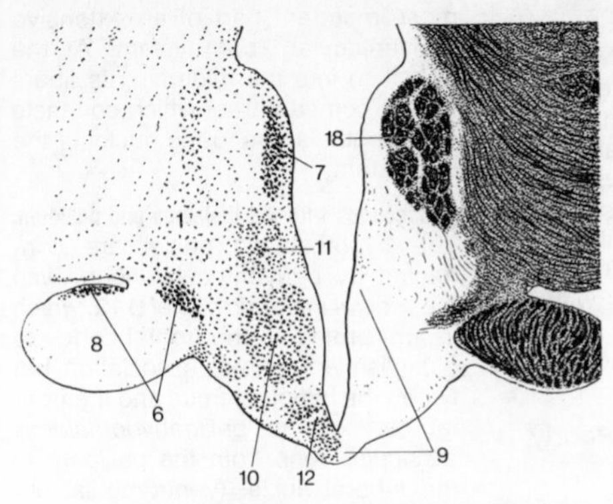

A Frontal section through the tuber cinereum (poorly myelinated hypothalamus)

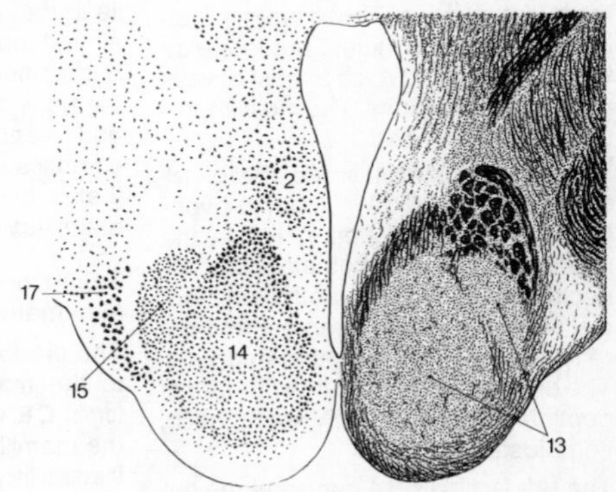

B Frontal section through the mamillary body (myelin-rich hypothalamus)

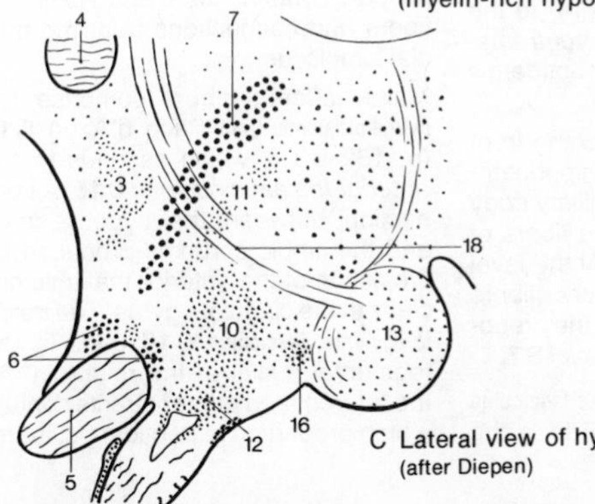

C Lateral view of hypothalamic nuclei (after Diepen)

Vascular Supply

The close association between the nervous and endocrine systems is revealed by the unusually rich vascular supply of individual hypothalamic nuclei. The *supra-optic nucleus* **A1** and the *paraventricular nucleus* **A2** are about six times as well vascularized as the remaining grey matter. Their nerve cells are in close contact with the capillaries, some of which are even enclosed within them (endocellular capillaries).

Fiber Relationships of the Poorly Myelinated Hypothalamus B

A large number of connections conduct olfactory, gustatory, viscerosensory and somatosensory impulses to the hypothalamus. These are generally not compact bundles, but loose divergent systems which almost always run in both directions. The most important are:

The **medial forebrain bundle (Medial telencephalic fasciculus)** (olfacto-hypothalamo-tegmental fibers) **B3** in the lateral part of the hypothalamus, connects almost all the hypothalamic nuclei with the olfactory centers and the reticular formation of the midbrain. The bundle contains a large number of peptidergic fibers (VIP, encephalin, somatostatin).

The **stria terminalis B4** curves along the caudate nucleus and joins the *amygdaloid body* **B5** (pp. 212, 214) to the *preoptic region* **B6** and the *ventromedial nucleus* **B7.** It is rich in peptidergic fibers (p. 218 B15, C).

The fibers of the **fornix B8** come from the pyramidal cells of the hippocampus **B9** and end in the mamillary body **B10.** A large number of the fibers of the fornix are peptidergic. At the level of the anterior commissure fibers branch off from the fornix to the preoptic region and the *tuberal nuclei* **B7.**

The **dorsal longitudinal fasciculus** (Schütz's tract) (p. 136B) **B11** is the most important part of an extensive periventricular fiber system. At the transition into the midbrain, its fibers form a compact tract which connects the hypothalamus to the nuclei of the brain stem.

Connections with the thalamus and pallidum. The hypothalamic nuclei are connected by periventricular fibers with the *medial thalamic nuclei* **B12,** which in turn project to the frontal cortex to establish an indirect association between the hypothalamus and the frontal cortex. The *pallidohypothalamic fasciculus* runs from the pallidum to the tuberal nuclei (ventromedial nucleus).

Commissures. The commissures which lie in the hypothalamic region contain almost no fibers from hypothalamic nuclei: fibers from the midbrain and the pons cross in the dorsal (Ganser) and ventral (Gudden) supra-optic commissures, and fibers from the subthalamic nuclei cross in the supramamillary commissure.

Fiber Connections of the Well Myelinated Hypothalamus C

The predominantly afferent pathways to the mamillary body **C10** are the **fornix C8** whose fibers mainly end in the mamillary body and the **peduncle of the mamillary body C13** from the tegmental nuclei of the midbrain, which supposedly contains taste and vestibular fibers and connections from the medial lemniscus.

Mainly efferent fibers comprise the **mamillothalamic tract** (Vicq d'Azyr) **C14** (p. 162 B31), which ascends to the *anterior thalamic nucleus* **C15.** A connection between the hypothalamus and the limbic cortex is produced by projection of the anterior thalamic nucleus to the cingulate gyrus. The **mamillotegmental fasciculus C16** ends in the tegmental nuclei of the midbrain. All the afferent and efferent tracts contain a high proportion of peptidergic fibers.

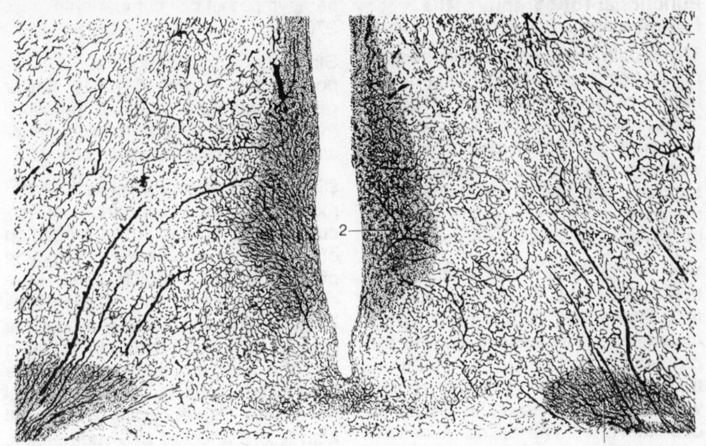

A Vascular supply of the hypothalamus in the rhesus monkey (after Engelhardt)

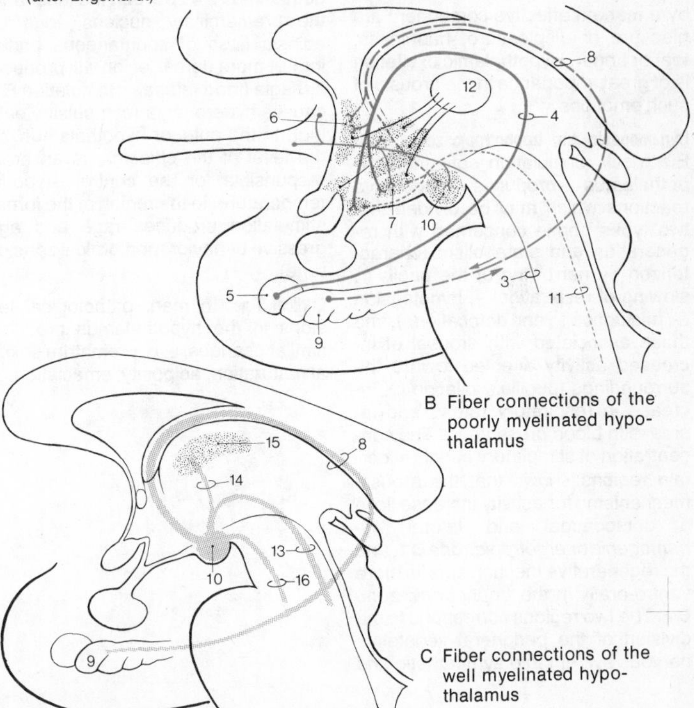

B Fiber connections of the poorly myelinated hypothalamus

C Fiber connections of the well myelinated hypothalamus

Functional Topography of the Hypothalamus

The hypothalamic centers influence all processes important for the maintenance of homeostasis and regulate the functions of the organs responding to momentary bodily stress: warmth, water and electrolyte balance, cardiac function, circulation and respiration, metabolism and the sleeping and waking rhythm are under their control.

Vital functions, such as eating, gastrointestinal activity and defecation, fluid intake and micturition are regulated from this region, just as are procreation and sexuality. Vital functions are initiated by bodily needs, which are perceived as hunger, thirst or sexual drive. The instinctive actions which allow the organism to survive and which are usually accompanied by a marked affective component are pleasure or displeasure, happiness, fear or anger. Hypothalamic excitation is of great importance in the arousal of such emotions.

Dynamogenic and trophotropic zones AB. Electrical stimulation of the hypothalamus produces vegetative reactions which may be divided into two types: those concerned with regeneration and metabolism (characterized by narrowing of the pupils ◊, slowing of respiration ~, hypotension ∪, micturition ⸾, and defecation ⁝), and those associated with arousal of increased activity directed toward the surroundings (pupillary dilation ○, increase in respiratory rate ⩘, and increase in blood pressure ∩). The concentration of stimulatory effects in certain regions shows that the arousal mechanisms for activity increase lie in a dorsocaudal and lateral *dynamogenic* or *ergotropic zone* **B1,** and the regenerative mechanisms lie more ventro-orally in the *trophotropic zone* **B2.** The two regions correspond to the division of the peripheral vegetative nervous system into sympathetic and parasympathetic components (p. 270).

Stimulation of the caudal hypothalamus in man **C** gives similar results: increase in blood pressure (red), pupillary dilation (blue), and increase in the respiratory rate (/ / /).

Stimulation and lesion experiments have demonstrated the importance of circumscribed regions in the regulation of certain vital processes: destruction of the tuber in the region of the infundibular nucleus **D3** in young animals leads to atrophy of the reproductive glands. Lesions between the chiasm and the tuber in infant rats, on the other hand, produce premature sexual maturation. Estrus cycling and sexual activity are affected by hypothalamic destruction. Lesions in the caudal hypothalamus **E4** between the tuber and the premamillary nucleus, lead to adipsia (loss of spontaneous drinking). A more dorsal lesion will produce aphagia (food refusal). Stimulation **F5** causes hyperphagia (compulsive eating). In the anterior hypothalamus, at the level of the chiasma, is an area responsible for the control of body temperature. In the region of the fornix stimulation produces rage and aggressive behavior (perifornical zone of rage).

Clinical Tips. In man, pathological lesions in the hypothalamus produce similar changes, e. g. premature sexual maturation, adiposity, emaciation.

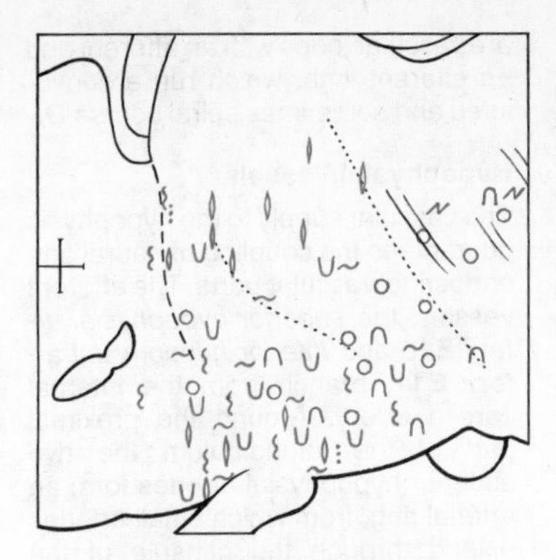

A Trophotropic and ergotropic zones of the diencephalon in the cat, stimulation chart (after Hess)

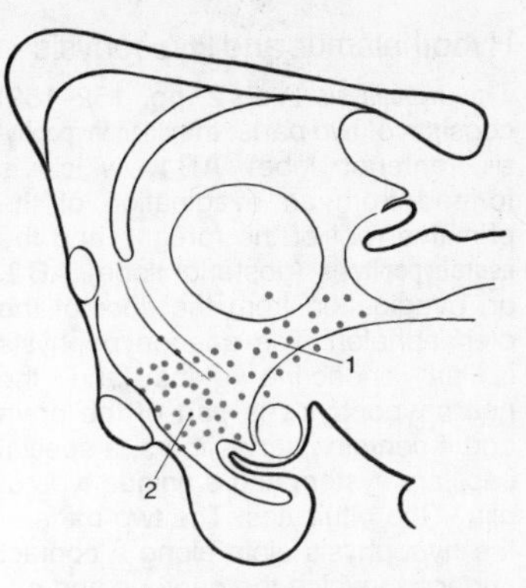

B The trophotropic and ergotropic zones in the cat, schematic

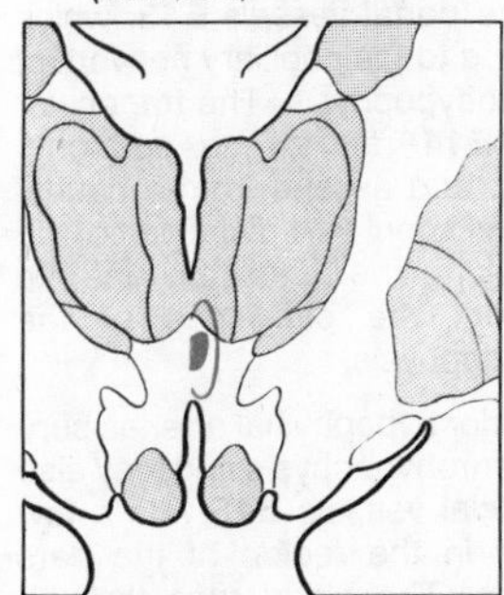

C Effects of stimulation in the caudal hypothalamus, man (after Sano, Yoshioka, Ogishiwa et al.)

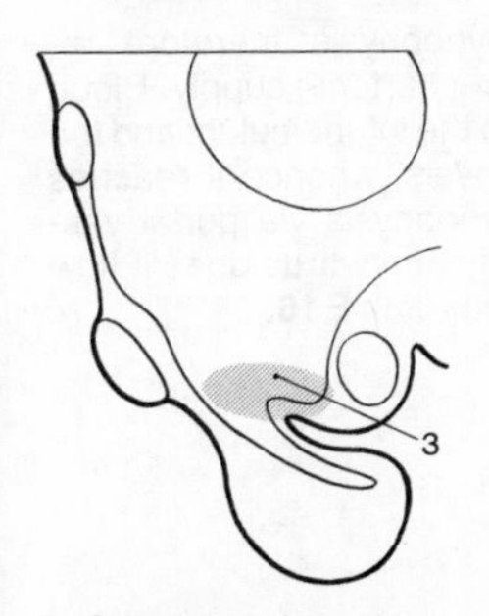

D Gonadotropic region in the cat (after Sawyer and Kawakami)

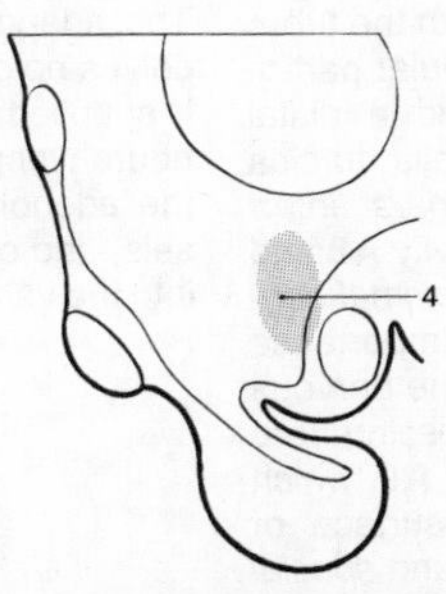

E Adipsia, experimental lesion in the rat (after Stevenson)

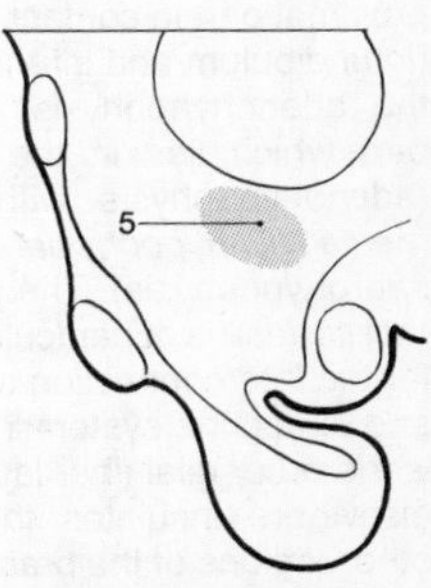

F Hyperphagia, stimulation experiment, cat (Brügge)

Hypothalamus and Hypophysis

The **hypophysis** (Vol. 2, pp. 152–159) consists of two parts: the **adenohypophysis** (anterior lobe) **AB1,** which is formed from an evagination of the primitive roof of the foregut, and the **neurohypophysis** (posterior lobe) **AB2,** an evagination from the floor of the diencephalon. The adenohypophysis is an endocrine gland and the neurohypophysis is part of the brain and it contains nerve fibers, a special capillary system and a unique type of glia – the pituicytes. The two parts of the hypophysis ajoin along a contact surface in which the nervous and endocrine-vascular systems are brought into contact with each other.

Infundibulum

(Hypophysial stalk) **A3.** The tuber cinereum tapers distally into the *fundibulum,* with the *infundibular recess* **A4.** The funnel shaped descent of the infundibulum where there is contact between the nervous and the endocrine system (neurohaemal zone) is also called the *median eminence of the tuber.* The adenohypophysis extends to the tuber as a thin layer that covers the anterior surface of the infundibulum (*infundibular part of the adenohypophysis* **A5**), and with a few islets of tissue **A6,** also its dorsal aspect. Thus, we can distinguish a proximal part in contact with the tuber (infundibulum and infundibular part of the adenohypophysis) and a distal part which lies in the sella turcica (adenohypophysis with *pars intermedia* **A7,** *hypophysial cavity* **A8** and neurohypophysis). The *proximal contact surface* is of particular importance for the interconnection of the nervous and endocrine systems. Missing here is the outer glial fiber layer **A9,** which elsewhere insulates the surface of other regions of the brain, and *special vessels* **C** enter from the adenohypophysis into the infundibulum. These are vascular loops with an afferent and an efferent limb, which run a convoluted and sometimes spiral course **D.**

Hypophysial Vessels

The vascular supply to the hypophysis guarantees the coupling of neural and endocrine-vascular parts. The afferent vessels, the *superior hypophysial artery* **E10** and *inferior hypophysial artery* **E11,** branch from the internal carotid artery. Around the proximal part of the infundibulum the two superior hypophysial arteries form an arterial ring, from which small arteries extend through the capsule of the adenohypophysis into the infundibulum and divide to form *special arteries* **E12.** Their efferent limbs collect in the *portal vessels* **E13,** which carry blood to the capillary network of the adenohypophysis. The trabecular arteries **E14** run to the adenohypophysis, and ascend from a caudal direction to supply the distal part of the infundibulum. The blood then runs into veins from the capillaries of the adenohypophysis.

Both inferior hypophysial arteries supply the neurohypophysis and they also form special vessels **E15** with a few branches in the region of the pars intermedia. The blood runs through short portal vessels into the sinusoids of the hypophysis.

The adenohypophysis, therefore, receives no direct arterial supply. Blood is supplied to the infundibulum and the neurohypophysis, whence it reaches the adenohypophysis via portal vessels, and only afterwards does it flow into the venous limb **E16.**

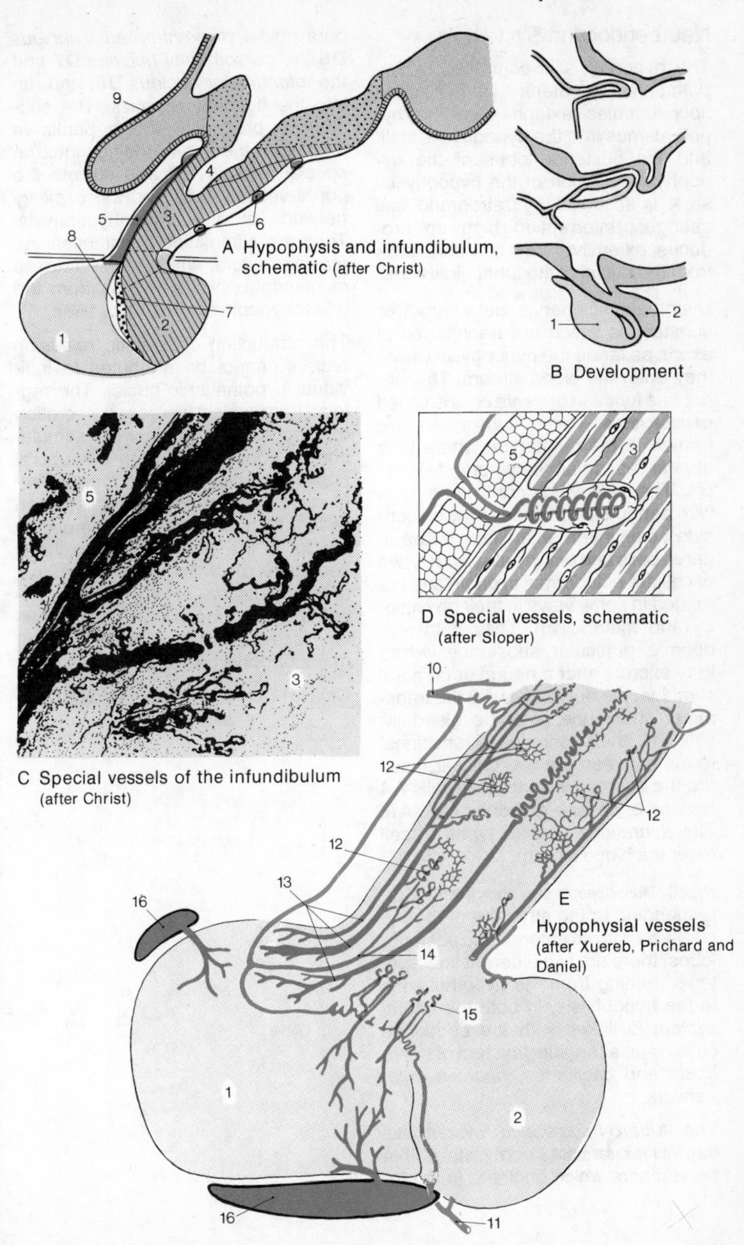

A Hypophysis and infundibulum, schematic (after Christ)

B Development

C Special vessels of the infundibulum (after Christ)

D Special vessels, schematic (after Sloper)

E Hypophysial vessels (after Xuereb, Prichard and Daniel)

Neuroendocrine System

The hypophysis is controlled by hypothalamic centers. Unmyelinated fiber bundles extend from the hypothalamus into the hypophysial stalk and the posterior lobes of the hypophysis. Section of the hypophysial stalk is followed by retrograde cell changes; interruption high up produces extensive loss of nerve cells from the nuclei of the tuber cinereum.

Hypothalamic nerve cells produce substances which are transported in axons as far as the hypophysis where they enter the blood stream. This endocrine function of nerve cells is called **neurosecretion.** These substances are formed in the perikaryon where they appear as small secretory droplets **B1.** The cells which are true neurons with dendrites and axons may be considered an intermediate form between nerve cells and gland cells. Both types of cell are ectodermal in origin and are related in some ways in their physiology and metabolism. They both produce a particular substance, which they excrete after a neural or humoral stimulus: the nerve cells **A2** the transmitter substance, and the glandular cells **A3** their secretion. Transitional forms between the two cellular types are the neurosecretory nerve cells **A4** and the endocrine glandular cells **A5.** The secretions of both types of cell enter the blood stream.

Hypothalamo-hypophysial fiber tracts. Corresponding to the structure of the hypophysis with its anterior and posterior lobes, there are two different fiber systems running from the hypothalamus to the hypophysis. In both the neural system is linked with the endocrine cells by an alternating system of nerve fibers and capillaries, *neurovascular network.*

The *tubero-infundibular system,* **tubero-infundibular tract D** consists of thin nerve fibers which originate in the tuberal nuclei, the *ventromedial nucleus* **D6** the *dorsomedial nucleus* **D7** and the *infundibular nucleus* **D8,** and run into the hypophysial stalk. The substances produced in the perikarya pass from the nerve endings into the special vessels **D9** and so into the portal vessels **D10** and the *capillary network* of the adenohypophysis. They are stimulating substances, *releasing factors,* which cause release of glandotrophic hormones from the adenohypophysis (Vol. 2, p. 153).

The production of specific *releasing factors* cannot be attributed to individual hypothalamic nuclei. The regions from which increased secretion may be elecited by electrical stimulation do not correspond to the tuberal nuclei. Stimulation of the preoptic region **C11** causes increased secretion of luteotrophic hormone. Stimulation caudal to the chiasm **C12** results in secretion of thyrotrophic hormone, and stimulation of the ventral hypothalamus (tuber cinereum to mamillary recess) **C13** causes secretion of gonadotrophic hormone.

Chiasm **CD14,** mamillary body **CD15.**

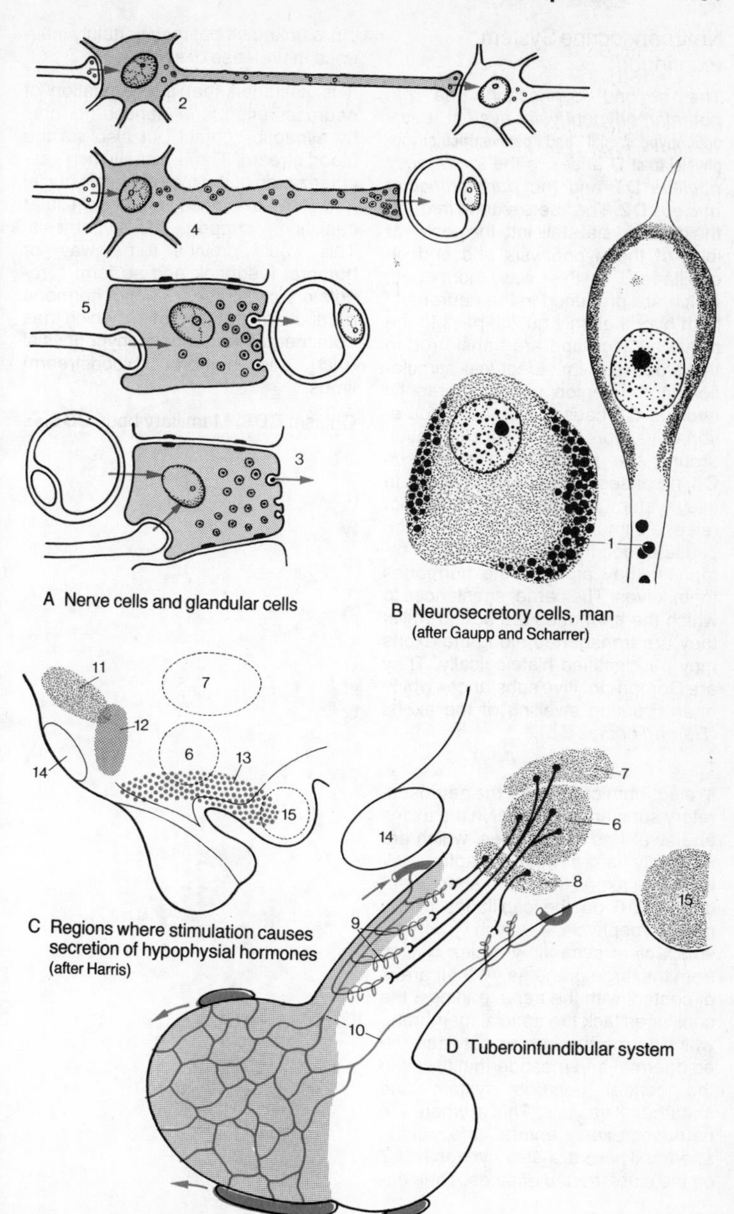

A Nerve cells and glandular cells

B Neurosecretory cells, man (after Gaupp and Scharrer)

C Regions where stimulation causes secretion of hypophysial hormones (after Harris)

D Tuberoinfundibular system

Neuroendocrine System (continued)

The second connection, the *hypothalamohypophysial system,* **supraopticohypophysial and paraventriculohypophysial tract D** arises in the *supra-optic nucleus* **D1** and the *paraventricular nucleus* **D2.** The fibers extend through the hypophysial stalk into the posterior lobe of the hypophysis and end on capillaries. In this way, hormones, which are produced in the neurons of both hypothalamic nuclei, pass to the nerve endings and are transferred to the blood stream. Electrical stimulation in the region of the supraoptic nucleus **C3** causes increased secretion of vasopressin, whilst stimulation around the paraventricular nucleus **C4** increases secretion of oxytocin. In this system the nerve cells do not release stimulating substances which cause endocrine cells to secrete hormones; they produce the hormones themselves. The carrier substances to which the hormones are bound whilst they are transported along the axons may be identified histologically. They are Gomori-positive substances which often produce swelling of the axons (*Herring bodies* **B5**).

In electronmicrographs, the neurosecretory substances appear in the axons and swellings as granules, which are markedly larger than synaptic vesicles. The axons form club-shaped endings **AD6** on the capillaries of the neurohypophysis, which contain small, clear synaptic vesicles distinct from the large granules. At their areas of contact with the nerve endings, the capillaries lack the special membrane which forms the border between the ectodermal and mesodermal tissue in the central nervous system and sheathe all vessels. This is where the neurosecretion enters the blood stream. There are also synapses **A7** on the club-shaped endings. Their origin is unknown but they certainly influence the release of secretions.

It is assumed that the regulation of neurosecretion is influenced not only by synaptic contact but also via the blood stream. The unusually rich vascularization of the hypothalamic nuclei and the existence of endocellular capillaries support this hypothesis. This would provide a pathway for humoral feedback and so form a regulatory circuit for control of hormone production and secretion which has both neural (supraopticohypophysial tract) and humoral (bloodstream) limbs.

Chiasm **CD8.** Mamillary body **CD9.**

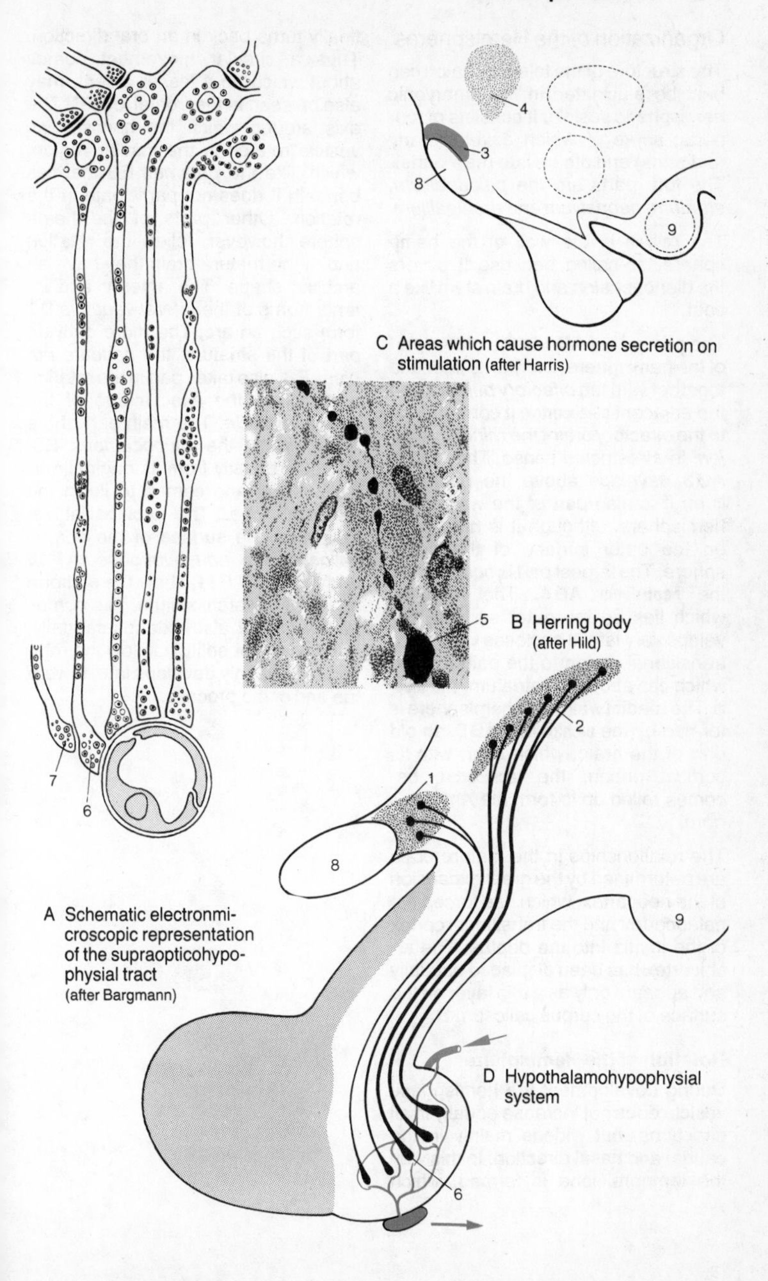

A Schematic electronmicroscopic representation of the supraopticohypophysial tract (after Bargmann)

B Herring body (after Hild)

C Areas which cause hormone secretion on stimulation (after Harris)

D Hypothalamohypophysial system

Organization of the Hemispheres

The structure of the telencephalon can best be elucidated in the embryonic hemispheric vesicle. It consists of four parts, some of which develop early (old parts) and others late (new parts). The four parts are the *paleopallium, striatum, neopallium* and *archipallium.*

The **pallium** is the wall of the hemisphere, so-called because it covers the diencephalon and brain stem like a coat.

The **paleopallium AB1** is the oldest part of the hemisphere. It forms its floor and together with the *olfactory bulb* **A2** and the adjacent **paleocortex,** it corresponds to the olfactory brain, the *rhinencephalon,* in a restricted sense. The **striatum AB3** develops above the paleopallium. It is also part of the wall of the hemisphere, although it is not visible on the outer surface of the hemisphere. The largest part is occupied by the **neopallium AB4.** The **neocortex,** which lies on the outer surface, develops very late. It encloses ventrally a transitional region to the paleocortex, which lies above the striatum, the **insula.** The medial wall of the hemisphere is formed by the **archipallium AB5,** an old part of the brain, which later, with its cortical ribbon, the **archicortex,** becomes rolled up to form the *Ammon's horn.*

The relationships in the mature brain are determined by the great expansion of the neocortex, which has forced the paleocortex and the transitional cortex of the insula into the depths. The archicortex has been displaced caudally and appears only as a thin layer on the surface of the corpus callosum **B.**

Rotation of the Hemisphere

During development the hemispheric vesicle does not increase equally in all directions, but widens mainly in the caudal and basal direction. In this way the temporal lobe is formed, which finally turns back in an oral direction. Thus a circular movement comes about which, to a lesser extent, may also be seen in the frontal lobe **C.** The axis around which the hemispheric vesicle rotates is the insular region, which, like the *putamen* **E6** that lies beneath it does not participate in the rotation. Other parts of the hemisphere, however, follow the rotation and in the mature brain they have an archlike shape. The anterior and inferior horns of the *lateral ventricle* **D7** form such an arc. The more outlying part of the striatum, the *caudate nucleus* **E8,** also takes part in the rotation and follows the exact curve of the lateral ventricle. The main part of the *archipallium,* the *hippocampus* **E9,** migrates basally from its original dorsal position and comes to lie in the temporal lobe. The archipallial remnant on the surface of the corpus callosum, the *induseum griseum* **F10** and the *fornix* **F11,** show the arciform shape of the archipallium. The *corpus callosum* **F12** also extends caudally, but it does not entirely follow the rotation since it only develops late, toward the end of the process.

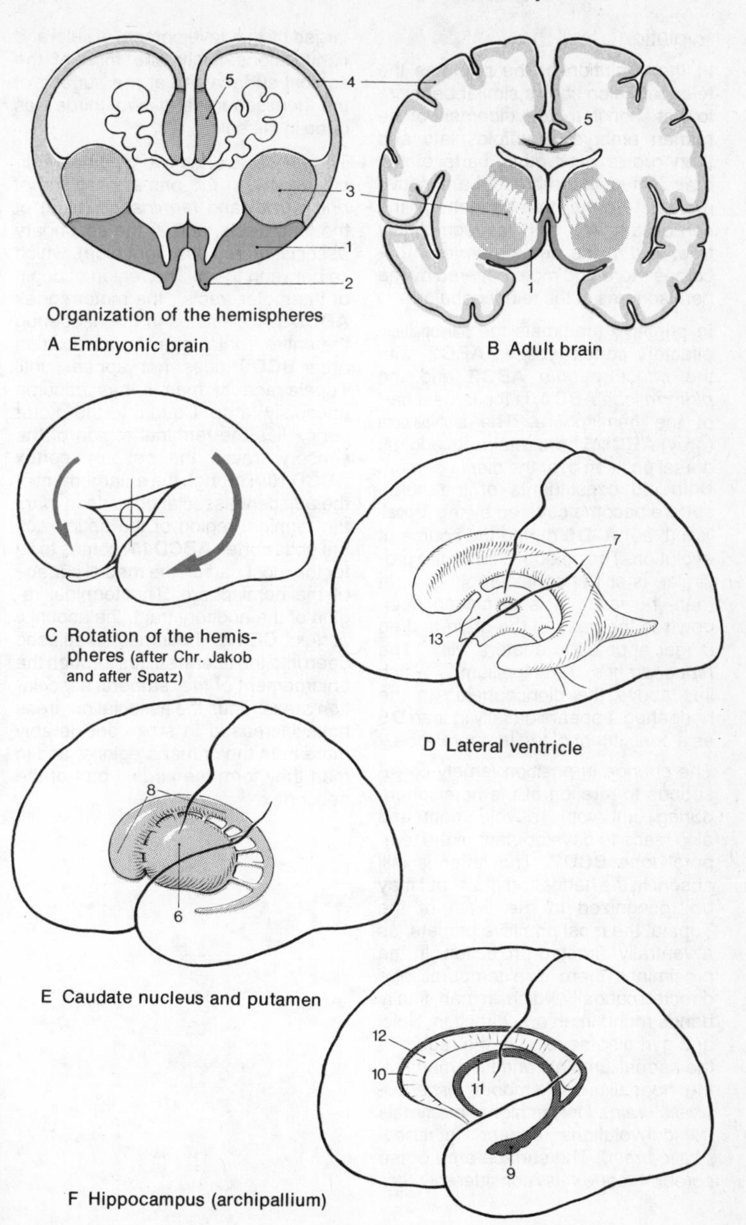

Organization of the hemispheres

A Embryonic brain

B Adult brain

C Rotation of the hemispheres (after Chr. Jakob and after Spatz)

D Lateral ventricle

E Caudate nucleus and putamen

F Hippocampus (archipallium)

Evolution

In the evolution of the primates the telencephalon shows similar behavior to that seen in the development of the human embryo: it unfolds late and then grows over other parts of the brain. Thus, in the brain of a primitive mammal (e. g. the hedgehog) the cerebellum **A1** still lies completely free, and in the primate series it becomes more and more covered by the hemispheres of the telencephalon.

In primitive mammals the **paleopallium** olfactory cortex (yellow) **ABC2,** with the *olfactory bulb* **ABC3** and the *piriform lobe* **ABC4,** is the largest part of the hemisphere. The **archipallium** (grey) **ABCD5** here retains its original dorsal position over the diencephalon. Both old constituents of the hemisphere become covered by the **neopallium** (blue) **A–D6** during the course of evolution. The paleopallium of the prosimiae is still of considerable size. In man, however, it is displaced deep down to the base of the brain and no longer appears in a lateral view. The *hippocampus* (archipallium), which lies above the diencephalon in the hedgehog, appears basally in man **D5** as a constituent of the temporal lobe.

The change in position largely corresponds to rotation of the hemisphere during embryonic development and also leads to development of the temporal lobe **BCD7.** The latter is still absent in the hedgehog brain, but may be recognized in the brain of the Tupaia, the most primitive primate, as a ventrally directed projection. In the prosimians there is a temporal lobe directed caudally which, in man, finally bends round in an oral direction. Sulci and gyri also develop in the region of the neopallium. In a primitive mammal the neopallium is smooth *(lissencephalic brain)*. Only in higher mammals do convolutions appear *(gyrencephalic brain)*. The surface area of the cerebral cortex is considerably enlarged by the development of sulci and convolutions. Only one third of the cortical surface lies at the surface of the hemisphere and two thirds lies deep in the sulci.

The neocortex has two types of cortical region: 1) the primary *regions of origin* (red) and *termination* (blue) of the long tracts, and 2) the secondary *association regions* (light blue), which lie between them. The region of origin of the motor tracts , the motor cortex **ABCD8,** constitutes in the hedgehog the entire frontal lobe. An association area **BCD9** does not appear until Tupaia and in man it has become unusually large. Caudal to the motor cortex lies the terminal region of the sensory tracts, the sensory cortex **ABCD10.** Through the enlargement of the adjacent association zone in man, the terminal region of the optic tract, the optic cortex **ABCD11,** comes to lie for the most part on the medial surface of the hemisphere. The terminal region of the auditory tract, the acoustic cortex **CD12,** becomes displaced deep into the lateral sulcus through the enlargement of the temporal association areas. Thus the association areas have increased in size considerably more than the primary regions, and in man they form the largest part of the neocortex.

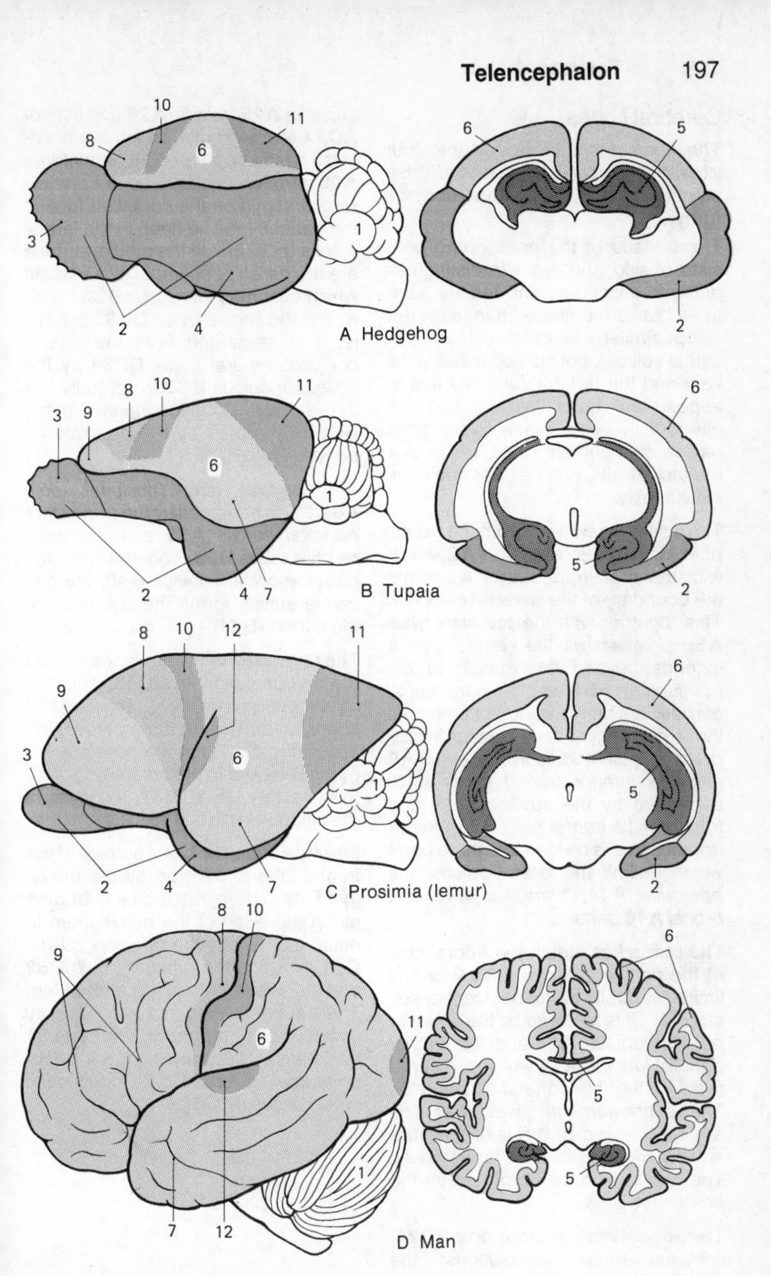

Evolution of the telencephalon (according to Edinger, Elliot Smith, Le Gros Clark)

Cerebral Lobes

The hemisphere is divided into four **lobes:** the *frontal* (red), *parietal* (blue striped), *temporal* (blue) and *occipital* (grey).

The surface of the hemisphere consists of **sulci** and **gyri**. We distinguish *primary, secondary* and *tertiary sulci.* In all brains the first primary sulci develop similarly (central sulcus, calcarine sulcus), but the secondary sulci vary and the tertiary sulci, the last to appear, are quite irregular and are different in every brain. Every brain has its own surface relief, which, like the facial traits, is an expression of its individuality.

The **frontal lobe** extends from the *frontal pole* **AC1** to the **central sulcus A2,** which with the *precentral sulcus* **A3** forms the boundary of the **precentral gyrus A4.** This together with the **postcentral gyrus A5,** is known as the **central region.** It extends beyond the *margin of the hemisphere* **AB6** onto the *paracentral gyrus* **B7.** The frontal lobe comprises three major gyri: the *superior frontal gyrus* **A8,** the *middle frontal gyrus* **A9** and the *inferior frontal gyrus* **A10,** separated by the *superior* **A11** and *inferior* **A12** *frontal sulci.* The inferior frontal gyrus is divided into three parts which delimit the lateral sulcus: the *opercular* **A14,** *triangular* **A15** and *orbital* **A16** *parts.*

The **parietal lobe** meets the *frontal lobe* at the postcentral gyrus A5 which is limited caudally by the *postcentral sulcus* **A17.** It is followed by the *superior parietal lobulus* **A18** and the *inferior parietal lobulus* **A19** which are separated by the *intraparietal sulcus* **A20.** The *supramarginal gyrus* **A21** lies around the end of the lateral fissure and the **angular gyrus A22** lies ventrally. The medial surface is formed by the *precuneus* **B23.**

The **temporal lobe** (*temporal pole* **AC24**) contains three convolutions: the *superior* **A25,** *middle* **A26** and *inferior* **AC27** *temporal gyri,* which are separated by the *superior* **A28** and *inferior* **A29** *temporal sulci.* The *transverse temporal gyri* on the dorsal surface of the parietal lobe lie deep in the lateral sulcus (p. 222). On the medial surface are the *parahippocampal gyrus* **BC30** which becomes the *uncus* **BC31** orally, and the *lingual gyrus* **BC32** caudally. It is separated from the *medial occipitotemporal gyrus* **BC34** by the *collateral sulcus* **BC33.** Ventrally lies the *lateral occipitotemporal gyrus* **BC35** separated by the *occipitotemporal sulcus* **BC36.**

The **occipital lobe** (*occipital pole* **ABC37**) is traversed by the *transverse occipital sulcus* **A38** and the deep **calcarine sulcus B39.** Together with the *parieto-occipital sulcus* **B40,** the calcarine sulcus forms the boundary of the *cuneus* **B41.**

The **cingulate gyrus B42** extends around the corpus callosum **BC43.** The *hippocampal sulcus* **B44** separates it caudally from the *dentate gyrus* **B45,** and orally it ends in the *paraterminal gyrus* **B46** and in the *subcallosal area* (parolfactory area) **B47.** *Isthmus of the cingulate gyrus* **B48.**

Base of the Brain. The basal surface of the frontal lobe is covered by the *orbital gyri* **C49.** The *gyrus rectus* **C50** runs along the edge of the hemisphere limited laterally by the *olfactory sulcus* **C51** in which the *olfactory bulb* **C52** and the *olfactory tract* are embedded. The tract is divided into two *olfactory striae,* which surround the *anterior perforated substance* **C53.** *Hippocampal sulcus* **C54,** *longitudinal cerebral fissure* **C55.**

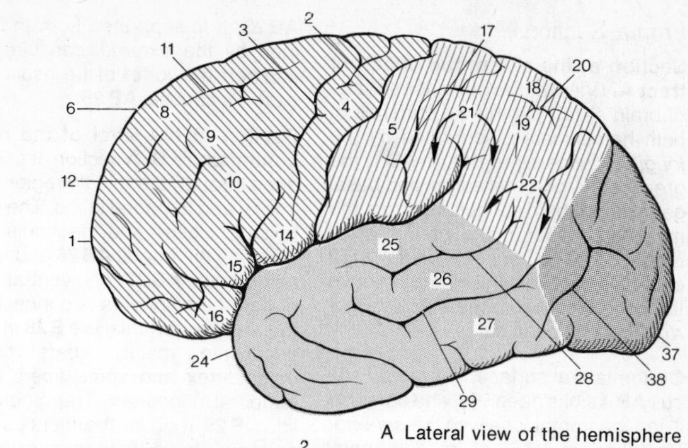

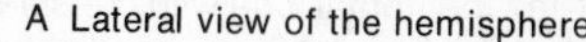
A Lateral view of the hemisphere

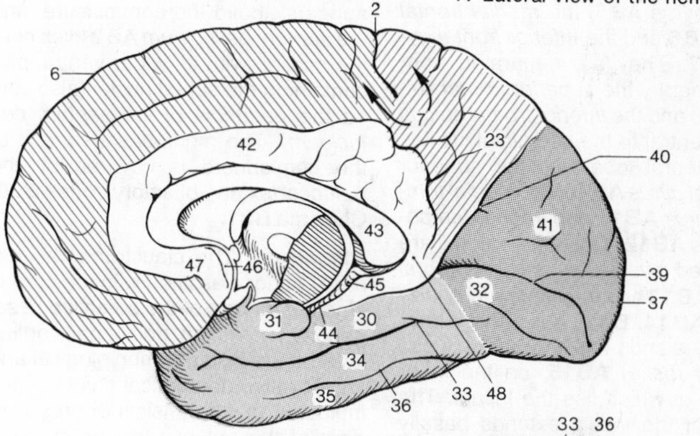

B Medial view of the hemisphere

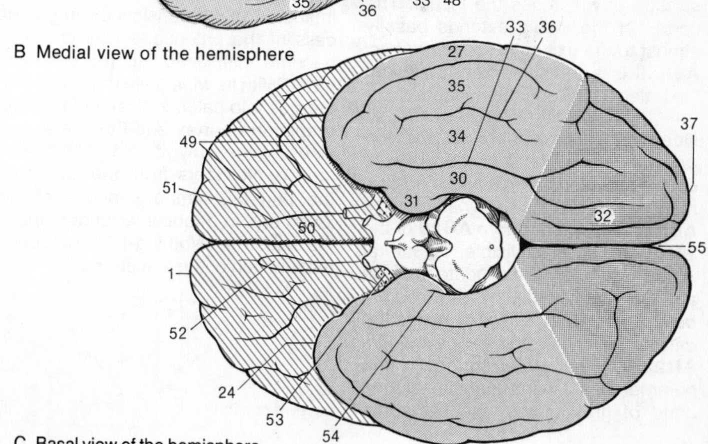

C Basal view of the hemisphere

Frontal Sections

Section at the exit of the olfactory tract A. (View of the posterior surface of brain slice.) On the cut surfaces of both hemispheres, separated by the *longitudinal cerebral fissure* **AB 1,** the grey matter, cerebral cortex and basal ganglia can be distinguished from the myelinated fiber mass of the white matter. The *corpus callosum* **AB 2** connects the two hemispheres. Above the corpus callosum the *cingulate gyrus* **AB 3** has been cut.

On the lateral surface, the *lateral sulcus* **AB 4** sinks deeply in, and dorsal to it lies the frontal lobe with the *superior frontal gyrus* **AB 5,** the *medial frontal gyrus* **AB 6** and the *inferior frontal gyrus* **AB 7.** They are separated from each other by the *superior frontal sulcus* **AB 8** and the *inferior frontal sulcus* **AB 9.** Ventral to the lateral sulcus lies the temporal lobe with the *superior temporal gyrus* **AB 10,** the *medial temporal gyrus* **AB 11** and *inferior temporal gyrus* **AB 12**. The temporal gyri are separated by the *superior temporal sulcus* **AB 13** and the *inferior temporal sulcus* **AB 14.** Deep down the lateral sulcus expands into the *lateral fossa (Sylvian fossa)* **AB 15,** on the inner surface of which lies the insula. The cortex of the insula extends basally almost to the exit of the *olfactory tract* **A 16.** It is a transition zone between the paleocortex and the neocortex.

Deep in the hemisphere lies the *corpus striatum,* which is divided into the *caudate nucleus* **AB 18** and the *putamen* **AB 19** by the *internal capsule* **AB 17.** The *anterior horn* **AB 20** of the lateral ventricle is visible. Its lateral wall is formed by the caudate nucleus and its medial wall by the *septum pellucidum* **AB 21,** which contains the *cavity of the septum pellucidum* **AB 22.** On the lateral surface of the putamen lies a narrow encapsulating band of grey matter, the *claustrum* **AB 23.** It is separated from the putamen by the *external capsule* **AB 24** and from the cortex of the insula by the *extreme capsule* **AB 25.**

Section at the level of the rostral commissure B. A section at this level goes through the middle region of the frontal and temporal lobes. The lateral fossa is closed and the insula is covered by the frontal **AB 26** and temporal opercula **AB 27.** The ventral region of both hemispheres is connected by the *anterior commissure* **B 28** in which decussate mainly fibers of the paleocortex and some fibers of the temporal neocortex. The globus pallidus **B 29** (part of the diencephalon) appears above the commissure, and the *septum pellucidum* **AB 21** lies near the midline with its wider ventral part containing the septal nuclei (also called the *peduncle of the septum pellucidum).* The mediobasal surface of the hemisphere is covered by the *paleocortex,* the olfactory cortex **B 30**. Chiasma **B 31.**

Claustrum. The claustrum was formerly either classified with the corpus striatum as part of the so-called basal ganglia, or as an additional cortical layer of the insula. Embryological and comparative anatomical studies have indicated that it consists of groups of cells of the paleocortex, which have become displaced during development. With its wide base the claustrum merges into paleocortical regions (the prepiriform cortex and the lateral nucleus of the amygdaloid). Unmyelinated nerve fibers from the cortex of the parietal, temporal and occipital lobes end in a topical arrangement in the claustrum. Nothing is known about the function of the claustrum.

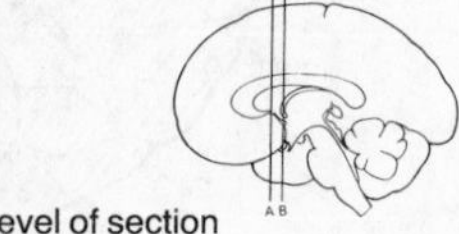

Level of section

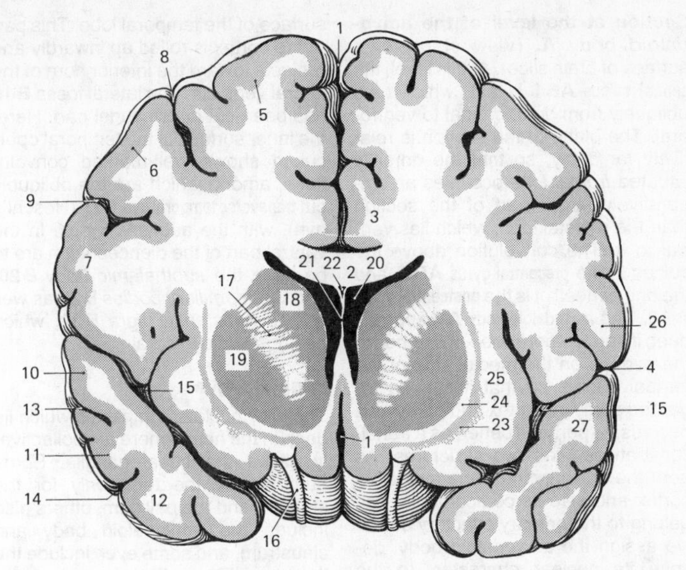

A Frontal section, emergence of the olfactory tract

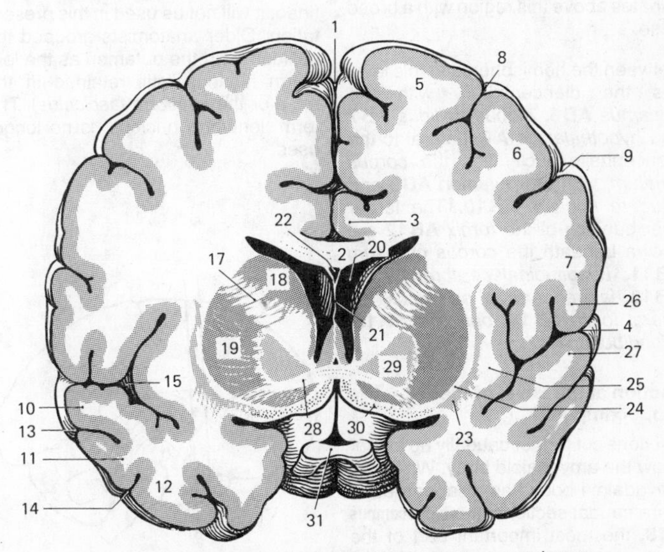

B Frontal section at the level of the anterior commissure

Section at the level of the amygdaloid body A. (View of posterior surface of brain slice). At this level, the **central sulcus AB1** is cut, which runs obliquely from dorsocaudal to ventro-oral. The plane of the section is relatively far orally, so that the dorsally situated *frontal lobe* occupies a comparatively larger part of the section than the parietal lobe, which lies ventral to it. The convolution above the sulcus is the **precentral gyrus AB2,** and the one beneath it is the **postcentral gyrus AB3.** The **amygdaloid body A4** appears deep in the temporal lobe. It extends to the surface on the medial side of the temporal lobe, so that it can be regarded partly as cortex and partly as a nucleus, or perhaps better, as a transition between the two structures. As both the surrounding periamygdaloid cortex and also its corticomedial half belong to the primary olfactory center we assign the amygdaloid body, despite its nuclear character, to the paleocortex. The *claustrum* **AB5** terminates above this region with a broad base.

Between the hemispheres at this level lies the diencephalon, with the *thalamus* **AB6,** *globus pallidus* **AB7** and *hypothalamus* **A8.** Lateral to the diencephalic nuclei lies the *corpus striatum,* with the *putamen* **AB9** and *caudate nucleus* **AB10.** The larger fiber bundles of the *fornix* **AB12** are shown beneath the *corpus callosum* **AB11.** *Longitudinal cerebral fissure* **AB13,** *lateral cerebral sulcus* **AB14,** *lateral fossa* **AB15,** *optic tract* **A16,** *infundibulum* **A17.**

Section at the Level of the Hippocampus B

Sections cut further caudally no longer show the amygdaloid body. When the amygdaloid body has disappeared in more caudal sections, the **hippocampus B18,** the most important part of the archicortex, appears on the medial surface of the temporal lobe. This part of the cortex is rolled up inwardly and projects toward the inferior horn of the lateral ventricle. The lateral fossa **B15** has been cut in its caudal part. Here, the inner surface of the temporal operculum shows well-marked convolutions, among which are the obliquely cut **transverse temporal gyri B19** (Heschl's gyri), with the auditory cortex. In the ventral part of the diencephalon are to be seen the *subthalamic body* **B20,** and the *mamillary bodies* **B21** as well as the *substantia nigra* **B22,** which belongs to the midbrain.

Basal Ganglia

The grey nuclear complexes which lie deep in the hemisphere are collectively called the basal ganglia. Some authors use the term only for the striatum and the pallidum, others also include the amygdaloid body and claustrum, and some even include the thalamus. Since the concept of the basal ganglia is vague and poorly defined, it will not be used in this presentation. Older anatomists grouped the pallidum and the putamen as the lentiform nucleus (still retained in the *ansa* or the *lentiform fasciculus).* The term 'lentiform nucleus' is no longer used.

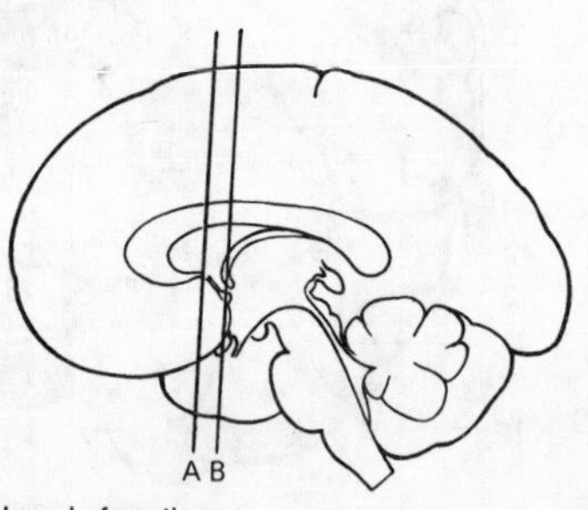

Level of section

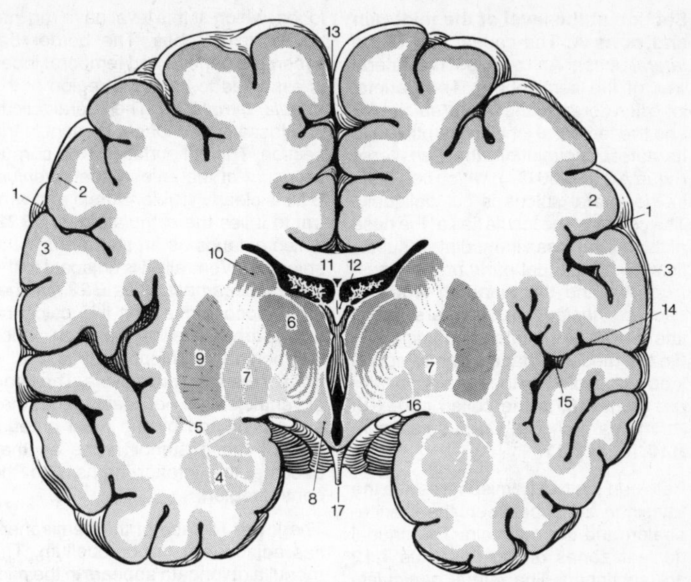

A Frontal section, level of the amygdaloid nucleus

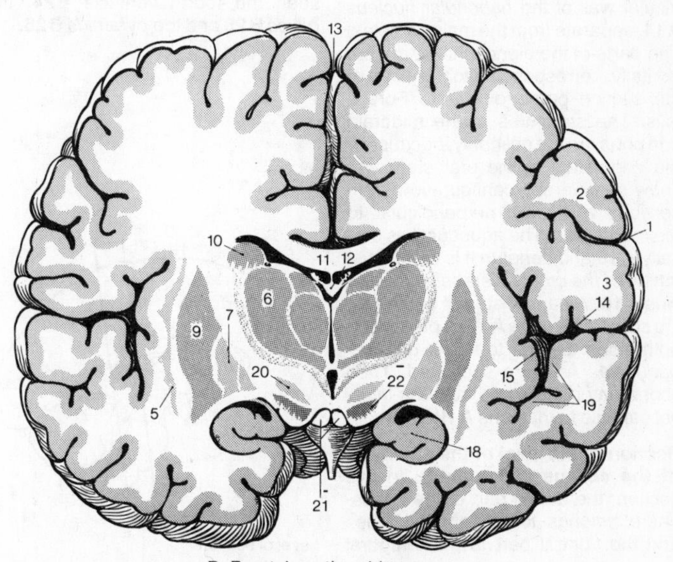

B Frontal section, hippocampus

Section at the level of the midbrain and pons A. The caudal part of the *lateral fossa* **A1** overlies the lateral wall of the hemisphere. The parietal lobe lies dorsal to the *lateral sulcus* **A2** and the temporal lobe lies ventral to it. Its dorsal convolution, the *transverse gyrus* **A3** (p. 236 C1), which lies deep in the lateral sulcus is cut obliquely. The cortex of the insula lies at the base of the lateral fossa immediately superficial to the caudal parts of the *claustrum* **A4** and the *putamen* **A5.** The *caudate nucleus* **A6** appears in the lateral wall of the lateral ventricle **A7.** On the medial surface of the temporal lobe, concealed by the *parahippocampal gyrus* **A8,** is the coiled cortex of *Ammon's horn* **A9.** Corpus callosum **A10**, fornix **A11.**

Between the two hemispheres lies the transition zone between the diencephalon and the midbrain. The caudal nuclear zones of the thalamus **A12** are sectioned. The *lateral geniculate body* **A13** lies medially on the ventricular wall of the *habenular nucleus* **A14,** separate from the main complex. The parts of the diencephalon are cut frontally, corresponding to the plane of the section perpendicular to Forel's axis. The structures of the midbrain and pons are cut obliquely, because of the angulation of the brainstem and show a different configuration from sections which are perpendicular to Meynert's axis. The aqueduct lies dorsally **A15** and beneath it is the *decussation of the cranial cerebellar peduncles* **A16.** A narrow strip of dark cells, the *substantia nigra* **A17,** is present on both sides. Lateral to it, the *cerebral peduncles* **A18** are sectioned. Then fibers may be followed from the internal capsule to the pons **A19.**

Section at the level of the splenium of the corpus callosum B. In this section, the dorsal part of the hemisphere belongs to the parietal lobes and the ventral part to the temporal lobes, which at this level have run into the occipital lobes. The border between the parietal and temporal lobes is assumed to be in the region of the *angular gyrus* **B20.** The lateral sulcus and fossa are no longer present in the section. The cut surface of the corpus callosum at the level of the *splenium* **B21** is clearly visible. Dorsal and ventral to it lies the *cingulate gyrus* **B22,** which extends as an arch around the splenium. Ventrally it is adjacent to the *parahippocampal gyrus* **B23.** Neither the hippocampus nor the calcarine sulcus are visible in the section, which lies behind the hippocampus and in front of the calcarine sulcus. The lateral ventricles are noticeably wide; they are constituted by the most anterior part of the posterior horns as they become the inferior horns and the central region.

The lower surface of the hemisphere lies adjacent to the cerebellum. The medulla oblongata appears in the middle and on its oblique section may be seen the fourth ventricle **B24,** the *olives* **B25** and the *pyramids* **B26.**

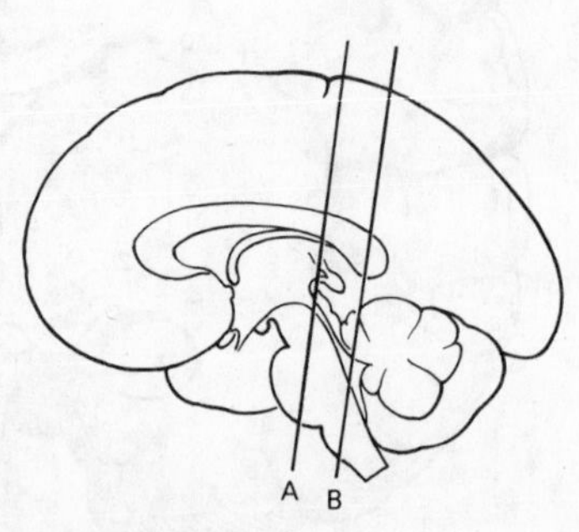

Level of section

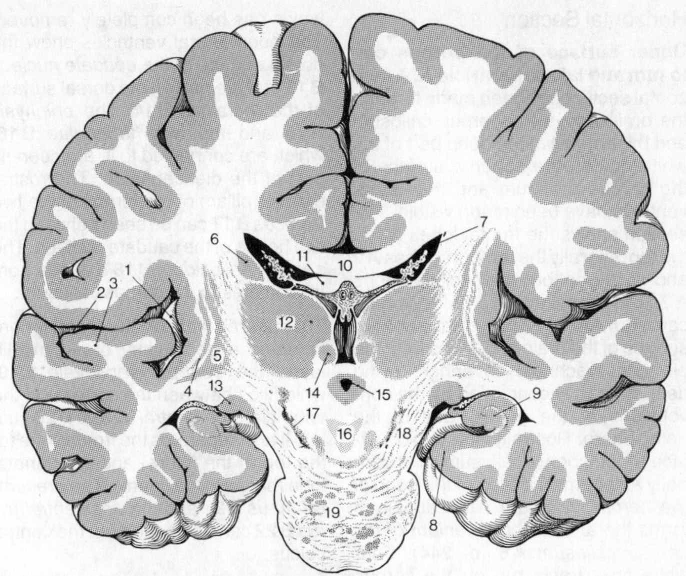

A Frontal section at the level of the midbrain and pons

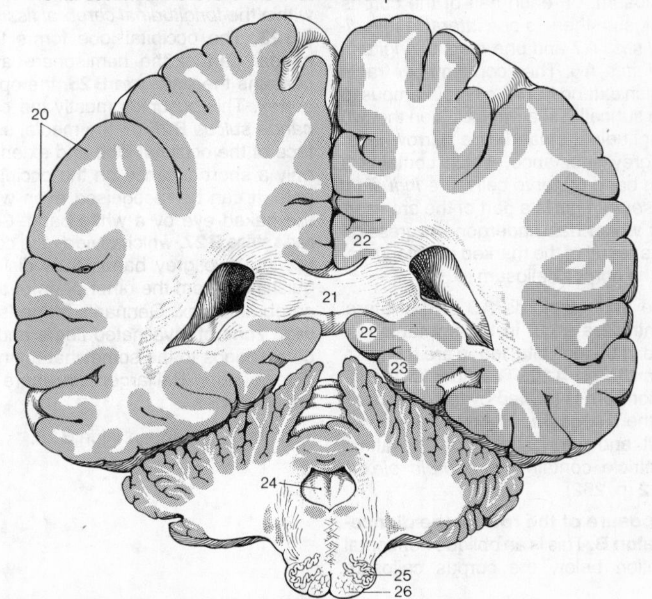

B Frontal section at the level of the splenium of the corpus callosum

Horizontal Section

Upper surface of the corpus callosum and lateral ventricle A. A horizontal section has been made through the brain above the corpus callosum and by removal of the deep part of the white matter, the superior surface of the corpus callosum and the lateral ventricle have been made visible. The section shows the frontal lobes **A1** at the top, laterally the temporal lobes **A2** and at the bottom the occipital lobes **A3.** The dorsal surface of the corpus callosum **A4** forms part of the superior surface of the brain and is free from the pia and arachnoid coverings. Lying deep down it is concealed by the convolutions of the medial wall of the hemisphere. Rostrally, the upper surface of the corpus callosum runs ventrally and forms the arch of the *genu of the corpus callosum* **A5;** caudally it forms the arch of the *splenium of the corpus callosum* **A6** (p. 244). Four white fiber tracts run on the corpus callosum: on each half of the corpus callosum there is one *lateral longitudinal stria* **A7** and one *medial longitudinal stria* **A8.** They contain fiber tracts which extend from the hippocampus to the subcallosal area. Between the two longitudinal striae lies a narrow layer of grey substance, which contains a thin band of nerve cells, the *indusium griseum.* This is a part of the archicortex which has undergone regression as a result of the marked development of the corpus callosum (p. 6).

The anterior horns **A9** of the lateral ventricle are cut in the frontal lobes, and in the occipital lobes are the posterior horns **A10.** The floor of the inferior horn is formed by the protrusion of the *hippocampus* **A11.** The central part and inferior horn of the lateral ventricle contain the *choroid plexus* **A12** (p. 262).

Exposure of the roof of the diencephalon B. This is an oblique horizontal section below the corpus callosum which has been completely removed. The open lateral ventricles show the dorsal surface of the *caudate nucleus* **B13** and medial to it the dorsal surface of the *thalamus* **B14.** The *epiphysis* **B15** and the two *habenulae* **B16,** which are connected to it, are seen as part of the diencephalon. The rostral course (pillars of the fornix) of the two *fornices* **B17** can be seen between the two heads of the caudate nucleus. The *septum pellucidum* **B18** extends from the fornices to the corpus callosum.

The lateral wall of the hemisphere contains a particularly broad myelin layer, the *centrum semiovale* **B19,** which lies between the cortex and the ventricle. The *central sulcus* **B20** cuts into it and separates the frontal lobe (at the top of the figure) and the parietal lobe (below in the figure). The *precentral gyrus* **B21** and the *postcentral gyrus* **B22** can be traced from the central sulcus.

The *cerebellum* **B24** is seen caudally within the *longitudinal cerebral fissure* **AB23.** The occipital lobe forms the caudal part of the hemisphere and contains the **striate area B25,** the optic cortex. This occupies mostly the *calcarine sulcus* **B26** on the medial surface of the occipital lobe and extends only a short distance on the occipital pole. It can be recognised even with the naked eye by a white band, **Gennari's stripe B27,** which divides the cortex into two grey bands thus distinguishing it from the other parts of the cerebral cortex. Gennari's stripe is a wide zone of myelinated fibers and it corresponds to the somewhat thinner outer band of Baillarger elsewhere in the neocortex.

Tectum of mesencephalon **B28.**

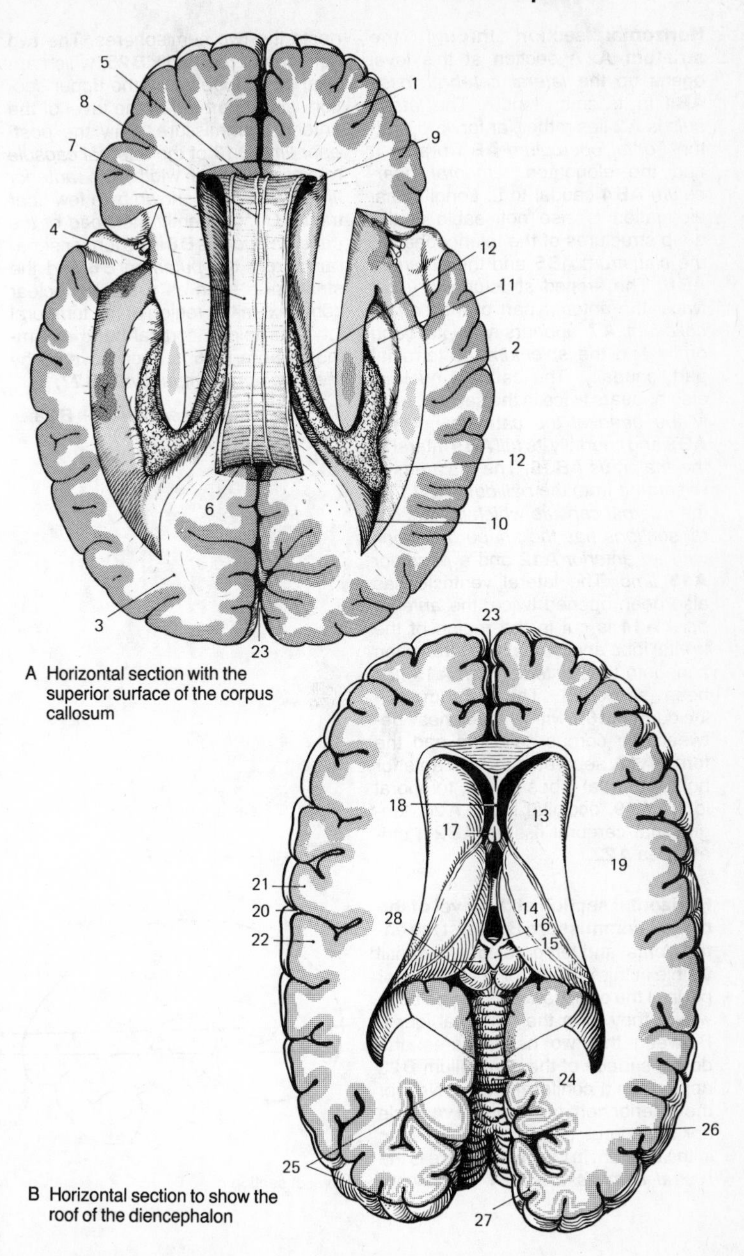

A Horizontal section with the superior surface of the corpus callosum

B Horizontal section to show the roof of the diencephalon

Horizontal section through the striatum A. A section at this level opens up the *lateral cerebral fossa* **AB 1** in its entire length. The *lateral sulcus* **A 2** lies rather far forward, with the *frontal operculum* **AB 3** oral to it and the elongated *temporal operculum* **AB 4** caudal to it. Longitudinal elongation is also noticeable in the deep structures of the telencephalon, the *claustrum* **AB 5** and the *putamen* **AB 6.** The arched structures are cut twice: the anterior part of the *corpus callosum* **A 7** appears with its *genu* orally and the *splenium,* its terminal part, caudally. The caudate nucleus also appears twice in the section, orally the *head of the caudate nucleus* **AB 8** and caudally its *tail* **AB 9** lateral to the *thalamus* **AB 10.** The thalamus is separated from the *pallidum* **AB 11** by the *internal capsule which in horizontal sections has the shape of a hook with an anterior* **A 12** and a *posterior* **A 13** *limb.* The lateral ventricle has also been opened twice: the *anterior horn* **A 14** is cut in the region of the frontal lobe and caudally the transition zone into the *posterior horn* **A 15** has been sectioned. The septum pellucidum **A 16,** which stretches between the corpus callosum and the fornix **A 17,** separates the two anterior horns. Frontal lobes **AB 18,** temporal lobes **A 19,** occipital lobes **A 20,** longitudinal cerebral fissure **AB 21,** striate area **A 22.**

Horizontal section at the level of the rostral commissure B. Whilst the entire frontal and temporal lobes are still visible in this section, only the anterior parts of the occipital lobes can be seen where they form the temporal lobes. Between the two hemispheres, the dorsal surface of the cerebellum **B 23** appears in a coniform shape. Neither the anterior horn of the lateral ventricle nor the corpus callosum can be seen in the section. In their place there is the *rostral commissure* **B 24** which connects the two hemispheres. The two *columns of the fornix* **B 25,** which appear close together in the upper section, are separated at the level of the anterior commissure. Only the *posterior limb* **B 13** of the *internal capsule* retains its usual width, the *anterior limb* **B 12** is only shown by a few fiber tracts. Consequently the *head of the caudate nucleus* **B 8** is no longer separated from the *putamen* **B 6** and the striatum is seen as a single nuclear complex. In the region of the temporal lobe the folded cortical band of Ammon's horn **B 26** is seen covered by the *parahippocampal gyrus* **B 27.**

Tectum of the mesencephalon **B 28.**

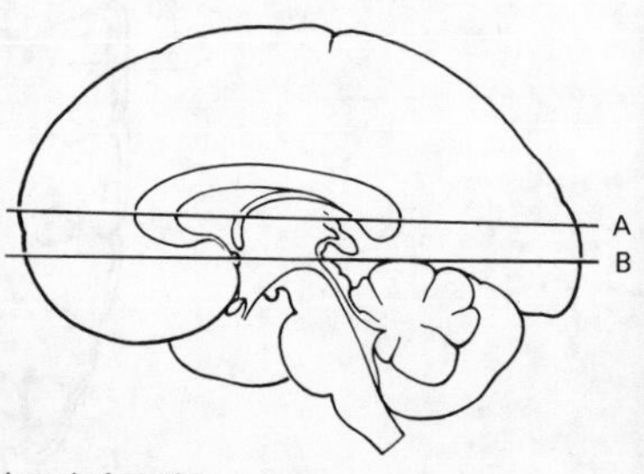

Level of section

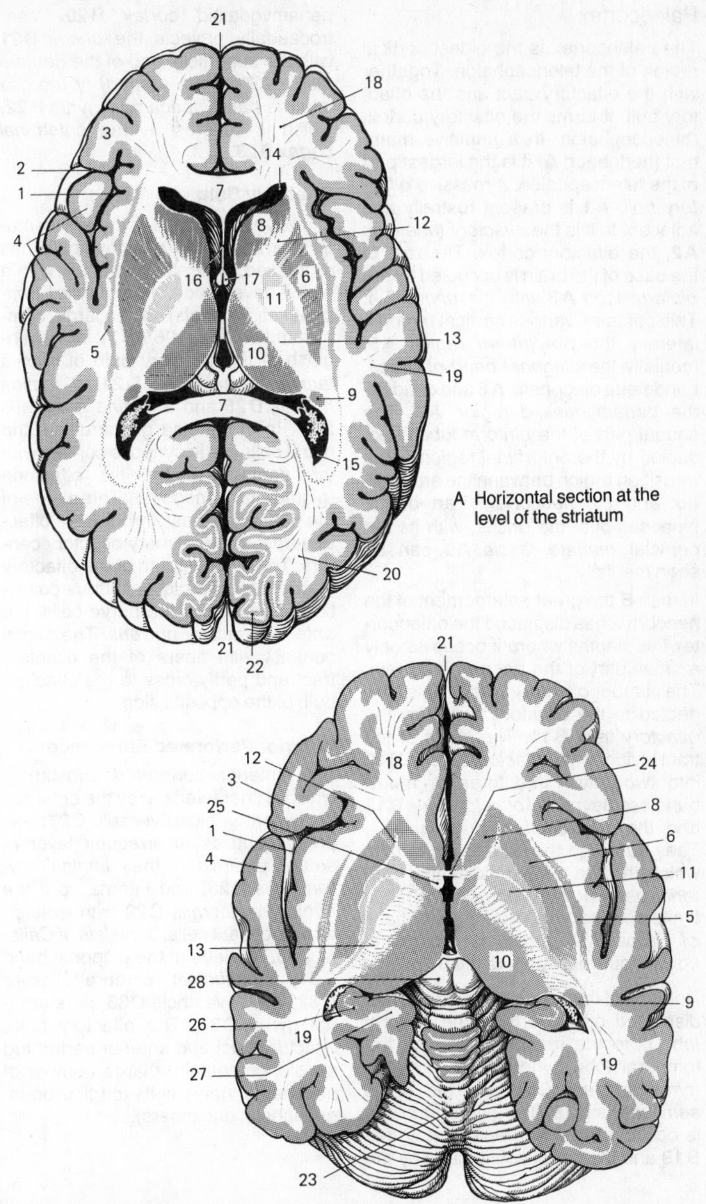

A Horizontal section at the level of the striatum

B Horizontal section at the level of the rostral commissure

Paleocortex

The paleocortex is the oldest cortical region of the telencephalon. Together with the olfactory tract and the olfactory bulb it forms the olfactory cortex, rhinencephalon. In a primitive mammal (hedgehog **A**) it is the largest part of the telencephalon. A massive *olfactory bulb* **A1** is present rostrally and adjacent to it is the *olfactory tubercule* **A2,** the olfactory cortex. The rest of the base of the brain is occupied by the *piriform lobe* **A3** with the *uncus* **A4.** This contains various cortical regions: laterally the *prepiriform region* **A5,** medially the *diagonal band of Broca,* bandeletta diagonalis **A6** and caudally the *periamygdaloid region* **A7.** The caudal part of the *piriform lobe* is occupied by the *entorhinal region* **A8,** a transition region between the archicortex and the neocortex. Part of the hippocampus, the uncus, with its superficial *dentate fascia* **A9** can be seen medially.

In man **B** the great enlargement of the neocortex has displaced the paleocortex into depths where it occupies only a small part of the base of the brain. The slender *olfactory bulb* **B10** is connected to the olfactory cortex by the *olfactory tract* **B11.** The fibers of the tract divide at the *olfactory trigone* **B12** into two (often into three or more) bundles: the *medial olfactory stria* **B13** and the *lateral olfactory stria* **B14.** They surround the *olfactory tubercle,* which in man has sunk deeply *(anterior perforated substance* **B15).** It is limited caudally by the *diagonal band of Broca* **B16,** which is thought to convey afferent fibers to the bulb.

Rotation of the hemisphere in man has displaced other parts of the *piriform lobe* onto the medial surface of the temporal lobe, where they form the *gyrus ambiens* **B17** and the *gyrus semilunaris* **B18.** The gyrus ambiens is occupied by the *prepiriform cortex* **B19** and the gyrus semilunaris by the periamygdaloid *cortex* **B20.** Ventrocaudally projects the *uncus* **B21** with the superficial end of the *dentate fascia (Giacomini's band).* It merges into the *parahippocampal gyrus* **B22,** which is covered by the *entorhinal cortex* **B23.**

Olfactory Bulb

In man, who belongs to the *microsmatic* mammals, the olfactory bulb has regressed, whereas mammals with a highly developed sense of smell *(macrosmatic* animals) have a large, complicated bulb (p. 196 AB3). We distinguish in the olfactory bulb of man a *lamina glomerulosa* **D24,** a *lamina mitralis* **D25** and a *lamina granularis* **D26.** In the lamina glomerulosa, the mitral cells are in synaptic contact with the endings of the fila olfactoria (glomerular synapses). The axons of the mitral cells pass through the olfactory tract to the primary olfactory centers. The whole length of the olfactory tract contains a discontinuous collection of medium sized nerve cells, the *anterior olfactory nucleus.* The axons connect with fibers of the olfactory tract and partly cross to the olfactory bulb of the opposite side.

Anterior Perforated Substance

The anterior perforated substance, which is characterized by the entrance of numerous blood vessels **C27,** has on the outside an irregular layer of small pyramids, the *lamina pyramidalis* **C28,** and internal to it the *lamina multiformis* **C29** with isolated groups of dark cells, the *islets of Calleja* **C30.** Nucleus of the diagonal band **C31,** longitudinal cerebral fissure **C32,** lateral ventricle **C33**, paraterminal gyrus **C34.** The olfactory bulb, olfactory tract and anterior perforated substance contain a large number of peptidergic nerve cells (corticoliberin, encephalin and others).

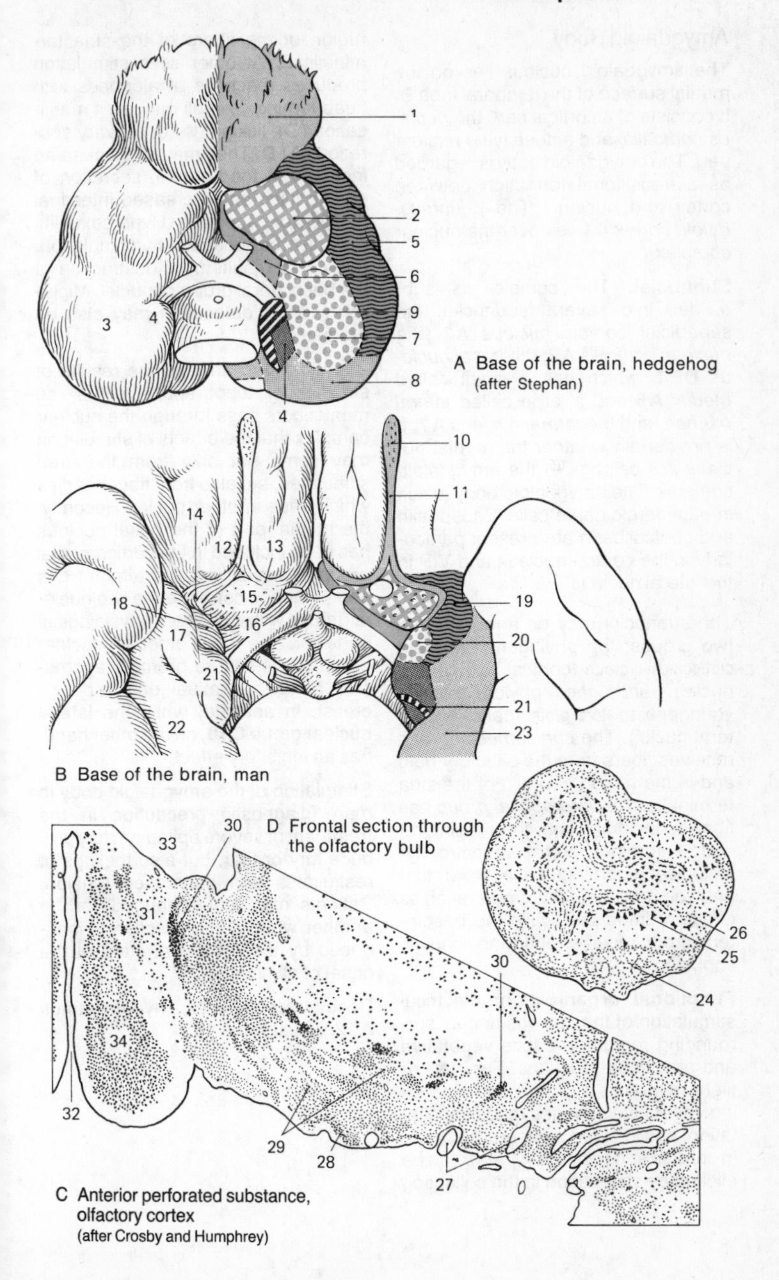

A Base of the brain, hedgehog
(after Stephan)

B Base of the brain, man

C Anterior perforated substance, olfactory cortex
(after Crosby and Humphrey)

D Frontal section through the olfactory bulb

Amygdaloid Body

The amygdaloid nucleus lies on the medial surface of the temporal lobe **B.** It consists of a cortical part, the *nucleus corticalis,* and a deep lying nuclear part. The amygdaloid body is regarded as a transitional formation between cortex and nucleus. The *periamygdaloid cortex* **A1** lies over the nuclear complex.

Subnuclei: The complex is subdivided into several subnuclei: the superficial *cortical nucleus* **A2,** the *nucleus centralis* **A3,** the *basal nucleus* **DE4,** which has a small celled *medial* **A5** and a large celled *lateral* **A6** *part,* and the *lateral nucleus* **A7.** It is not certain whether the medial nucleus **A8** belongs to the amygdaloid complex. The amygdaloid body is rich in peptidergic nerve cells. Encephalin and corticoliberin are present particularly in the central nucleus and VIP in the lateral nucleus.

The subnuclei may be arranged into two groups: the phylogenetically old **corticomedial group** (cortical and central nuclei) and the phylogenetically younger **basolateral group** (basal and lateral nuclei). The *corticomedial group* receives fibers from the olfactory bulb and is the region of origin of the stria terminalis. The *laterobasal group* has fiber connections with the prepiriform and entorhinal cortex. Electrophysiological recordings have shown that only the corticomedial group receives olfactory impulses, while the basolateral group receives optic and acoustic impulses.

Functional organization: electrical stimulation of the nucleus and its surrounding regions produce vegetative and emotional reactions. Fury (●) or flight response (○), with corresponding vegetative changes (pupillary dilatation, rise in blood pressure, increase in cardiac and respiratory rate), can be elicited by stimulation in the collecting region of the fibers of the stria terminalis **E.** At other sites stimulation produces reactions of alertness with head turning. Stimuli may elicit mastication (○), licking (●) or salivary secretion (▲) **D.** They may also cause an increase in food intake, secretion of gastric juice and increased intestinal motility or appetite. Hypersexuality may also be produced by stimulation, but it also results from destruction of the basolateral group of nuclei. Micturition (Δ) or defecation may also be produced.

It is difficult to arrange the results of stimulation topographically as so many fibers pass through the nuclear complex that the effects of stimulation may come not only from the area stimulated, but also from fiber bundles which arise in other nuclei. Recently, the medial part of the basal nucleus has been included in the corticomedial group of nuclei, and an attempt has been made to correlate the two nuclear groups with a different physiological behavior: the corticomedial nuclear group **C9** is said to promote aggressive behavior, sexual drive and increase in appetite, while the lateral nuclear group **C10,** on the other hand, has an inhibitory effect.

Stimulation of the amygdaloid body in man (diagnostic precaution in the treatment of severe epilepsy) may produce fury or fear, but also feelings of restfulness and decreased tension. Patients may feel "changed" or "in another world". The reaction is influenced by the emotional state at the onset of stimulation.

Optic tract **A–E11,** hypothalamus **A12,** claustrum **A13.**

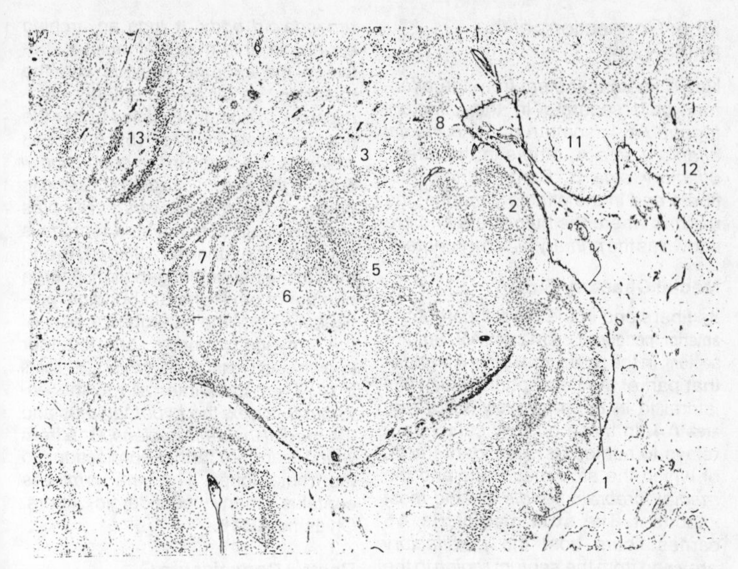

A Organization of the amygdaloid nucleus, frontal section, semischematic

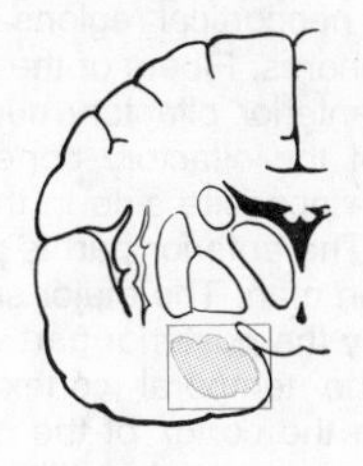

B Plane of the section

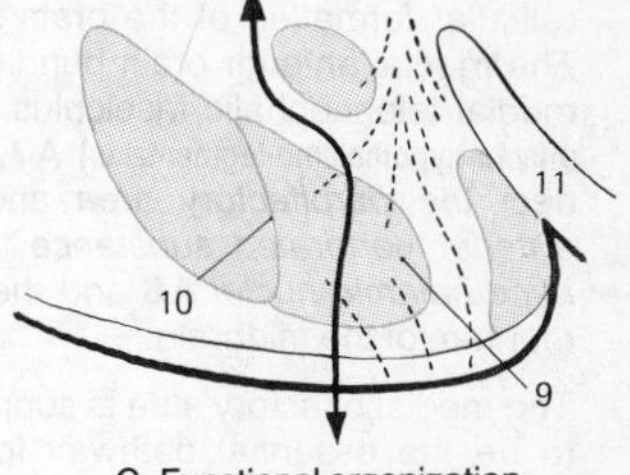

C Functional organization
(after Koikegami)

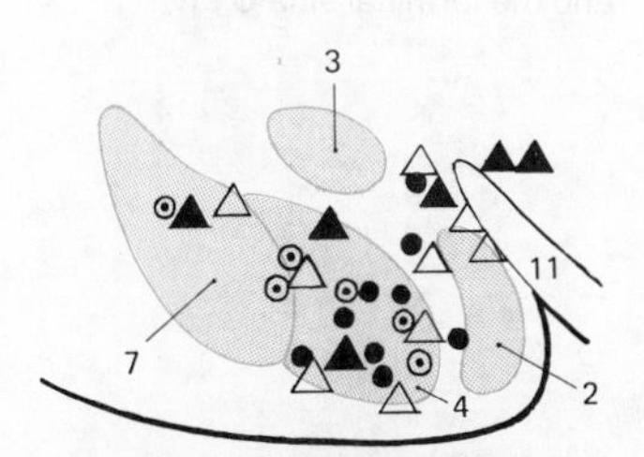

D Vegetative reactions following experimental stimulation in the cat
(after Ursin and Kaada)

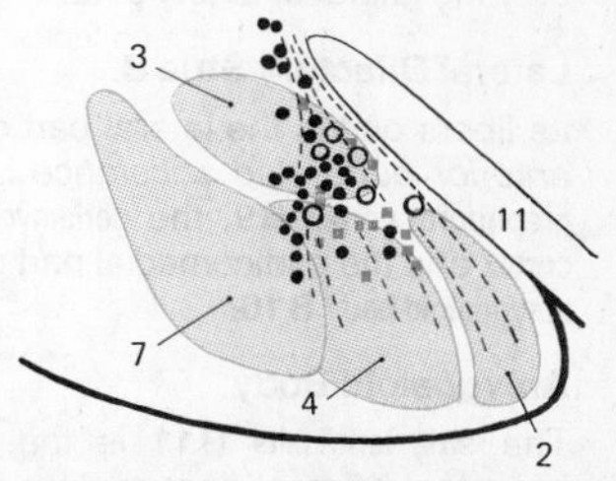

E Menace and flight reactions following experimental stimulation in the cat
(after de Molina and Hunsperger)

Fiber Connections of the Paleocortex

Nerve fibers that stem from the *olfactory bulb* separate in the *olfactory trigone* **AB1**: some run in the *medial olfactory stria* **A,** others in the *lateral olfactory stria* **B.** Some of the bulb fibers end in the olfactory trigone; the regions where the bulb fibers end are known as the *primary olfactory centers.*

Medial Olfactory Stria A

Its fibers end in the medial part of the **anterior perforated substance A2,** in the **septum,** in the **paraterminal gyrus A3** and in that part of the hemisphere which lies lateral to it, the **parolfactory (subcallosal) area A4.** In those primary centers are relays to numerous secondary tracts, of which the *stria medullaris* and the *medial forebrain bundle* are the most important. The **stria medullaris A5** carries fibers from the *parolfactory area* and from the *septum region* to the *habenular nuclei* **A6** from which the habenulotegmental tracts lead to the reticular formation of the brain stem. The medial anterior brain bundle, the medial telencephalic fasciculus (**fibrae olfacto-hypothalamo-tegmentales) A7,** connect the *parolfactory area* and the *anterior perforated substance* to the *hypothalamic nuclei* **A8** and the tegmentum of the midbrain.

The medial olfactory stria is supposed to be the essential pathway for the sense of smell. However, in animals anosmia does not result from cutting only the lateral olfactory stria.

Lateral Olfactory Stria B

Its fibers end in the lateral part of the *anterior perforated substance* in the **prepiriform cortex B9,** the **periamygdaloid cortex** and the corticomedial part of the **amygdaloid body B10.**

Amygdaloid Body

The **stria terminalis B11** is the most important efferent fiber system of the *amygdaloid body.* It runs an arching course in the marginal sulcus, between the caudate nucleus and the thalamus, as far as the anterior commissure. In doing this, it lies under the thalamostriate vein (p. 159 B14, 163 AB2). Its fibers end in the *septal nuclei* **B12,** in the *preoptic region* **B13** and in nuclei of the *hypothalamus.* Bundles of fibers pass from the stria terminalis over into the stria medullaris **B5** and then to the habenular ganglion. A large number of other efferent bundles, particularly from the laterobasal part of the amygdaloid body, are together known as **ventral amygdalofugal fibers B14.** They pass inter alia to the *entorhinal cortex,* the *hypothalamus* and the *medial thalamic nucleus,* **B15** from which there are further connections to the frontal lobes. The stria terminalis contains many peptidergic fibers. Chiasma **AB16.**

Rostral Commissure C

The commissure joins the palaeocortical and the neocortical regions of the two hemispheres. Fibers of the olfactory tract (anterior olfactory nucleus) **C17** and of the olfactory cortex **C2** cross to the opposite side in the anterior part. The anterior part is poorly developed in man. The major section is formed by the posterior part where fibers of the temporal cortex **C18** mostly from the cortex of the medial temporal gyrus, cross. In addition, the posterior part also receives crossed fibers from the amygdaloid body **C10** and the terminal stria **C11.**

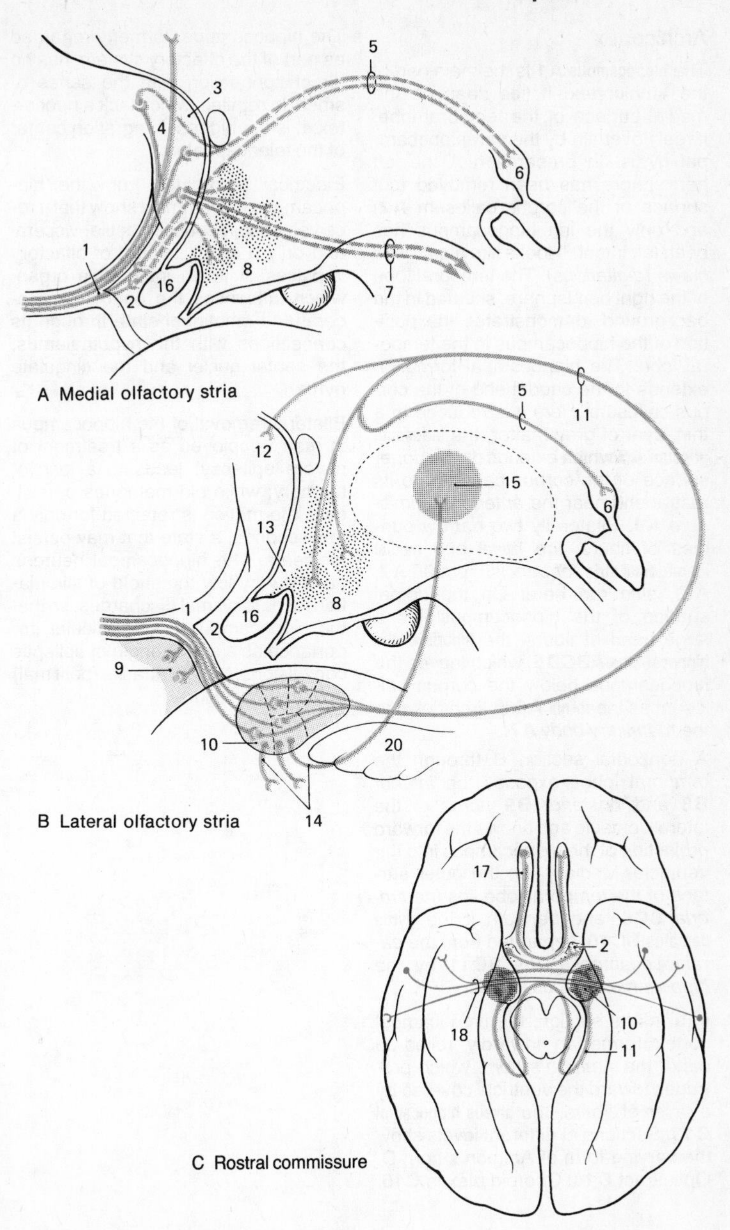

A Medial olfactory stria

B Lateral olfactory stria

C Rostral commissure

Archicortex

The **hippocampus A1** is the main part of the archicortex. It lies deep on the medial surface of the temporal lobe, largely overlain by the parahippocampal gyrus. In preparation **A** the left hemisphere has been removed (cut surface of the corpus callosum **A2**) and only the left hippocampus has been left intact. It looks like a paw with claws *(digitations)*. The temporal lobe of the right hemisphere, situated in the background, demonstrates the position of the hippocampus in the temporal lobe. The hippocampal formation extends to the caudal end of the corpus callosum. Here it is reduced to a thin layer of grey matter, the **induseum griseum A3** which extends on the upper surface of the corpus callosum to its rostral end near the *anterior commissure* **A4.** Bilaterally two narrow bundles of fibers, the **lateral** and **medial longitudinal stria** *(of Lancisi)* (p. 206 A7, A8), also run here. On the dorsal surface of the hippocampus lies a large band of fibers, the **fimbria of the hippocampus ABCD5,** which leaves the hippocampus below the corpus callosum as the **fornix A6** to arch down to the *mamillary body* **A7.**

A horizontal section **B** through the temporal lobes exposes the *inferior* **B8** and *posterior* **B9** *horns* of the lateral ventricle and shows the forward projection of the hippocampus into the ventricle. Medially, on the outer surface of the temporal lobe lies the *fimbria* **BC5**, and beneath it the **gyrus dentatus BC10,** separated from the *parahippocampal gyrus* **BC11** by the *hippocampal sulcus* **BC12.**

In a frontal section, the hippocampal cortex forms an inwardly rolled-up band, the *Ammon's horn,* which protrudes toward the ventricle covered by a layer of fibers, the **alveus hippocampi C13.** Sections at different levels show the varying form of Ammon's horn D. Optic tract **C14.** Choroid plexus **C15.**

The hippocampus, formerly regarded as part of the olfactory system, has no direct connection with the sense of smell. In reptiles, which lack a neocortex, it is the highest integration center of the telencephalon.

Electrical recordings from the hippocampus of mammals show that it receives optic, acoustic, tactile, visceral and only a small amount of olfactory impulses. It is an integrative organ, which influences the endocrine, visceral and emotional state through its connections with the hypothalamus, the septal nuclei and the cingulate gyrus.

Bilateral removal of the hippocampus in man (employed as a treatment of severe epilepsy) leads to a loss of memory; while old memories persist, new information is retained for only a few seconds, a state that may persist for years. The hippocampal neurons have a very low threshold of stimulation for paroxysmal discharges, so that the hippocampus is of particular importance as a site of origin of epileptic convulsions, twilight attacks (petit mal) and memory failure.

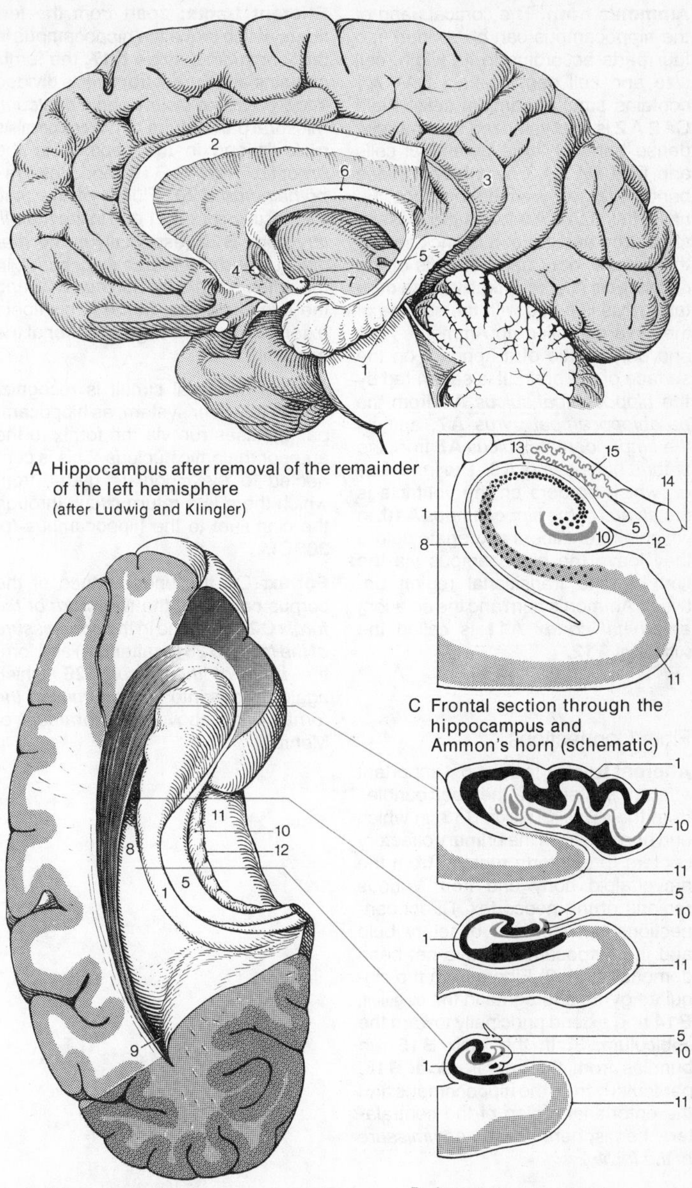

A Hippocampus after removal of the remainder of the left hemisphere
(after Ludwig and Klingler)

C Frontal section through the hippocampus and Ammon's horn (schematic)

B View of the hippocampus from above
(after Sobotta)

D Ammon's horn, sections at different levels

Ammon's horn. The cortical band of the hippocampus can be divided into four parts according to its width, cell size and cell density: field CA1 **A1** contains small pyramidal cells. Field CA2 **A2** is characterized by a narrow, dense band of large pyramidal cells and field CA3 **A3** by a broad, loose band of large pyramidal neurons. Finally, field CA4 **A4** forms the loosely-structured end zone. It is enclosed by the narrow, dark band of cells of the **dentate gyrus A5 (fascia dentata).** The dentate gyrus is fused with the surface of the inwardly rolled-up Ammon's horn and only a little of it appears on the surface of the brain. It is separated by the *hippocampal-sulcus* **A6** from the *parahippocampal gyrus* **A7,** and by the *fimbriodentate sulcus* **A8** from the *fimbria hippocampi* **A9.** The inner layer, which borders on the ventricle is the *alveus of the hippocampus* **A10,** in which efferent fibers aggregate before they leave the hippocampus via the fimbria. The transitional region between Ammon's horn and the adjacent *entorhinal cortex* **A11** is called the **subiculum A12.**

Fiber Connections

Afferent tracts: 1) the most important afferent system are the fiber bundles from the **entorhinal region B13,** in which end bundles from the primary olfactory centers (prepiriform region), from the amygdaloid body and from various regions of the neocortex. Direct connections between the olfactory bulb and the hippocampus have not been demonstrated. 2) Fibers from the cingulate gyrus aggregate in the **cingulum B14** and extend principally toward the subiculum. 3) In the **fornix B15** run bundles from the *septal nuclei* **B16,** particularly from the hippocampus and the entorhinal region of the contralateral hemisphere (via the *commissure of the fornix).*

Efferent tracts: apart from the few fibers which leave the hippocampus in the *longitudinal striae* **B17,** the fornix contains all efferent tracts. It is divided into a *precommissural* and a *postcommissural* part. Fibers of the **precommisural fornix** end in the septum, in the *preoptic region* **B19** and the *hypothalamus* **B20**. Fibers of the **postcommissural fornix B21** end in the *mamillary body* **B22** (especially in the *medial nucleus of the mamillary body),* in the *anterior thalamic nucleus* **B23** and the *hypothalamus.* Some fornix fibers extend to the central grey matter of the midbrain.

A large neuronal circuit is recognizable in this fiber system, as hippocampal impulses run via the fornix to the anterior thalamic nucleus. This is connected to the cingulate gyrus, from which there is a return circuit through the cingulum to the hippocampus (p. 306C).

Fornix. On the undersurface of the corpus callosum, the two *crura of the fornix* **C24** combine in the *commissure of the fornix* **C25** (psalterium) and form the *body of the fornix* **C26,** which again divides into two *columns of the fornix* **C27** above the foramen of Monro.

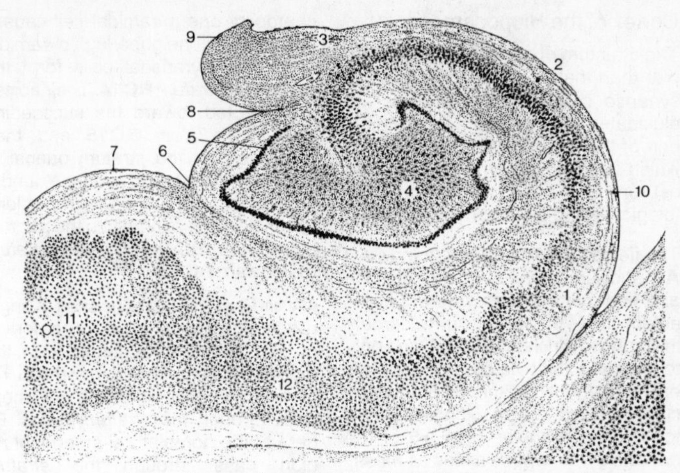

A Ammon's horn, frontal section through the hippocampus

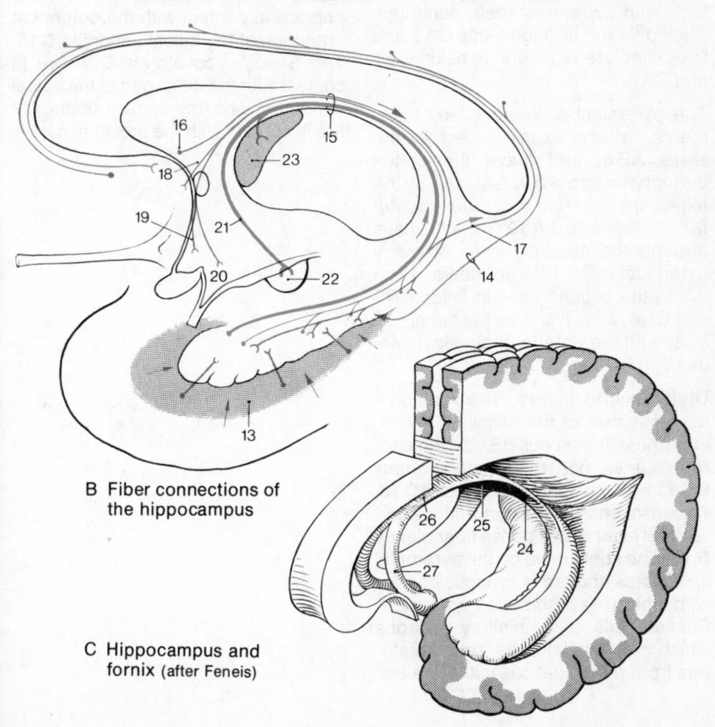

B Fiber connections of the hippocampus

C Hippocampus and fornix (after Feneis)

Cortex of the Hippocampus

The structure of the archicortex is simpler than that of the neocortex, so its synaptic connections are easier to elucidate. The hippocampal cortex is one of the few regions of the brain in which inhibitory and stimulating neurons have been identified both histologically and electrophysiologically.

The fields CA1 **A1,** CA2 **A2** and CA3 **A3** show differences in organization and fiber connections. Afferent fibers enter the Ammon's horn, largely through the **perforating tract A4,** and only to a small extent via the alveus. They end on dendritic branches of the **pyramidal cells A5.** Certain fibers extend toward the *granular cells* **A6** of the dentate gyrus, whose axons as *mossy fibers* **A7** are likewise in synaptic contact with pyramidal cell dendrites. They only run through fields CA3 and CA4, and are not found in fields CA1 and CA2.

The pyramidal cells are efferent elements, whose axons collect in the **alveus AB8,** and leave the cortex through the **fimbria A9.** Given off by the axons are recurrent collaterals *(Schaffer collaterals)* **A10** which pass through the dendritic trees of many pyramidal cells. Efferent fibers which run to the septum arise in fields CA3 and CA4, while fibers for the mamillary body and the anterior thalamic nucleus stem from area CA1.

Division into layers. The Ammon's horn consists of the following layers: innermost, the *alveus* **B8** with the efferent fibers, then the *stratum oriens* **BC11** with multipolar **basket cells BC12,** whose axons split up and fill the pyramidal layer with a dense fiber plexus **B13.** The fibers envelop the pyramidal cell bodies and are in synaptic contact with them (axosomatic synapses). Basket cells are inhibitory neurons which are stimulated by axon collaterals from pyramidal cells, and the discharge of one pyramidal cell causes inhibition in neighboring pyramidal cells. The pyramidal cells form the *pyramidal stratum* **BC14.** Their apices are directed toward the succeeding *stratum radiatum* **BC15** and their bases toward the stratum oriens. In both directions arborize dense dendritic plexuses. Branches from the long apical dendrite extend into the *stratum lacunosum* **BC16** and the *stratum moleculare* **BC17.**

Afferent fibers from various sites of origin run in different layers. A large portion of the commissural fibers from the hippocampus and the entorhinal region of the contralateral hemisphere pass through the *stratum oriens.* Fibers of the homolateral entorhinal region pass through the *stratum lacunosum* and *stratum moleculare* and are in contact with the outermost branches of the apical dendrites **C18.** The Schaffer collaterals **C10** are in contact with the distal part of the apical dendrites, and the mossy fibers **C7** are in contact with the proximal part.

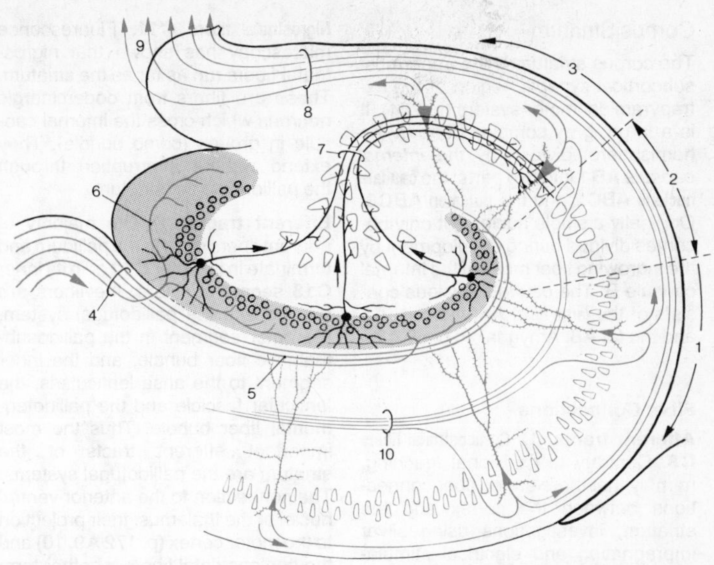

A Organization of the hippocampus
(after Cajal)

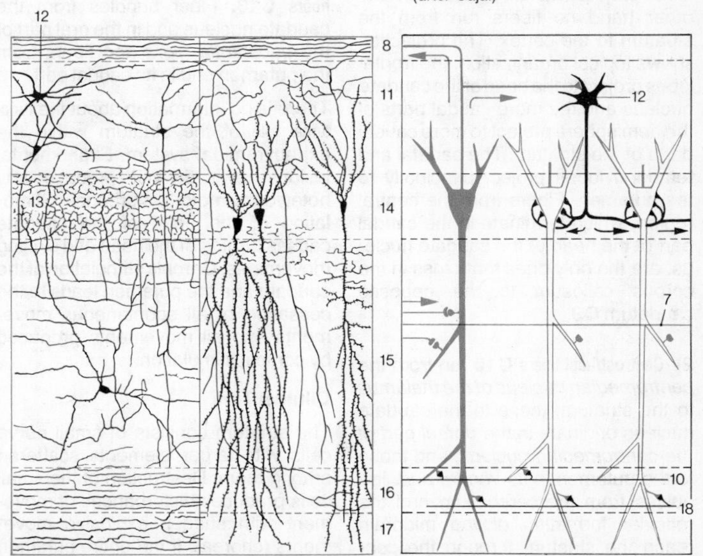

B Structure of the hippocampal cortex, silver impregnation (after Cajal)

C Neuron synapses
(after Anderson, Blackstad and Lömo)

Corpus Striatum

The corpus striatum is the uppermost subcortical synaptic region of the extrapyramidal motor system (p. 284). It is a large, grey complex deep in the hemisphere, divided by the *internal capsule* **AB1** into two parts: the **caudate nucleus ABC2** and the **putamen ABC3.** Originally a single nucleus, it only becomes divided during development by the ingrowing fiber mass of the internal capsule B. The caudate nucleus consists of the **large head A4,** the **body A5** and the **tail A6.** Amygdaloid body **A7.**

Fiber Connections

Afferent tracts. 1) **Corticostriatal fibers C8.** Contrary to traditional teaching, namely that there are no fiber connections between the cortex and the striatum, investigations using silver impregnation and electrical stimulation have shown that all the cerebral lobes project onto the striatum. On the other hand no fibers run from the striatum to the cortex. The projection shows topical organization: the frontal lobes project to the head of the caudate nucleus and the more caudal parts of the hemisphere project to more caudal parts of the striatum. The parietal and temporal lobes project principally to the putamen. Fibers from the central region, which terminate in the caudal part of the head of the caudate nucleus, are the only ones that cross in the corpus callosum to the opposite caudatum **C9.**

2) **Centrostriatal fibers C10** run from the *centromedian nucleus of the thalamus* to the striatum, those to the caudate nucleus originate in the *dorsal part of the centromedian nucleus,* and those to the putamen in its *ventral part.* Impulses from the cerebellum and the reticular formation of the midbrain reach the striatum through the centromedian nucleus.

Nigrostriatal fibers C11. Fluorescence microscopy has shown that nigrostriatal fibers run as far as the striatum. These are fibers from dopaminergic neurons which cross the internal capsule in groups (comb bundle). They extend without interruption through the pallidum to the striatum.

Efferent tracts: 1) The majority of efferent fibers pass to the pallidum and terminate in its outer **C12** or in its inner **C13** segments. Here, the fibers are relayed into the pallidofugal system, the outer segment in the pallidosubthalamic fiber bundle, and the inner segment to the ansa lenticularis, the lenticular fascicle and the pallidotegmental fiber bundle. Thus the most important efferent tracts of the striatum are the pallidofugal systems. The connection to the anterior ventral nuclei of the thalamus, their projection to the motor cortex (p. 172 A9, 10) and the corticostriatal fibers together form a large circle of neurons. 2) **Strionigral fibers C13.** Fiber bundles from the caudate nucleus end in the oral part of the *substantia nigra* and fibers from the putamen end in its caudal part.

There is no information about the **functional role** of the striatum within the extrapyramidal system. Experimental lesions in it do not produce any noteworthy motor disturbance. Stimulation of the head of the caudate causes turning of the head and circling movements. Chronic stimulation of the caudate and the putamen leads to the cessation of all spontaneous movements and all movements produced by cortical stimulation.

Clinical Tips

The striatum consists of small nerve cells with larger elements scattered among them. Destruction of the small cells produces disturbances of movement with sudden explosive movements (chorea). If the entire putamen is destroyed this type of hyperkinesia does not occur.

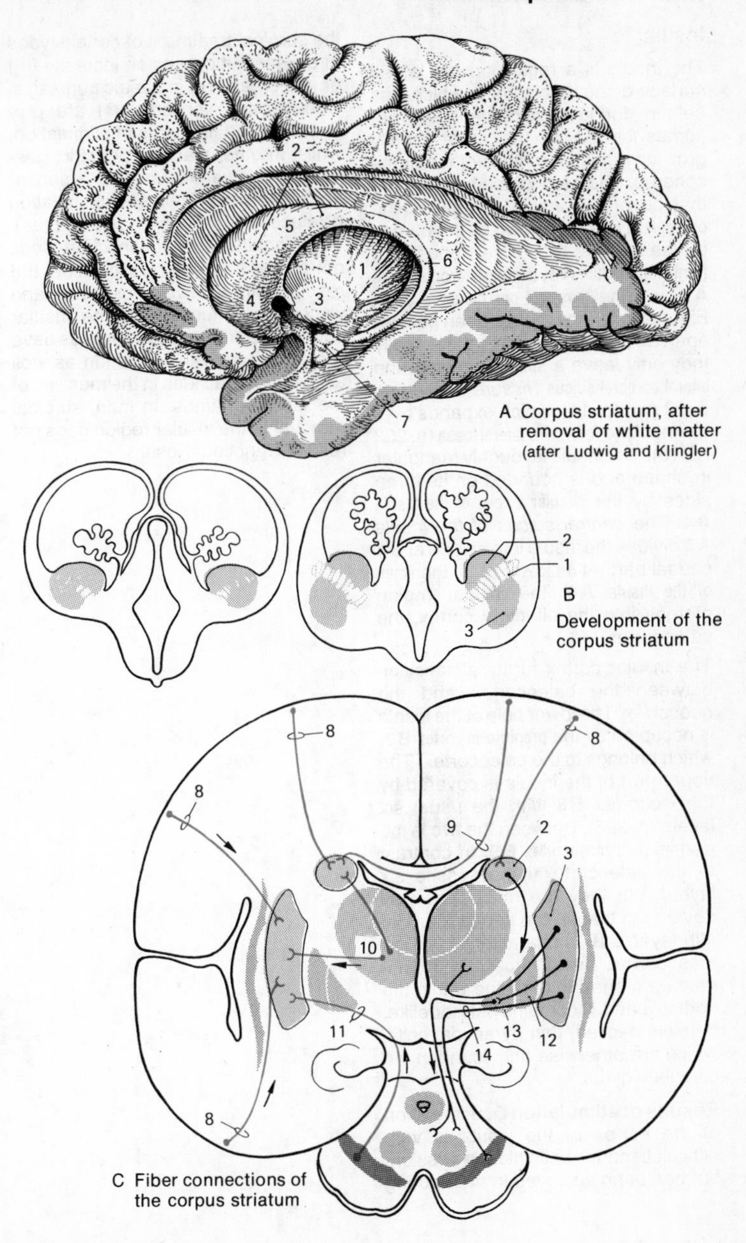

A Corpus striatum, after removal of white matter (after Ludwig and Klingler)

B Development of the corpus striatum

C Fiber connections of the corpus striatum

Insula

The insula is a region on the lateral surface of the hemisphere, which lags behind during development and becomes buried by the more rapidly growing adjacent regions of the hemispheres. The parts of the hemisphere that cover the insula are called the **opercula.** They are known by the names of the adjacent cerebral lobes, the **frontal operculum A 1,** the **parietal operculum A2** and the **temporal operculum A3.** In Fig. **A** the opercula have been pushed apart to reveal the insula. Normally they only leave a slitlike opening, the **lateral cerebral sulcus** (*fissure of Sylvius,* p. 10 A4) (p. 11), which expands over the insula to form the **lateral fossa** (p. 202 AB 15). The insula is roughly triangular in shape and is bounded on its three sides by the **circular sulcus of the insula A4.** The *central sulcus of the insula* **A5** divides the insula into an oral and a caudal part. At its lower pole, the **limen of the insula A6,** the insular region merges into the olfactory cortex, the paleocortex.

The insular cortex forms a transition between the paleocortex and the neocortex. The lower pole of the insula is occupied by the **prepiriform cortex B 7,** which belongs to the paleocortex. The upper part of the insula is covered by the isocortex **B8** with the usual six layers (p. 226). Between the two is the transitional **mesocortex B9.** In contrast to the paleocortex it possesses six layers, but they are much less well developed than in the neocortex. The fifth layer **C10** is characteristic of the mesocortex, being especially prominent as a narrow, dark line within the cortical band. It contains pallisadelike, densely packed, thin pyramidal cells, which are otherwise only found in the cingulate gyrus.

Results of stimulation D. Stimulation of the cortex of the insula is very difficult because of its hidden position, but has been achieved in man during the surgical treatment of certain types of epilepsy. It causes an increase (+) or a decrease (–) in gastric peristagis. Nausea and vomiting (●) are produced at certain points of stimulation, whilst in other places epigastric, gastric (×) or lower abdominal (○) sensations result. At certain sites stimulation produces sensations of taste (● blue). Although the stimulation chart does not permit a topical classification, the results point to a viscerosensory and visceromotor function of the insular cortex. Experiments on monkeys have resulted in salivary secretion as well as in motor reactions in the muscles of the face and limbs. In man, surgical removal of the insular region does not result in functional loss.

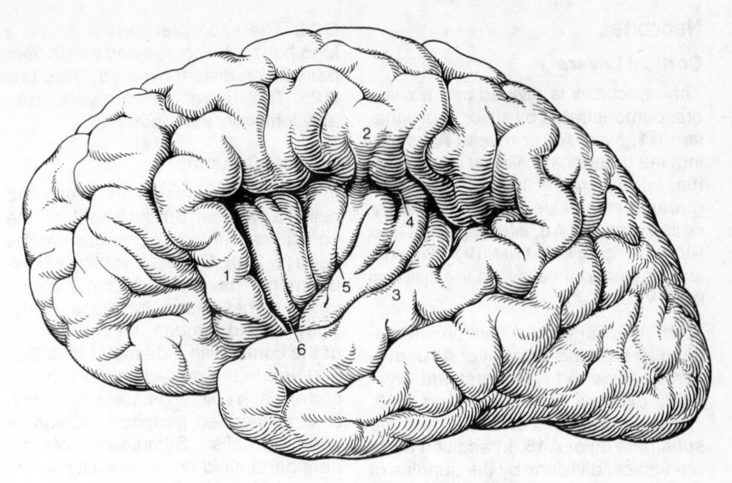

A Insula, the opercula have been pulled apart
(after Retzius)

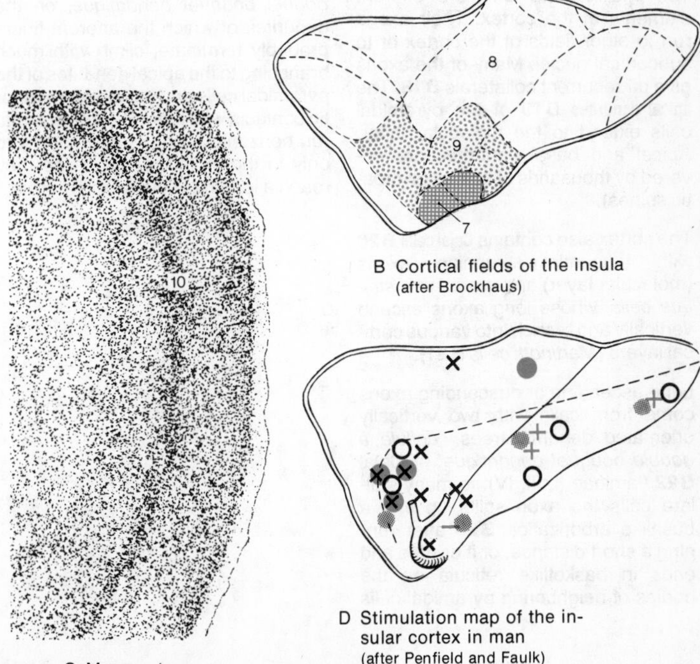

B Cortical fields of the insula
(after Brockhaus)

C Mesocortex

D Stimulation map of the insular cortex in man
(after Penfield and Faulk)

Neocortex

Cortical Layers

The neocortex is divided into six layers, demonstrated by silver impregnation **A1,** Nissl **A2** or myelin **A3** staining: the outermost **molecular layer** (I) **A4,** the **external granular layer** (II) **A5** with granular or stellate cells, the **outer pyramidal layer** (III) **A6, internal granular layer** (IV) **A7, ganglionic layer** (V) **A8** with large pyramidal cells and the **multiform layer** (VI) **A9.**

In the myelin-stained section we distinguish the *tangential layer* **A10,** *dysfibrous layer* **A11,** *suprastriate layer* **A12,** *external Baillarger band* **A13,** *internal Baillarger band* **A14** and the *substriate layer* **A15.** In addition there are vertical divisions by the bundles of *radial fibers* **A16.**

Cortical Neurons

The **pyramidal cells B17** are efferent elements of the cortex. Their axons run to other fields of the cortex or to subcortical nuclei. Many of the axons give off recurrent collaterals **B18.** The **apical dendrites B19** of the pyramidal cells extend to the molecular layer. Apical and basal dendrites are covered by thousands of spines (sympatic spines).

The cortex also contains **Cajal cells B20** with tangentially running axons (molecular layer) and *granular* or *stellate cells,* whose long axons ascend vertically and branch into various cortical layers *(Martinotti cells* **B21).**

Long ascending or descending axons come from cells with two vertically orientated dendritic trees, *'cellule à double bouquet dendritique'* of Cajal **B22** (laminae II, III, IV). In many stellate cells the axon splits up into a bushlike arborization **B23** after running a short distance, or it divides and ends in basketlike reticula on the bodies of neighboring pyramidal cells **B24.** The axon branches may run a long horizontal course and end in fiber baskets on distant pyramidal neurons **B25.** The *basket cells* are very probably inhibitory interneurons.

Vertical Columns

In the sections stained for myelin and cells, a vertical arrangement of the cortical cells into cell columns can be seen (p. 240). Electrophysiological experiments have shown that the columns of cells form functional units. They extend through the entire width of the cortex with a diameter of 300 to 500 μm. In the projection fields of the cortex each column is associated with a circumscribed peripheral region of sensory cells. Stimulation of the peripheral field always elicits a response from the entire column. The long vertical stellate cell axons are the substrate for the functional unity of the columns. The axons of the *cellule à double bouquet dendritique,* on the dendrites of which the afferent fibers probably terminate, climb with much branching to the apical dendrites of the pyramidal cells and have many synaptic contacts with them **B26.** Axons that run horizontally are usually short and only in the molecular layers do they reach a length of several millimeters.

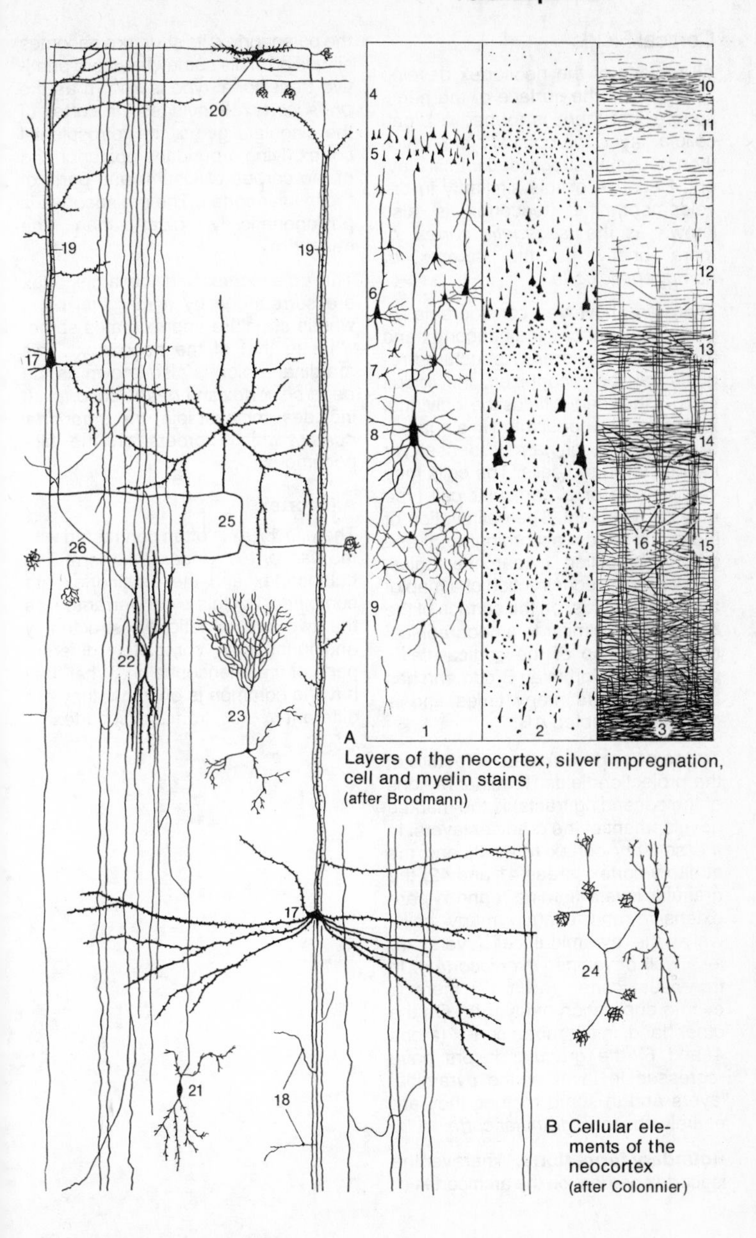

A Layers of the neocortex, silver impregnation, cell and myelin stains (after Brodmann)

B Cellular elements of the neocortex (after Colonnier)

Cortical Fields

All regions of the neocortex develop similarly: on the surface of the hemisphere first there is formed a broad cellular layer, the *cortical plate,* which then becomes subdivided into six layers. Because of this similarity in development, the neocortex is also known as the *isogenetic cortex,* or more briefly as the **isocortex** or *homogenetic cortex.*

There are, however, considerable regional variations in the neocortex and a number of differently structured regions can be recognized, the **cortical fields.** In the various fields the individual layers may be constituted in a number of ways: wide or narrow, densely or loosely arranged cells. The cells may vary in size or a particular cell type may be dominant. The delimitation of individual fields according to these criteria is called cytoarchitectonics, and it permits construction of a map of the cortical fields, on the surface of the hemisphere, similar to a geographical map. The map of the cortical fields produced by Korbinian Brodmann has been confirmed many times and is generally accepted **AB.**

Types of cortex. A special feature of the projection fields (terminal regions of the ascending tracts) is the marked development of the granular layers. In the sensory cortex (area 3) and the auditory cortex (areas 41 and 42) the granular layers (lamina II and IV) are extensive and contain many cells, whilst the pyramidal cell layers are less well developed *(koniocortex).* In the visual cortex (area 17) there is even a duplication of layer IV. On the other hand, in the motor cortex (areas 4 and 6) the granular layers have regressed in favor of the pyramidal layers and in some regions they are entirely absent *(agranular cortex).*

Boundary formations. Wherever the isocortex borders on the archicortex or the paleocortex, its structure becomes simplified. This somewhat more primitive transitional type is known as the *proisocortex.* It includes the cortex of the cingulate gyrus, the retrosplenial cortex (lying around the posterior end of the corpus callosum) and parts of the insular cortex. The proisocortex is phylogenetically older than the neocortex.

The paleocortex and the archicortex are surrounded by a marginal band whose structure approximates somewhat to that of the neocortex. This marginal region is also known as the *periarchicortex* and *peripaleocortex.* It includes, for example, the *entorhinal cortex,* which borders on the hippocampus.

Allocortex

The *allocortex* is often contrasted with the *isocortex.* Under this term the paleocortex and the archicortex are combined, but this is not warranted, as the two are genetically, structurally and functionally completely different parts of the telencephalon. What they have in common is only that they are different (ἄλλος) from the isocortex.

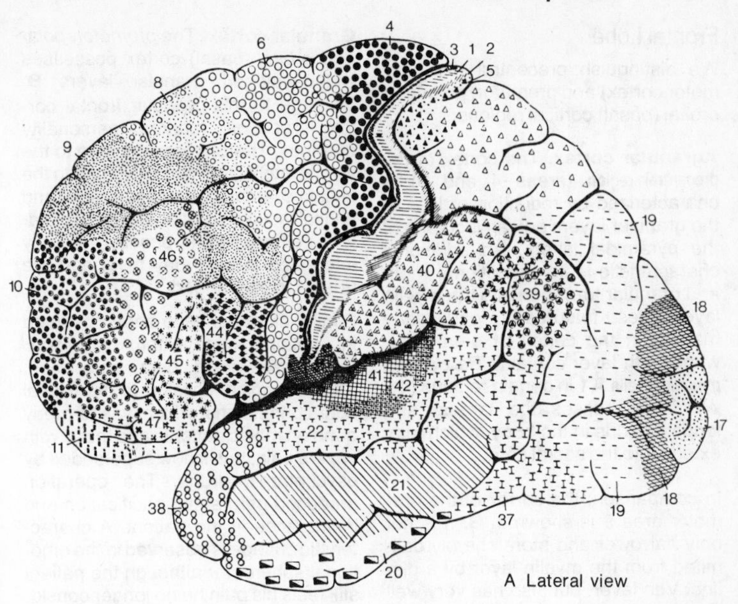

A Lateral view

Cortical fields of the hemisphere
(after Brodmann)

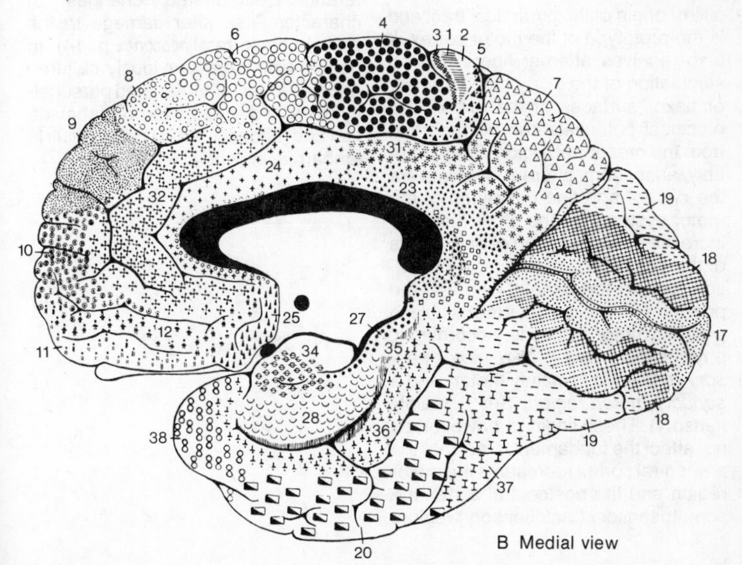

B Medial view

Frontal Lobe

We distinguish precentral (the true motor cortex) and premotor, polar and orbital (basal) cortical regions.

Agranular cortex. The cortex of the **precentral region** (areas 4 and 6) is characterized by reduction or loss of the granular layers and an increase of the pyramidal layers. An additional characteristic is its exceptional width and granular transition into the myelin layer. These features are particularly marked in the cortex of area 4 **A.**, whose Vth layer contains *Betz's* **giant pyramidal cells A 1** in certain regions (4 γ ▲). They posses the largest and longest axons in the body, which may extend into the sacral cord.

In comparison, the cortex of the premotor area 9 is shown in **B.** It is not only narrower and more sharply delimited from the myelin layer by a distinct VIth layer, but also has very well developed granular layers (II and IV).

The agranular cortex is the principal site of origin of the pyramidal tract and is the prototype of the motor cortex. It also receives afferent fibers and on stimulation of the skin of the extensor or flexor surfaces of the extremities electrical potentials can be recorded from the precentral region. They probably arise from afferent systems for the control and fine regulation of the motor system. On the other hand, by increased stimulation in certain points of the postcentral region (somatosensory cortex) the parietal lobe and the premotor frontal region, it is possible to produce motor reactions. Accordingly, physiologists speak of a *motor sensory region* Ms I (mainly motor) and a *sensory-motor region*, Sm I (mainly sensory). These findings, however, do not affect the fundamental fact that the precentral cortex represents the motor region and the postcentral cortex the somatosensory (tactosensory) region.

Granular cortex. The *premotor, polar* and *orbital* (basal) cortex possesses well-developed granular layers **B.** Damage to this granular frontal cortex produces marked personality changes. There is less damage to the formal intellectual capacity, than to the initiative, ambitious, concentrative and critical faculties. Patients show a childish, selfsatisfied euphoria, are only interested in everyday trivia and are unable to plan ahead.

Similar changes have been observed in patients on whom *prefrontal leucotomies* have been performed. This is surgical division of the frontal fiber connections, which was formerly done on manic patients and those with severe pain, but is now superseded by psychotropic drugs. The operation produced permanent pacification and indifference in the patient. A characteristic change is observed in the emotional sphere, for although the patient still feels his pain he no longer considers it troublesome. To pain, previously unbearable, the patient becomes indifferent. Deep-seated changes of character arise after damage to the orbital cortex (**basal neocortex** p. 14). In people who were previously cultured and had a well-differentiated personality there is disintegration of behavior, tact and modesty, which could result in embarrassing social offences.

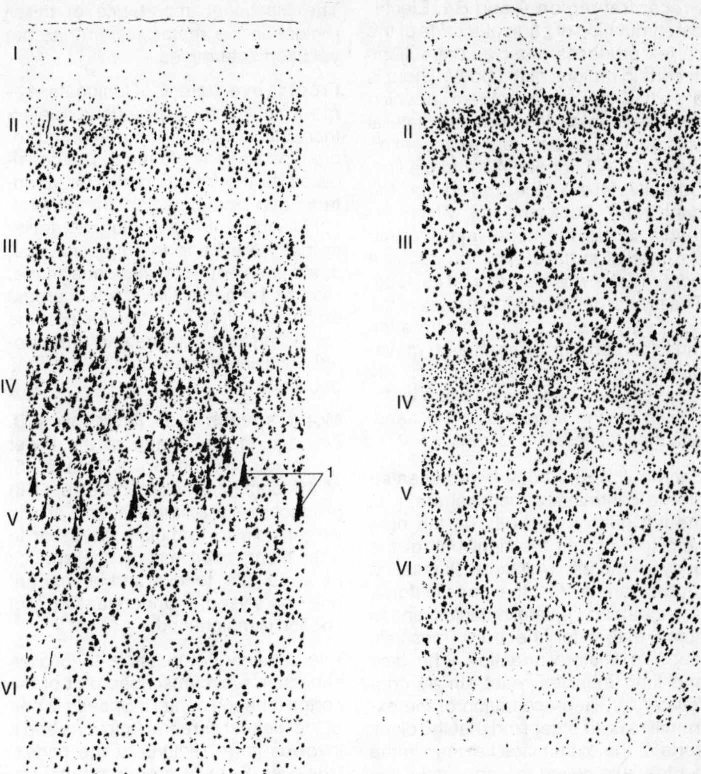

B Area 9

A Area 4, motor cortex

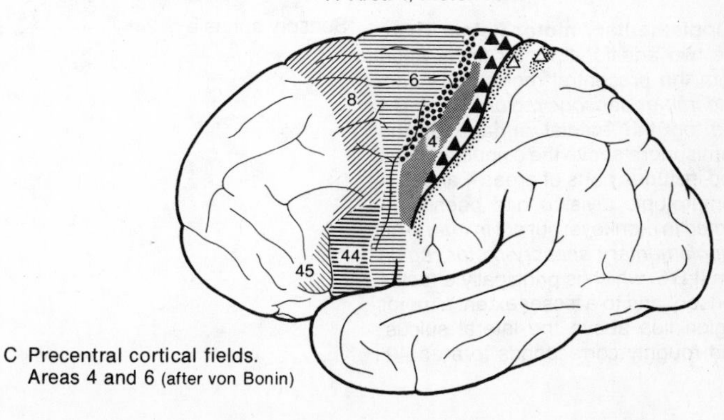

C Precentral cortical fields.
Areas 4 and 6 (after von Bonin)

Precentral region A and B1. Electrical stimulation of particular parts of the cortex produces muscle contractions in certain regions of the body. There is a somatotopic arrangement in which the head region lies above the lateral sulcus, the lowest part representing the throat **A2,** tongue **A3** and the lips. Dorsally follows the region for the hand, arm, trunk and leg, which extends across the upper margin onto the medial surface. This produces a *homunculus* which stands on its head. The areas for the individual parts of the body vary in size: the parts where the muscles perform differentiated movements are represented by particularly large areas. The largest region is occupied by the fingers and the hand, and the smallest by the trunk.

Each half of the body is represented on the contralateral hemisphere, i. e. the left half of the body on the right hemisphere and the right half of the body on the left. Unilateral stimulation of the areas for the masticatory-, laryngeal- and palatal muscles, and to some extent for the trunk muscles, leads to bilateral reaction. The face and limb muscles react strictly contralaterally. Representation of the extremities is so organized that the distal parts of the extremities lie deep in the central sulcus and the proximal parts are represented more rostrally on the precentral convulution B1.

Supplementary motor fields. There are two additional motor fields apart from the precentral region. The *second motor-sensory region* Ms II **B4** lies on the medial surface of the hemisphere above the cingulate gyrus and includes parts of areas 4 and 6. A somatotopic division has been confirmed in monkeys, but not in man. The *supplementary sensory-motor region* Sm II **B5,** which is principally a tactile sensory and to a lesser extent a motor region lies above the lateral sulcus, and roughly corresponds to area 40. The functional importance of these regions in the motor system has not yet been determined.

Frontal eye field C. Conjugate eye movements may be produced by electrical stimulation of the precentral region, particularly of area 8. It is considered as the frontal center for voluntary eye movements. In general, stimulation causes conjugate deviation of the eyes to the opposite side, sometimes with simultaneous movement of the head. The fibers of area 8 do not end directly on the nuclei of the ocular muscles, but instead are probably relayed in the interstitial nucleus of Cajal.

Motor speech area (Broca's area) D. Damage in the region of the lower frontal convolution (areas 44 and 45) of the dominant hemisphere (p. 246) produces **motor aphasia.** Patients are no longer able to formulate words or to utter them, even though the speech muscles (lips, tongue and larynx) are not paralyzed. Understanding of speech is maintained.

It is certainly not possible to localize speech in a circumscribed part of the cortex (speech center). Speech is one of the highest cerebral functions and it involves large regions of the cortex. However, Broca's area is a most important synaptic region in the complex neuronal basis of speech.

Sensory aphasia p. 246.

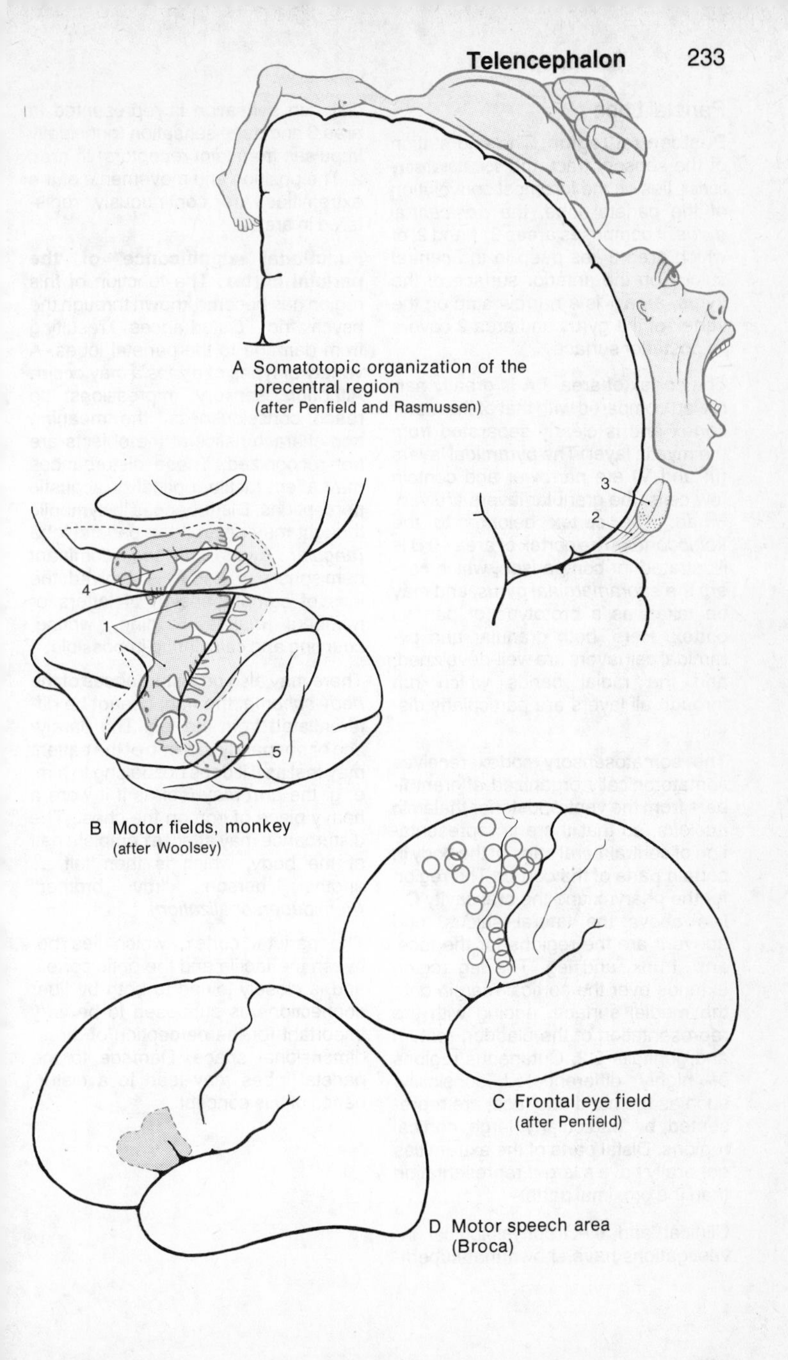

A Somatotopic organization of the precentral region
(after Penfield and Rasmussen)

B Motor fields, monkey
(after Woolsey)

C Frontal eye field
(after Penfield)

D Motor speech area
(Broca)

Parietal Lobe

Postcentral region. The end station of the sensory tract, the **somatosensory cortex,** lies on the foremost convolution of the parietal lobe, the *postcentral gyrus.* It comprises areas 3, 1 and 2, of which area 3 lies deep in the central sulcus on the anterior surface of the gyrus, area 1 is a narrow strip on the vertex of the gyrus and area 2 covers its posterior surface.

The cortex of area 3 **A** is greatly narrowed compared with that of the motor cortex and is clearly separated from the myelin layer. The pyramidal layers (III and V) are narrower and contain few cells, the granular layers are very broad. The cortex belongs to the *koniocortex.* The cortex of area 40 **B** is illustrated for comparison, which covers the *supramarginal gyrus,* and may be taken as a prototype of parietal cortex. Here, both granular and pyramidal cell layers are well-developed, and the radial bands which run through all layers are particularly distinct.

The somatosensory cortex receives somatotopically organized afferent fibers from the ventroposterior thalamic nucleus, so that there is representation of contralateral parts of the body in certain parts of the cortex. The region for the pharynx and the oral cavity **C1** lies above the lateral sulcus, and above it are the regions for the face, arm, trunk, and leg. The leg region extends over the cortical margin onto the medial surface, ending with the representation of the bladder, rectum and genitalia **C2.** Cutaneous regions of highly differentiated sensibility, such as the hand and face, are represented by particularly large cortical regions. Distal parts of the extremities generally have a larger representation than the proximal parts.

Clinical and electrophysiological investigations have shown that superficial skin sensation is represented in area 3 and deep sensation (principally impulses from joint receptors) in area 2. The position and movements of the extremities are continuously registered in area 2.

Functional significance of the parietal cortex. The function of this region has become known through the psychiatric disturbances resulting from damage to the parietal lobes. A variety of types of *agnosia* may occur. Although sensory impressions do reach consciousness, the meaning and characteristics of the objects are not recognized. These disturbances may affect tactile, optical or acoustic perceptions. Disturbances in symbolic thought may occur if the parietal lobe *(angular gyrus)* of the dominant hemisphere (p. 246 A) is involved: the loss of comprehension of letters or numbers makes reading, writing, counting and calculating impossible.

There may also be disturbances of the *body schema:* the right cannot be differentiated from the left. The paralyzed or nonparalyzed limb of the patient may feel as if it does not belong to him, e. g. the arm may feel as if it were a heavy piece of iron on the chest. The disturbance may affect a complete half of the body, which is then felt as another person, "my brother" *(hemidepersonalization).*

The parietal cortex, which lies between the tactile and the optic cortex, and is closely joined to both by fiber connections, is supposed to be very important for the perception of three-dimensional space. Damage to the parietal lobes may lead to a disturbance of this concept.

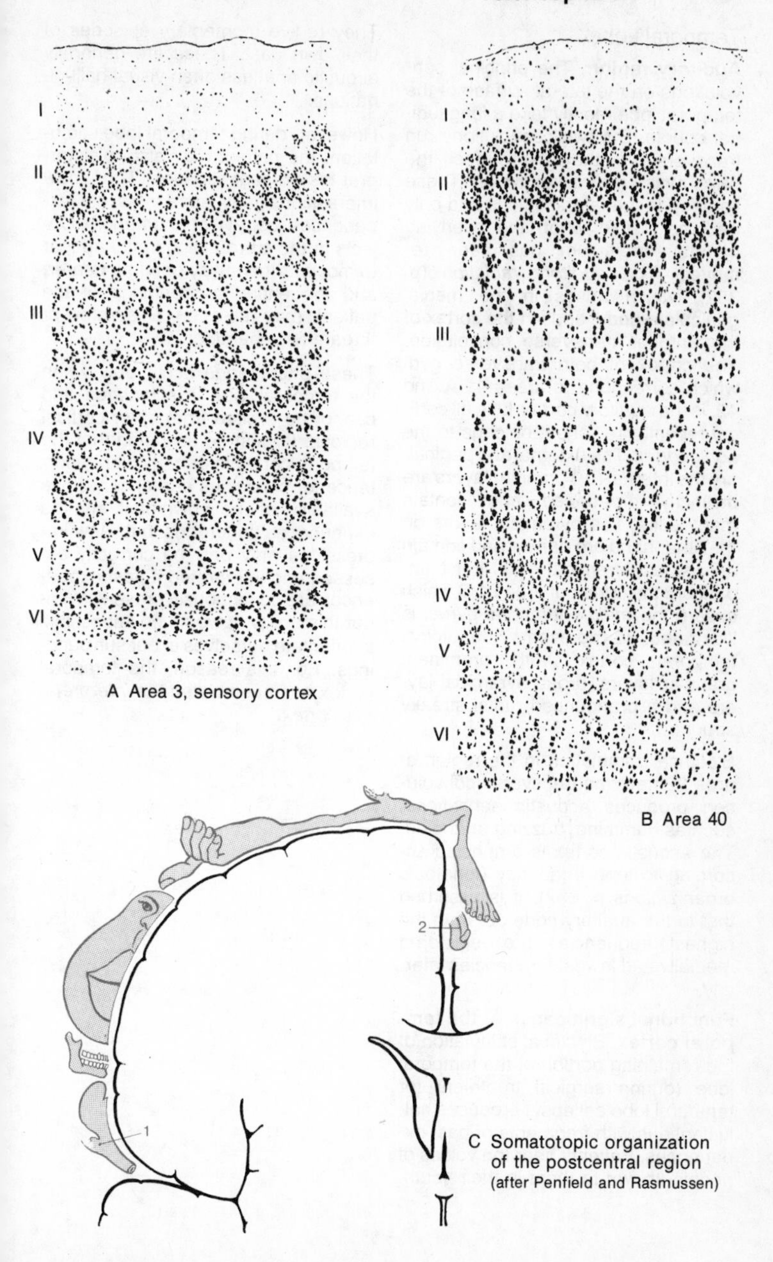

A Area 3, sensory cortex

B Area 40

C Somatotopic organization of the postcentral region (after Penfield and Rasmussen)

Temporal Lobe

Auditory region. The principal convolutions on the lateral surface of the temporal lobe mostly take a longitudinal course. But two convolutions run transversely on the dorsal surface, the **Heschl's transverse temporal gyri C 1.** These lie deep in the lateral sulcus and only become visible when the overlying parietal operculum has been removed. The *auditory radiation* (p. 352), which arises in the medial geniculate body, ends in the cortex of the anterior transverse convolution. The cortex of both transverse gyri, which correspond to areas 41 **A** and 42, is the *auditory cortex.* Like all cortical receptor regions it belongs to the *koniocortex.* The external and particulary the internal (IV) granular layers are distinctly broadened and contain many cells. The pyramidal layers on the other hand are narrow and contain only small pyramidal cells. For comparison, the cortex of area 21 **B** which covers the *middle temporal gyrus,* is illustrated. It represents the typical temporal cortex, with prominent granular layers, broad pyramidal layers and a pronounced radial streaky structure.

Electrical stimulation in the region of area 22 near the transverse convolution produces acoustic sensations, such as humming, buzzing or ringing. The acoustic cortex is organized according to tone frequency (tonotopic organizations p. 352). It is assumed that in the auditory cortex of man the highest frequencies are registered medially and lowest frequencies laterally.

Functional significance of the temporal cortex. Electrical stimulation of the remaining portion of the temporal lobe (during surgical treatment for temporal lobe epilepsy) produces hallucinations with fragments of past experiences. Patients hear the voices of people known to them in their youth. They re-live momentary episodes of their own past. These are generally acoustic and less often visual hallucinations.

However, during temporal lobe stimulation, the patient may also misinterpret his present situation. Thus, new impressions may appear as old experiences *(déja-vu phenomenon).* Objects in the surroundings may appear to move further away or come nearer, and the entire surroundings of the patient may take on a sinister or threatening character.

These phenomena occur only when the temporal lobe is stimulated and cannot be elicited from other cortical regions. It must be assumed that the temporal cortex is of particular importance for conscious and unconscious availability of one's own past, and the experiences undergone in it. New impressions may only be correctly assessed and interpreted if past experiences are continuously present. Without this ability, it would be impossible to orientate ourselves in our surroundings. For this reason, the temporal cortex has been called the *interpretative cortex.*

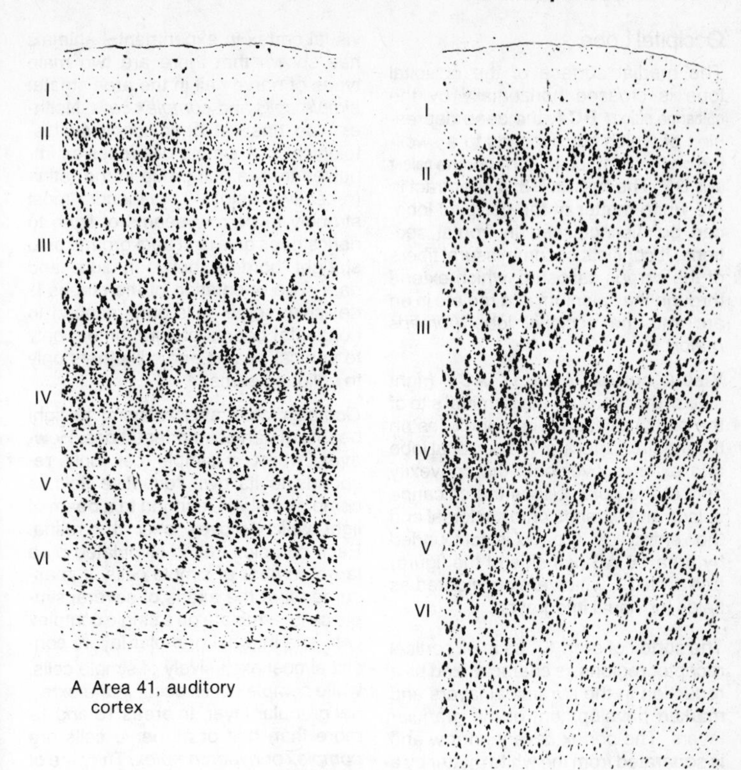

A Area 41, auditory cortex

B Area 21

1

C Transverse temporal gyri, Heschl's transverse gyri

Occipital Lobe

The medial surface of the occipital lobe is crossed horizontally by the **calcarine sulcus BC1,** the deep depression of which corresponds to a swelling on the ventricular surface, the **calcar avis B2.** The **tapetum B3,** a fiber tract in the white matter of the occipital lobe, can be distinguished in frontal sections. It consists of commissural fibers of the corpus callosum, which extend through the splenium and radiate in an arc into the occipital lobe *(forceps major).*

Visual cortex. Area 17 **AC4** (right side of the figure) is the terminal site of the optic radiation. The cortex lies on the medial surface of the occipital lobe and extends over onto the convexity only at the pole. It lines the calcarine sulcus and extends onto its dorsal and the ventral lips. Area 17 is surrounded by area 18 **A5** (left side of the figure) and area 19, which are considered as fields for visual integration.

The cortex of area 17, as in all cortical receptor regions, is characterized by a reduction in the pyramidal layers and marked development of the granular layers. The cortex is very narrow and is separated from the white matter by a richly cellular lamina VI. The internal granular layer (IV) is divided by a hypocellular zone (IVb), which corresponds to the *Gennari stripe* in sections stained for myelin. In this sparsely celled zone lie very large cells, the giant stellate or *Meyner's cells.* The two cell-rich layers of the inner granular layer (IVa and IVc) contain very small granule cells. They belong to the most densely cellular layers of the entire cerebral cortex. Area 18 has a homogenous granular layer of large granular cells. Area 19 forms a transition to the parietal and temporal cortex.

Functional organization. Electrophysiological investigation of the visual cortex in experimental animals has shown that there are two main types of nerve cell in the area striata: *simple cells* and *complex cells.* Neither type has yet been identified histologically. A simple cell receives impulses from a cell group in the retina (receptive field). It responds most strongly to narrow beams of light, to dark stripes on a light background or to straight borders between light and dark. The orientation of the stripes is decisive; some cells only respond to horizontal beams of light, others only to vertical beams and still others only to oblique beams.

Complex cells also respond to light beams of particular orientations. However, whilst a simple cell only responds to its own receptive field, a complex cell will respond to a beam of light which moves over the retina. Each complex cell is stimulated by a large number of simple cells. It is assumed that the axons of several simple cells terminate on a single complex cell. The internal granular layers consist almost exclusively of simple cells, while complex cells occur in the external granular layer. In areas 18 and 19 more than half of all nerve cells are *complex* or *hypercomplex.* They are of particular importance for shape and pattern recognition.

Electrical stimulation of the visual cortex produces a sensation of light or flashes. Stimulation of areas 18 and 19 is supposed to produce also figures and shapes. In addition there is deviation of gaze *(occipital eye center).* The eye movements controlled by the occipital lobes are purely reflex ones, in contrast to the voluntary movements which are controlled by the frontal eye center.

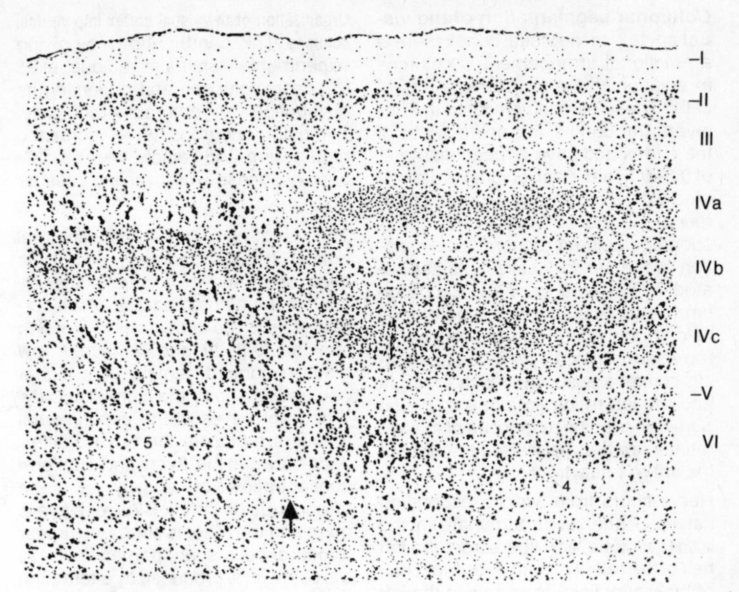

A Area 17 (after Brodmann)

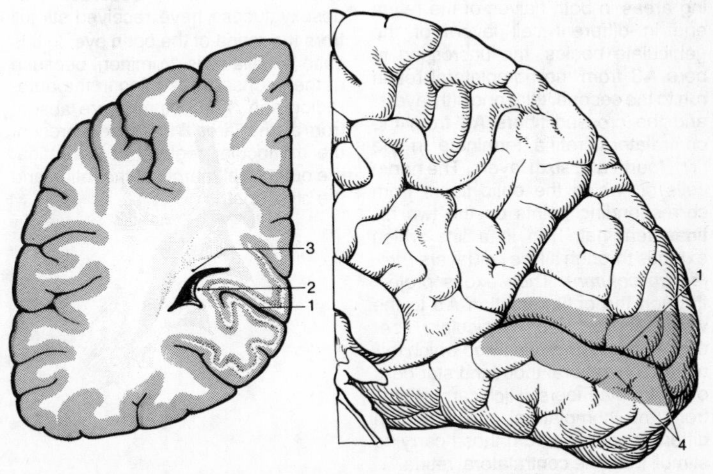

B Frontal section of occipital lobes C Medial section with area 17

Columnar segmentation of the visual cortex. In the visual cortex there is a functional arrangement in columns in addition to the structural layers of cells. These extend vertical to the cell layers throughout the entire width of the cortex and have a mean diameter of 0.3–0.5 mm. Each column is associated with a circumscribed area of the retina. All the neurons in the corresponding column respond to stimulation of the sensory cells in its associated retinal area (p. 226). Each column is associated with the peripheral field in only one of the two retinas. In the visual cortex **A1** there are alternating columns associated with the right and the left retina *(columns of ocular dominance)*. The responses of the two retinas are thus separated throughout the entire visual tract.

Nerve fibers from two corresponding halves of the retina terminate in the **lateral geniculate body:** fibers from the left half of the retina of both eyes (right half of the visual field) terminate in the left geniculate body **A2,** and fibers from the right half of both retinas (left half of the visual field) end in the right geniculate body. The fibers from corresponding areas in both halves of the retina end in different cell layers of the geniculate bodies: the uncrossed fibers **A3** from the homolateral retina run to the second, third and fifth layers, and the crossed fibers **A4** from the contralateral retina terminate in the first, fourth and sixth layers. The nerve cells on which the optic fibers from corresponding points on the two retinas terminate run in a line which extends through all the cell layers *(projection column)*. Their axons project through the optic radiation **A5** to the visual cortex. Each geniculate fiber divides into numerous twigs which terminate on several thousand star cells of lamina IV. Fibers which carry stimuli from the homolateral retina end in different columns from those carrying stimuli from the contralateral retina.

Organization of the visual cortex into vertical columns. This can be shown by giving experimental animals radio-labelled ^{14}C-desoxyglucose and observing its distribution autoradiographically. Stimulated neurons have increased metabolism and rapidly take up ^{14}C-desoxyglucose whilst resting cells do not.

The visual cortex of an experimental animal (rhesus monkey) with both eyes open shows a striped appearance in the autoradiograph which corresponds to the known cell layers **B6:** laminae I, II, III and V have a low glucose content, lamina VI a higher one and lamina IV the highest activity. When both eyes of the experimental animal are closed, there is no difference between the layers and a uniform concentration of activity is seen throughout the cortex **B7.** If one eye is open and the other is shut, a columnar appearance is seen, perpendicular to the cell layers, produced by alternating dark and pale columns **B8.** The shut eye is represented by the pale columns, whose neurons have not taken up any labelled glucose. The dark columns with newly absorbed ^{14}C-desoxyglucose have received stimuli from the retina of the open eye. In this case lamella IV is prominent because of the intense blackening in the autoradiograph. The columns are absent from a small area **B9.** These represent the monocular regions of the retina, the outermost margin of the retina and the blind spot.

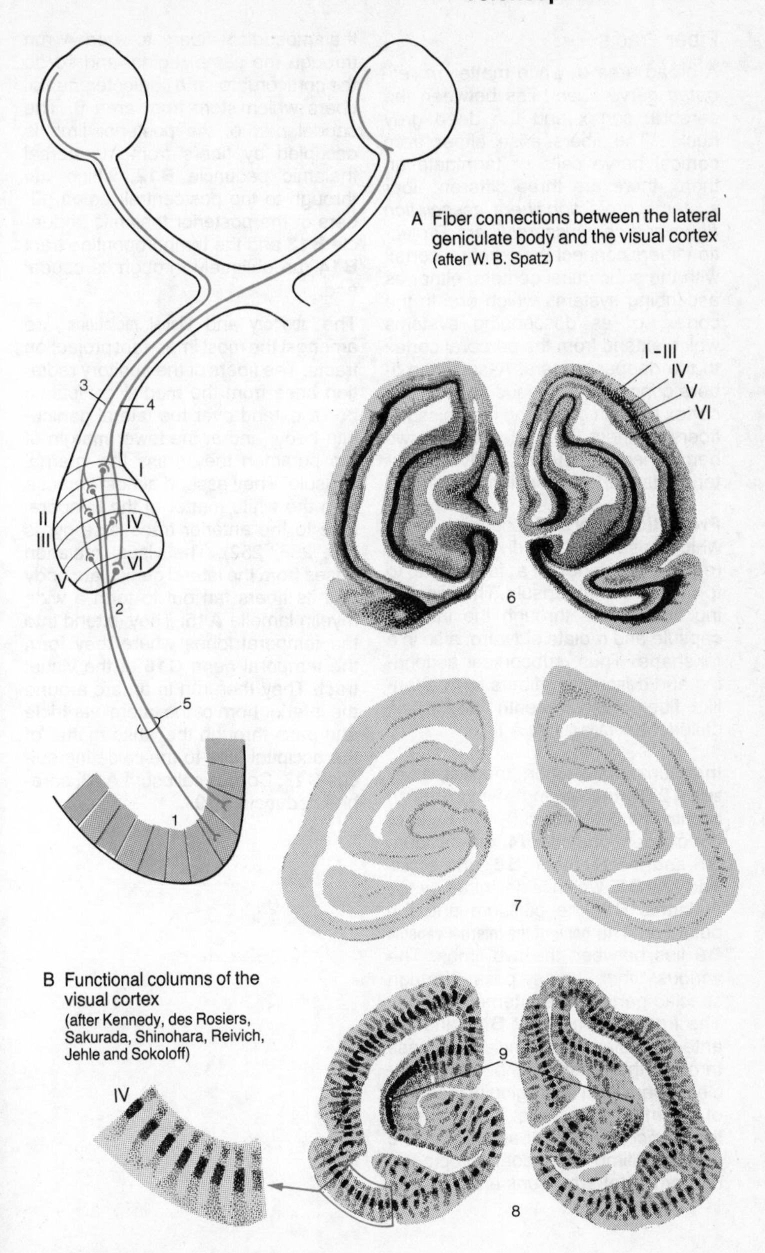

A Fiber connections between the lateral geniculate body and the visual cortex (after W. B. Spatz)

B Functional columns of the visual cortex (after Kennedy, des Rosiers, Sakurada, Shinohara, Reivich, Jehle and Sokoloff)

Fiber Tracts

A broad area of white matter *(myelinated nerve fibers)* lies between the cerebral cortex and the deep grey nuclei. The fibers arise either from cortical nerve cells or terminate on them: there are three different fiber systems: *projection fibers, association fibers* and *commissural fibers.* Projection fibers connect the cerebral cortex with the subcortical centers, either as ascending systems which end in the cortex, or as descending systems which extend from the cerebral cortex to the deeper centers. Association fibers connect the various parts of the cortex to each other and commissural fibers connect the cortex of the two hemispheres: they are really only interhemispherical association fibers.

Projection fibers. Descending tracts which arise from the diverse cortical regions combine in a fan-shape to form the internal capsule. The ascending fibers pass through the internal capsule and radiate outward, also in a fanshape. Thus, subcortical ascending and descending fibers form a fan-like fiber mass beneath the cortex, called the **corona radiata A1.**

In a horizontal section, the **internal capsule A2 B** forms an angle which has an **anterior limb B3,** limited by the head of the caudate nucleus **B4,** the pallidum **B5** and the putamen **B6,** and a **posterior limb B7,** which is delimited by the thalamus **B8,** the pallidum and the putamen. The **genu of the internal capsule B9** lies between the two limbs. The various fiber tracts pass through specific parts of the internal capsule. The frontopontine tract **B10** and the anterior thalamic peduncle **B11** pass through the anterior limb. Corticonuclear fibers lie in the region of the genu of the internal capsule. The fibers of the *corticospinal tract* adjoin it in the posterior limb in a somatotopic arrangement of arm, trunk and leg. The thalamocortical fibers to area 4 run through the same region, and so do the corticorubro- and corticotegmental fibers which stem from area 6. The caudal part of the posterior limb is occupied by fibers from the dorsal thalamic peduncle **B12,** which run through to the postcentral region. Fibers of the posterior thalamic peduncle **B13** and the temporopontine tract **B14** run obliquely through its caudal end.

The **auditory** and **visual radiations** are amongst the most important projection tracts. The fibers of the auditory radiation arise from the medial geniculate body, extend over the lateral geniculate body, and at the lower margin of the putamen they cross the internal capsule. They ascend almost vertically in the white matter of the temporal lobe to the anterior transverse gyrus (pp. 236, 252). The visual radiation arises from the lateral geniculate body and its fibers fan out to form a wide myelin lamella **A15.** They extend into the temporal lobes where they form the temporal genu **C16** of the visual tract. They then run in an arc around the inferior horn of the lateral ventricle and pass through the white matter of the occipital lobe to the calcarine sulcus **C17.** Corpus callosum **A18,** cerebral peduncle **A19.**

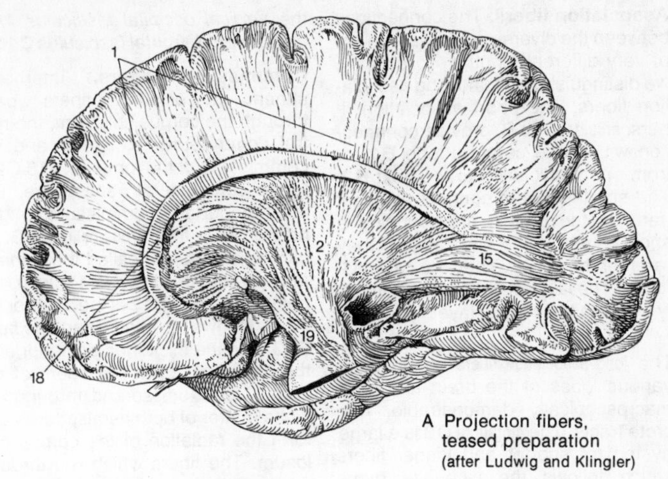

A Projection fibers, teased preparation
(after Ludwig and Klingler)

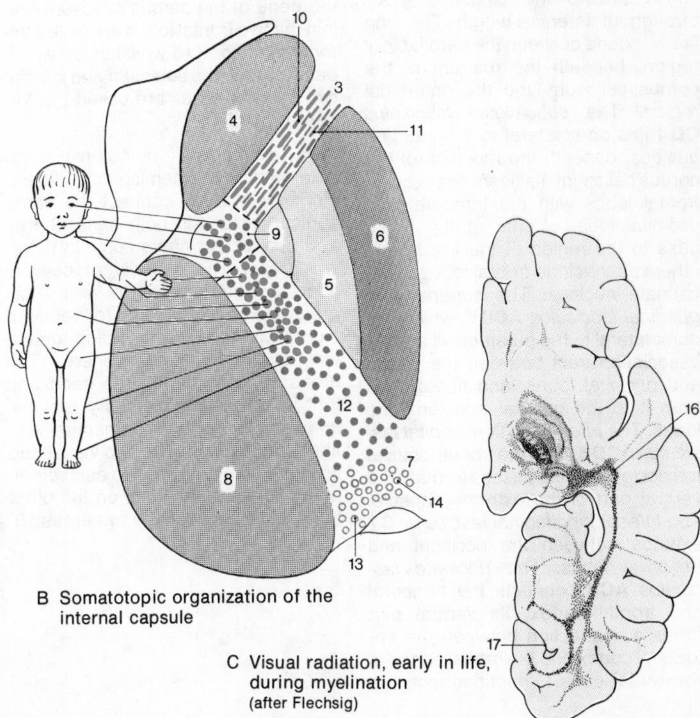

B Somatotopic organization of the internal capsule

C Visual radiation, early in life, during myelination
(after Flechsig)

Association fibers. The connections between the diverse cortical areas are of very different lengths. To simplify, we distinguish short and long association fibers. The short association fibers, **arcuate fibers** B, provide connnection within one cerebral lobe **B1,** or from one convolution to the next **B2.** The shortest fibers connect adjacent parts of the cortex; after running a short course in the white matter they re-enter the cortex. They are called U-fibers. The layer of U-fibers lies directly beneath the cortical layer.

The **long association fibers** connect the various lobes of the brain and form macroscopically demonstrable, discrete tracts. The **cingulum D3** is a large system of shorter and longer fibers which follows the cingulate gyrus throughout its entire length. The long fibers extend between the parolfactory region, beneath the rostrum of the corpus callosum and the entorhinal region. The *subcallosal fasciculus* **CD4** lies dorsolateral to the caudate nucleus, beneath the radiation of the corpus callosum. Its fibers connect the frontal lobes with the temporal and occipital lobes. Some of the fibers pass to the region of the insula and others connect the frontal lobes to the caudate nucleus. The *superior longitudinal fasciculus* **ACD5,** which lies dorsolateral to the putamen, is a large association tract between the frontal and occipital lobes, and fibers pass from it to the parietal and temporal lobes. The *inferior fronto-occipital fasciculus* **ACD6,** from the frontal lobes to the occipital lobes, passes through the ventral part of the extreme capsule. The *inferior longitudinal fasciculus* **C7** extends between the occipital and temporal lobes. The *uncinate fasciculus* **AC8** connects the temporal and frontal cortex. Its ventral part forms a connection between the entorhinal cortex and the orbital cortex of the frontal lobes. Other fiber tracts are the *vertical occipital fasciculus* **AC9** and the *orbitofrontal fasciculus* **C10.**

Commissural fibers. Interhemispheric association fibers pass through the corpus callosum, the rostral commissure (p. 214C) and the commissure of the fornix (p. 218 C25) to the contralateral hemisphere. The most important commissure of the neocortex is the **corpus callosum E.** Its curved oral part is called the *genu of the corpus callosum* **E11** with the pointed *rostrum* **E12** at the anterior tip. There follows the middle part, the *body* **E13,** and the thickened posterior end, the *splenium* **E14.** The fibers of the corpus callosum spread through the white matter of both hemispheres and form the radiation of the corpus callosum. The fibers which run through the genu of the corpus callosum and join the two frontal lobes are called the *forceps minor* **F15** whilst those which pass through the splenium and join the two occipital lobes are called the *forceps major* **F16.**

We distinguish homotopic and heterotopic interhemispheric fibers. *Homotopic fibers* connect the same cortical regions in both hemispheres and *heterotopic fibers* connect different areas. The majority of callosal fibers are homotopic. Not all cortical areas are interconnected to the same extent with the corresponding area in the contralateral hemisphere. The areas concerned with the hand and foot in both somatosensory regions, for example, have no interhemispheric fiber connections. The two visual cortices are also not joined to each other. Many fiber connections, on the other hand, exist between the two areas 18.

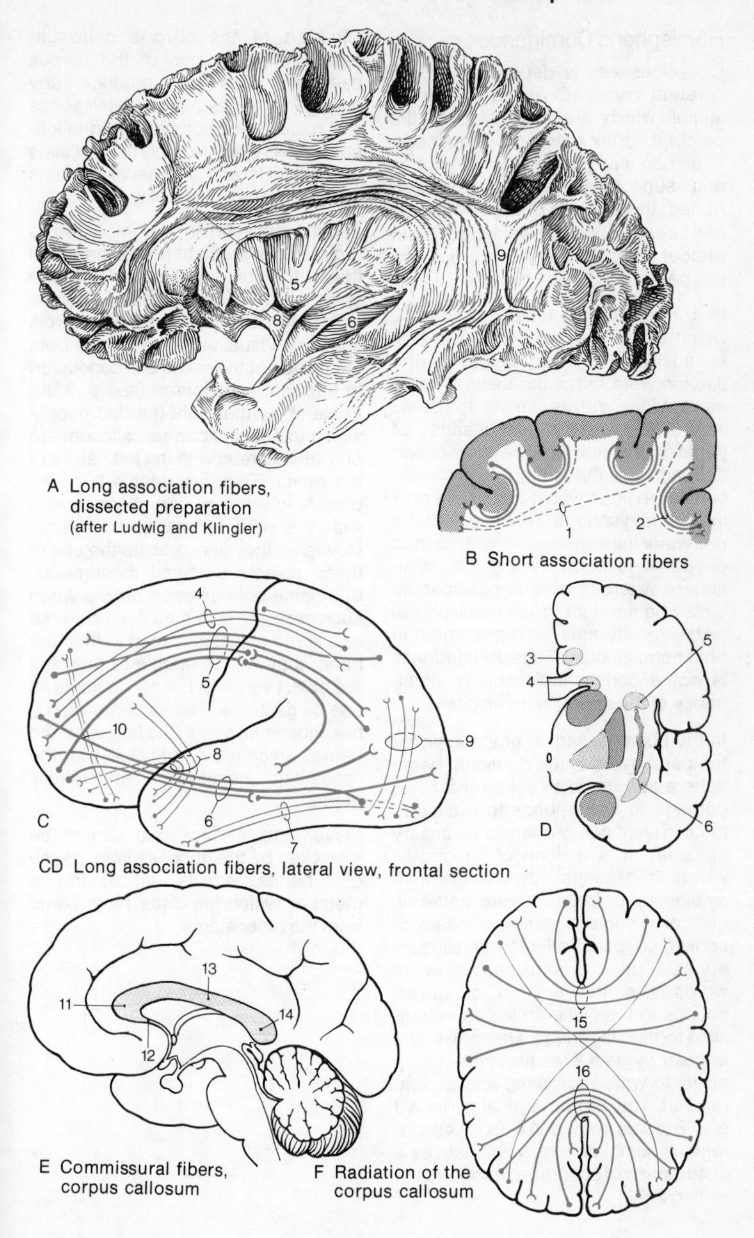

A Long association fibers, dissected preparation (after Ludwig and Klingler)

B Short association fibers

C

D

CD Long association fibers, lateral view, frontal section

E Commissural fibers, corpus callosum

F Radiation of the corpus callosum

Hemispheric Dominance

Consciousness is dependent on the cerebral cortex. Only those sensory stimuli which are conducted to the cerebral cortex reach consciousness. If the connections between the cortex and subcortical centers are interrupted, there is loss of consciousness and later an apparant waking state but without consciousness *(apallic syndrome)*.

Only man has the faculty of speech, which as internal speech is the presupposition for thought, just as the spoken word forms the basis of communication, and as writing transmits information across thousands of years. In the individual person speech depends on the integrity of certain cortical regions which usually lie only in one hemisphere. This is called the *dominant hemisphere;* in right-handed people it is normally the left hemisphere, whereas in left-handed people it may be the right or left hemisphere, or the faculty may be represented in both hemispheres. Thus, handedness is not a certain indication of dominance of the opposite hemisphere.

In the posterior region of the superior temporal gyrus of the dominant hemisphere lies **Wernicke's speech center A1,** damage to which leads to a disturbance in word comprehension (sensory aphasia). It is a zone of integration which is essential for the constant availability of learned word patterns, and for the interpretation of heard or spoken speech. Patients with sensory aphasia utter a senseless flow of words and the speech of others sounds to them like an incomprehensible foreign language. Damage to the *angular gyrus* **A2** results in loss of the ability to write *(agraphia)* and to read *(alexia).* The supramarginal gyrus **A3** and Broca's region **A4** (p. 232) are regions stimulation of which causes a disturbance of spontaneous speech or writing **A5.**

Division of the corpus callosum (split brain). Division of the corpus callosum does not produce any change in personality or intelligence and patients are completely unnoticeable in everyday life. Only special tests of the tactile and the visual systems will reveal any abnormality **B.**

Touch sensation in the left hand is registered in the right hemisphere and that of the right hand in the left hemisphere (in right-handed people it is the dominant hemisphere which controls speech). Visual stimuli which hit both left halves of the retina are conducted to the right hemisphere (see p. 328). Tests show that right-handed people with a transected corpus callosum are only able to read with the left halves of the retina. They are unable to name objects which they can only perceive with the right halves of the retina. However, they can indicate the use of these objects by hand movements. The same phenomenon occurs when such persons have their eyes covered and some object is placed in their left hand: they will be unable to describe the object verbally, but can indicate its use by gestures. Things perceived by the right hand, or with the left half of the retina, which are associated with the "speaking" hemisphere are readily named.

Movements of one limb cannot be mimicked by the opposite limb, as the one hemisphere is unable to remember which impulses have come from the other side.

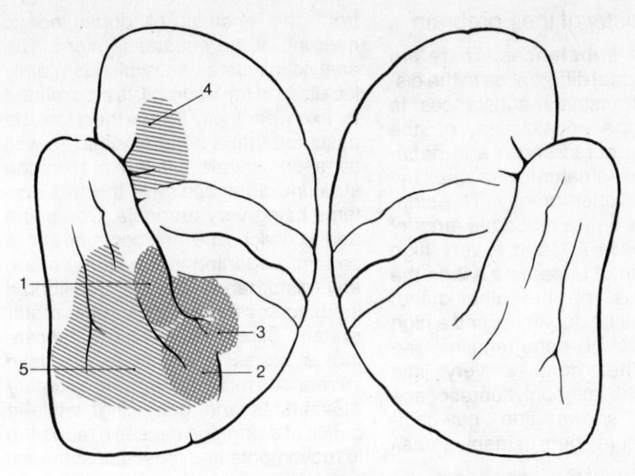

A Speech and writing centers in a right-handed subject

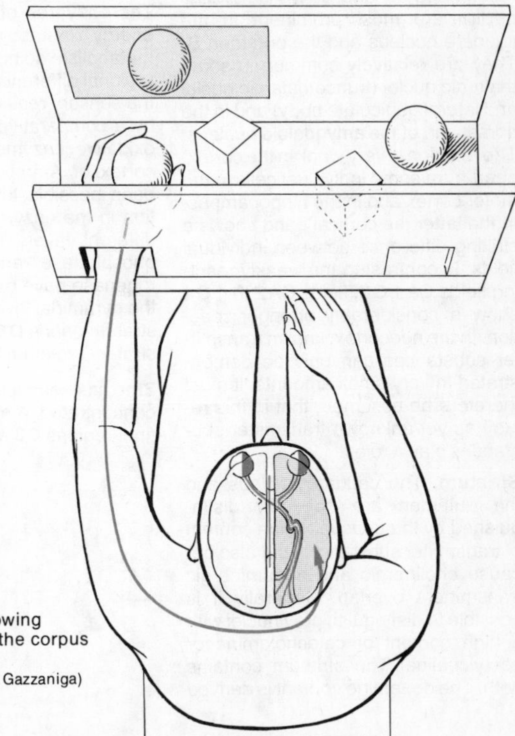

B Reactions following transection of the corpus callosum
(after Sperry and Gazzaniga)

Histochemistry of the Forebrain

Transmitter substances. There are marked regional differences in the distribution of transmitter substances. In particular, the localization of the **catecholamines** noradrenalin and dopamine in the striatum and the hypothalamus varies greatly. The striatum contains only a moderate amount of *noradrenalin* **A1** and a very high concentration of *dopamine* **A2.** In the hypothalamus, on the other hand, there is very little dopamine and a high concentration of noradrenalin. The cerebral cortex contains very little catecholamine, although fluorescence microscopy shows fine catecholaminergic nerve fibers in many areas.

Cholinergic neurons, which are demonstrated by the **acetylcholinesterase reaction** are most prominent in the caudate nucleus and the putamen **B.** They are relatively common in some thalamic nuclei (truncothalamic nuclei and lateral geniculate body) and in the dorsal part of the amygdaloid nucleus. The reaction is very weak in the cortex, apart from some individual cells in the motor cortex and in the hippocampus. In the latter the cortical band shows a striking difference between individual fields: in contrast to the weakly reacting fields CA1 **C3,** fields CA2 to 4 **C4** show a considerably stronger reaction. In the neocortex, known transmitter substances can only be demonstrated in very small amounts, it must therefore be assumed, that in this region as yet unknown transmitter substances play a role.

Striatum. The caudate nucleus and the putamen are not only distinguished by the unusually high content of transmitter substances, but also because cholinergic and dopaminergic transmitters overlap. Generally it is possible to distinguish grey nuclei with a high content of catecholamine or acetylcholine. The striatum contains both. The dopamine content is derived from the endings of dopaminergic neurons of the substantia nigra. The acetylcholinesterase, which is mainly localized in the neuropil, is also related to axon endings. It can therefore be assumed that a dopaminergic as well as a cholinergic system end in the striatum, antagonism of the two systems being very probable. Dopamine and acetylcholine are contained in a certain proportion in the striatum and any disturbance of this equilibrium leads to severe changes in the motor system. Clinical observations indicate that an excess of dopamine results in increased mobility with involuntary movements and grimacing, while a deficit of dopamine causes a reduction in movements and muscular stiffness, as in *Parkinson's disease.*

The **enzymes** of cellular respiration, energy metabolism and structural metabolism do not show such marked regional differences. In the neocortex, the sensory regions are characterized by a comparatively higher content of oxidative enzymes. It is only in the cortex of Ammon's horn that it has been possible to demonstrate variation in the enzyme distribution in the different layers of the cortex. Acid phosphatase and succinic dehydrogenase have been demonstrated in the pyramidal layer **D5** ATPase in the stratum oriens **D6** and aldolase in the stratum radiatum **D7.**

Zinc has been found to be particularly concentrated in endings in the mossy fiber regions **C3** and **C4** (p. 220).

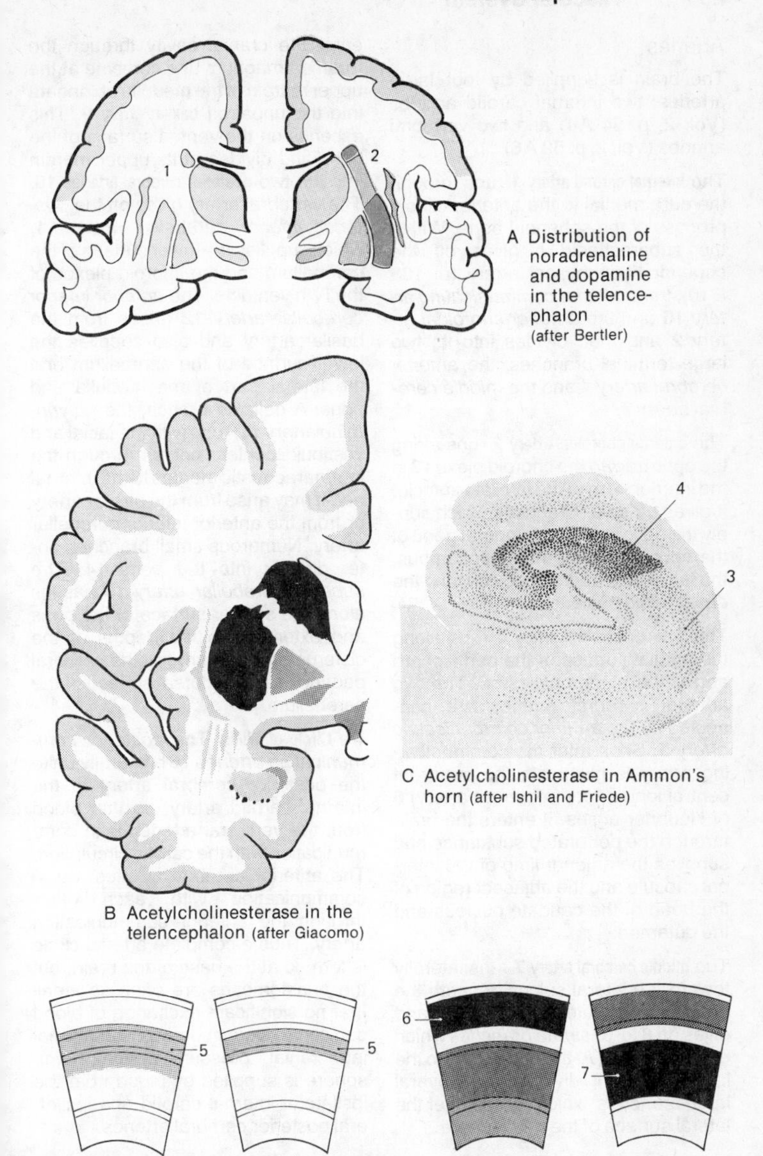

A Distribution of noradrenaline and dopamine in the telencephalon (after Bertler)

B Acetylcholinesterase in the telencephalon (after Giacomo)

C Acetylcholinesterase in Ammon's horn (after Ishil and Friede)

Acid phosphatase Succinic dehydrogenase ATPase Aldolase

D Enzyme pattern of the cortex of hippocampus H3 (after Fleischhauer and after Lowry)

Arteries

The brain is supplied by four large arteries: two internal carotid arteries (Vol. 2, p. 54 A4) and two vertebral arteries (Vol. 2, p. 52 A6).

The **internal carotid artery 1** runs through the dura medial to the anterior clinoid process of the sphenoid bone. Within the subarachnoid it gives off the superior hypophyseal artery (p. 188 E10), the *posterior communicating artery* **16** and the *anterior choroidal artery* **2** and then divides into its two large terminal branches, the *anterior cerebral artery* **4** and the *middle cerebral artery* **7.**

The **anterior choroidal artery 2** runs along the optic tract to the choroid plexus **3** in the inferior horn of the lateral ventricle. It gives off small branches which supply the optic tract, the temporal genu of the optic radiation, the hippocampus, the tail of the caudate nucleus, and the amygdaloid nucleus.

The **anterior cerebral artery 4** runs along the medial surface of the hemisphere above the corpus callosum. The two anterior cerebral arteries are connected by the *anterior communicating artery* **5.** Soon after the communicating artery is given off, the recurrent central long artery (recurrent artery) **6** of Heubner arises. It enters the brain through the perforated substance and supplies the anterior limb of the internal capsule and the adjacent region of the head of the caudate nucleus and the putamen.

The **middle cerebral artery 7** runs laterally toward the lateral sulcus beneath the anterior perforated substance, and gives off 8 to 10 *striate branches* which enter the brain. At the entrance into the lateral fossa it divides into several large branches, which spread over the lateral surface of the hemisphere.

The two **vertebral arteries 8,** which arise from the paired subclavian arteries, enter the cranial cavity through the foramen magnum and combine at the upper margin of the medulla oblongata into the unpaired **basilar artery 9.** This ascends on the ventral surface of the pons and divides at its upper margin into the two **posterior cerebral arteries 10.** The vertebral artery gives off the *posterior inferior cerebellar artery* **11,** which supplies the lower surface of the cerebellum and the choroid plexus of the IVth ventricle. The *anterior inferior cerebellar artery* **12** arises from the basilar artery and also supplies the lower surface of the cerebellum and the lateral part of the medulla and pons. A delicate branch, the *labyrinthine artery* **13,** runs with the facial and vestibulocochlear nerves through the internal acoustic meatus to the internal ear. It may arise from the basilar artery or from the anterior inferior cerebellar artery. Numerous small branches enter directly into the *pons* **14.** The *superior cerebellar artery* **15** passes along the supper surface of the pons and extends into the depths of the cisterna ambiens around the cerebral peduncles to the dorsal surface of the cerebellum.

The Circle of Willis. The *posterior communicating arteries* **16** bilaterally unite the posterior cerebral artery to the internal carotid artery, so that blood from the vertebral arteries is in communication with the carotid circulation. The anterior cerebral arteries are in communication with each other through the anterior communicating artery. Thus a complete arterial circle is formed at the base of the brain, but the anastomoses are often so small that no significant exchange of blood can occur. Under conditions of normal intracranial pressure, each hemisphere is supplied by blood from the ipsilateral internal carotid and ipsilateral posterior cerebral arteries.

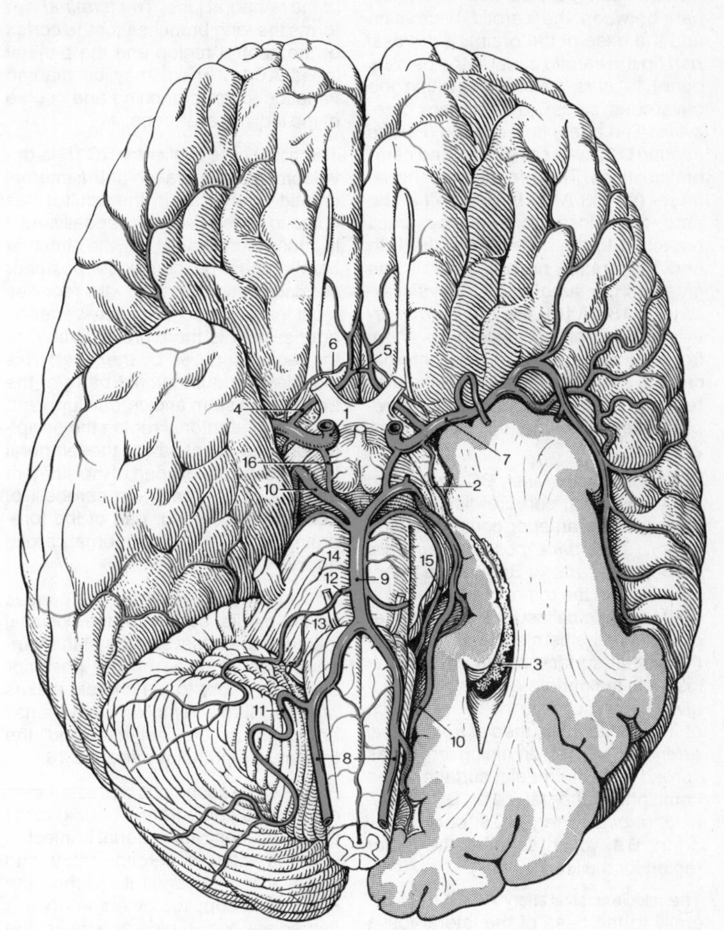

Vessels at the base of the brain

Internal Carotid Artery

The **internal carotid artery C1** may be divided into a *pars cervicalis* (cervical part between the carotid bifurcation and the base of the brain), a *petrosal part* (in the carotid canal in the petrous bone), a *pars cavernosa* (within the cavernous sinus), and a *pars cerebralis.* The artery is S-shaped (*carotid syphon* **C2**) in its cavernous and cerebral sections. The inferior hypophysial artery (p. 188 A11) is given off in the pars cavernosa, followed by small branches to the dura and to the IVth and Vth cranial nerves. After it has given off the *superior hypophysial artery* (p. 188 A10), the *ophthalmic artery* and the *anterior choroidal artery* from its cerebral part, the internal carotid artery divides into two large terminal branches, the *anterior cerebral artery* and the *middle cerebral artery.*

The **anterior cerebral artery BC3** turns into the longitudinal cerebral fissure after giving off the anterior communicating artery. The *pars postcommunicalis (pericallosal artery)* **BC4** of the artery runs around the rostrum and the genu of the corpus callosum **B5** to the medial wall of the hemisphere, and along the dorsal surface of the corpus callosum to the parieto-occipital sulcus. It gives off branches to the basal surface of the frontal lobe (*medial frontobasal artery* **B6**). The remaining branches spread over the medial surface of the hemisphere *frontal rami* **BC7,** *callosomarginal artery* **BC8,** *paracentral artery* **B9,** which supplies the motor region for the limbs.

The **middle cerebral artery AC10** runs laterally to the base of the lateral fossa and divides into several sets of branches. The artery is divided into three parts: from the *sphenoidal part* arise the *central arteries* (fine branches to the striatum, thalamus and internal capsule), the *insular part* gives off the short *insular arteries* **C11,** which supply the insular cortex with the *lateral frontobasal artery* **A12** and the *temporal arteries* **A13** to the cortex of the temporal lobe. The *terminal part* forms the long branches for the cortex of the central region and the parietal lobes **AC14.** There may be marked variation in the branching and course of the individual arteries.

The **posterior cerebral artery BC15** is developmentally a branch of the internal carotid artery, but in the adult it has come to lie relatively far caudally. As it is only connected to the internal carotid artery by the slender posterior communicating artery, it receives most of its blood from the vertebral arteries and is therefore attributed to the region supplied by the latter. This includes the subtentorial parts of the brain (brain stem and cerebellum) and in the supratentorial region the occipital lobe the basal part of the temporal lobe and the caudal part of the striatum and thalamus (tentorium cerebelli p. 268). All the anterior part of the forebrain is supplied by the internal carotid artery.

The posterior cerebral artery arborises on the medial surface of the occipital lobe and the basal surface of the temporal lobe. It gives off the posterior choroidal artery to the choroid plexus of the IIIrd ventricle and small branches to the striatum and the thalamus. Ophthalmic artery **C16.**

Figure **C** is a schematic representation of a carotid angiogram. For diagnostic purposes, contrast material is injected into the internal carotid artery and within a few seconds it reaches the entire area supplied by the artery. An immediate x-ray picture shows the arterial tree. When the film is read it must be remembered that all the vessels are shown in one plane.

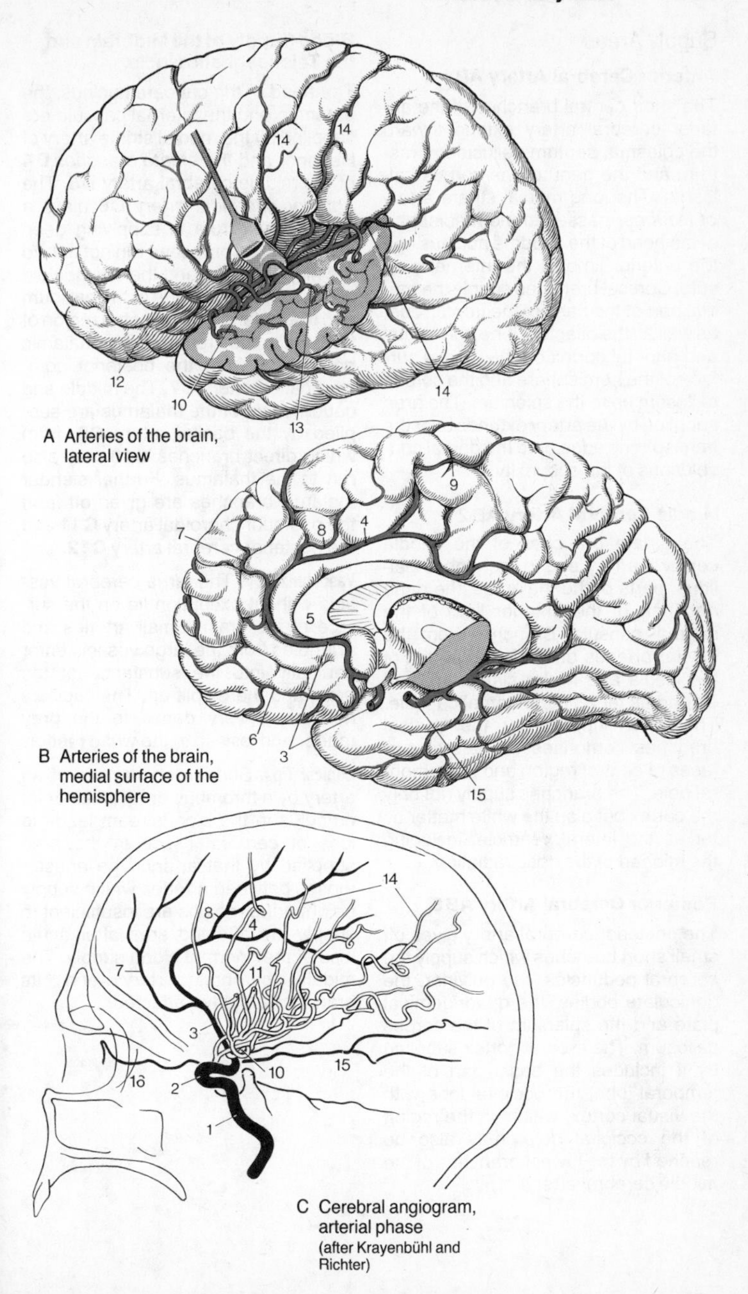

A Arteries of the brain, lateral view

B Arteries of the brain, medial surface of the hemisphere

C Cerebral angiogram, arterial phase (after Krayenbühl and Richter)

Supply Areas

Anterior Cerebral Artery AB 1

The short central branches of the anterior cerebral artery extend toward the chiasma, septum pellucidum, rostrum and the genu of the corpus callosum. The long medial striate *artery of Heubner* passes to the medial part of the head of the caudate nucleus and the anterior limb of the internal capsule. Cortical branches supply the medial part of the base of the frontal lobe, as well as the olfactory lobe, the frontal and parietal cortex on the medial surface of the hemisphere and the corpus callosum up to the splenium. The area supplied by the artery extends over the hemispheric edge onto the dorsal convolutions of the convexity.

Middle Cerebral Artery AB 2

The *striate branches* of the middle cerebral artery end in the globus pallidus, parts of the thalamus, the genu and part of the anterior limb of the internal capsule. Branches from the insular arteries divide in the cortex of the insula and in the claustrum, and extend as far as the external capsule. The area supplied by the cortical branches comprises the lateral surfaces of central region and the temporal pole. The branches supply not only the cortex but also the white matter as far as the lateral ventricle, including the midpart of the optic radiation.

Posterior Cerebral Artery AB 3

The posterior cerebral artery gives off small short branches which supply the cerebral peduncles, the pulvinar, the geniculate bodies, the quadrigeminal plate and the splenium of the corpus callosum. The area of cortex supplied by it includes the basal part of the temporal lobe, the occipital lobe with the visual cortex, which, in the region of the occipital pole, may also be reached by the lowest branches of the middle cerebral artery.

Blood Supply of the Midbrain and the Telencephalic Nuclei

The head of the caudate nucleus, the putamen and the internal capsule are supplied by the medial striate artery of Heubner and the *striate branches* **D4** of the middle cerebral artery **D5.** The *anterior choroidal artery* **C6** plays a very variable role in supplying deep structures. Its branches run not only to the hippocampus and the amygdaloid nuclei, but also to parts of the pallidum and the thalamus. The rostral region of the thalamus receives a thalamic branch **C8** from the posterior communicating artery **C7.** The middle and caudal parts of the thalamus are supplied by the basilar artery **C9,** from which direct branches **C10** may also run to the thalamus. Further slender thalamic branches are given off from the posterior choroidal artery **C11** and the posterior cerebral artery **C12.**

Vascularization. The large cerebral vessels without exception lie on the surface of the brain. Small arteries and arterioles from the large vessels enter vertically into the substance of the brain and there split up. The capillary network is very dense in the grey matter and less so in the white matter.

Clinical Tips. Sudden obstruction of an artery by a thrombus or by an air or fat embolus in the bloodstream leads to loss of cerebral tissue in the area supplied by that artery. The anastomoses between arteries which supply neighboring regions are insufficient to supply the affected area after acute loss of their normal blood supply. The middle cerebral artery and its branches are often affected.

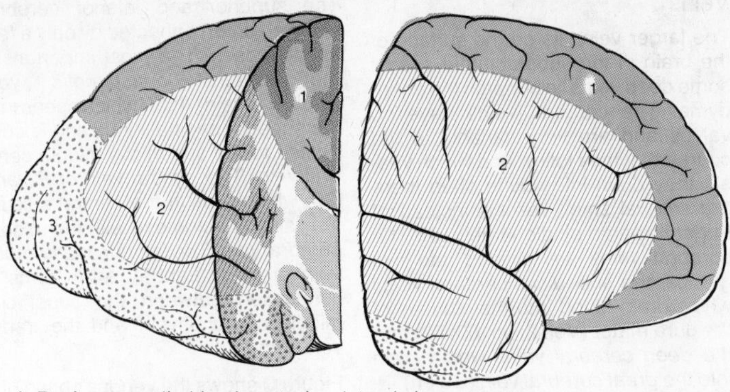

A Regions supplied by the arteries, lateral view

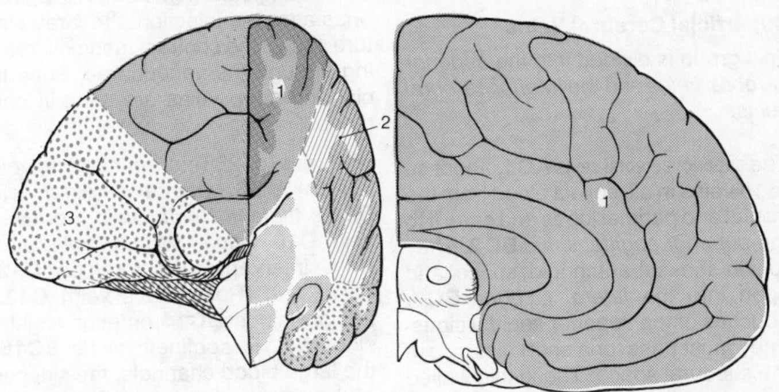

B Regions supplied by the arteries, medial view

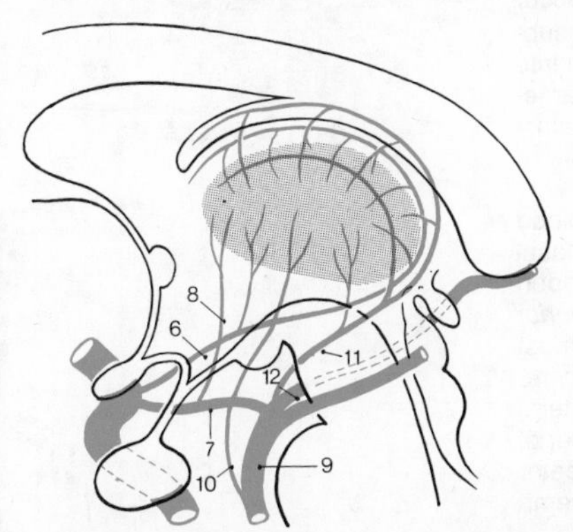

C Arterial supply of the thalamus
(after Van den Bergh and Vander Eeken)

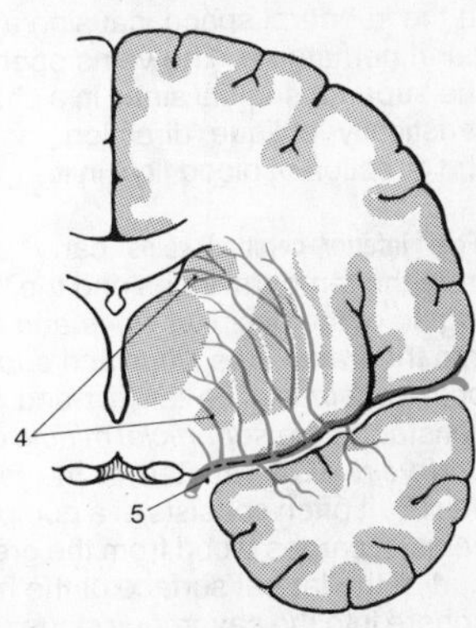

D Arterial supply to the striatum

Veins

The larger veins lie on the surface of the brain in the subarachnoid space; some deep veins lie beneath the ependyma. The cerebral veins have no valves and are very variable in the course that they follow and which vessel they drain into. Quite often there are several small vessels instead of the anticipated single large vein. The cerebral veins are divided into two groups: the *superficial cerebral veins* which drain blood into the sinuses of the dura mater (Vol. 2, pp. 60, 62) and the *deep cerebral veins* which drain into the *great cerebral vein* (Galen).

Superficial Cerebral Veins

This group is divided into the *superior cerebral veins* and the *inferior cerebral veins.*

The **superior cerebral veins AC1,** about 10 to 15 veins in all, collect blood from the frontal and parietal lobes and carry it to the *superior sagittal sinus* **BC2.** They run in the subarachnoid space and open into the *lateral lacunae* **BC3,** pouches of the superior sagittal sinus. They must pass for a short distance in the subdural space. These thin walled veins readily become torn in head injuries and bleeding may then occur in the subdural space, causing a subdural hematoma. The veins open into the superior sagittal sinus in a characteristically oblique direction, against the direction of blood flow in it.

The **inferior cerebral veins** carry blood from the temporal lobes and the basal region of the occipital lobes and open into the *transverse sinus* and *superior petrosal sinus.* The largest and most constant is the *superficial middle cerebral vein* **AC4** situated in the lateral sulcus. It often consists of a number of veins. It carries blood from the greater part of the lateral surface of the hemisphere into the cavernous sinus.

The superior and inferior cerebral veins are interconnected by only a few anastomoses. The most important is the *superior anastomotic vein* (Trolard's vein) **AC5,** which opens into the superior sagittal sinus and is connected to the superficial middle cerebral vein. The *central vein of Roland* **C6,** which runs in the central sulcus, may also anastomose with the middle cerebral vein. The *inferior anastomotic vein* (vein of Labbé) **AC7** forms a connection between the superficial middle cerebral vein and the transverse sinus.

Figure **C** shows the venous phase of a carotid angiogram (p. 252). A few seconds after the injection, the x-ray picture shows the contrast medium draining through the venous tree. Superficial and deep veins are seen in one plane.

Deep veins (p. 258): *great cerebral vein* (Galen) **BC8,** *internal cerebral vein* **BC9,** *thalamostriate vein* (terminal vein) **C10,** *vein of septum pellucidum* **C11,** interventricular foramen **C12,** *basal vein* (Rosenthal's vein) **C13,,** straight sinus **BC14,** inferior sagittal sinus **BC15,** confluent sinus **BC16;** the large blood channels, the sinuses of the dura mater see Vol. 2, pp. 60, 62.

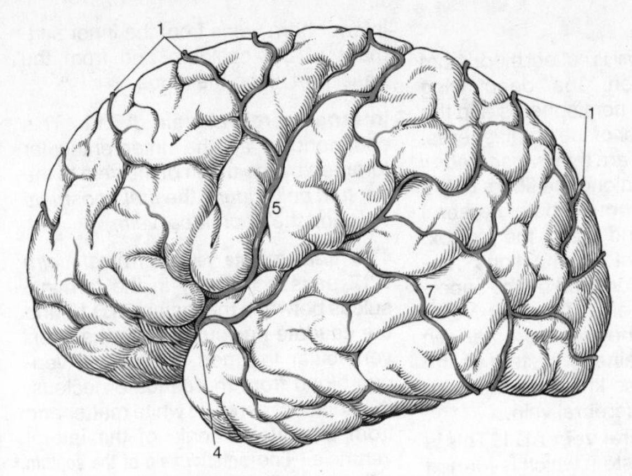

A Veins of the brain, lateral view

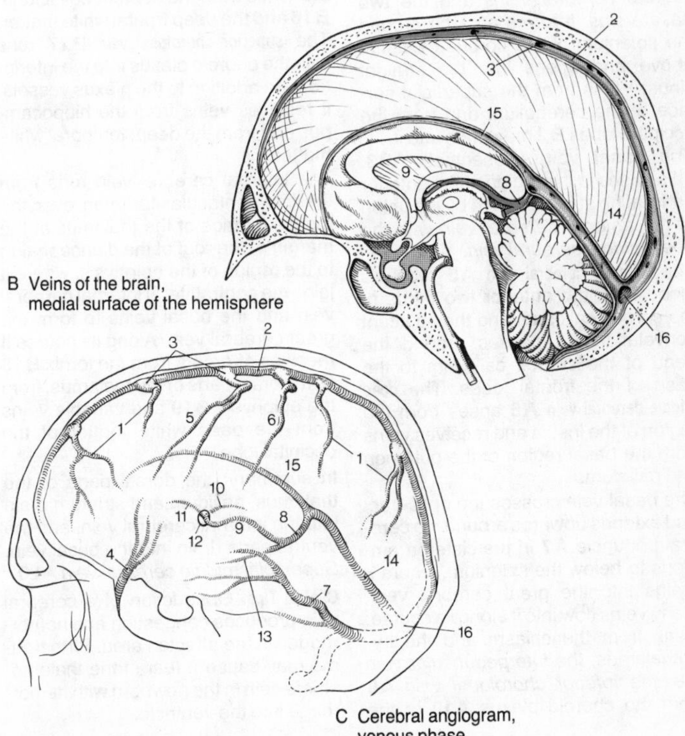

B Veins of the brain, medial surface of the hemisphere

C Cerebral angiogram, venous phase (after Krayenbühl and Richter)

Deep Veins

The **deep cerebral veins** collect blood from the diencephalon, the deep lying structures of the hemispheres and the deep white matter of the hemispheres. In addition there are thin *transcerebral veins* which run along the fibers of the corona radiata from the outer cerebral white matter and from the cortex. These represent connections between the areas drained by the superficial and deep veins. The deep veins drain blood into the *great cerebral vein* (Galen). The drainage system of the deep veins is also known as the system of the deep cerebral vein.

The great cerebral vein AB1. This is a short vascular stem which is formed by the junction of four veins, the two *internal cerebral veins* and the two *basal veins.* It forms an arch around the splenium of the corpus callosum above and opens into the straight sinus. Veins from the superficial surface of the cerebellum and from the occipital lobes **B2** may drain into it.

The basal vein (Rosenthal) **AB3.** This arises in the region of the anterior perforating substance **A4** by the union of the *anterior cerebral vein* with the *deep middle cerebral vein.*

The **anterior cerebral vein A5** receives blood from the anterior two thirds of the corpus callosum and the adjacent convolutions. It passes around the genu of the corpus callosum to the base of the frontal lobes. The **deep middle cerebral vein A6** arises from the region of the insula and receives veins from the basal region of the putamen and pallidum.

The basal vein crosses the optic tract and extends upwards around the cerebral peduncle **A7** in the cisterna ambiens to below the splenium, where it drains into the great cerebral vein. Many veins flow into it along its course: veins from the chiasm and the hypothalamus, the *interpeduncular vein* **A8,** the *inferior choroideal vein* **A9** from the choroid plexus **A10** of the inferior horn veins from the inner segment of the pallidum and from the basal part of the thalamus.

Internal cerebral vein AB11. This commences at the interventricular foramen by the union of the vein of the *septum pellucidum,* the *thalamostriate vein* and the *choroideal vein.*

The **thalamostriate vein** *(terminal vein)* **B12** passes rostrally in the terminal sulcus between the thalamus **B13** and the caudate nucleus **B14** to the interventricular foramen. It receives venous blood from the caudate nucleus, the adjacent cerebral white matter and from the lateral angle of the lateral ventricle. The **anterior vein of the septum pellucidum B15** receives venous branches from the septum pellucidum **B16** and the deep frontal white matter. The **superior choroideal vein B17** runs with the choroid plexus into the inferior horn. In addition to the plexus vessels, it receives veins from the hippocampus and from the deep, temporal white matter.

The internal cerebral vein runs from the interventricular foramen over the medial surface of the thalamus at the margin of the roof of the diencephalon to the region of the epiphysis, where it joins the contralateral internal cerebral vein and the basal veins to form the great cerebral vein. Along its course it receives afferents from the fornix **B18** from dorsal parts of the thalamus, from the epiphysis **B19** and variable veins from the deep white matter of the occipital lobe.

In summary: the dorsal parts of the thalamus, pallidum and striatum drain into the internal cerebral vein, and the ventral parts drain into the basal vein. *Superficial middle cerebral vein* **A20.**

Clinical Tips. Obstruction of a cerebral vein produces congestion and hemorrhages in the affected area. Birth trauma may cause a tear in the thalamostriate vein in the newborn with hemorrhage into the ventricle.

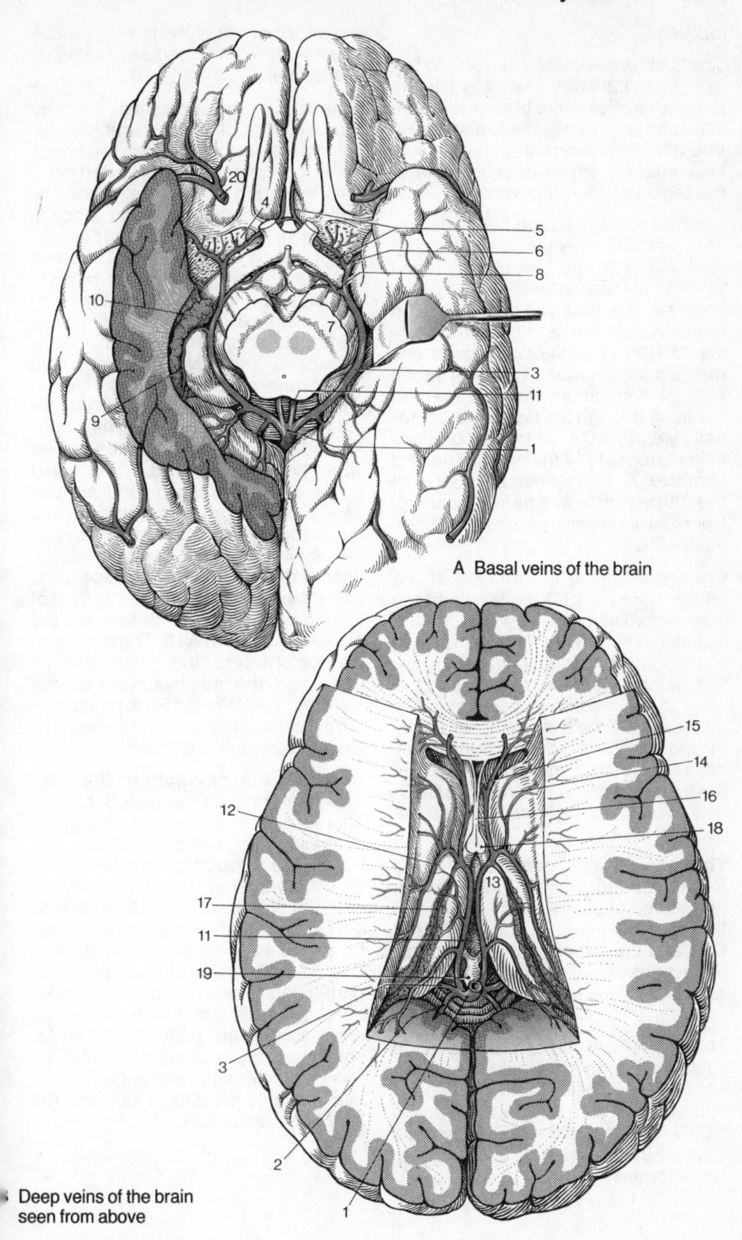

A Basal veins of the brain

B Deep veins of the brain seen from above

Survey

The CNS is completely surrounded by *cerebrospinal fluid.* This also fills the internal cavities of the brain, the ventricles, so that it is possible to distinguish internal and external cerebrospinal fluid spaces, which communicate in the region of the IVth ventricle.

Internal cerebrospinal fluid spaces. The ventricular system consists of four ventricles: two lateral ventricles (I and II) **A1** in the telencephalic hemispheres, the IIIrd ventricle **ABC2** in the diencephalon, and the IVth ventricle **ABC3** in the hindbrain (pons and medulla oblongata). The two lateral ventricles communicate with the IIIrd ventricle through the **interventricular foramen** *(Monro)* **AC4,** which lies on each side in front of the thalamus. The IIIrd ventricle in turn communicates with the IVth ventricle by a narrow opening, the **cerebral aqueduct** *(aqueduct of Sylvius)* **ABC5.**

Corresponding to the rotation of the hemisphere (p. 194) the **lateral ventricle** is semicircular in shape, with a caudally directed spur. We distinguish several regions: the *anterior horn* **BC6** in the frontal lobe, bordered laterally by the head of the caudate nucleus, medially by the septum pellucidum and dorsally by the corpus callosum; the *narrow central part* (cella media) **BC7** above the thalamus; the *temporal horn* **BC8;** and the *occipital horn* **BC9** in the occipital lobe.

The lateral wall of the **IIIrd ventricle** is formed by the thalamus with the interthalamic substance **C10** (p. 10) and the hypothalamus. The *optic recess* **C11** and the *infundibular recess* **C12** project anteriorly, and the *suprapineal recess* **C13** and the *pineal recess* **C14** caudally.

The **IVth ventricle** forms a tent-shaped space over the rhomboid fossa between the cerebellum and the medulla and extends as a long *lateral recess* **BC15** on both sides. Each recess ends in the foramen of Luschka, the *lateral opening of the IVth ventricle.* At the attachement of the inferior medullary velum lies the *median aperture of Magendie.*

External cerebrospinal fluid spaces. The external cerebrospinal fluid space lies between the two layers of the leptomeninges. It is limited internally by the pia mater and externally by the arachnoid (subarachnoid space p. 268 A13). It is narrow over the convexity of the brain and only becomes enlarged at the base of the brain in certain areas, where it forms the cerebrospinal fluid *cisterns.* While the pia mater adheres closely to the outer surface of the CNS, the arachnoid membrane spans the sulci, indentations and fossae so that over deeper indentations larger spaces are formed, the **subarachnoid cisterns,** filled with cerebrospinal fluid. The largest space is the *cerebellomedullary cistern* **A16** between the cerebellum and the medulla. The *interpeduncular cistern* **A17** lies in the angle between the floor of the diencephalon, the cerebral peduncles and the pons, and in front of it, in the region of the chiasm lies the *chiasmatic cistern* **A18.** The surface of the cerebellum, the quadrigeminal plate and the epiphysis border the *cisterna ambiens* **A19** (superior cistern) traversed by a wide-meshed network of connective tissue.

Cerebral fluid circulation. The cerebrospinal fluid is produced by the choroid plexus. It flows from the lateral ventricles into the IIIrd ventricle and from there through the aqueduct into the IVth ventricle. There it enters the external cerebrospinal fluid space through the median and lateral foramina of the IVth ventricle. Drainage of CSF into the venous circulation occurs partly via the arachnoid villi (p. 268), which protrude into the venous sinus, and partly at the point of exit of the spinal nerves, where there is a transition into the dense venous plexus and into the nerve sheaths (a route into the lymphatic circulation).

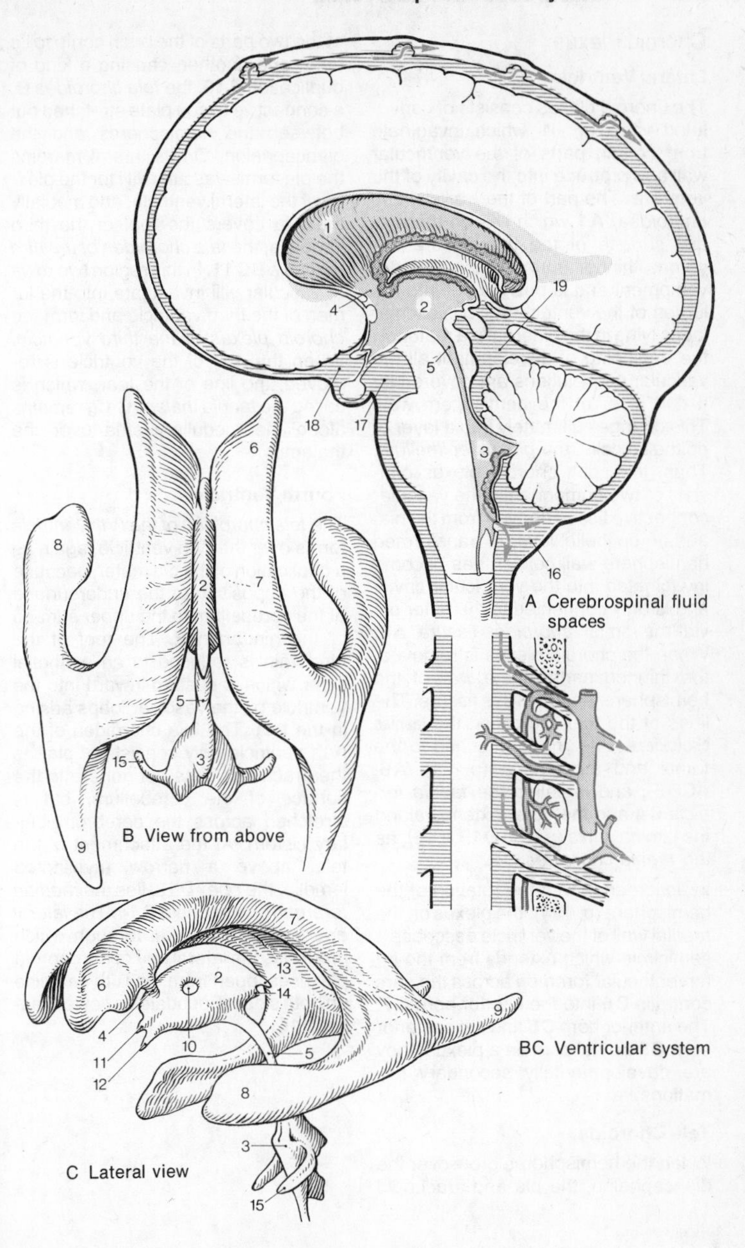
1
2
3
4
5
16
17
18
19
A Cerebrospinal fluid spaces
6
7
8
9
15
B View from above
10
11
12
13
14
C Lateral view
BC Ventricular system

Choroid Plexus

Lateral Ventricle

The choroid plexus consists of convoluted vascular villi which invaginate from certain parts of the ventricular wall and protrude into the cavity of the ventricle. The part of the wall *(lamina choroidea)* **A1** which lies on the medial surface of the hemisphere becomes thinner during embryonic development and is pushed out into the lumen of the ventricle **A2** by vascular loops lying in the external pia mater. At the beginning of development all the vascular convolutions are covered by a thin layer of the hemisphere wall. This changes ultimately into a layer of epithelial cells, the *plexus epithelium.* Thus, the adult choroid plexus consists of two components: the vascular connective tissue derived from the pia, and an epithelial layer of transformed hemisphere wall cells. It has become invaginated into the ventricular cavity and is only connected to the outer pia via the small *choroidal fissure* **A3.** When the choroid plexus is removed, the thinned parts of the wall of the hemisphere tear off at this fissure. The lines of the tear are called the **taeniae choroideae.** One line is attached to the fornix and the fimbria (p. 216 A6, ACD5), and is called the *taenia fornicis* **C4** and the other extends along the lamina affixa (p. 158 D15, E16), as the *taenia choroidea* **C5.**

In accordance with the rotation of the hemisphere (p. 194), the plexus on the medial wall of the ventricle describes a semicircle which extends from the interventricular foramen across the pars centralis **C6** into the inferior horn **C7.** The anterior horn **C8** and the posterior horn **C9** do not contain a plexus; they are developmentally secondary formations.

Tela Choroidea

When the hemispheres grow over the diencephalon, the pia and arachnoid of the two parts of the brain come to lie upon one another, causing a kind of duplication **A10,** the *tela choroidea* **B,** a connective tissue plate stretched out between the hemispheres and the diencephalon. On its lateral margins the pia forms vascular villi for the plexus of the lateral ventricle, and medially the tela covers the roof of the third ventricle, the *tela choroidea of the IIIrd ventricle* **BC11.** In this region two rows of vascular villi invaginate into the lumen of the third ventricle and form the *choroid plexus of the third ventricle.* When the roof of the ventricle is removed, the line of the tear, which is called the *taenia thalami* **C12,** remains along the medullary stria over the thalamus.

Fourth Ventricle

The *tela choroidea of the IVth ventricle* forms over the IVth ventricle, again as a duplication of the pia mater, because of the apposition of the undersurface of the cerebellum to the upper surface of the hindbrain **E.** The roof of the hindbrain is reduced to an epithelial layer which is pushed inward into the ventricle by the vascular loops arising in the tela. The tela choroidea of the IVth ventricle only consists of pia, as the arachnoid does not adhere to the surface of the cerebellum, but is stretched across the cerebromedullary cistern. At the attachment of the tela, above a narrow myelinated lamella, the *obex* **D13,** lies the *median foramen* of Magendie **D14.** The *lateral apertures* of Luschka, through which protrudes the lateral end of the choroid plexuses, open from the IVth ventricle on both sides (Bochdalek's flower basket **D15**).

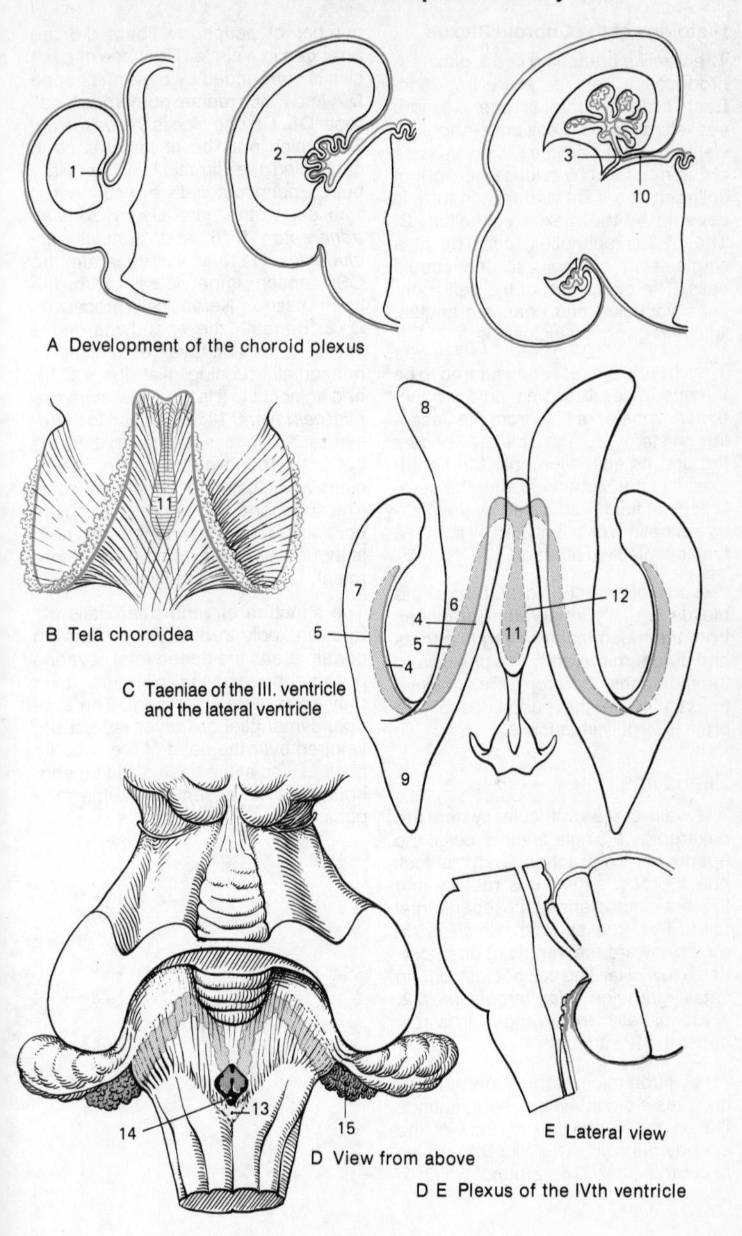

A Development of the choroid plexus

B Tela choroidea

C Taeniae of the III. ventricle and the lateral ventricle

D View from above

E Lateral view

D E Plexus of the IVth ventricle

Histology of the Choroid Plexus

The treelike branching of the plexus **A** produces a very large free surface. Each branch contains one or more vessels, arteries, capillaries and thin-walled venous cavities. The vessels are surrounded by a loose network of collagen fibers **B1** and this, in turn, is covered by the plexus epithelium **B.** The plexus epithelium consists of a single layer of finely ciliated cuboid cells. The cytoplasm of the cells contains vacuoles and coarse granules, lipid and glycogen inclusions.

The choroid plexus is considered to be the site of cerebrospinal fluid production. A transfer of fluid from the vascular system of the plexus occurs through its epithelium into the ventricles. It is not certain whether the cerebrospinal fluid is secreted by the plexus epithelium or is dialyzed by it, i. e., a type of selective filtration.

Like the pia-arachnoid and dura, the plexus is richly innervated (branches from the trigeminal and vagus nerves and autonomic fibers). The plexus and the meninges, therefore, are sensitive to pain, while the substance of the brain is largely insensible.

Ependyma

The walls of the ventricular system are covered by a single layer of cells, the ependyma **C.** Each ependymal cell has a process that runs radially into the brain substance, the ependymal fiber. The free surface, which is directed toward the ventricle, often carries a few cilia. The cell bodies contain small granules *blepharoblasts* **C2,** which usually are arranged in a row beneath the surface.

An electron micrograph shows irregular, vesicle-containing evaginations **D3** on the ventricular surface of the ependymal cells. The cilia **D4** contain a central fiber **D5,** around which a number of peripheral fibers **D6** are arranged in a circle. The base of each cilia is surrounded by a granular zone **D7,** into which radiate numerous small roots **D8.** On one side is the *basal foot* **D9,** which may be of importance in controlling the direction of the ciliary beat. Ependymal cells are connected with each other laterally by *zonulae adherentes* **D10** and *zonulae occludentes* **D11,** which separate the CSF space imperviously from the brain tissue. Nerve cell processes **D12.** Beneath the ependyma lies a sparsely cellular layer of radially or horizontally running glial fibers **C13,** and adjacent to it is the *subependymal glial cell layer* **C14.** In addition to astrocytes, it contains transitional forms between ependymal cells and astrocytes and clusters of small dark cells. The subependymal zone, in which the continuous replenishment of glial cells is thought to occur, contains therefore mainly undifferentiated types of glia.

The structure of ventricular walls differs markedly in different regions. In certain areas the ependymal covering or the subependymal glial fiber layer may be completely absent. The subependymal glial cell layer is best developed over the head of the caudate nucleus and at the base of the anterior horn, but is absent over the hippocampus.

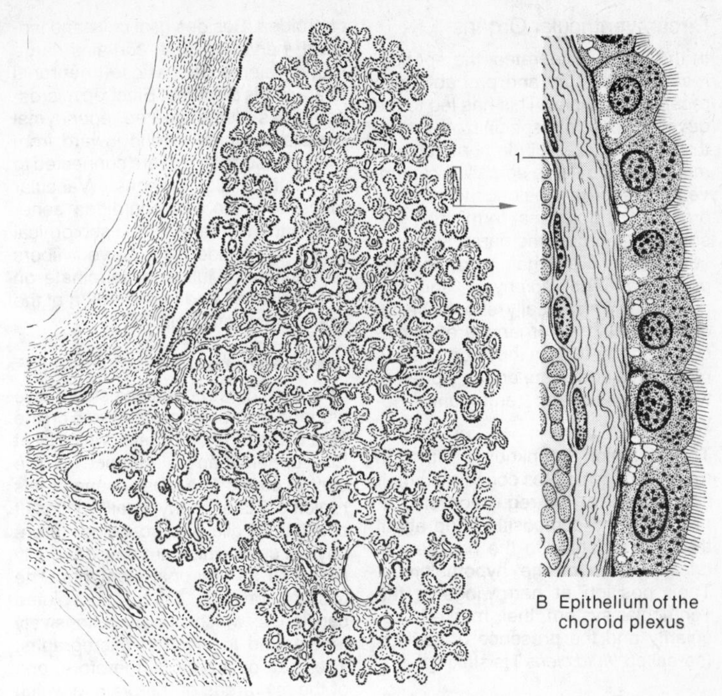

A Choroid plexus

B Epithelium of the choroid plexus

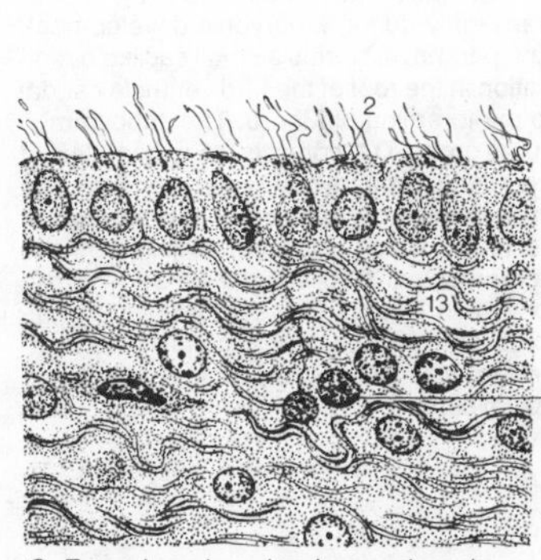

C Ependymal and subependymal layers

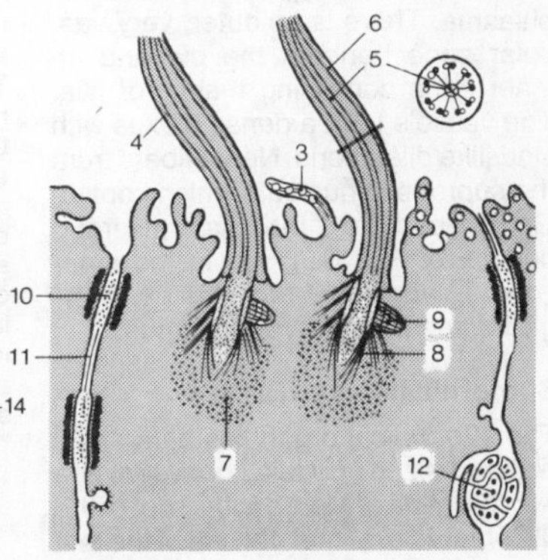

D Ependymal cell, electron microscopical scheme (Brightman and Palay)

Circumventricular Organs

In the lower vertebrates, the ependyma has secretory and probably also receptor functions. This has led to the development of specialized structures, which are still demonstrable in mammals. To these so-called *circumventricular organs* belong the vascular organ of the lamina terminalis, the subfornical organ, the paraphysis, the subcommissural organ, and the area postrema. The epiphysis and hypophysis, which really should be included with these organs, are not discussed here. In man these organs are regressive and some are only present temporarily during embryonic development.

Their function is unknown, although there are hypotheses concerning their importance for the regulation of CSF pressure and composition and about their relationships to the neuroendocrine system of the hypothalamus. Their position at narrowings in the ventricular system, their marked vascularity and the presence of cavities (so-called *'fluid clefts'*) is striking.

Vascular Organ of the Lamina Terminalis AD1

This lies in the lamina terminalis, the rostral end of the IIIrd ventricle, between the rostral commissure and the chiasma. There is an outer, very vascular zone beneath the pia and an inner zone consisting mainly of glia. The vessels form a dense plexus with sinuslike dilatations. Nerve fibers from the supraoptic nucleus, which contain Gomori-positive material (Herring bodies, p. 192 B5), run in the inner zone. In addition it receives peptidergic fibers from the hypothalamus.

Subfornical Organ BD2

The subfornical organ lies as a small pinhead-sized nodule between the two foramina of Monro in the roof of the IIIrd ventricle, at the oral end of the tela choroidea. Besides glial cells and isolated nerve cells it contains large round elements, whose neuronal character is disputed. Electron microscopy has demonstrated ependymal canaliculi, which extend inward from the outer surface and are connected to the intercellular spaces. Vascular loops from the tela choroidea penetrate into the interior of the subfornical organ. Peptidergic nerve fibers (somatostatin, luliberin) terminate on the capillaries and in the region of the ependymal canaliculi.

Area Postrema CD3

The area postrema consists of two slender, symmetrical structures on the floor of the rhomboid fossa, which lie at the funnel shaped entrance to the central canal. The loose tissue in this region contains many small cavities. It consists of glia and so-called parenchymal cells, which are now generally regarded as neuronal elements. The tissue contains many convoluted capillaries, which appear extensively fenestrated in electron micrographs. The area postrema is, therefore, one of the few areas of the brain in which the blood brain barrier is readily permeable.

Paraphysis and Subcommissural Organ

In man these two structures only appear transitorily during embryonic development. The paraphysis forms a small saclike evagination in the roof of the IIIrd ventricle caudal to the foramina of Monro. The subcommissural organ **D4** consists of a complex of cylindrical ependymal cells beneath the epithalamic commissure. They produce a secretion which does not disperse in the cerebrospinal fluid but thickens to form a long, thin thread, Reissner's fiber. In animals, in which the central canal is not obliterated, this fiber extends into the lower spinal cord.

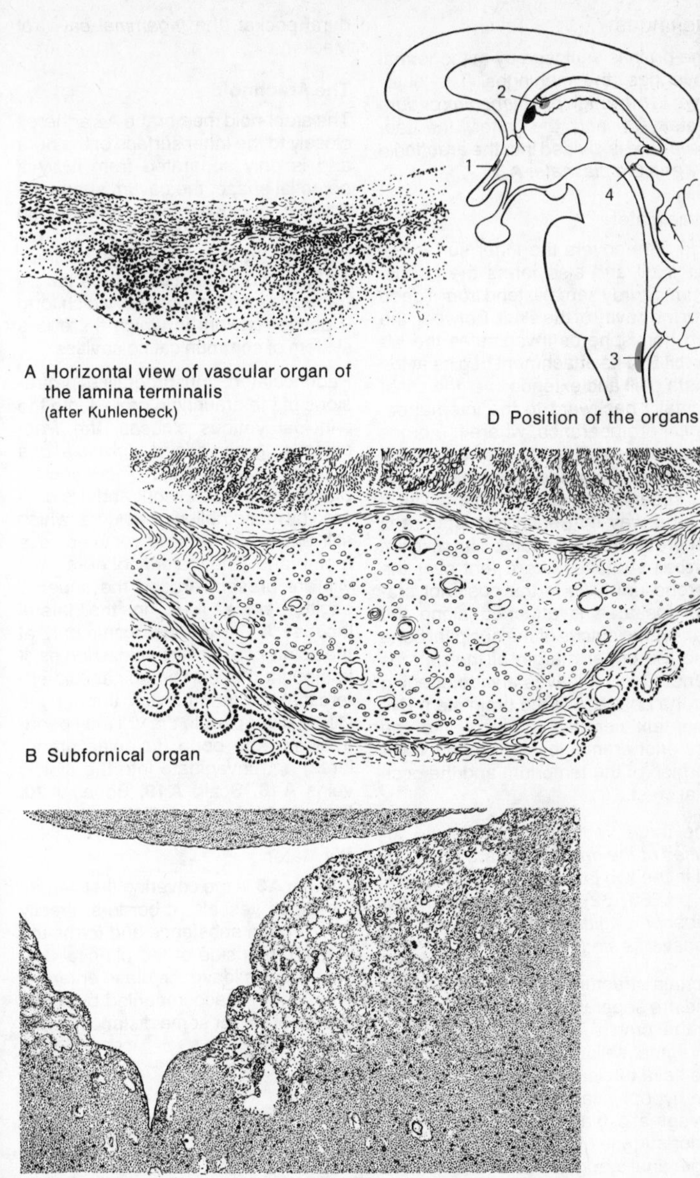

A Horizontal view of vascular organ of the lamina terminalis (after Kuhlenbeck)

B Subfornical organ

C Area postrema

D Position of the organs

Meninges

The brain is enclosed by mesodermal coverings, the meninges. The outer layer is the tough *pachymeninx* or *dura mater* **A1** and the inner, the *leptomeninx,* is divided into the *arachnoid* **A2** and the *pia mater* **A3.**

Dura Mater

The dura covers the inner surface of the skull and also forms the periosteum. Sturdy septa extend from it deep into the cavity of the skull. Between the two hemispheres invaginates the **falx cerebri B4.** Its attachment begins at the crista galli and extends over the crista frontalis backward to the internal occipital protuberance, where it merges into the *tentorium cerebelli* **B5,** which extends to both sides. The falx divides the superior part of the cranial cavity in such a way that each hemisphere is secured in its own space. The **tentorium cerebelli** stretches like a tent over the cerebellum lying in the posterior cranial fossa. It is attached along the transverse sulcus of the occipital bone and the upper margin of the petrous bone. Orally it leaves a wide opening for the passage of the brain stem **B6.** The falx cerebelli projects into the posterior cranial fossa from the lower surface of the tentorium and the occipital crest.

The large venous channels, the *sinuses of the dura mater,* are embedded in the two laminae of the dura (Vol. 2, pp. 60, 62). Section through the *superior sagittal sinus* **B8** and the *transverse sinus* **B9.**

Certain structures lie in dural pockets and are separated from the remainder of the cranial cavity. Thus, the diaphragma sellae **B10** stretches over the sella turcica, and is penetrated by the hypophysial stalk through the diaphragmatic hiatus **B11.** On the anterior surface of the petrous bone, the trigeminal ganglion is enclosed in a dural pocket, the *trigeminal cave* (of Meckel).

The Arachnoid

The arachnoid membrane A2 adheres closely to the inner surface of the dura and is only separated from it by a potential space, the cavum subdurale **A12.** It encloses the subarachnoid space, which contains the cerebrospinal fluid, the *cavum subarachnoidale* **A13,** and is connected to the pia by trabeculae **A14** and septa, which form a dense network in which there is a system of communicating cavities.

Pedunculated, mushroomlike protrusions of the arachnoid project into the principal venous sinuses, the *Pacchioni's granulations (granulations arachnoideales)* **A15.** They consist of an arachnoid network and are covered by mesothelium. The dura, which still encloses them, is reduced to a membrane. The majority of arachnoid villi are present around the superior sagittal sinus **A16,** in the lateral lacunae **A17,** and less commonly at the points of exit of the spinal nerves. It is thought that cerebrospinal fluid enters the venous circulation through the villi. In older people the villi may penetrate into the bone (foveolae granulares) and invaginate into the diploic veins **A18.** Scalp **A19,** Bones **A20,** Diploe **A21** (Vol. 1, p. 20 D5).

Pia Mater

The pia **A3** is the covering that carries the blood vessels. It borders directly on the brain substance and forms the mesodermal side of the pia-glial barrier. Vessels leave the pia to enter the brain and are accompanied by a thin sheath of pia for some distance.

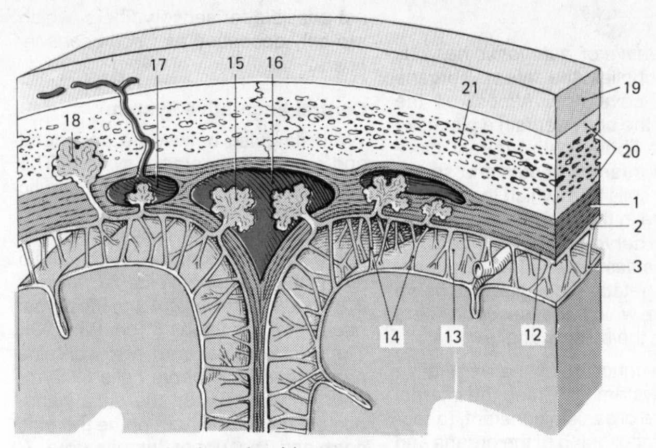

A Meninges and subarachnoid space

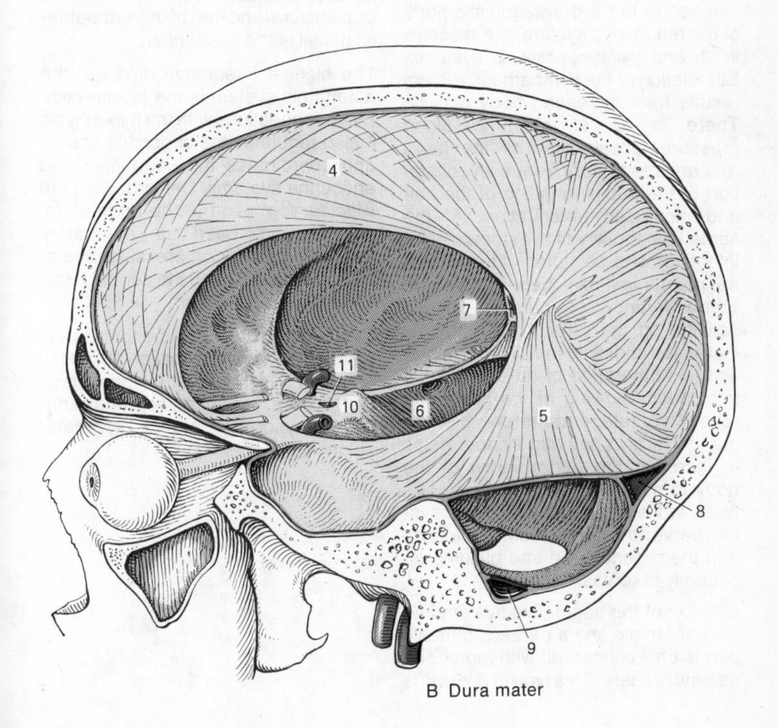

B Dura mater

Survey

The *vegetative* or *autonomic nervous system* supplies the internal organs and their coverings. Almost all the tissues of the body contain a plexus of fine nerve fibers, both afferent (sensory) and efferent (motor and secretory). The cells from which the efferent fibers arise in the body form scattered clusters of nerve cells surrounded by a connective tissue covering. The autonomic (vegetative) ganglia. The nerve cells from which the sensory fibers arise lie in the spinal ganglia.

The main function of the vegetative nervous system is to keep the internal state of the organism constant, to regulate the functioning of the organs and to cope with variations in its environment. Regulation is obtained by coordination of the two antagonistic parts of the autonomic system, the **sympathetic 1** and **parasympathetic 2** systems. Stimulation of the sympathetic system results from increased body activity. There is an elevation in blood pressure, acceleration of the heart rate and respiratory frequency, dilatation of the pupils, bristling of the hair and increased perspiration. At the same time motility of the stomach and intestines and secretion of the intestinal glands is diminished. If there is a preponderance of parasympathetic activity, on the other hand, there is increased intestinal motility and secretion, a furtherance of defecation and micturition, slowing of the heart and respiratory rate, and narrowing of the pupil. The sympathetic system helps to increase the performance of the body in stress and emergency situations, while the parasympathetic system serves the metabolism of the body and the regeneration and building up of bodily reserves.

Division of the vegetative system into a sympathetic and a parasympathetic part is only concerned with motor and secretory fibers. This type of division is not possible for sensory fibers, which are only described as *viscerosensory fibers.*

Central Vegetative System

We distinguish between a peripheral and a central vegetative system. The central cell groups of the sympathetic and parasympathetic system lie in different regions. The parasympathetic nerve cells form nuclei in the brain stem: the *Edinger-Westphal nucleus* **3,** the *salivatory nuclei* **4** and the dorsal nucleus of the vagus **5** (pp. 98, 108). The sacral spinal cord also contains parasympathetic nerve cells **6.** Sympathetic neurons, on the other hand, occupy the lateral horn of the thoracic cord and the upper lumbar cord **7.** Within the CNS the localization of the parasympathetic nuclei is therefore *craniosacral* and that of the sympathetic nuclei is *thoracolumbar.*

The highest integrative organ of the autonomic system is the **hypothalamus.** By its connections with the hypophysis it also regulates the endocrine glands and coordinates the vegetative and endocrine systems. Cell groups in the *reticular formation* of the brain stem, (which control heart rate, respiratory frequency and blood pressure; see p. 138) also take part in the central regulation of organ function.

Sympathetic trunk **8,** superior cervical ganglion **9,** stellate ganglion **10,** celiac ganglion **11,** superior mesenteric ganglion **12,** inferior mesenteric ganglion **13,** hypogastric plexus **14,** greater splanchnic nerve **15.**

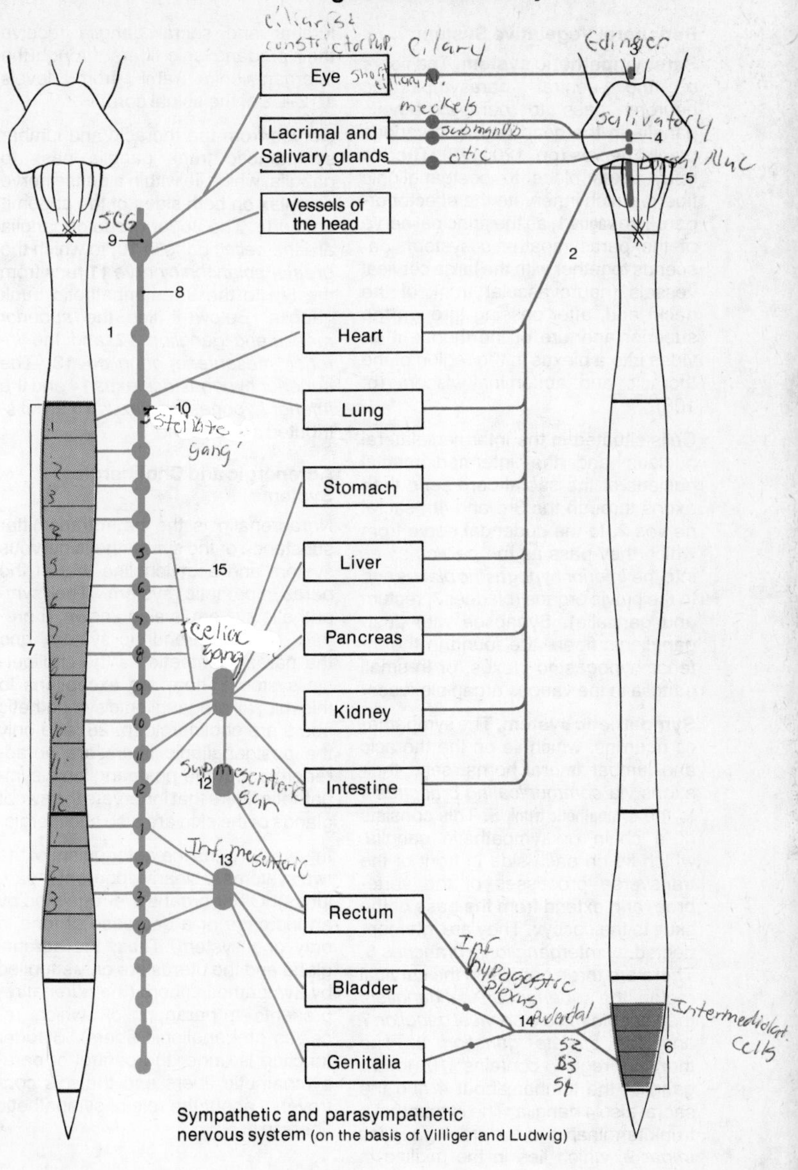

Sympathetic and parasympathetic
nervous system (on the basis of Villiger and Ludwig)

Peripheral Vegetative System

Parasympathetic system. The fibers of the central parasympathetic neurons pass to parasympathetic ganglia in the head region in various cranial nerves (pp. 120, 122). There a relay takes place to postganglionic fibers, which innervate the effector organ. The **vagus 1,** as the principal nerve of the parasympathetic system, descends together with the large cervical vessels (neurovascular trunk of the neck) and, after passing through the superior aperture of the thorax, it divides into a plexus in the region of the thoracic and abdominal viscera (p. 108).

Cells situated in the intermediolateral nucleus and the intermediomedial nucleus of the sacral cord send their axons through the 3rd and 4th sacral nerves **2,** to the pudendal nerve from which they pass as the *pelvic nerves* into the *inferior hypogastric plexus* and to the pelvic organs (bladder **3,** rectum and genitalia). Synapses with postganglionic fibers are found in the inferior hypogastric plexus, or in small ganglia in the various organ plexuses.

Sympathetic system. The sympathetic neurons, which lie on the thoracic and lumbar lateral horns, send their axons via *communicating branches* **4** to the **sympathetic trunk 5.** This consists of a chain of sympathetic ganglia, which lie on each side in front of the transverse processes of the vertebrae, and extend from the base of the skull to the coccyx. They are interconnected by *interganglionic branches* **6.** There are three ganglia in the cervical region, the *superior cervical ganglion,* the variable *middle cervical ganglion* **7** and the *stellate ganglion* **8.** The thoracic region contains 10 to 11 ganglia, the lumbar about 4 and the sacral also 4 ganglia. The sympathetic trunk terminates in the small *ganglion impar* **9,** which lies in the midline in front of the coccyx. The cervical, lower lumbar and sacral ganglia receive their preganglionic fibers through the interganglionic rami from levels T12–L2 of the spinal cord.

Nerves from the thoracic and lumbar sympathetic trunk ganglia pass to ganglia, which lie within a dense nerve complex on both sides of the abdominal aorta. The upper group of ganglia are the *celiac ganglia* **10,** to which the *greater splanchnic nerve* **11** runs from the 5th to the 9th sympathetic trunk ganglia. Below it lies the *superior mesenteric ganglion* **12** and the *inferior mesenteric ganglion* **13.** The *superior hypogastricplexus* **14** and the *inferior hypogastric plexus* **15** are distributed in the pelvis.

Adrenergic and Cholinergic Systems

Noradrenalin is the neurotransmitter substance of the sympathetic nervous system and acetylcholine that of the parasympathetic system. The sympathetic system is also known, therefore, as the adrenergic system, and the parasympathetic as the cholinergic system. There are exceptions to this rule; all preganglionic sympathetic fibers are cholinergic (p. 28) and only the postganglionic fibers are noradrenergic, but the postganglionic sympathetic fibers that innervate the sweat glands of the skin are also cholinergic.

In some organs the antagonism of the two systems is clearly apparent (heart, lungs). Other organs are regulated by an increase or a decrease of tone in only one system. Thus, the adrenal gland and the uterus are only supplied by sympathetic fibers (the adrenal represents a paraganglion which receives preganglionic fibers). Bladder function is under the control of parasympathetic fibers and there is controversy about the role of sympathetic fibers in it.

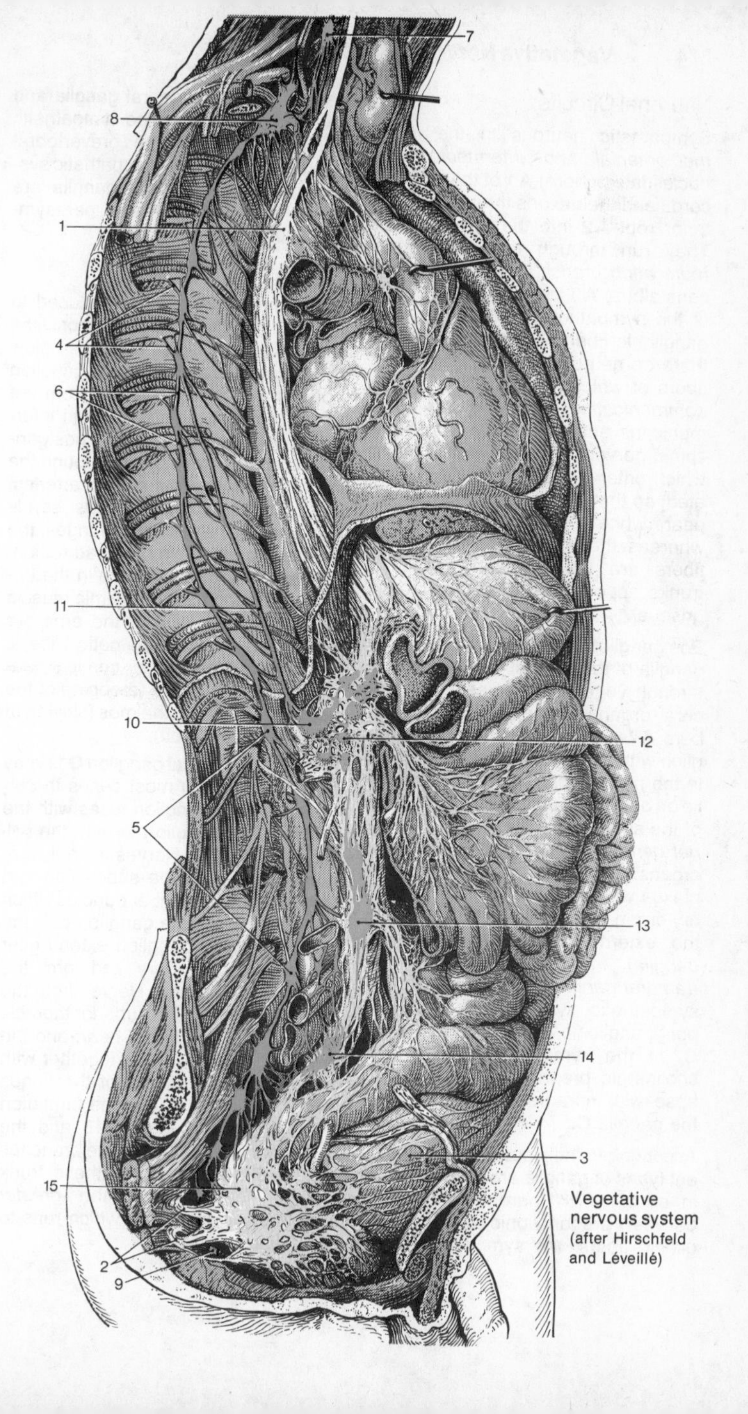

Vegetative nervous system (after Hirschfeld and Léveillé)

Neuronal Circuits

Sympathetic neurons in the *intermediomedial* and *intermediolateral nuclei* (lateral horn) **A1** of the thoracic cord send their axons through the anterior root **A2** into the spinal nerve. They run through the *white communicating branch* (ramus communicans albus) **A3** to reach the ganglion of the sympathetic trunk **A4** as preganglionic fibers. Some of them end there on neurons, the postganglionic fibers of which run through the *grey communicating branch* (ramus communicans griseus) **A5** back into the spinal nerve; the preganglionic fibers which enter the ganglion are myelinated, so that the connecting trunk appears white (r. communicans albus), whereas the efferent postganglionic fibers are unmyelinated and their trunks appear grey (r. communicans griseus).

Postganglionic fibers **A6** from the ganglia of the sympathetic trunk pass through vegetative nerves to the visceral organs. Some preganglionic fibers **A7** may pass through the ganglion without synapsing and terminate in the *prevertebral ganglia* **A8,** which lie on both sides of the aorta. A number of the smaller and the smallest *terminal ganglia* **A9** lie within the interior organs. They form part of the nerve plexus which extends throughout every organ, and are to be found both in the external coverings *(extramural ganglia)* and within the organ *(intramural ganglia).* Whilst in the parasympathetic system both pre- and post-ganglionic fibers are cholinergic **B,** in the sympathetic system the cholinergic preganglionic fibers synapse with noradrenergic neurons in the ganglia **C.**

According to their location three different types of ganglia are distinguished, in all of which there are synapses between preganglionic and postganglionic fibers: the sympathetic trunk ganglia, the prevertebral ganglia and the terminal ganglia. The sympathetic trunk ganglia and the prevertebral ganglia belong to the sympathetic system, while the terminal ganglia are largely but not exclusively parasympathetic.

Sympathetic Trunk

The cervical ganglia are reduced to three in number; the uppermost, the **superior cervical ganglion D10** lies below the base of the skull near the ganglion nodosum. It receives fibers from the upper thoracic regions through interganglionic branches. Its postganglionic fibers form a plexus around the internal and external carotid arteries. The *internal carotid plexus* sends branches to the cranial meninges, the eye and to glands in the head region. The superior tarsal muscle in the upper eyelid and the ophthalmic muscle in the posterior wall of the orbit are innervated by sympathetic fibers. Damage to the superior cervical ganglion results in ptosis (drooping of the upper lid) and enopthalmos (sinking of the eye into the orbit).

The middle cervical ganglion **D11** may be absent and in most cases the inferior cervical ganglion fuses with the first thoracic ganglion to form the **stellate ganglion D12.** Nerves from it form plexuses around the subclavian and vertebral arteries. Fiber bundles which connect the stellate ganglion with the medial cervical ganglion extend over the subclavian artery and form the *ansa subclavia* **D13.** Nerves from the cervical **D14** and the superior thoracic ganglia **D15** run to the heart and the hila of the lungs where, together with parasympathetic fibers of the vagus nerve, they take part in the formation of the *cardiac plexus* **D16** and the *pulmonary plexus* **D17.** The branches of the 5th to 9th sympathetic trunk ganglia join to form the *greater splanchnic nerve* **D18,** which runs to the celiac ganglia.

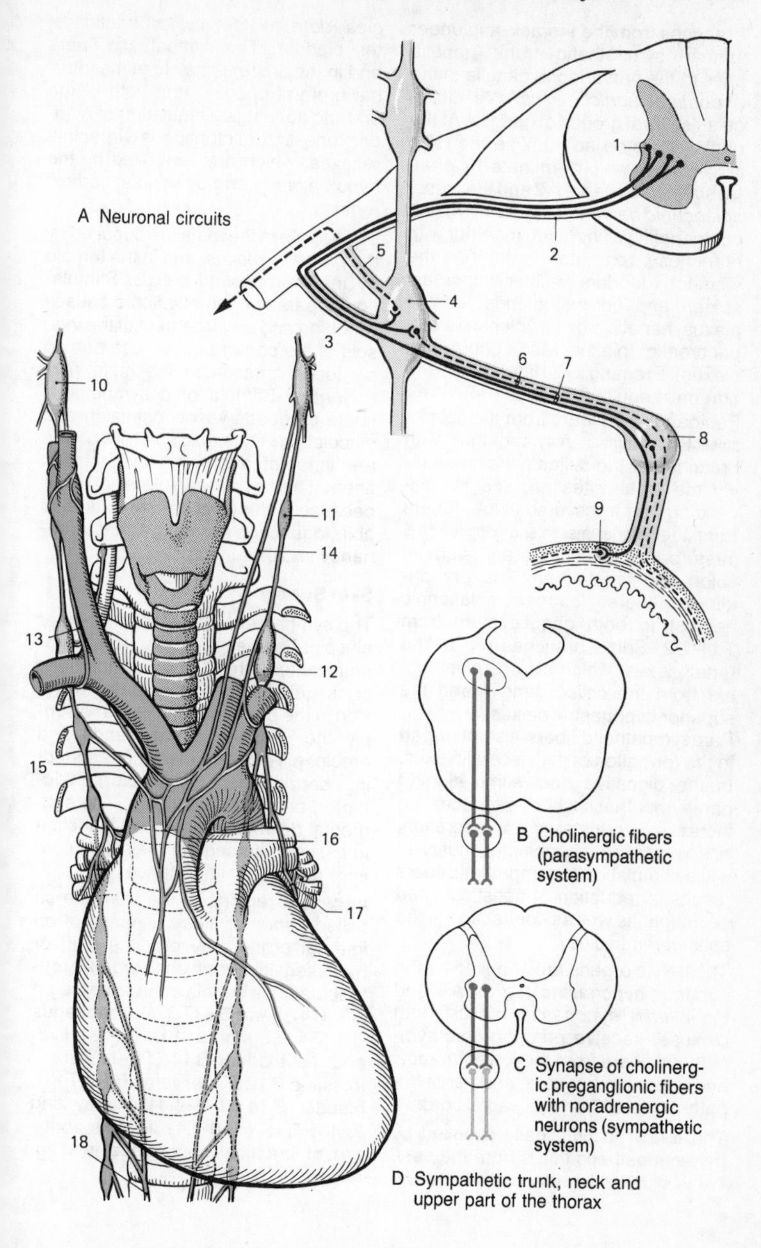

B Cholinergic fibers (parasympathetic system)

C Synapse of cholinergic preganglionic fibers with noradrenergic neurons (sympathetic system)

D Sympathetic trunk, neck and upper part of the thorax

Branches from the thoracic and upper lumbar sympathetic trunk ganglia pass to the prevertebral ganglia of the *abdominal aortic plexus.* We distinguish several groups of ganglia. At the origin of the celiac trunk lie the **celiac ganglia A1,** in which terminate the *greater splanchnic nerve* **A2** and the *lesser splanchnic nerve* **A3** (T9–T11). Their postganglionic fibers run together with branches from the aorta to the stomach, duodenum, liver, pancreas, spleen and adrenal glands (gastric plexus, hepatic plexus, splenic plexus, pancreatic plexus and suprarenal plexus). Preganglionic fibers run to the adrenal medulla (Vol. 2, p. 164).

Postganglionic fibers from the **superior mesenteric ganglion A4,** together with branches of the celiac ganglion, supply the small intestine, and the ascending and transverse colon. Fibers from the **inferior mesenteric ganglion A5** pass to the descending and sigmoid colon and the rectum. The preganglionic fibers (lumbar splanchnic nerves) to both ganglia stem from T11–L2. Some branches run to the renal plexus, which also receives fibers from the celiac ganglia and the superior hypogastric plexus.

Parasympathetic fibers also take part in the formation of the visceral plexus. In the digestive tract stimulation of parasympathetic fibers produces an increase in peristalsis and secretion, and relaxation of sphincter muscles, while stimulation of sympathetic fibers results in reduction of peristalsis and secretion, as well as contraction of the sphincter muscles.

The pelvic organs are supplied by the *superior hypogastric plexus* **A6** and the *inferior hypogastric plexus.* Both plexuses receive preganglionic sympathetic fibers from the lower thoracic and upper lumbar cord, and parasympathetic fibers from the sacral cord.

The **bladder** is principally supplied by parasympathetic fibers from the *vesical plexus,* which innervate the muscles *(detrusor)* serving contraction of the bladder. The sympathetic fibers end in the smooth muscle at the internal urethral opening and both of the ureteric openings. Regulation of bladder tone and micturition is via spinal reflexes, which are controlled by the hypothalamus and by various cortical regions.

The **genitalia** in the male are supplied by the *prostatic plexus,* and in the female by the *uterovaginal plexus.* Stimulation of parasympathetic fibers causes widening and engorgement of the vessels of the corpora cavernosa and so produces erection in the male *(Nn. erigentes).* Stimulation of sympathetic fibers produces vasoconstriction and ejaculation. The muscles of the uterus are innervated only by sympathetic fibers. Their functional role is not clear, because a denervated uterus is also able to function normally during pregnancy and parturition.

Skin Supply

The sympathetic fibers which re-enter all spinal nerves via (grey) rami communicantes from the sympathetic trunk ganglia (p. 274A) pass to the skin in the peripheral nerves, and supply the vessels, sweat glands and erector pili muscles in the corresponding dermatomes (vasomotor, sudomotor, pilomotor functions). The segmental delimitation of loss of function in cases of damage to the spinal cord is of diagnostic importance.

In certain regions of the skin, the so called *zones of Head,* disease of an internal organ may produce pain or hypersensitivity and each organ is represented in a certain zone: diaphragm **B7** (C4), heart **B8** (T3, 4), esophagus **B9** (T4, 5), stomach **B10** (T8), liver and gall bladder **B11** (T8–11), small intestine **B12** (T10), colon **B13** (T11), bladder **B14** (T11–L1), kidney and testes **B15** (T10–L1). Head's zones are of practical importance in diagnosis.

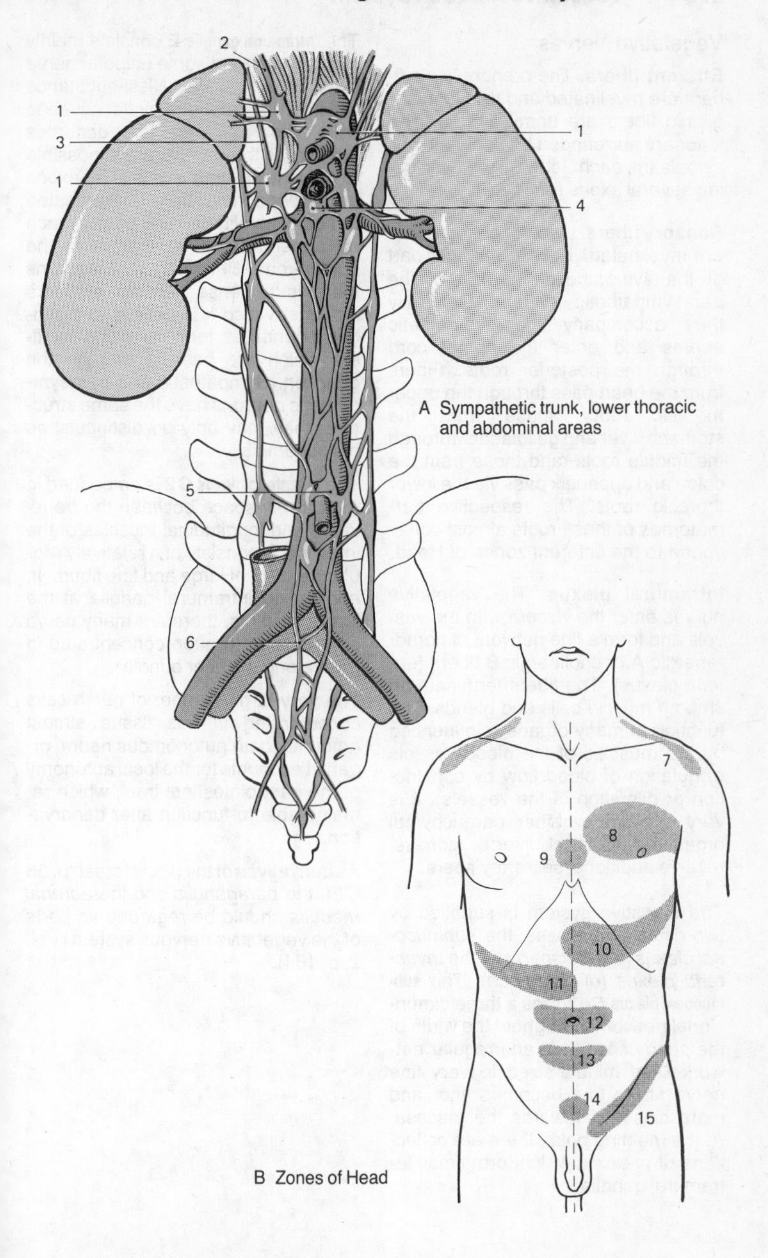

A Sympathetic trunk, lower thoracic and abdominal areas

B Zones of Head

Vegetative Nerves

Efferent fibers. The preganglionic fibers are myelinated and the postganglionic fibers are unmyelinated. The latter are surrounded by Schwann cell cytoplasm, each Schwann cell enclosing several axons (p. 34 A8).

Sensory fibers. Viscerosensory fibers are myelinated. They are neither part of the sympathetic nor part of the parasympathetic system. Generally they accompany the sympathetic nerves and enter the spinal cord through the posterior roots. Fibers from the heart pass through the upper thoracic roots, those from the stomach, liver and gallbladder through the middle roots, and those from the colon and appendix pass via the lower thoracic roots. The respective dermatomes of these roots almost correspond to the different zones of Head.

Intramural plexus. The vegetative nerves enter the viscera with the vessels and form a fine network of noradrenergic **A** or cholinergic **B** fibers (enteric plexus). The fibers terminate on smooth muscle cells and glands. The function of many organs is influenced by the muscles of the blood vessels (regulation of blood flow by constriction or dilatation of the vessels). It is very doubtful whether parenchymal organs, such as the liver or kidneys, receive additional secretory fibers.

The digestive system is supplied by two different plexuses: the *submucosal plexus* (of *Meissner*) and the *myenteric plexus* (of *Auerbach*). The **submucosal plexus C1** forms a three dimensional network throughout the width of the submucosa. It is an irregular network **D** of middle-sized to very fine nerve fibers that becomes finer and more closeknit towards the mucosa. At the meeting point, there are collections of neurons which form small intramural ganglia.

The **intramural ganglia E** consists mainly of multipolar and some unipolar nerve cells with granular Nissl-substance which are surrounded by flat covering cells. Their numerous long dendrites are thin, so they are usually impossible to distinguish from axons. The axons are exceptionally thin, unmyelinated or poorly myelinated and often branch off a dendrite rather than from the perikaryon. Between the neurons there is a dense network of fibers amongst which it is difficult to distinguish dendrites, terminal axons and fiers which are passing through the ganglion. Sympathetic and parasympathetic neurons have the same structure; they may only be distinguished histochemically.

The **myenteric plexus C2** is embedded in the narrow space between the transverse and longitudinal muscles of the intestine. It consists of a relatively regular network of large and fine fibers. In addition to intramural ganglia at the junction points, there are many nerve cells which are often concentrated in rows along the fiber bundles.

The very large number of nerve cells concentrated in this tissue almost amounts to an autonomous neural organ. It accounts for the local autonomy of the gastrointestinal tract, which remains able to function after denervation.

As derivatives of the neural crest (p. 56 C2), the *paraganglia* and the *adrenal medulla* should be regarded as parts of the vegetative nervous system (Vol. 2, p. 164).

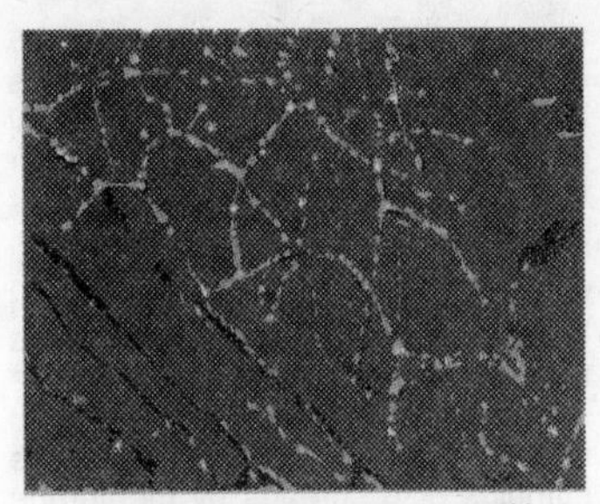

A Sympathetic fibers demonstrated by fluorescence microscopy

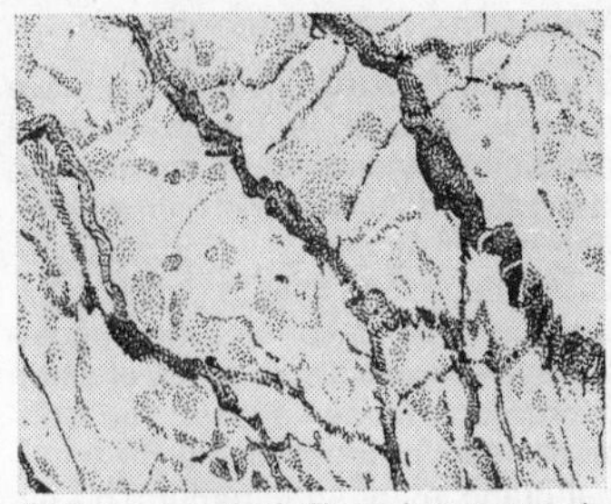

B Parasympathetic fibers demonstrated by the acetylcholinesterase reaction

AB Nerve fibers in the heart muscle of the rat (after Schiebler and Winckle)

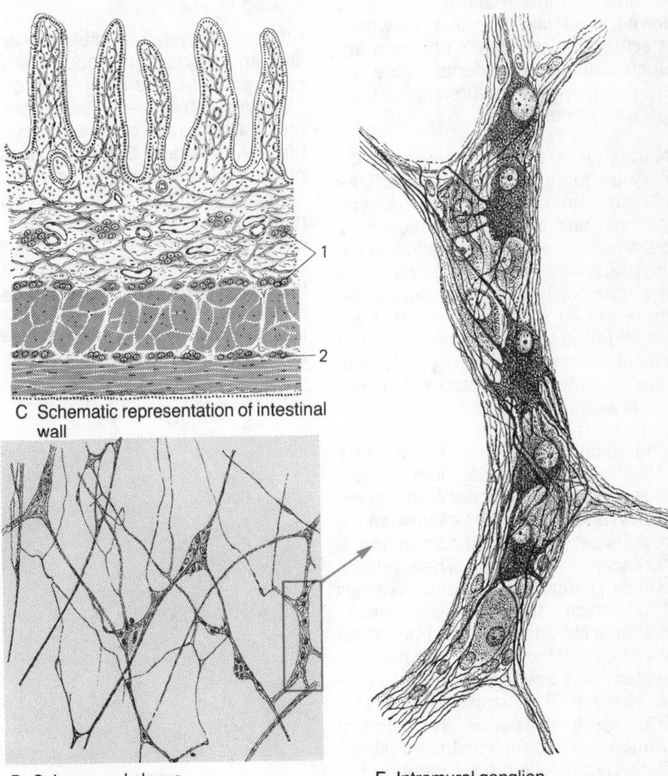

C Schematic representation of intestinal wall

D Submucosal plexus

E Intramural ganglion

Vegetative Nerves (continued)

Vegetative neuron A. The vegetative nervous system is formed from a number of individual elements, the vegetative neurons *(neuron theory).* The *continuity theory,* which was postulated for a long time to describe the terminal arborizations of the vegetative system, was disproven by electron microscopic observations. The continuity theory suggested that the terminal branches of an intramural plexus formed a network *(terminal reticulum)* in which the processes of different neurons were continuously interwoven both with each other and with the innervated muscle and gland cells to form a plasmatic syncytium. However, electron microscopy did not show any such continuous elements. There are, however, special features in the postganglionic neurons.

Numerous axo-axonal synapses **D1** may be found along the nerve fiber bundles, not only amongst the sympathetic and parasympathetic fibers, but also between sympathetic and parasympathetic fibers. In the region of the terminal branches of the axons, there are no special structures, such as are found on the motor end plates of striated muscle. There are only varicose enlargements **ABCD2** of the terminal axonal branches.

The axonal swellings may produce indentations in the smooth muscle cells and may even penetrate them. However, they usually lie between the muscle cells without direct membrane contact as in a synapse. The swellings contain clear and granular vesicles **C3,** which resemble presynaptic boutons. Noradrenalin, the transmitter substance of the sympathetic nervous system, has been found in the granular vesicles. The Schwann cell sheath **B4,** which surrounds the terminal branches is absent from the swellings. Correspondingly, the adjacent part of the wall of the smooth muscle cell **BC5** lacks a basement membrane **B6.** This is clearly the site of transmission of stimuli from the vegetative nerve fiber to the smooth muscle cell.

The vesicles contained in the axonal swellings discharge into the intercellular space **C7.** The transmitter diffuses through the tissue and transmits the stimulus to a large number of smooth muscle cells. It is probable that the stimulus also spreads by membrane contact between the muscle cells. The smooth muscle cells are closely associated through *gap junctions* **B8** where the basement membrane is lacking.

Efferent vegetative fibers supply smooth muscle and gland cells (secretory fibers). The innervation of the gland resembles to a considerable extent that of muscle cells. It is assumed that the receptors **C9** on the superficial surface of the innervated cell react specifically with the appropriate transmitter substance. Receptors on the outer surface of the cell membrane activate one of the enzymes lying on the inner surface of the membrane, adenylate cyclase **C10.** The adenylate cyclase system then stimulates the production of secretion by the gland cell.

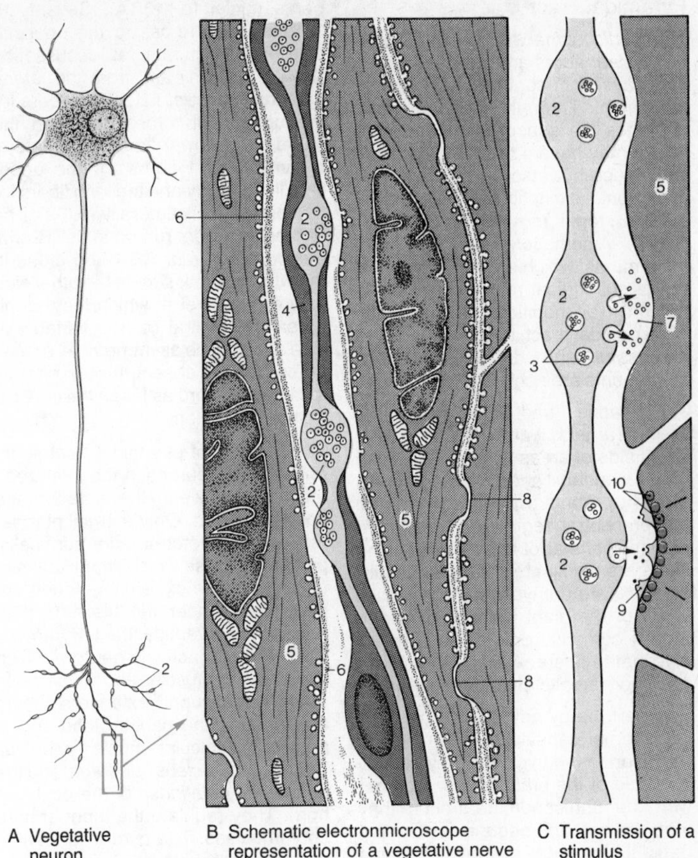

A Vegetative neuron

B Schematic electronmicroscope representation of a vegetative nerve fiber

C Transmission of a stimulus

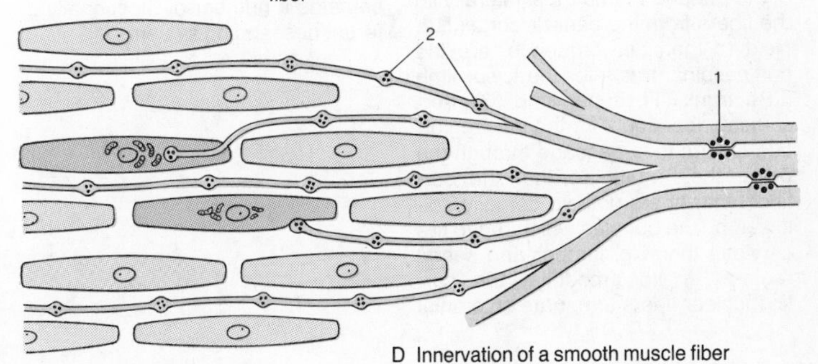

D Innervation of a smooth muscle fiber

Pyramidal Tract

The *corticospinal tract* and the *corticonuclear fibers* are regarded as the pathways for control of voluntary movement. Through them the cortex regulates the subcortical motor centers. It may have a retarding or inhibitory effect, but also produces a constant tonic excitation through which sudden, rapid movements are promoted. Automatic and stereotyped movements, which are under the control of subcortical motor centers, are thought to be modified by the influence of pyramidal tract impulses, so that purposeful and directly controlled movements are achieved.

The pyramidal tract fibers arise in the precentral fields 4 and 6 **A 1**, in parietal lobe fields of areas 3, 1 and 2, and in the secondary sensorimotor region (area 40). About two thirds stem from the precentral region and one third from the parietal lobe. Only 60% of the fibers are myelinated and the remaining 40% are unmyelinated. The thick fibers of the giant pyramidal cells in area 4 comprise only 2 to 3% of all myelinated fibers. All the other fibers arise from smaller pyramidal cells.

Fibers of the pyramidal tract traverse the internal capsule (p. 242). At the transition to the midbrain they occupy the base of the brain where, together with the corticopontine tracts, they form the cerebral peduncles. The pyramidal tract fibers occupy the middle of the peduncles and most laterally lie the fibers from the parietal cortex **B 2**. Next to them, in sequence, are the corticospinal tracts for the lower limb (LS), trunk (T), upper limb (C), and corticobulbar fibers for the face region **B 3**. During their passage through the pons, they rotate so that the corticobulbar fibers now lie dorsally, followed by the bundles which run to the cervical, thoracic, lumbar and sacral regions. In the medulla, the corticonuclear fibers terminate on cranial nerve nuclei (p. 132 A). Seventy to 90% of fibers cross to the opposite side in the pyramidal decussation **AB 4** (p. 52 A 1) and from the *lateral corticospinal tract* **AB 5.** The fibers for the upper limb cross dorsal to the fibers to the lower limb. In the lateral pyramidal tract fibers for the upper limb lie medially and the long fibers for the lower limb lie laterally (p. 52). The uncrossed fibers run on in the *ventral corticospinal tract* **AB 6** and cross in the white commissure to the other side only at the level at which they terminate. The ventral tract is variable in size and may be asymmetrical, or may be completely absent. It only extends in the spinal cord as far as the cervical or thoracic region.

The majority of pyramidal tract fibers end on interneurons in the intermediate zone between the anterior and posterior horns. Only a small number reaches the motor anterior horn cells. principally those which supply the distal parts of the extremities. They are particularly under the discrete influence of the pyramidal tract. Pyramidal tract impulses activate neurons which supply flexor muscles and inhibit those neurons that supply extensors. Fibers that arise in the parietal lobe end in posterior funicular nuclei *(nucleus gracilis* and *nucleus cuneatus)* and the *substantia gelatinosa* of the posterior horn. They regulate the input of sensory impulses. The pyramidal tract is therefore not a single motor tract but contains a number of functionally different descending systems.

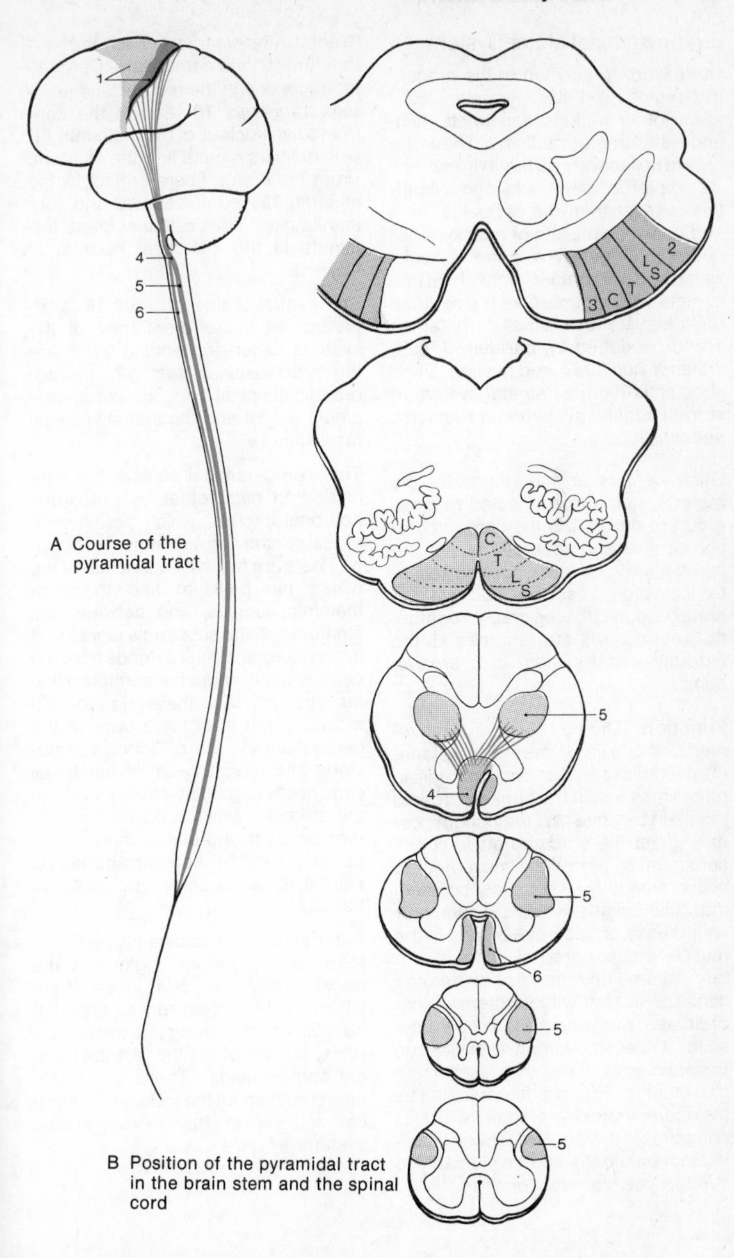

A Course of the pyramidal tract

B Position of the pyramidal tract in the brain stem and the spinal cord

Extrapyramidal Motor System

Definition. In addition to the precentral region and the pyramidal tract many other regions and tracts influence the motor activity. They are known collectively as the *extrapyramidal motor system.* Phylogenetically this is older than the pyramidal tract and unlike it consists of multisynaptic neurons. Originally a group of nuclei which had a characteristically high iron content was regarded as the extrapyramidal system: the striatum (putamen **1** and caudatum **2**), pallidum **3,** subthalamic nucleus **4,** red nucleus **5** and substantia nigra **6** *(striatal system* or *extrapyramidal system in a restricted sense).*

Other centers of importance for the motor system are connected with this group of nuclei, but they are integration centers and not motor nuclei: the cerebellum **7,** thalamic nuclei **8,** reticular formation, vestibular nuclei **9** and certain cortical regions. Together these structures are regarded as the *extrapyramidal system in a broader sense.*

Function. During voluntary movements of one limb, there is simultaneous stimulation of muscle groups in other limbs and in the trunk, in order to maintain balance and the upright posture under the changed static conditions, and to permit the movement to occur smoothly. These concomitant muscular actions, which are often only an increase or decrease of tone in the muscle groups, are not under voluntary control and are not performed consciously, but without them no coordinated movement would be possible. Other movements performed unconsciously are the associated movements. They are all controlled by the extrapyramidal system, which is comparable to a servomechanism acting independently and involuntarily to support voluntary movements.

Tracts. Afferent tracts reach the system through the cerebellum. Cerebellar tracts end in the red nucleus (*dentatorubral tract* **10**) and in the centromedian nucleus of the thalamus **11,** whose fibers extend into the striatum. From the cortex fibers project to the striatum **12,** red nucleus **13** and substantia nigra **14.** Vestibular fibers terminate in the interstitial nucleus of Cajal **15.**

The central tegmental tract **16** is regarded as the efferent tract of the system. Other descending tracts are the reticulospinal tract **17,** the rubroreticulospinal tract, the vestibulospinal tract **18** and the interstitiospinal fasciculus **19.**

The extrapyramidal centers are connected to each other by numerous neuronal circuits, which permit reciprocal control and adjustment of activity. There are two-way connections between the pallidum and the subthalamic nucleus, and between the striatum and the substantia nigra **20.** A large neuronal circuit extends from the cerebellum through the centromedian nucleus of the thalamus to the striatum, and from there back to the cerebellum via the pallidum, red nucleus and olive **21.** Other functional circuits are formed by cortical fibers to the striatum, which produce a recurrent circuit to the cortex through the pallidum and the anterior and lateral ventral thalamic nuclei (p. 180A, p. 222C).

The frontal and occipital visual fields, together with certain regions of the parietal and temporal lobes, from which complex mass movements can be elicited by strong electrical currents, are known as the 'extrapyramidal cortical fields'. There is still disagreement about the inclusion of cortical regions in the extrapyramidal system.

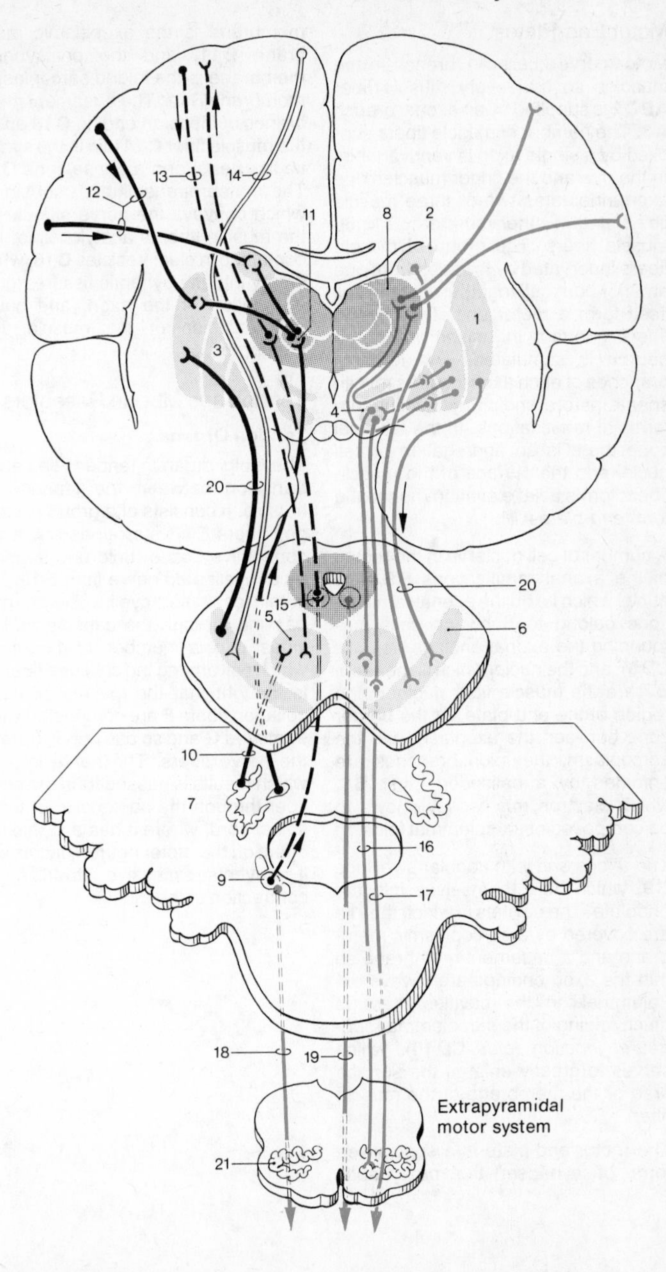

1
2
3
4
5
6
7
8
9
10
11
12
13
14
15
16
17
18
19
20
21
Extrapyramidal motor system

Motor End-Plates

Motor nerve fibers **A1** branch in the muscles so that every muscle fiber **ABC2** is supplied by an axonal branch **A3.** The number of muscle fibers supplied by a single axon is very variable. In the eye and the finger muscles one axon innervates two or three muscle fibers and in other muscles 50 to 60 muscle fibers. The group of muscle fibers innervated by the axon from one anterior horn cell together with the cell itself form a motor unit. The muscle fibers contract in unison when the neuron is stimulated. The terminal branches of each axon loses its myelin sheath before ending by splitting into terminal ramifications. In the terminal zone there is an aggregation of cell nuclei and the surface of the muscle fibers forms a flat elevation (hence the term *end-plate* **A4**).

A number of cell nuclei lie in the region of the axonal ramifications **B5.** The nuclei which lie on the axonal ramifications belong to Schwann cells surrounding the axonal endings (*teloglia* **CB6**), and the nuclei which lie beneath **B7** are the muscle fiber nuclei in the region of the end-plate. At the border zone between the axoplasm and the sarcoplasm, the axon branches are enfolded by a palisaded layer **B8,** which electron microscopy shows to be composed of sarcolemmal folds.

The axons end with nodular swellings **C9,** which bury themselves into the endplate. The cavities in which they lie are covered by a sarcoplasmic membrane and a basement membrane so that the axon endings are always extralemmal. In the cavities there is much folding of the sarcolemma (*subneural junction folds* **CD10**), which serves to greatly enlarge the surface area of the membrane of the muscle fiber.

The motor end-plate is a specialized form of synapse; the presynaptic membrane is the axoplasmic membrane **D11,** and the postsynaptic membrane is the folded sarcoplasmic membrane **D12.** The basement membranes of the axon ending **C13** and of the muscle fiber **C14** join in the synaptic fissure to form a dense zone **D15.** The neurotransmitter substance which conveys the nerve stimulus to the muscle fiber is acetylcholine. It is contained in clear vesicles **C16,** which empty into the synaptic fissure during stimulation of the axon, and cause depolarisation of the muscle fiber membrane.

Tendon and Muscle Receptors

Tendon Organs

The Golgi organ in tendons lies at the transition between the tendon and muscle. It consists of a group of collagen fibers **E17,** surrounded by a thin connective tissue tunic and supplied by a myelinated nerve fiber **E18.** The latter loses its myelin sheath after passing through the capsule and divides into a number of branches, which coil among the collagen fibers. It is thought that the loosely arranged collagen fibers **F** are compacted when stretched **G** and so cause pressure on the nerve fibers. The nerve impulse which results is passed from the nerve fiber through the posterior root to the spinal cord, where it has an inhibitory effect on the motor neuron. In this way it prevents excessive stretching or contraction of the muscle.

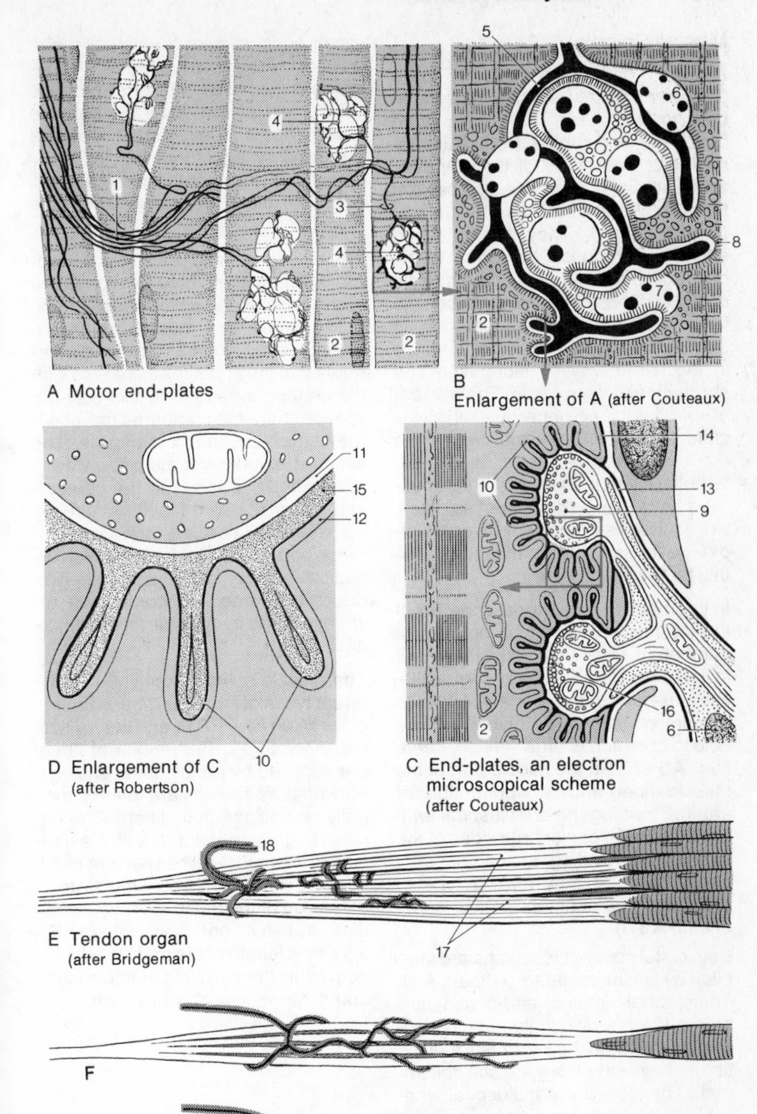

A Motor end-plates

B Enlargement of A (after Couteaux)

C End-plates, an electron microscopical scheme (after Couteaux)

D Enlargement of C (after Robertson)

E Tendon organ (after Bridgeman)

F

G

F G Tendon organ in a relaxed (F) and contracted (G) muscle (after Bridgeman)

Muscle Spindle

A muscle spindle consists of 5 to 10 fine, transversely striated muscle fibers (*intrafusal fibers* **A1**), surrounded by a fluid-filled connective tissue capsule **A2.** The fibers of the spindles, which are up to 10 mm long, lie parallel to the other fibers of the muscle *(extrafusal fibers).* They are attached either to the muscle tendon or to the connective tissue poles of the muscle capsule. Since the intrafusal fibers lie in the same longitudinal direction as the extrafusal fibers, they are affected in the same way by elongation and shortening of the muscle. The number of spindles in each individual muscle is quite variable. Those muscles which are concerned with fine, highly differentiated movements (finger muscles) have a large number of spindles, while muscles which perform only simple movements (trunk muscles) contain far fewer spindles.

In the middle or equatorial zone **A3** of the intrafusal fiber, which contains a number of cell nuclei, myofibrils are lacking; this part is not contractile. Only the two distal parts **A4,** which contain cross-striated myofibrils, are able to contract. A large sensory nerve fiber **A5** ends in the middle part of the muscle fiber and is wound spirally around, forming the *annulospiral ending* **AC6B.** On one or both sides of the annulospiral ending a finer sensory nerve fiber **A7** may form a type of umbelliform attachment (*flower-spray ending* **A8D**).

Both distal contractile regions are supplied by fine motor fibers (γ-fibers **A9**). Their small motor end-plates are marked by poorly developed junctional folds, but as in the extrafusal muscle fibers their end-plates are still epilemmal. The sensory annulospiral endings, on the other hand, lie beneath the basement membrane of the muscle fiber **C10** and are thus hypolemmal. The γ-fibers for these specialized motor endings stem from small motor cells in the anterior horn, the γ-neurons, whose impulses produce contraction of the distal parts of the muscle fibers. The resultant stretch of the sensory equatorial part not only stimulates the annulospiral ending, but also alters the sensitivity of the spindle.

The muscle spindle is a stretch receptor, which is stimulated by elongation of the muscle and which ceases to be active when the muscle contracts. During stretching of a muscle, the frequency of the impulses increases with the change in length. In this way the spindle transmits information about the current length of the muscle. The stimuli are not only transmitted via the spinocerebellar tracts to the cerebellum, but are also transmitted directly via reflex collaterals to the large anterior horn cells. Their stimulation by sudden stretching of the muscle produces an immediate contraction response of the muscle (stretch reflex p. 44).

The muscle spindle contains two different types of intrafusal fibers: the *nuclear chain fiber* **EF11** and the *nuclear bag fiber* **EF12.** Both types of fibers are supplied by annulospiral endings. Flowerspray endings are found principally on nuclear chain fibers. The nuclear bag fibers react to active extension of the muscle, while nuclear chain fibers register the continuing state of extension of the muscle. Muscle spindles transmit not only information about the length of the muscle but also convey to the cerebellum information about the velocity of contraction.

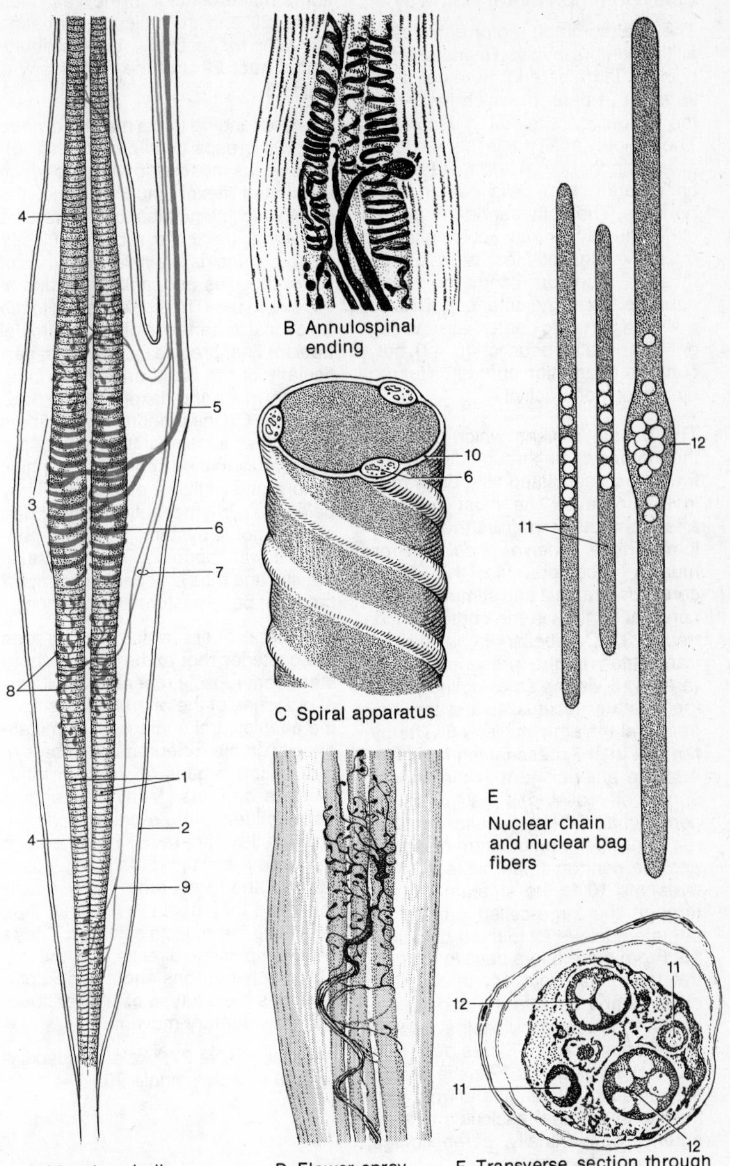

A Muscle spindle

B Annulospinal ending

C Spiral apparatus

D Flower-spray ending

E Nuclear chain and nuclear bag fibers

F Transverse section through a muscle spindle

Final Common Motor Pathway

The final common motor pathway for all the centers involved in motor activity is the large anterior horn cell **1** and its axon (α-neuron) which innervates the voluntary skeletal musculature. The majority of all tracts that run to the anterior horn do not terminate directly on anterior horn cells but on interneurons. These influence the neuron either directly, or may act by inhibiting or activating reflexes between the muscle receptors and the motor neurons. Thus, the anterior horn is not a simple synaptic center, as was suggested in the introduction (p. 44), but a complex integration apparatus for regulation of motor activity.

The central regions which influence the motor system through descending tracts are associated with each other in many ways. The most important afferent tracts stem from the cerebellum. These receive impulses from muscle receptors via the spinocerebellar tracts **2** and stimuli from the cerebral cortex via the corticopontine tracts **3.** Cerebellar impulses are transmitted to the precentral cortex (area 4) **6** via the small-celled part of the dentate nucleus **4,** and the ventrolateral thalamic nucleus **5.** The pyramidal tract **7,** descending from area 4 to the anterior horn in the pons **8,** sends off collaterals back into the cerebellum. Other cerebellar impulses run through the emboliform nucleus **9** and the centromedian nucleus of the thalamus **10** to the striatum **11,** and through the large-celled part of the dentate nucleus **12** to the red nucleus **13.** From there fibers pass in the central tegmental tract **14,** through the olive **15** and back to the cerebellum, and in the rubroreticulospinal tract **16** to the anterior horn. Fibers pass from the nucleus globosus **17** to the interstitial nucleus of Cajal **18** and from it via the interstitiospinal fasciculus **19** to the anterior horn. Finally, cerebellofugal fibers make contact in the vestibular nuclei **20** and the reticular formation **21** with, respectively, the vestibulospinal tract **22** and the reticulospinal tract **23.**

The descending tracts may be divided into two groups according to their effects on the muscles: one group which stimulates flexor muscles and the other which stimulates extensors. The pyramidal tract and rubroreticulospinal tract principally activate neurons of flexor muscles and inhibit neurons of the extensors. This corresponds to the functional importance of the pyramidal tract for fine, precise movements, particularly of the hand and finger muscles, in which the flexors play a decisive role. On the other hand, the fibers of the vestibulospinal tract and of the reticular formation of the pons inhibit flexors and activate extensors. They belong to a phylogenetically older part of the motor system, which is concerned with opposing the effects of gravity, and thus is of particular importance for body posture and balance.

Peripheral fibers which run through the posterior root to the anterior horn stem from muscle receptors. Collateral branches of the afferent fibers from the annulospiral endings **24** terminate directly on the α-neurons; the fibers of the tendon organs **25** end on intermediate neurons. Many descending tracts influence the α-neurons via the spinal reflex apparatus. These end on the large α-cells and the small γ-cells **26.** As the γ-neurons have a low threshold they begin to fire first, thus activating the muscle spindles. These send impulses directly to the α-neurons. γ-neurons and muscle spindles thus have a type of 'starter' function for voluntary movement.

Accessory olive **27,** skeletal musculature **28** muscle spindle **29.**

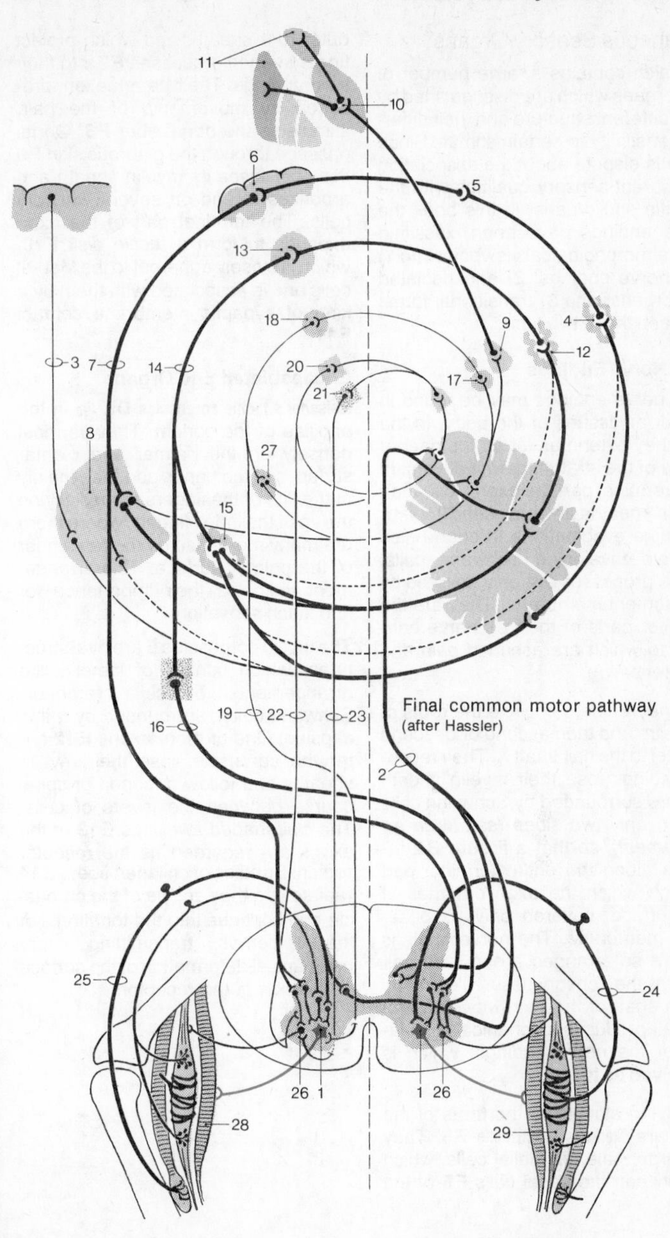

Final common motor pathway
(after Hassler)

Cutaneous Sensory Organs

The skin contains a large number of end organs which are distinguished by their different structure and their different sensitivity to certain stimuli. Since there is dispute about the association of different sensory qualities with the specific end organs, in this book the nerve endings have been classified from a morphological viewpoint into 1) free nerve endings; 2) encapsulated end organs; and 3) transitional forms between these two types.

Free Nerve Endings

Free nerve endings may be found in almost all tissues of the body. In the skin they extend as far as the deeper layers of the stratum germinativum. In their terminal part the axons lose their myelin sheaths and may send knot- or fingerlike evaginations through holes in their enveloping Schwann cells. These processes are only covered by a basement membrane. They are the receptor parts of the free nerve endings, to which are ascribed pain and cold sensation.

Thin nerve fibers encircle the roots of the hairs and then ascend or descend parallel to the hair shaft **A.** Their receptor endings lose their myelin sheath and are surrounded by Schwann cells **B1** on only two sides *(sandwich arrangement),* so that a fissure **B2** remains along the entire terminal part through which the axon emerges at the surface, covered only by basement membrane. The nerve endings **C3** are so arranged around the hair **C4** that the sensory fissure lies radial to the shaft of the hair. Movement of the hair produces mechanical stimulation of the nerve endings, which is perceived as touch.

Also lying alongiside the roots of the hairs are **Merkel's touch cells F5.** They are large, pale, epithelial cells, which lie between the basal cells **F6** of the outer root sheath, and which project finger shaped processes **F7** into their surroundings. Their deformation, produced by movements of the hair, stimulates the nerve fiber **F8.** On its passage through the glabrous skin **F9** this fiber loses its myelin sheath and arborizes to end on several sensory cells. The terminal part of the axon expands to form a *tactile disk* **F10,** which is closely adherent to the Merkel cells and is associated with them by a kind of synaptic membrane contact **F11.**

Encapsulated End Organs

Meissner's tactile corpuscles DE lie in the papillae of the corium. They lie most densely on the palmar and plantar surfaces of the hands and feet, mainly at the tips of the fingers (especially on the tip of the index finger), where there are many more than on the remainder of the palmar surface. This arrangement indicates their importance for fine touch sensation.

The tactile corpuscles **E** are oval structures, which consist of lamella like arrangements of cells (probably Schwann cells), surrounded by a thin capsule. One or more axons **E12** enter the corpuscle, lose their myelin sheaths and follow a looped or spiral course between the layers of cells. The clubshaped swellings **E13** of the axons are regarded as the receptor region. Bundles of collagen fibers **E14** radiate into the capsule of the corpuscle and continue into the tonofibrils of the epidermis, transmitting every mechanical deformation of the surface of the skin of the receptors.

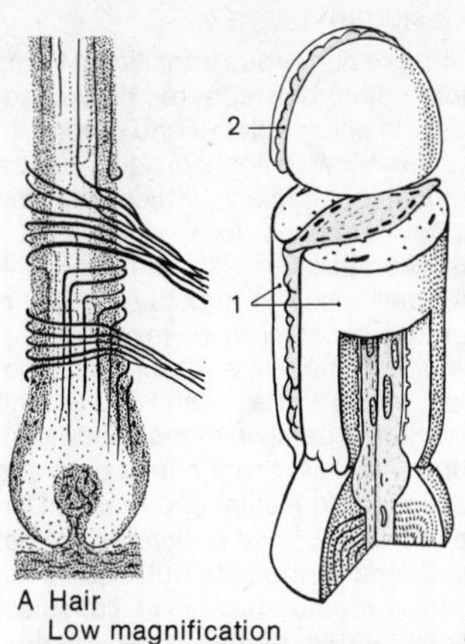

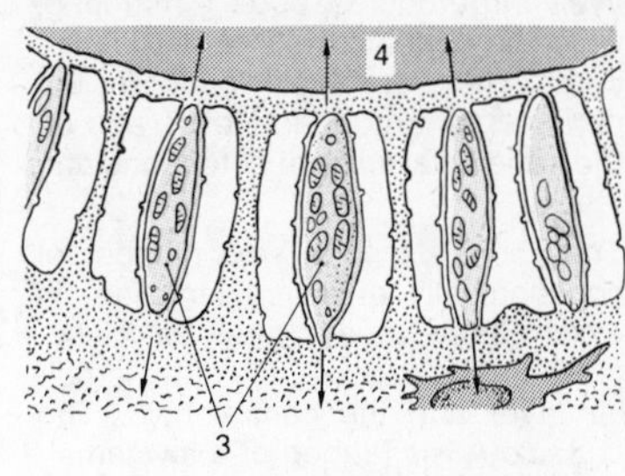

C Arrangement of the nerve endings on hairs (after Andres and v. Düring)

B Electron microscopical scheme of a nerve ending (after Andres)

A Hair
Low magnification

ABC Free nerve endings related to hairs

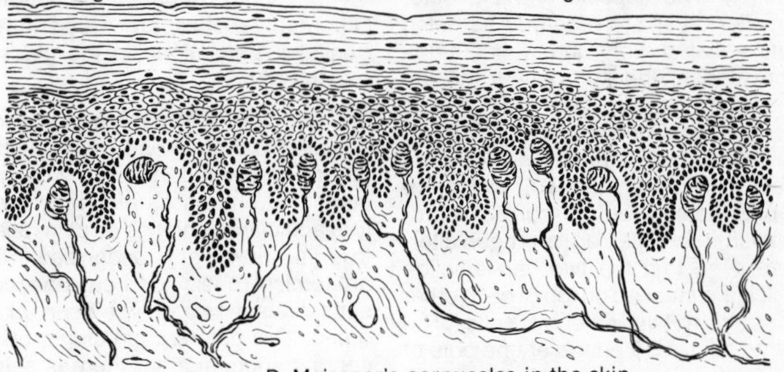

D Meissner's corpuscles in the skin

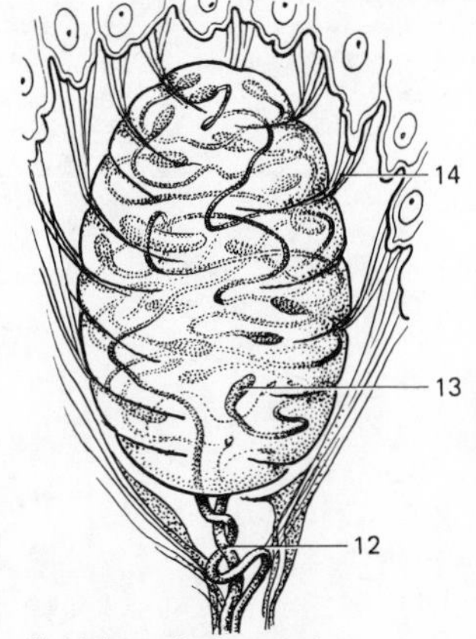

E Meissner's corpuscle

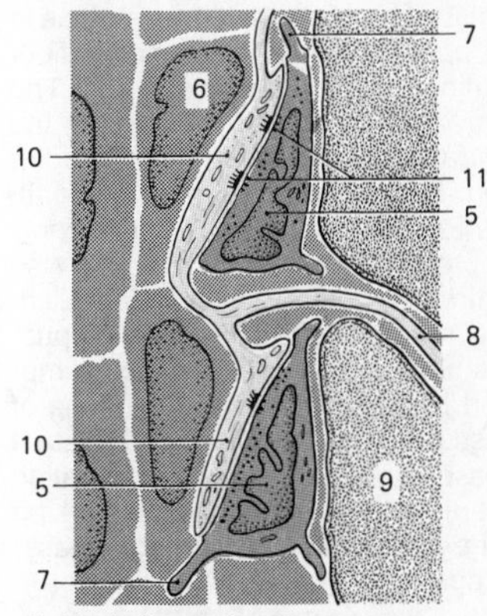

F Merkel's tactile disks as seen by electron microscopy (after Andres and v. Düring)

The **Vater-Pacini corpuscles ABC** are relatively large bodies, up to 4 mm long, which lie beneath the skin in the subcutis. They are also found in the periosteum, around the joints and on the superficial surface of tendons and fascia.

They consists of a large number of concentrically arranged lamellae, in which three layers may be distinguished; the capsule, the outer lamellar layer and the central bulb. The *capsule* **A1** is formed of a few lamellae, which are reinforced by connective tissue fibers. The *outer lamellar layer* **AC2** consists of closed circular protoplasmic lamellae **B3,** separated by noncommunicating fluid-filled spaces. The capsule and the outer lamellar layer are regarded as being differentiated from the perineurium. The densely-packed layers of the fluid-free *central bulb* **ABC4** are formed from Schwann cells. The central bulb consists of two symmetrical stacks of halflamellae **B5,** which are separated by a radial fissure. The protoplasmic lamellae are arranged alternately, so that the two half-lamellae which are stacked on top of each other arise from two different cells. The nerve fiber **AC6** enters at one pole of the corpuscle and travels in the middle of the central bulb to its end. The myelin sheath is lost at the entrance to the central bulb and the nerve fiber terminates in a clublike swelling. The unmyelinated part of the axon in the central bulb **ABC7** is the receptor area. Perineurium **A8,** myelin sheath **A9.** Pacinian corpuscles act not only as pressure receptors, but are also particularly sensitive to vibration. Electrical recordings from isolated corpuscles have shown that they are stimulated by deformation and by lifting of pressure, but not by constant pressure. As very sensitive vibration receptors they respond to the shaking of the floor caused by someone passing by.

Transitional Types

There are numerous transitional forms intermediate between free nerve endings and encapsulated end organs. In all of them nerve fibers divide to form a terminal complex in which the fine axonal branches form a knot or a bushlike structure with nodular terminal swellings. The structures are surrounded by a more or less well developed connective tissue capsule. They include the oval Krause end bulbs in the papillae of the corium, the round Golgi-Mazzoni corpuscles and the elongated Ruffini end organs **D** in the subcutis. These structures are not only found in the skin but also in the mucous membranes, joint capsules, the coverings of the large arteries. Different varieties are found in the genitalia, particularly in the glans **E** and the clitoris.

The attribution of particular sensory qualities to certain end organs is disputed. The importance of the corpuscular end organs as mechanoreceptors and of free endings as pain receptors is, however, recognized. The anatomical basis for heat and cold sensations has not yet been elucidated. The different structures of receptors suggests that there is selection of stimuli by them.

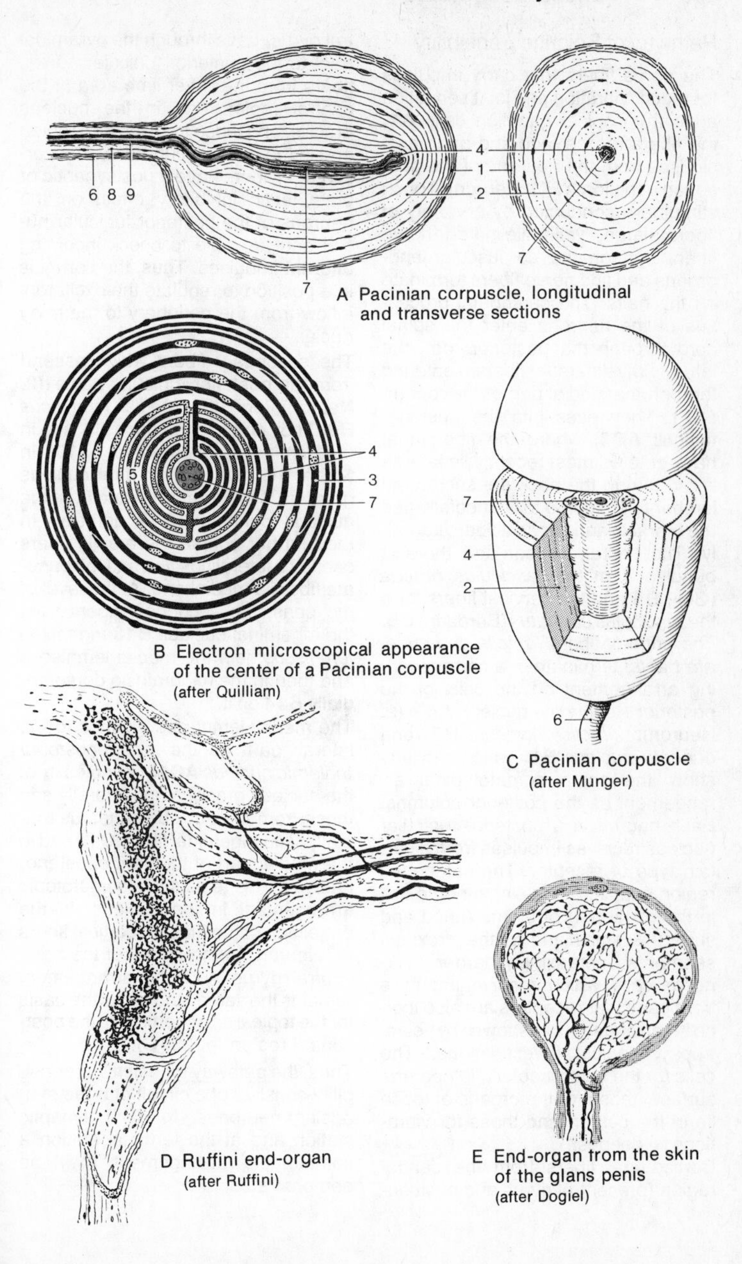

A Pacinian corpuscle, longitudinal and transverse sections

B Electron microscopical appearance of the center of a Pacinian corpuscle (after Quilliam)

C Pacinian corpuscle (after Munger)

D Ruffini end-organ (after Ruffini)

E End-organ from the skin of the glans penis (after Dogiel)

Pathway of Epicritic Sensibility

The nerve fibers that carry impulses for touch, vibration and joint sensation arise from spinal ganglion cells **A1,** the fibers for the face and paranasal sinuses from the neurons of the gasserian ganglion **A2** (1. Neuron). Touch stimuli are conducted by a variety of fibers: large, well-myelinated nerve fibers end on the corpuscular end-organs and fine nerve fibers terminate on the hairs. The centripetal processes of the neurons enter the spinal cord through the posterior root, the large, myelinated fibers entering through the medial part of the root (p. 56F). They pass into the posterior funiculi **AB3,** where the fibers that have entered most recently lie laterally, so that in the neck the sacral and lumbar fibers come to lie medially and the thoracic and cervical fibers laterally. The sacral, lumbar and thoracic bundles form the *fasciculus gracilis* (Goll) **B4** and the cervical fibers form the *fasciculus cuneatus* (Burdach) **B5.** The primary fibers *(gracile and cuneate tracts)* terminate in a corresponding arrangement on the cells of the posterior column nuclei **A6** (2. Neuron), *gracile nucleus* **B7** and *cuneate nucleus* **B8,** which similarly show the same somatotopical arrangement as the posterior columns. Each neuron in a posterior funicular nucleus receives impulses from a certain type of receptor. The cutaneous region supplied by a single nerve cell in the distal part of a limb (hand and finger) is small and in the proximal segments increasingly larger. The nerve cells which receive impulses from particular receptors are also topically arranged, as shown by electrophysiological investigations. The cells for the hair receptors lie near the surface, those for the organs of touch lie in the center, and those for vibrations lie deepest.

Corticofugal fibers from the central region (precentral gyrus and postcentral gyrus) pass through the pyramidal tract to the posterior funicular nuclei, fibers from the lower limb area of the central region end in the nucleus gracilis, and those from the arm region end in the nucleus cuneatus. The corticofugal fibers have a postsynaptic or presynaptic inhibitory effect on the neurons of the posterior funicular nuclei and are able to check incoming afferent impulses. Thus, the cortex is in a position to regulate the excitatory inflow from the periphery to the relay nuclei.

The secondary fibers which ascend from the posterior funicular nuclei (2. Neuron) form the *medial lemniscus* **B9.** They cross to the opposite side in the *lemniscal decussation* **B10,** in which the fibers from the nucleus gracilius **B11** lie ventrally and from the nucleus cuneatus **B12** dorsally. In their further course the gracile fibers occupy a lateral position and the cuneate fibers a medial one. At the level of the pons, the secondary fibers from the trigeminal nucleus **B13** *trigeminal lemniscus,* join the medial lemniscus and then in the midbrain lie dorsomedially **B14** on it.

The medial lemniscus extends to the lateral part of the *ventroposterior thalamic nucleus* **AC15;** bundles from the nucleus gracilius end laterally and those from the nucleus cuneatus medially. The trigeminal fibers **C16** end in the medial part of the ventroposterior nucleus. This results in a somatotopic arrangement in the nucleus. In the projection of the thalamocortical fibers (3. Neuron) to the cortex of the postcentral gyrus **A17** the arrangement remains the same, and forms the basis for the topical arrangement of the postcentral region (p. 234C).

Thus, the pathway for epicritical sensibility consists of a circuit of three synapsing neurones. In each synaptic station and at the terminal station a somatotopic arrangement may be demonstrated.

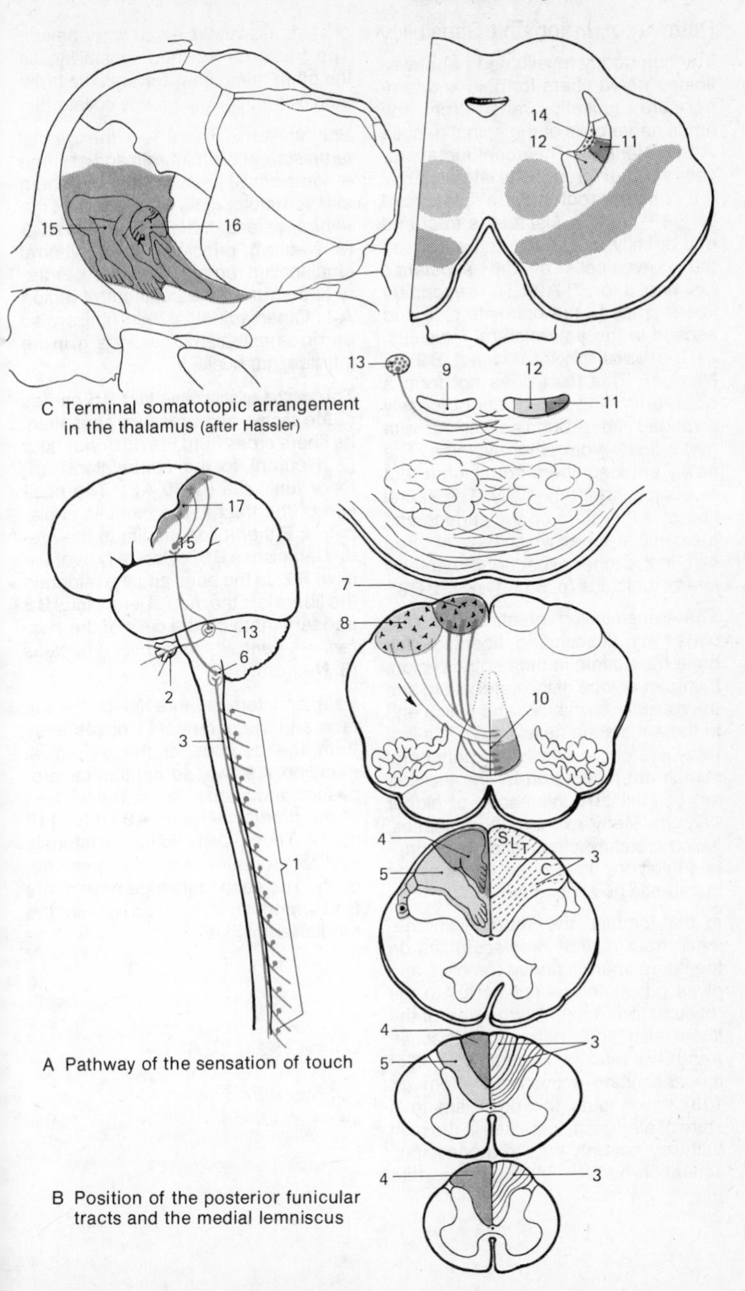

C Terminal somatotopic arrangement in the thalamus (after Hassler)

A Pathway of the sensation of touch

B Position of the posterior funicular tracts and the medial lemniscus

Pathway of Protopathic Sensibility

The fine poorly myelinated or unmyelinated nerve fibers for pain and temperature sensation arise from the small nerve cells of the spinal ganglia **A1** (1. Neuron). Their centripetal processes enter through the lateral part of the posterior root into the spinal cord (p. 56F), fork in *Lissauer's tract* and end directly, or via an interneuron, on the nerve cells of the *substantia gelatinosa* (p. 50 A2). The secondary fibers cross to the opposite side and ascend in the anterolateral funiculus, as the **lateral spinothalamic tract B2** (2. Neuron). This tract does not form a discrete trunk but consists of loosely arranged fiber bundles mixed with nerve fibers from other systems. The newly entering fibers from higher root levels join ventromedially. The sacral fibers lie therefore on the surface and the cervical fibers, which have entered last, in the inner aspect of the anterolateral funiculus (p. 50 A1, p. 132 B8).

The transmission of stimuli is regulated by descending fibers, which have their origin in the central region, the anterior lobe of the cerebellum and the reticular formation, and which end in the substantia gelatinosa. This is a relay system in which the peripheral stimuli are modulated by the facilitating or inhibitory influences of higher centers. Many axo-axonal synapses, which are characteristic for presynaptic inhibition, have been found in the substantia gelatinosa.

In the medulla the lateral spinothalamic tract *(spinal lemniscus)* lies on the lateral margin above the olive and gives off numerous collaterals to the reticular formation. A large part of the fibers also ends here *(spinoreticular tract)*. The reticular formation is part of the ascending activation system (p. 138) which puts the organism in a state of alertness. Impulses in the pain pathway not only lead to a conscious sensation but also increase the state of alertness via the reticular formation. The tracts for epicritic sensibility, on the other hand, pass through the brain stem without giving off any collaterals.

Spinothalamic fibers join the *medial lemniscus* in the midbrain and assume a dorsolateral position. Many of them end somatotopically on the cells of the *ventroposterior thalamic nucleus* **AC3** (3. Neuron), principally in the ventral small-celled region. From there, tertiary fibers run to the postcentral region **A4.** Other spinothalamic fibers also end in other thalamic nuclei, e. g. in the intralaminar nuclei.

The **ventral spinothalamic tract B5** carries crude tactile and pressure sensation. Its fibers cross from the posterior horn (2. Neuron) to the contralateral anterior funiculus (p. 50 A3). The position of the tract in the medulla is disputed; Either it lies medial to the medial lemniscus **B6,** or laterally over the olive **B7.** In the pons and the midbrain the fibers join the medial lemniscus **B8** and terminate on the cells of the *posterior ventral thalamic nucleus* (3. Neuron).

Pain and temperature fibers for the face and the paranasal sinuses arise from the neurons of the trigeminal ganglion **A9**, whose centripetal processes terminate in the *spinal nucleus of the trigeminal nerve* **AB10** (p. 116 BC4). There, pain fibers lie laterally and temperature fibers lie more medially. The secondary trigeminal fibers **B11** *(trigeminal lemniscus)* join the medial lemniscus.

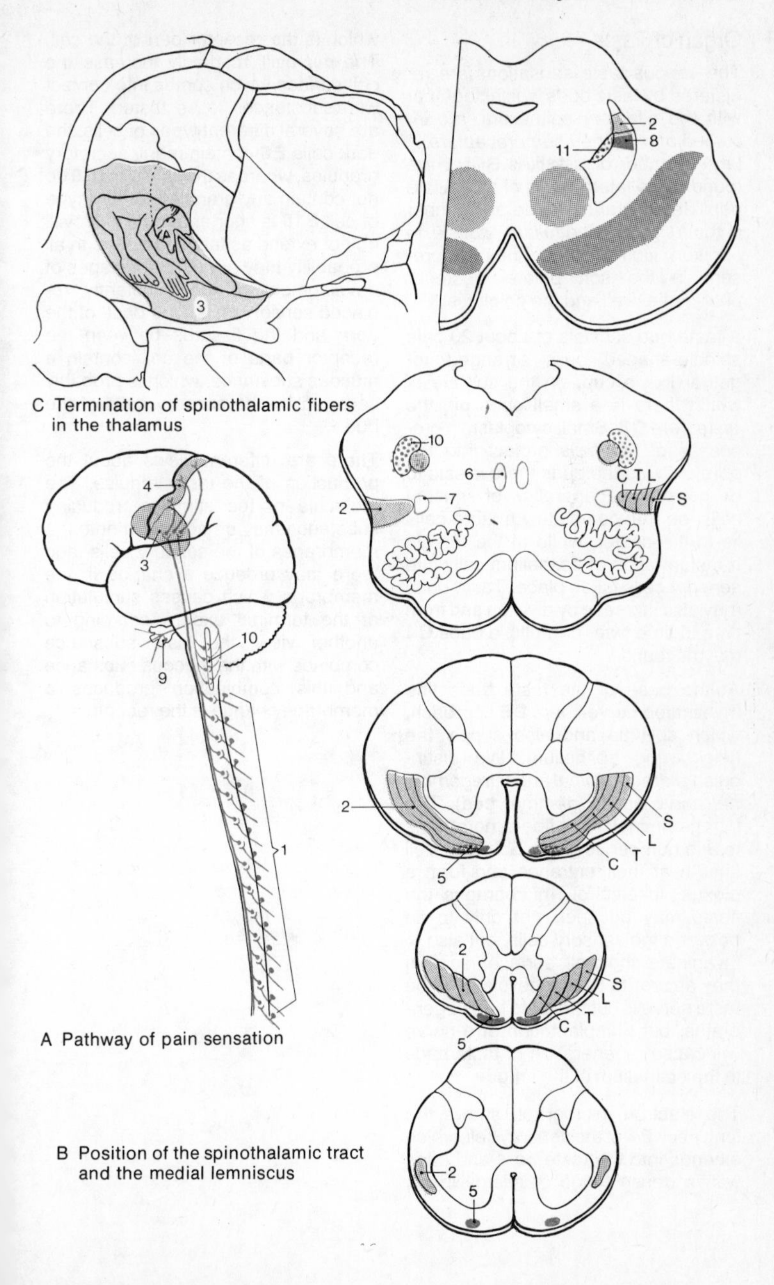

C Termination of spinothalamic fibers in the thalamus

A Pathway of pain sensation

B Position of the spinothalamic tract and the medial lemniscus

Organ of Taste

The various taste sensations are registered by taste buds which, together with the olfactory epithelium, are included among the *chemoreceptors.* A large number of **taste buds BCD 1** are found in the lateral walls of the vallate papillae **AB 2,** and a moderate number in the *fungiform* and *foliate papillae.* In addition, isolated taste buds are present in the soft palate, posterior pharyngeal wall and the epiglottis.

A taste bud consists of about 20 pale spindle-shaped cells arranged together like a cup, on the surface of which there is a small open pit, the *taste pore* **C 3.** Small cytoplasmic processes of the cells project into the pore. Around the buds there is said to be constant regeneration of sensory cells, so that completely mature cells lie in the center, while at the margin transformation of epithelium cells into sensory cells takes place. Taste buds may also increase by division and from time to time twin- or multiple buds **C 4** may be found.

At the base of the taste buds fine myelinated nerve fibers **D 5** approach, which split up and also supply the neighboring epithelium. We distinguish *extragemminal* and *intragemminal* nerve fibers (gemma: bud). The intragemminal nerve fibers, normally 2 to 3 in number, lose the Schwann cell sheath at their entrance and form a plexus. In electron micrographs the fibrils may be seen not only to lie between the sensory cells, but also to invaginate the cell surface, so that they appear to lie intracellularly. If the taste nerve is cut the taste bud degenerates, but reimplantation of a nerve will cause regeneration of taste buds in the epithelium of the tongue.

The electron micrograph shows the long neck **E 6** of the sensory cell, which extends into the taste pore and ends with a dense fringe of microvilli **E 7,** which is the receptor part of the cell. The microvilli markedly increase the cell surface which comes into contact with substances to be tasted. There are several different types of cells: the dark cells **E 8** contain many secretory granules, whereas the light cells **E 9** do not contain any granules. A third type of cell **E 10** is shorter and its microvilli do not extend as far as the pore. In all probability they are different stages of development of the constantly replaced sensory cells. The base of the pore and the fissures between the receptor parts of the cell contain a mucous substance, which is probably secreted by the sensory cells of the bud.

There are different views about the production of the taste impulse. The molecule of the stimulus-producing substance may be adsorbed onto the membranes of the sensory cells, and there may produce a change in the membrane which causes stimulation of the terminal axon. According to another view, the taste substance combines with the mucous substance and this combination produces a membrane change in the receptors.

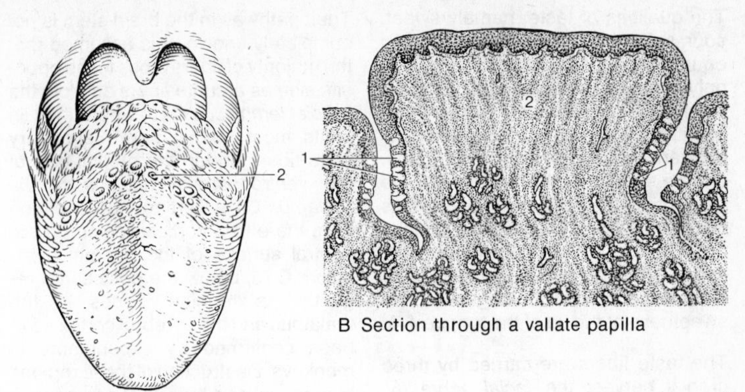

A Tongue with papillae

B Section through a vallate papilla

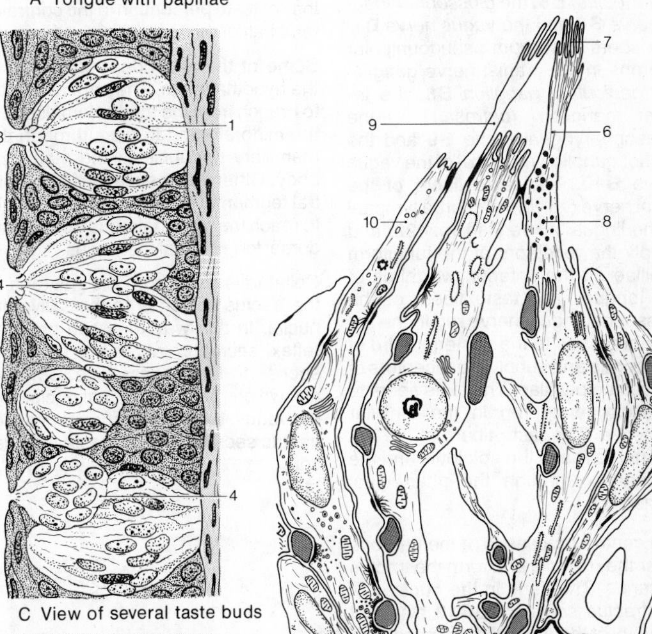

C View of several taste buds

E Electron micrograph
(after Popoff)

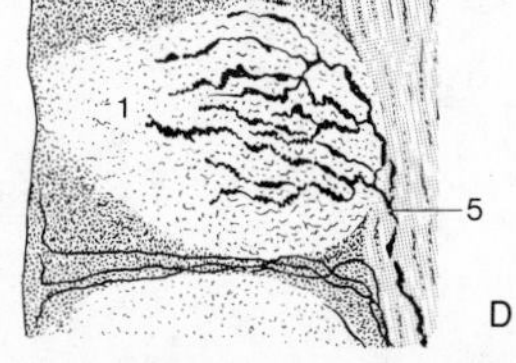

D Nerve endings

CDE Taste buds

The qualities of taste, namely sweet, sour, salty and bitter, are not detected equally by all taste buds. Some buds only react to sweet or only to sour, whereas others register two or three qualities. There are no morphological differences between any of the buds. On the surface of the tongue too, the perception of the individual qualities varies. Sour is perceived particularly on the lateral margin of the tongue **A1**, salt on the margins and the tip **A2**, bitter at the base of the tongue **A3** and sweetness at the tip of the tongue **A4.**

The taste fibers are carried by three cranial nerves: the *facial nerve (N. intermedius)* **B5,** the *glossopharyngeal nerve* **B6** and the *vagus nerve* **B7.** The fibers arise from pseudounipolar neurons in the cranial nerve ganglia, the *geniculate ganglion* **B8,** the inferior ganglion *(petrosal)* of the glossopharyngeal nerve **B9** and the inferior ganglion *(nodose)* of the vagus nerve **B10.** The taste fibers of the facial nerve run via the chorda tympani to the lingual nerve **B11** (p. 122) and supply the receptors of the fungiform papillae on the anterior two thirds of the tongue. The taste fibers of the glossopharyngeal nerve run in the lingual branches to the posterior third of the tongue and supply the receptors of the vallate papillae. The taste fibers to the soft palate run in the tonsillar branches. The taste fibers of the vagus nerve reach the epiglottis and the epipharynx through the pharyngeal branches.

The central processes of the neurons enter the medulla and form the tractus solitarius. They end in the *nucleus of the tractus solitarius* **BC12** at about the level of the entry of the nerves. The nucleus enlarges in this region and contains a cell complex, which is also known as the gustatory nucleus.

The secondary taste fibers arise from the nucleus of the tractus solitarius. Their pathway in the brain stem is not completely known. It is assumed that the majority of fibers cross to the opposite side as arcuate fibers and join the *medial lemniscus* **C13,** where they lie in its most medial part. Secondary taste fibers end in the medial part of the *ventroposterior nucleus of the thalamus* **C14.** Tertiary fibers extend from there to a cortical region on the ventral surface of the parietal operculum **C15,** below the postcentral region. The terminal zones in the thalamus and the cerebral cortex have been confirmed by experiments in monkeys. Destruction of these regions in man, caused by disease, produces loss of taste perception in the contralateral half of the tongue.

Some of the secondary fibers run to the hypothalamus. They are assumed to branch from the medial lemniscus in the midbrain and to pass through the mamillary peduncle to the mamillary body. Other fibers synapse in the ventral tegmental nucleus and are thought to reach the hypothalamus through the dorsal longitudinal fasciculus.

Collaterals pass from the neurons of the tractus solitarius to the salivatory nuclei. In this way they can produce reflex secretion of saliva as a response to taste sensations. Collaterals, which run to the dorsal nucleus of the vagus, form a reflex connection for gastric secretion due to taste stimulation.

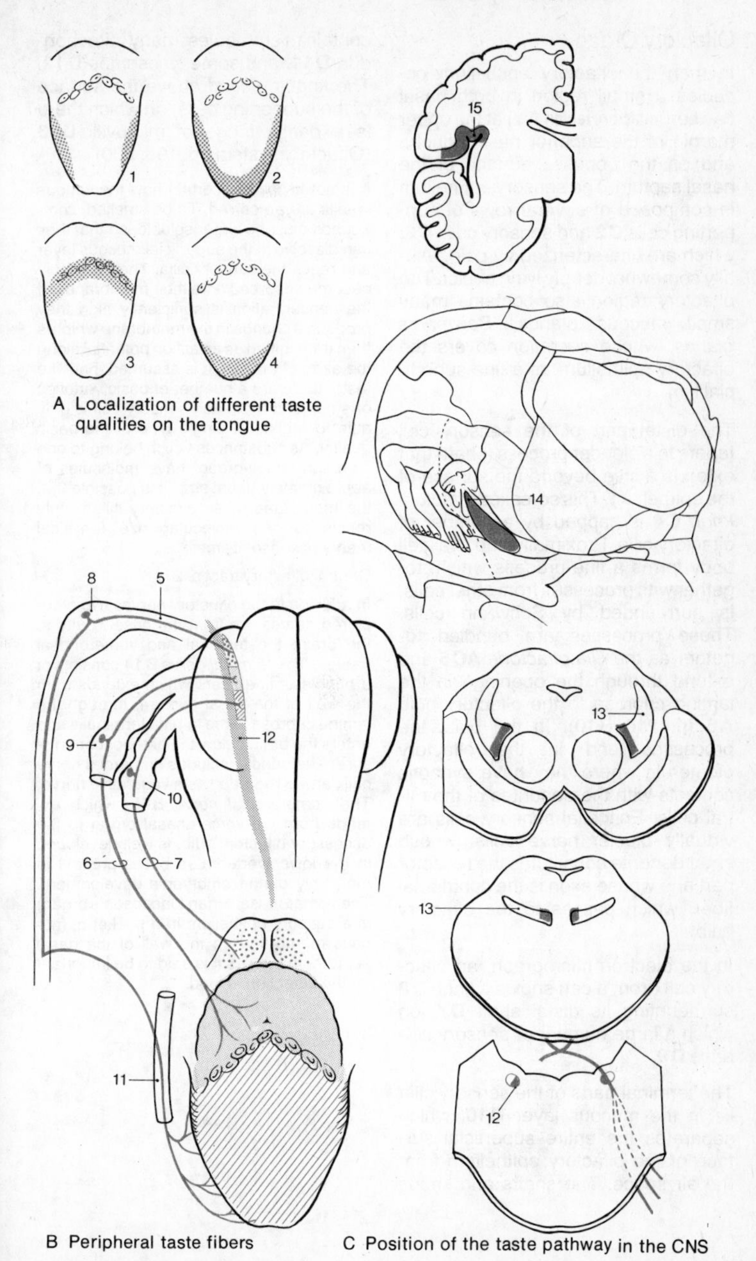

A Localization of different taste qualities on the tongue

B Peripheral taste fibers

C Position of the taste pathway in the CNS

Olfactory Organ

In man the olfactory epithelium occupies a small region in both nasal cavities (**olfactory region A1**) at the upper margin of the superior nasal concha and on the opposite surface of the nasal septum. The sensory epithelium is composed of several rows of supporting cells **C2** and sensory cells **C3**, which are characterized by pale, usually somewhat deeply lying nuclei. The olfactory region also contains many small mucous glands, *Bowman's glands,* whose secretion covers the olfactory epithelium as a fine superficial film.

The distal part of the sensory cell tapers to a slender process (shaft) that extends a little beyond the surface of the epithelium. This so-called *olfactory knob* **C4** is capped by a number of olfactory cilia. Proximally, the oval cell body forms a fine process which, together with processes from other cells, is surrounded by Schwann cells. These processes are bundled together as the *filia olfactoria* **AC5** and extend through the openings in the lamina cribrosa to the olfactory bulb **A6** (p. 210 B10). In the bulb, the processes end in the olfactory glomeruli, where they have synaptic contacts with the dendrites of the mitral cells. Epithelial sensory cells are virtually bipolar nerve cells, whose short dendrite constitutes the receptor part and whose axon is the centripetal fiber, which passes to the olfactory bulb.

In the electron micrograph, an olfactory cell (from a cat) shows a knob **D8** surmounting its distal shaft **D7,** on which a large number of *sensory cilia* arise **D9.**

The terminal parts of the sensory cilia lie in the mucous layer **D10,** which separates the entire superficial surface of the olfactory epithelium from the air space. The shafts and knobs contain microtubules, many mitochondria **D11,** and some lysosomes **D12.** The knobs extend above the surface of the supporting cells, on which there is a dense fringe of microvilli **D13** (Olfactory system pp. 196, 200).

It is not known for certain how the various smells are perceived. To be smelled, compounds must be water soluble, so that they can dissolve in the superficial mucous layer and reach the sensory cilia. They probably become adsorbed onto their membranes. If the concentration is sufficiently high they produce a change in the membrane which is then transmitted as an action potential along the axon of the cell. It is assumed that, like taste, there are a number of basic varieties of smell and that each sensory cell is only able to register a single type of the basic quality. As substances which belong to one common 'smell-group' have molecules of approximately equal size, it is possible that the membrane of an olfactory cilium only reacts to one molecular size (physical theory of sense of smell).

Central olfactory tract p. 214A.

In addition to the olfactory nerves two other paired nerves run from the nasal cavity to the brain: the terminal and vomeronasal nerves. The *terminal nerve* **B14** consists of a bundle of fine fibers which extends from the wall of the nasal septum, through the lamina cribrosa to the lamina terminalis and enters the brain below the anterior commissure. The bundle includes numerous nerve cells and is regarded as a vegetative nerve. The *vomeronasal nerve* **B15**, which extends from the vomeronasal organ to the accessory olfactory bulb, is well developed in the lower vertebrates but is present in man only during embryonic development. The vomeronasal organ (Jacobson's organ) is a sensory epithelium in a pocket of mucous membrane on the wall of the nasal septum. In reptiles it is said to be important for the detection of food.

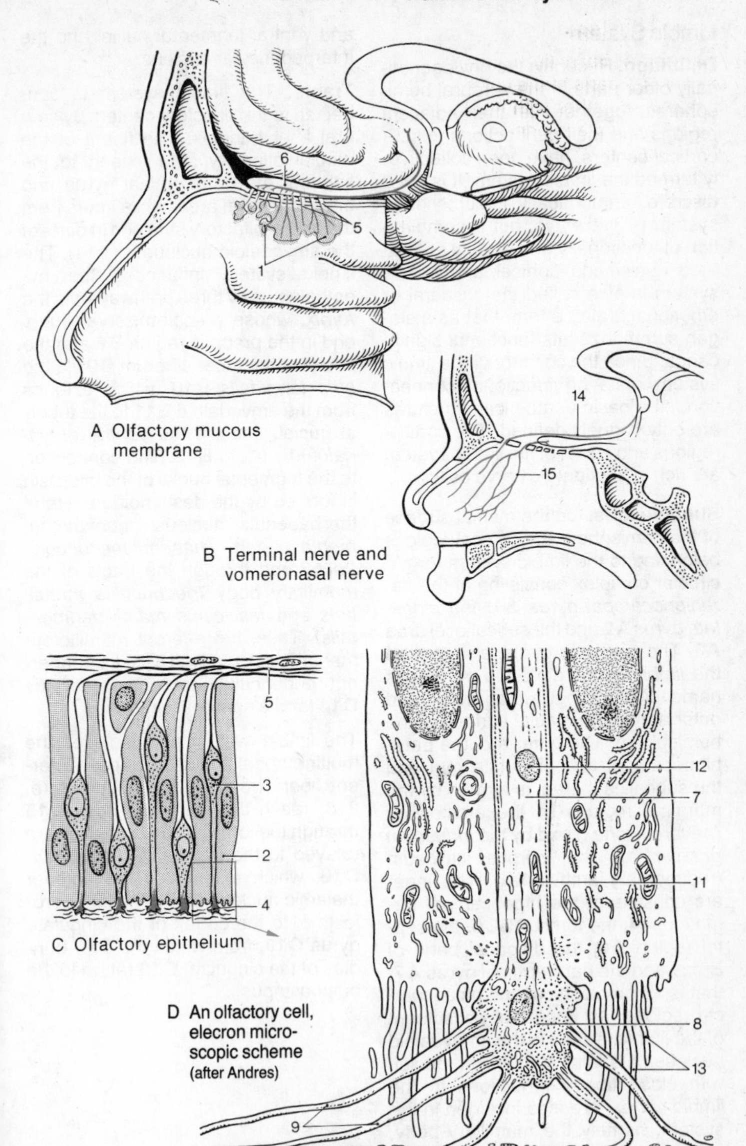

A Olfactory mucous membrane

B Terminal nerve and vomeronasal nerve

C Olfactory epithelium

D An olfactory cell, elecron microscopic scheme (after Andres)

Limbic System

Definition. Recently, the phylogenetically older parts of the cerebral hemispheres, together with their adjacent regions and their connections to subcortical centers, have been collectively termed the limbic system. It is not a discrete, anatomically circumscribed system of pathways, but a combination of functionally and closely associated nuclei and cortical areas. The system is also called the 'visceral or emotional brain', a term that as a slogan summarizes its functional significance. Since the concept of the limbic system rests on functional connections, the basic anatomical structures are only vaguely defined. The cortical regions and nuclei of the limbic system are rich in peptidergic nerve cells.

Subdivisions. On the medial surface of the hemisphere the cortical regions belonging to the limbic system form a circular complex consisting of the *parahippocampal gyrus* **A1**, the *cingulate gyrus* **A2** and the subcallosal area **A3.** The cingulate gyrus, also termed the *limbic gyrus,* gave the system its name. We distinguish an inner and outer arc on the medial surface of the hemisphere. The outer arc, the parahippocampal and cingulate gyri and the subcallosal area, is formed by the marginal regions of the archicortex *(periarchicortex)* and by the *induseum griseum* of the corpus callosum (rudimentary archicortex). The inner arc consists of the hippocampal formation **A4,** the fornix **ABC5,** the septal region **A6,** the diagonal band of Broca and the paraterminal gyrus **A7,** that is, of archicortical and paleocortical regions. An important constituent of the limbic system is the amgydaloid nucleus. Certain subcortical nuclei, with close fiber connections to the limbic cortex, are also included in the system, namely the mamillary body, anterior thalamic nucleus, habenular ganglion and in the midbrain the dorsal and ventral tegmental nuclei and the interpeduncular nucleus.

Tracts. The limbic system is connected to the olfactory centers by several fiber bundles. The fibers of the medial olfactory stria extend to the septum, the paraterminal gyrus and the subcallosal area, while fibers from the lateral olfactory stria end in parts of the amygdaloid nucleus (p. 214). The limbic system influences the hypothalamus by three pathways: via the *fornix,* whose precommissural fibers end in the preoptic region **B8** and the nuclei of the tuber cinerum **B9** via the *stria terminalis* **B10,** which extends from the amygdaloid **B11** to the tuberal nuclei, and via the *ventral amygdalofugal fibers* **B12.** The connection to the tegmental nuclei of the midbrain is formed by the descending tracts of the habenular nucleus *(habenulotegmental* and *habenulopeduncular tracts)* and through the tracts of the mamillary body *(pedunculus mamillaris* and *fasciculus mamillotegmentalis).* Thus, the efferent mamillotegmental fasciculus **C13** and the afferent pedunculus corporis mamillaris **C14** form a neuronal circuit.

The limbic system also includes the multineuronal circuit of Papez. Efferent fibers from the hippocampus (p. 218) reach the mamillary body **C15** through the fornix, where impulses are relayed to the bundle of Vicq d'Azyr **C16,** which extends to the anterior thalamic nucleus **C17.** The latter projects onto the cortex of the cingulate gyrus **C18,** from which the fiber bundles of the cingulum **C19** return to the hippocampus.

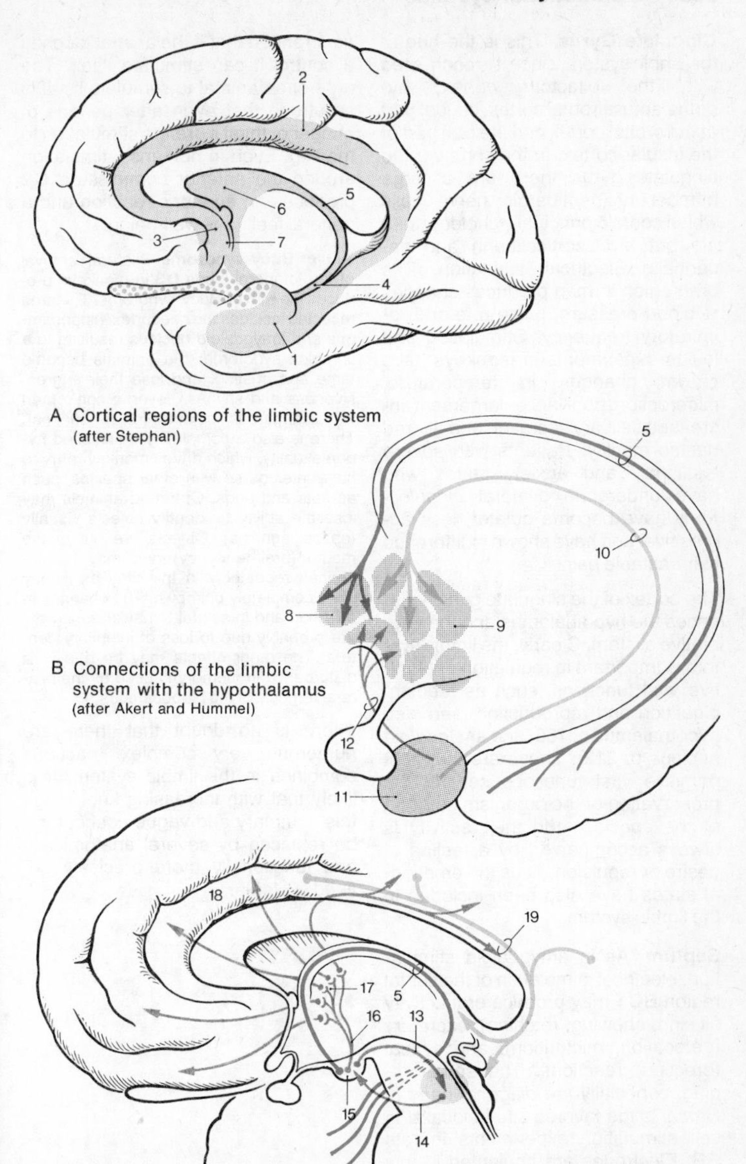

A Cortical regions of the limbic system
(after Stephan)

B Connections of the limbic system with the hypothalamus
(after Akert and Hummel)

C Neuronal circuit of Papez
(after Akert and Hummel)

Cingulate Gyrus. This is the hub of the limbic system, since it is connected with the olfactory cortex, hypothalamus, frontal cortex, caudal part of the orbital cortex and the oral part of the insular cortex. In the cortex of the cingulate gyrus there are a large number of peptidergic nerve cells which contain principally cholecystokinin but also corticoliberin and encephalin. Electrical stimulation of its oral region in man produces changes in blood pressure, pulse rate and respiratory frequency. Stimulation and lesion experiments in monkeys have shown changes in temperature, piloerection, pupillary enlargement, increased salivary secretion and altered gastric motility. Patients with severe excitatory and anxiety states who have undergone bilateral *cingulectomy* have become quieter, less aggressive, and have shown indifference to intractable pain.

The cortex of the cingulate gyrus influences the hypothalamus and the vegetative system. Clearly, the limbic system is important in regulation of primitive vital functions, such as feeding, digestion and reproduction (see also hypothalamus p. 186, and amygdaloid nucleus p. 212). These are the most primitive vital functions serving the preservation of the organism and that of the species, and their activity is always accompanied by a feeling of desire or repulsion. Thus the emotional states have also been included in the limbic system.

Septum. As in amygdaloid stimulation, electrical stimulation of the septal region **BC1** may produce oral activity (licking, chewing, retching), excretory (defecation, micturition), and sexual (erection) reactions. The septal region, especially the *diagonal band of Broca,* is the favored site in localizing self stimulation experiments in rats **AB.** Electrodes are implanted in this area in animals and the apparatus is so arranged that if the animal touches a control it can stimulate itself. The urge of an animal to stimulate itself is so strong that even after periods of hunger or thirst it prefers stimulation to feeding. Even in humans, stimulation around the anterior commissure **C2** produces an euphoric reaction and a general feeling of well-being.

Klüver-Bucy-syndrome. Bilateral removal of the temporal lobes **D3** in monkeys produces the Klüver-Bucy-syndrome. The area resected includes the neocortex, hippocampus and amygdaloid nucleus resulting in a complex syndrome; the animals become tame and trusting, and lose their aggressiveness and shyness, even if confronted with dangerous objects (e. g. snakes). There is also a completely uninhibited hypersexuality, which drives monkey to try to have intercourse with other species, such as cats and dogs. Operated animals may lose the ability to identify objects visually (optical agnosia). Objects are put in the mouth (oral tendency) and may be examined repeatedly in this way as if they were completely unknown. The changes in affection and in sexual behavior especially are probably due to loss of inhibitory centers. The other effects may be due to a disturbance in memory caused by the bilateral removal of the hippocampus.

There is no doubt that there are numerous very complex reactions combined in the limbic system. It is likely that with increasing knowledge this summary and vague concept will be replaced by several anatomically and functionally more precisely defined systems.

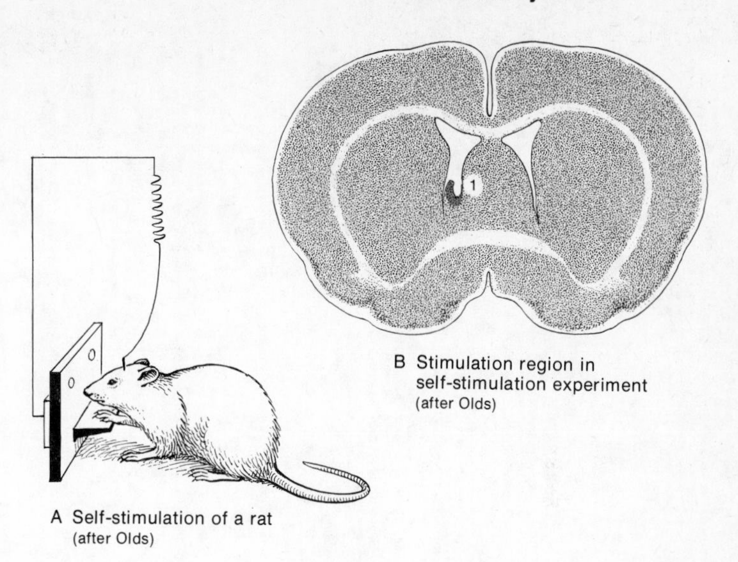

A Self-stimulation of a rat
(after Olds)

B Stimulation region in self-stimulation experiment
(after Olds)

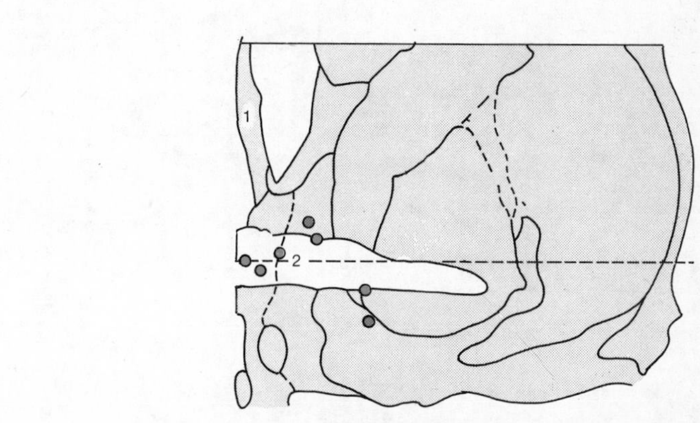

C Stimulation points for euphoric reactions in man
(after Schaltenbrand, Spuler, Wahren and Wilhelmi)

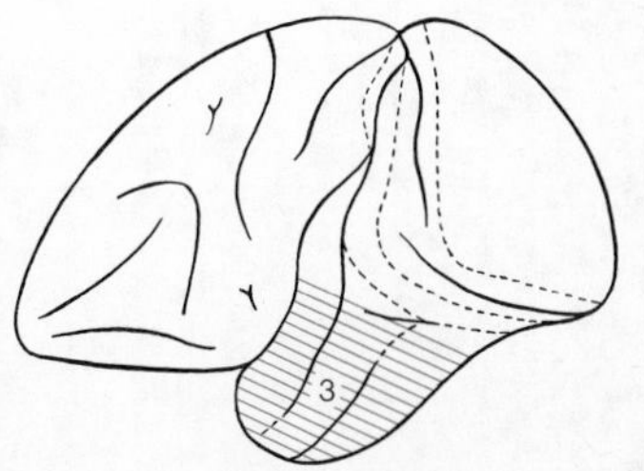

D Removal of the temporal lobe in the Klüver-Bucy syndrome
(after Klüver and Bucy)

Sensory Organs

Eyelids and the Orbits

Eyelids. The eyeball is embedded in the orbit and is covered by the eyelids. The *upper lid* **A1** and the *lower lid* **A2** border the *palpebral fissure,* which ends at the *medial angle of the eye* **A3** in a prolongation surrounding the *lacrimal caruncle* **A4.**

In the Mongolian race the upper lid continues medially as a fold of skin *(plica palpebronasalis)* onto the lateral surface of the nose (Mongolian fold). The fold is present as a transitory structure in infants and is called the *epicanthus.*

The lids are strengthened by scalelike, dense connective tissue plates, consisting of collagen fibers, the *superior tarsus* **B5** and *inferior tarsus* **B6,** which are attached by the *lateral* and *medial palpebral ligaments* **B7** to the margins of the orbits. The eyelids contain the elongated *tarsal glands,* the meibomian glands **C8,** which extend throughout the entire height of the lid. Their secretion prevents the tears from flowing over the edge of the lids. They open on the posterior edge of the lid margin, the *limbus posterior.* Several rows of *eyelashes* **C10** arise from the anterior margin, *limbus anterior* **AC9.** The inner wall of the lids is covered by the **tunica conjunctiva C11,** which at the *fornix of the conjunctiva* **C12** extends onto the anterior surface of the eyeball. The *superior tarsal muscle* **C13** and the *inferior tarsal muscle* **C14** (of smooth, unstriated fibers and innervated by sympathetic nerves) are attached to the tarsus. These assist in regulating the size of palpebral fissure. The lids are closed by the *orbicularis oculi muscle* **C15** (facial nerve, Vol. 1, p. 330 A20). The upper lid is elevated by the *levator palpebrae superioris muscle* **BC16** (oculomotor nerve) whose origin is at the upper margin of the optic canal. Its superficial tendinous extension **C17** penetrates into the subcutaneous connective tissue of the upper lid, while the deep tendon plate **C18** is attached to the upper margin of the tarsus.

Lacrimal apparatus. The **lacrimal gland B19** lies above the lateral corner of the eyelid and is subdivided by the tendon of the levator palpebrae superioris muscle into an *orbital* **B20** and a *palpebral* **B21** *part.* At the fornix of the conjunctiva the excretory ducts of the gland give off the lacrimal secretion, which keeps the anterior surface of the eye constantly moist, and collects eventually in the *lacus lacrimalis* of the medial corner of the eye. There, at the inner surface of each lid, is a small opening, the *puncta lacrimalia* **B22,** which leads into the *lacrimal canal* **B23.** These first either run upward or downward and then form a right angle, join and open into the *lacrimal sac* **B24,** which extends from the *nasolacrimal duct* **B25** to the inferior meatus of the nose. Blinking not only produces a uniform moistening of the outer surface of the eyeball, but it also exerts a suction effect on the outflow of tear secretion by alternately expanding and contracting the nasolacrimal duct.

Orbits. The orbital cavity, which is covered by periosteum *(periorbita)* **C26,** is filled with fatty tissue, the *corpus adiposum orbitae* **C27,** in which the eyeball **C28,** the optic nerve **C29** and the ocular muscles **C30** are embedded. At the anterior margin of the orbit the fat is bordered by the *orbital septum* **C31.** The fat is separated from the eyeball by a connective tissue capsule, the *vagina bulbi* **C32,** which encloses a narrow fissure, the *episcleral space* **C33.** Bones of the orbit **C34.**

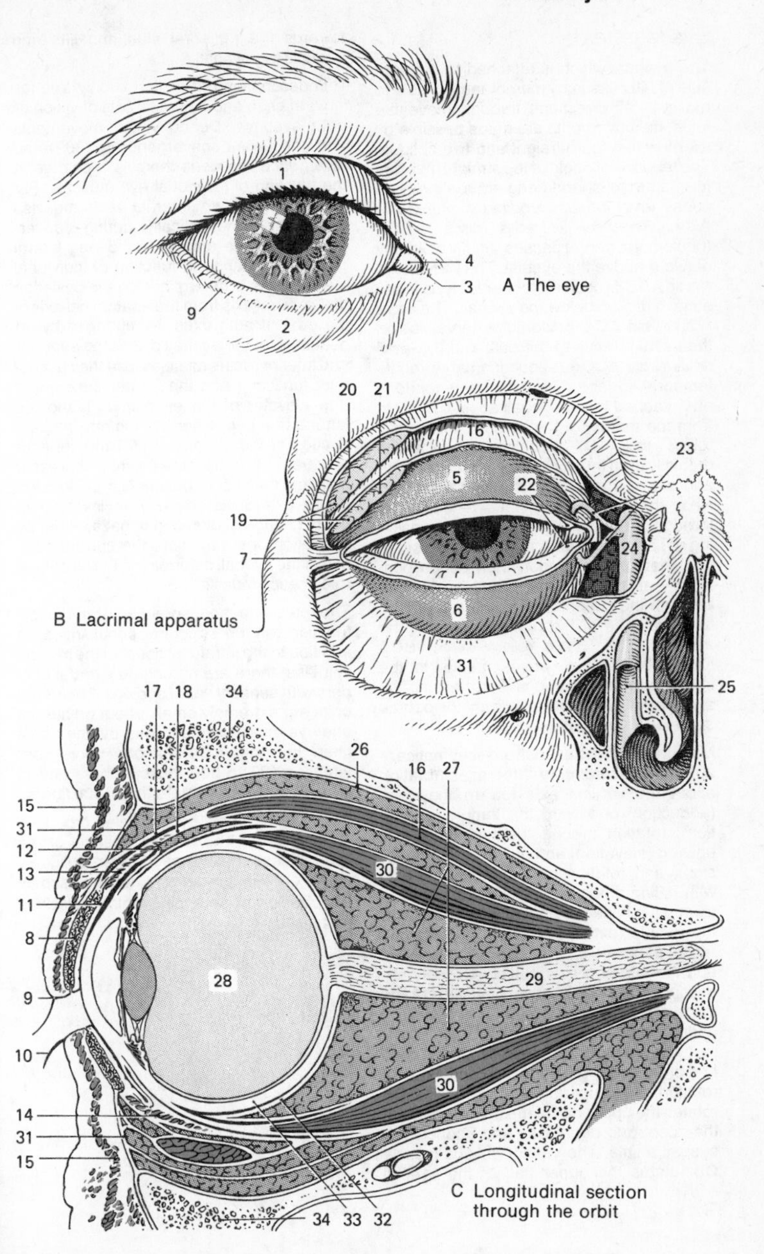
1
4
3
9
2
A The eye
20
21
16
23
5
22
19
24
7
6
B Lacrimal apparatus
31
25
17
18
34
26
16
27
15
31
12
13
30
11
8
28
29
9
10
30
14
31
15
34
33
32
C Longitudinal section through the orbit

Eye Muscles

The eyeball, which is attached to the capsule of fatty tissue by membranes, is able to move in all directions through moveable folds. Its movements are made possible by six muscles, four straight and two oblique. The tendons of origin of the straight muscles form a funnel-shaped ring around the optic canal, the *anulus tendineus communis* **AB1.** The **superior rectus muscle ABC2** (oculomotor nerve) passes slightly obliquely outward above the eyeball. The **inferior rectus muscle ABC3** (oculomotor nerve) runs in the same direction below the eyeball. The **medial rectus muscle AC4** (oculomotor nerve) lies on the nasal surface of the bulb and the **lateral rectus muscle ABC5** (abducens nerve) on its temporal surface. The flat muscle tendons are attached to the eyeball about 0.5–1 cm from the margin of the cornea. The **superior oblique muscle AC6** (trochlear nerve) arises medially from the body of the sphenoid bone and extends almost to the orbital margin. Near the orbital margin its tendon runs through the **trochlea A7,** a broad loop consisting of fibrous cartilage and lined by a synovial sheath. The tendon then bends posteriorly at an acute angle and is attached beneath the superior rectus muscle on the temporal side of the upper surface of the eyeball. The **inferior oblique muscle BC8** (oculomotor nerve) arises medial to the infra-orbital margin and runs to the temporal surface of the bulb. *Levator palpebrae superioris muscle* **B9.**

Movements of the eyeball are schematically arranged according to three axes: rotation about the vertical axis toward the nose (adduction) or toward the temple (abduction); rotation around the horizontal axis upward (elevation) and downward (depression), and rotation around a sagittal axis with rolling of the upper half of the eyeball toward the nose (internal rotation) or toward the temple (external rotation). The medial rectus muscle **C4** produces only adduction and the lateral rectus muscle **C5** only abduction. The superior rectus muscle **C2** lifts the eyeball and additionally produces slight adduction and internal rotation. The inferior rectus muscle **C3** depresses the eyeball and produces slight adduction and external rotation. The superior oblique muscle **C6** rotates the upper half of the eyeball towards the nose, and depresses and abducts the eyeball a little. The inferior oblique muscle **C8** rotates the upper half of the eyeball towards the temporal side and lifts and abducts the eyeball a little.

This description of function is only true for forward gaze and a parallel axis of vision of both eyeballs. During visual movements and concurrent convergence (p. 332) and divergence reactions there is a change in the function of individual eye muscles. For example, the two internal recti muscles which act synergistically during convergence, become antagonists during lateral gaze. The change in function of individual eye muscles is determined by the deviation of the visual axis from the anatomical axis of the orbit. If both axes are changed by an obduction of the eyeball of 23°, the superior and inferior rectus muscles lose their accessory functions and the former becomes a true elevator of the eyeball **D11** and the latter a true depressor. During maximal adduction of the eyeball of 50°, the superior oblique muscle becomes a true depressor **E13** and the inferior oblique a true elevator. Every extraocular muscle is involved in every change of direction of gaze, either by relaxation or contraction, the current position of the eyeball determining the functional role of each muscle.

The precision and speed of the muscle function rely on structural peculiarities. In addition to the intrafusal fibers of the muscle spindles, there are numerous extrafusal fibers with sensory spiral endings. The motor units are extremely small: about 6 fibers of one eye muscle are supplied by one nerve fiber. In finger muscles, one nerve fiber supplies 100 to 300 muscle fibers and in other muscles often more than 1500 fibers.

Paralysis of single eye muscles produces double vision. The relative position of the two images – next to each other, obliquely above or obliquely below – may be used by the clinician to determine which muscle is paralysed.

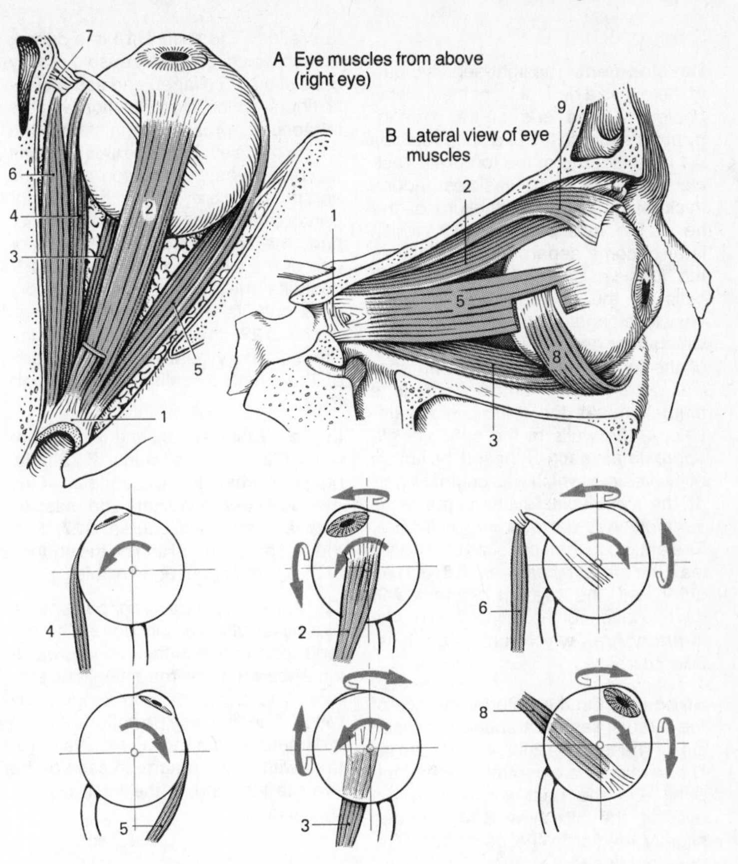

A Eye muscles from above (right eye)

B Lateral view of eye muscles

C Manner of functioning of the eye muscles of the right eye

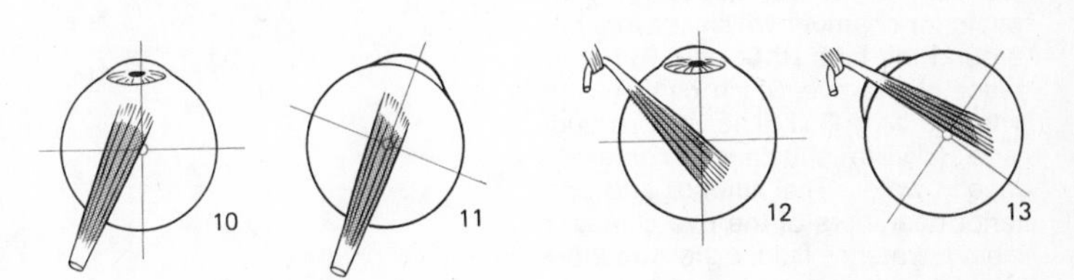

D Superior rectus muscle of the right eye, 10 forward gaze, 11 abduction of 23°

E Superior oblique muscle of the right eye, 12 forward gaze, 13 adduction of 50°

Survey

Development. The light-sensitve part of the eye is a derivative of the diencephalon. At the end of the first intrauterine month the two *optic vesicles* **A1** evaginate from the forebrain vesicle **A2.** The optic vesicles induce thickenings of the epithelium of the head, the *lens placodes* **A3,** which later become separated bilaterally as the *lens vesicles* **A4.** The epithelial cells on the posterior wall of each vesicle elongate into *lens fibers* **A5,** which later develop into the main part of the lens. The cells of the anterior wall of the vesicle remain as the lens epithelium. As the *optic cup* **A6** develops, the walls of the optic vesicle approximate each other and the lumen of the vesicle, which was originally part of the ventricular system, the *optic ventricle* **A7,** narrows into a fissure. The optic cup then consists of an inner leaf, the *stratum cerebrale* **A8,** and an outer leaf, the *stratum pigmenti* **A9,** the two layers of the retina. The *hyaloid artery* **A10,** which runs to the lens, later regresses.

Structure. On the anterior surface of the eyeball lies the transparent **cornea B11.** The lens **B12** lies behind the **iris B13,** which has a central opening, the pupil. On the posterior wall of the eyeball, the **optic nerve B14** emerges slightly medial to the optic axis. The eye contains three cavities: 1) the anterior chamber **B15** bordered by the cornea, the iris and the lens; 2) the posterior chamber which lies in a ring around the lens **B16;** and, 3) the interior of the eye which contains the *vitreous body* **B17.** The vitreous body is a jellylike substance, containing mostly water. The anterior and posterior chambers of the eye contain a clear watery fluid, the aqueous humour.

The wall of the eyeball consists of three layers: the sclera, the uvea and the retina. The **sclera B18** is a dense, taut connective tissue capsule mainly consisting of collagen and some elastic fibers. which in conjunction with the intraocular pressure maintains the shape of the eyeball. The **uvea** contains the blood vessels and forms the iris and the **ciliary body B19** in the anterior part of the eyeball and in the posterior part, the **choroid B20**. The posterior part of the **retina**, the *pars optica* **B21,** contains the light sensitive sensory cells and its anterior part, the *pars caeca* **B22**, the pigment epithelium. The boundary between the two parts of the retina is called the **ora serrata B23.**

In the eyeball we distinguish an anterior **B24** and a posterior **B25** pole, between which lies the *equator of the eye* **B26.** Some vessels and muscles take a *meridional course* **B27,** i. e. they follow a curved line on the surface of the eyeball from pole to pole.

The anterior and posterior parts of the eye have different functions. The anterior part contains the image projecting apparatus, the refracting lens system. The posterior part proper, the retina. The eye may be compared with a camera, which has a lens system in front with a diaphragm, in case of the eye the iris, and at the back a light-sensitive film.

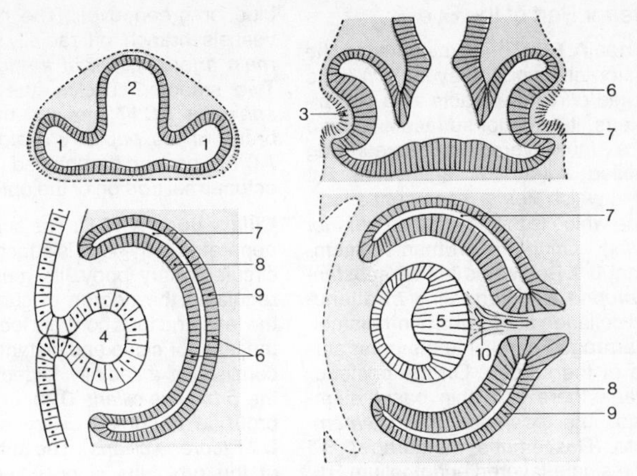

A Development of the eye

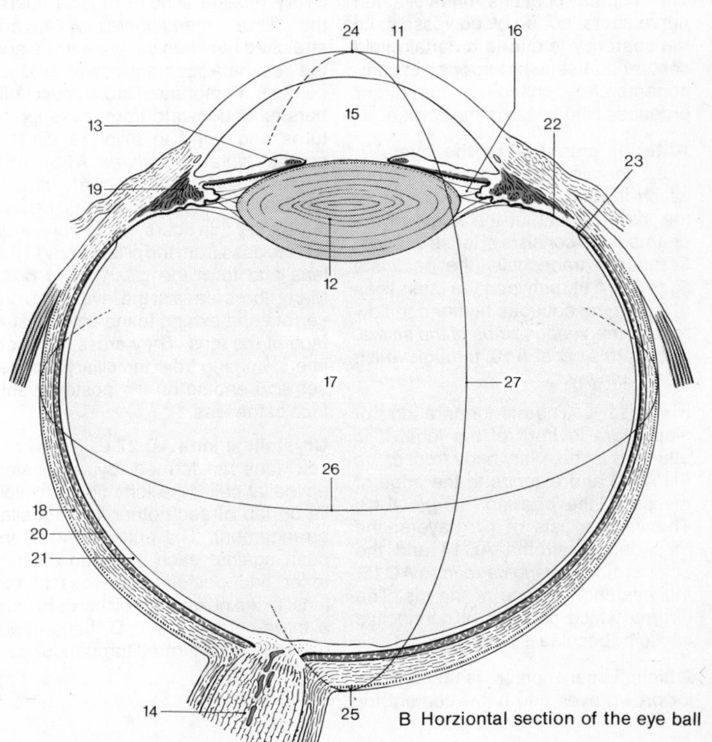

B Horziontal section of the eye ball

Anterior Part of the Eye

Cornea A 1 B. The cornea bulges like a watch-glass on the eyeball. With its marked curvature it acts as a collecting lens. Its anterior surface is formed by a multilayered, nonkeratinizing stratified squamous epithelium **B2** under which lies a basement membrane which is followed by an *anterior limiting lamina* (Bowman's membrane) **B3.** Below this lies the *substantia propria* **B4,** consisting of attenuated collagen fibers which form lamellae arranged parallel to the outer surface of the cornea. On the posterior surface there is a thin basal membrane, the *posterior limiting membrane* (Descemet's membrane) **B5** and a single-layered endothelium **B6.** The cornea contains unmyelinated nerve fibers, but no blood vessels. Its transparency is due to a certain fluid content and state of turgor of its components. Any change in the turgor produces cloudiness of the cornea.

Anterior chamber of the eye A7. This contains the aqueous humor which is produced by the vessels of the iris. The wall of the *angle of the chamber* **A8** consists of loose connective tissue trabeculae (the *pectinate ligament* **A9),** between the interstices of which the aqueous humor can flow toward the *venous sinus of the sclera,* Schlemm's canal **A10,** through which it drains away.

Iris A 11 C. The iris forms a kind of diaphragm in front of the lens. It is attached to the ciliary body *(root of the iris* **A12)** and extends to the edge of the pupil, the *pupillary margin* **A13.** The iris consists of two layers; the mesodermal stroma **AC14** and the ectodermal *pars iridica retinae* **AC15,** the posterior surface of the iris. The stroma which consists of connective tissue trabeculae is pigmented.

A high pigment content is responsible for brown eyes and a low content for blue or green eyes. The numerous vessels branch off radially from the *main arterial circle of the iris* **AC16.** Two smooth muscles, the *pupillary sphincter* **AC17** and the thin membrane of the *pupillary dilator muscle* **AC18,** have differentiated from the ectodermal portion of the optic cup.

Ciliary body A 19 D. The suspensory apparatus of the lens is attached to the circular ciliary body. Its musculature regulates the degree of curvature of the lens and thus controls focussing of the lens for close and distant vision. It consists of a radially folded surface, the *orbiculus ciliaris* **D20,** from which protrude about 80 *ciliary processes* **D21** *(corona ciliaris).* The anterior part of the orbiculus is occupied by the ciliary muscle, whose muscle fibers, the *fibrae meridionales* **A22,** are stretched between the ora serrata and the pectinate ligament as well as Descemet's membrane. Radial muscle fibers extend inward from these structures and bend to follow a circular course *(fibrae circulares* **A23).** The posterior surface of the ciliary body is covered by the *pars ciliaris retinae* **A24.** Very thin fibers, *fibrae zonulares* **AD25,** pass from the ciliary body to the lens and form the *ciliary zone* **D26.** Many fibers leave at the level of the ora serrata and extend to the anterior surface of the lens. They cross the short fibers, coming from the ciliary processes and ending on the posterior surface of the lens.

Crystalline lens AD 27 E. The biconvex lens is formed by elongated epithelial cells, the lens fibers, which lie on top of each other in a lamellar arrangement. The ends of the fibers push against each other on the anterior and posterior surfaces of the lens. In the newborn their borders form a three-cornered star **D.** Lens fibers continue to be formed throughout life.

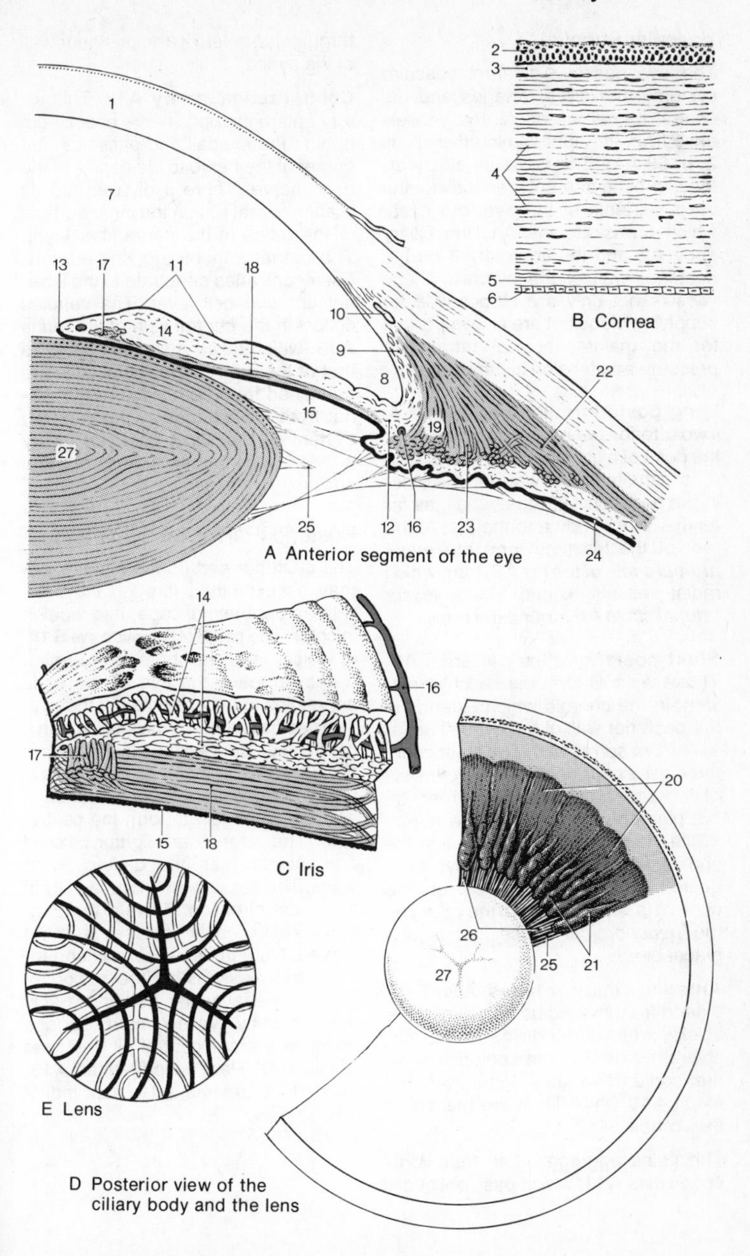

B Cornea

A Anterior segment of the eye

C Iris

E Lens

D Posterior view of the ciliary body and the lens

Vascular Supply

The eye has two different vascular systems: the ciliary arteries and the central retinal artery. All the vessels arise from the ophthalmic artery (Vol. 2, p. 58 A4). The posterior ciliary arteries are the afferent branches for the vascular tunic of the eye, the uvea, which forms the iris **A1,** the ciliary body **A2** and the choroid **A3** on the posterior wall of the eyeball. These vessels not only are responsible for supplying blood but are also important for the maintenance of intraocular pressure and tension in the eyeball.

Long posterior ciliary arteries A4. Two arteries penetrate the sclera near the point of exit of the optic nerve. One runs in the temporal wall and the other in the nasal wall of the eyeball, as far as the ciliary body and the iris. At the root of the iris they form the *major arterial circle of the iris* **A5,** from which radial vessels extend to the *lesser arterial circle* **A6** around the pupil.

Short posterior ciliary arteries A7. These vessels form the vascular network in the choroid which extends in the posterior wall of the eyeball as far as the *ora serrata* **A8.** The inner zone, the *lamina choroidocapillaris,* consists of very wide capillaries and borders on the pigment epithelium of the retina. While the pigment epithelium and the choroidocapillaris are firmly connected, the outer surface of the choroid is separated from the sclera by the *perichoroidal space* and is displaceable.

Anterior ciliary arteries A9. They extend from the rectus muscles to the sclera, where they divide in the episcleral tissue and the conjunctiva. In the conjunctiva they form *marginal loops* **A10** which lie in the margin of the cornea.

The veins aggregate into four *vorticose veins* **A11,** which pass obliquely through the sclera in the posterior wall of the eyeball.

Central retinal artery A12. This artery enters the optic nerve about 1 cm behind the eyeball and passes in the center of the nerve to the papilla of the optic nerve. There it divides into its branches that run on the inner surface of the retina in the nerve fiber layer. The retinal arteries are *end-arteries.* Their capillaries penetrate to the internal granular cell layer. The venules collect in the *central vein of the retina* **A13,** which takes a similar course to that of the artery. The visual cells are nourished from both sides of the retina: from the exterior by the capillary system of the short posterior ciliary arteries, and internally from the central artery.

Posterior Part of the Eye (Fundus)

The posterior part (fundus) of the eye may be examined through the pupil with an ophthalmoscope. It is reddish in color. The **papilla of the optic nerve B14,** in which all nerve fibers of the retina collect to leave the eye as the optic nerve, lies in the nasal half of the eye. The papilla is a light, whitish disk with a central flat depression, the *cavity of the disk* **B15.** In the papilla the central artery divides into several branches and the veins join to form the central vein. The arteries are lighter colored and thinner than the darker, wider veins. The vessels run more radially in the nasal direction. Toward the temporal region their course is more curved. Numerous vessels extend toward the **macula B16,** the region of greatest visual acuity. Its transversely oval, slightly yellow-tinted surface contains a small central pit, the **fovea centalis B17.** Pars optica retinae **A18,** pars ciliaris retinae **A19,** pars iridica retinae **A20.**

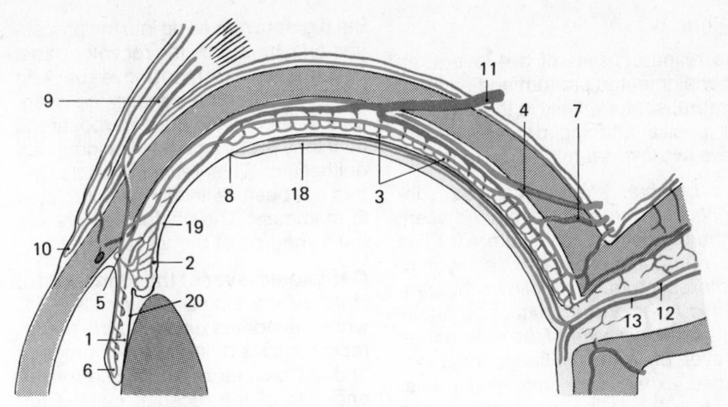

A Vascular supply of the eye

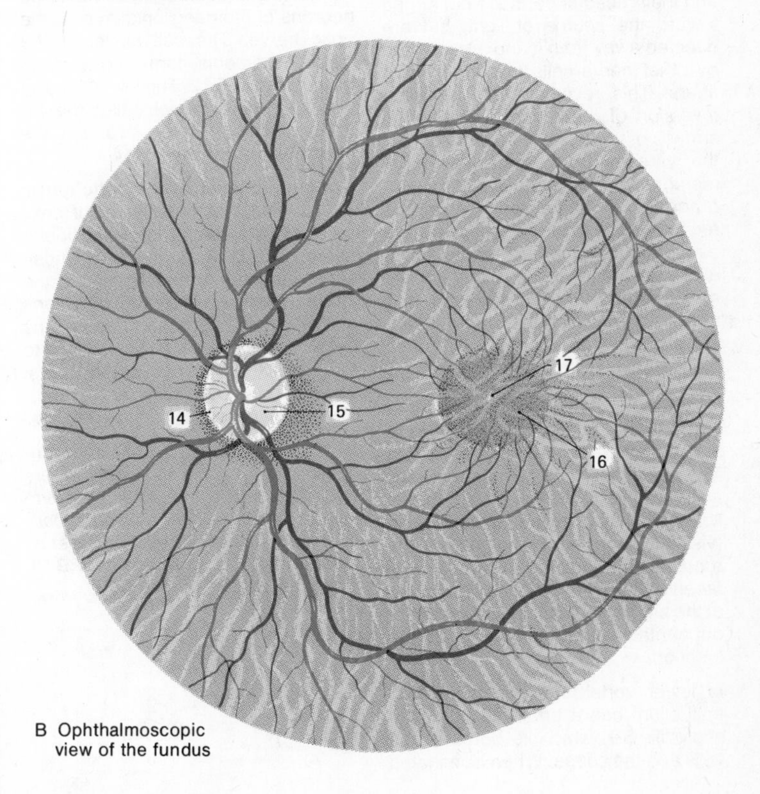

B Ophthalmoscopic view of the fundus

Retina

The retina consists of two layers: an outer pigmented stratum and an inner cerebral stratum. Only in the region of the papilla and the ora serrata are these two layers firmly grown together.

The *cerebral stratum* can be subdivided into three cell layers. Adjacent to the pigmented epithelium **AB1** lies the *neuroepithelial layer,* a layer of photoreceptors. Next follows the *ganglionic layer of the retina,* and finally the *ganglionic layer of the optic nerve,* a layer of large multipolar neurons, whose axons form the optic nerve. Thus, in the retina the sensory cells and their receptor parts are not turned toward the source of light, but are directed away from it and are covered by other nerve cells and their nerve fibers. This reverse order is called *inversion* of the retina. The internal surface of the retina is separated from the vitreous body by a basement membrane, the *internal limiting membrane* **B2.** In the neuroepithelium the receptor parts of the sensory cells are separated from the nuclear parts by a glial membrane, the *external limiting membrane* **B3.** Between the two membranes stretch the long *supporting glial cells of Müller* **B4** with their elongated, leaflike processes.

Neuro-epithelial layer AB5. The neuroepithelium contains two types of sensory cell, **rods B6** and **cones B7.** The duality theory assumes that rods are sensitive to light and dark in dim light, and the cones are color sensitive (vision in bright light). The nuclei of the receptor cells form the *external granular layer* **A8.** The rods and cones end at the pigmented epithelium, and without contact with it they are unable to function.

In lower vertebrates the pigmented epithelium has a brushlike border of microvilli **B9**, which lie between the rods and the cones. When illuminated the pigment moves up into the processes and surrounds the receptor parts. In dim light the pigment moves back. In some species the sensory cells are able to stretch and during exposure to light they press against the pigmented epithelium. This *retinomotor system* has not been definitely demonstrated in mammals. The photoreceptors are the 1. neuron of the visual pathway.

Ganglionic layer of the retina AB10. Here lie the bipolar relay cells **B11,** whose dendrites extend to the photoreceptor cells (2. neuron), where they end with their terminal branches on the end-bulb of the receptor cells. Their axons make contact with the large neurons of the ganglionic layer of the optic nerve. The cell nuclei of the bipolar elements form the *internal granular layer* **A12.** Their synapses lie in zones free of nuclei, called the *external plexiform layer* **A13** and the *internal plexiform layer* **A14.**

Ganglionic layer of the optic nerve AB15. The innermost layer of cells consists of a row of large multipolar nerve cells **B16** (3. neuron), whose short dendrites form synapses in the internal plexiform layer with the axons of the bipolar neurons. Their axons extend in the nerve fiber layer **AB17** to the papilla as unmyelinated fibers. They form the optic nerve and eventually end in the lateral geniculate body.

Association cells. Transverse connections within the individual layers are produced in the outer plexiform layer by axons of the *horizontal cells* **B18** and in the inner plexiform layer by dendrites of the *amacrine cells* **B19**. Choroidocapillary lamina **A20.**

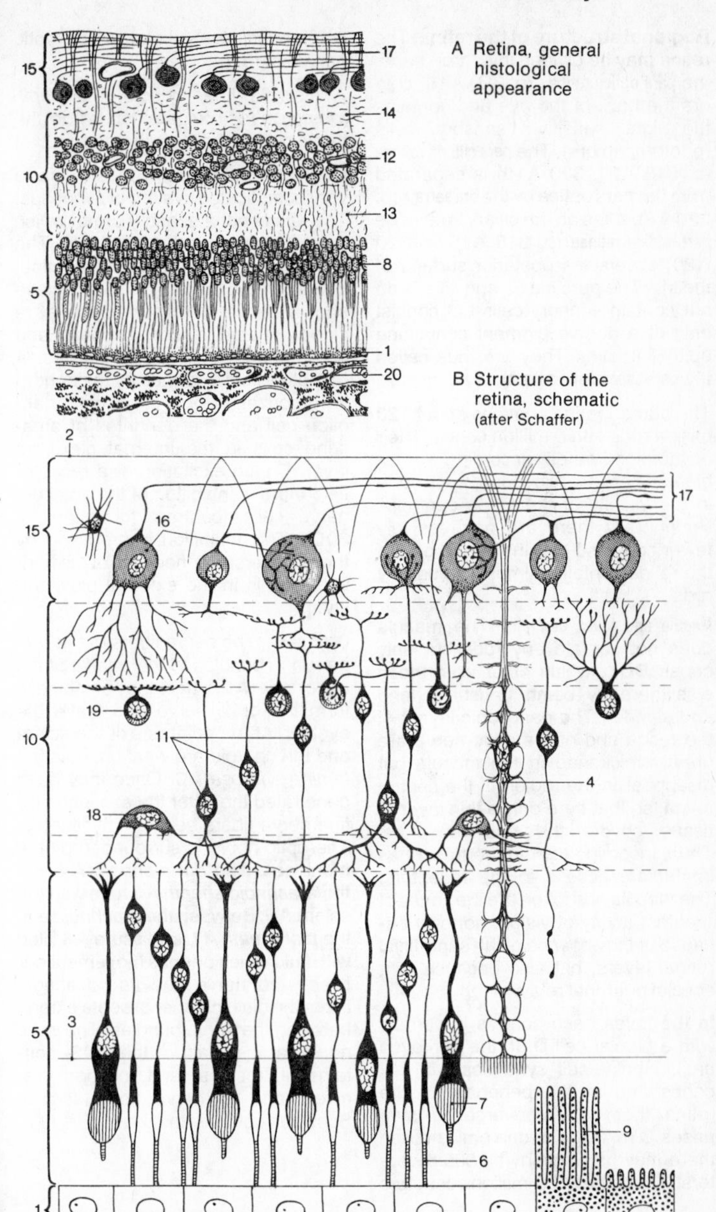

A Retina, general histological appearance

B Structure of the retina, schematic (after Schaffer)

Regional structure of the retina. The retina may be divided into three parts: the **pars optica retinae** (p. 320 A18) covers the back of the eye and contains the light sensitive sensory cells (photoreceptors). The **pars ciliaris retinae** (p. 318 A24, 320 A19) is separated from the pars optica by the **ora serrata** (p. 320 A8). It lies on the ciliary body. The **pars iridica retinae** (p. 318 A15, p. 320 A20) covers the posterior surface of the iris. The pars iridica and ciliaris do not contain sensory cells but consist only of a double pigment containing epithelial layer. They are thus called the **pars caeca** (blind part).

The human retina contains about 120 million rods and 6 million cones. Their distribution varies according to regions. There are two rods to each cone in the vicinity of the fovea centralis. Peripherally there are progressively fewer cones so that the lateral part of the retina contains almost exclusively rods.

Macula and fovea centralis. The macula does not contain any rods but only cones **BD1,** which differ from those elsewhere by being unusually long and slender. The ganglionic layers of the retina and of the optic nerve are greatly thickened in the macula but disappear in the region of the central fovea, so that here only a thin layer of tissue covers the photoreceptors. Thus, incoming light in the fovea has immediate access to the receptors. The macula and fovea are the zones of greatest acuity of vision, not only because of the absence of the superficial retinal layers, but also because of a special neuronal relay.

In the fovea each cone is in contact with a bipolar cell **D.** In the perifovea each bipolar cell synapses with six cones and at the periphery of the retina there are convergence synapses **C** (p. 30). For one optic neuron there may be as many as 500 receptors. Altogether 130 million receptors correspond to only one million optic nerve fibers.

A discrete fiber bundle, the *papillomacular fasciculus,* extends from the nerve cells of the macula to the papilla.

Neuronal Synapses. The retina has a very complicated system of synapses. The eye is not a sensory organ which merely transmits light stimuli. The stimuli instead are refined in the retina. Electrophysiological studies have shown that groups of sensory cells are closely related to *receptive fields* and act together as functional entities. A receptive field is formed by the dendritic arborization of a large optic ganglion cell and the dendrites of amacrine cells in the internal plexiform layer. During excitation of a receptor field there is inhibition of the adjacent nerve cells (contrast formation, p. 30D). This inhibition is produced by the transverse connections of the horizontal cells in the external plexiform layer.

Optic nerve. The nerve fibers of the retina run in bundles to the *papilla of the optic nerve* **A2,** where they combine to form the optic nerve and leave the eyeball. At the point of exit the sclera and the choroid are weakened as the *lamina cribrosa* **A3.** Once they have penetrated the latter the exceptionally fine nerve fibers become myelinated. The optic nerve is surrounded by the meninges. The *dural sheath* **A4** and the *arachnoid sheath* **A5** fuse with the sclera **A6.** Between the *arachnoid* and the *pial sheath* **A7** is a fissure **A8** filled with fluid which permits movement between the nerve and its coverings. From the pia a number of septa extend between the nerve bundles. The optic nerve, a fiber tract of the CNS, contains astrocytes and oligodendrocytes, and its nerve fibers lack a Schwann cell sheath.

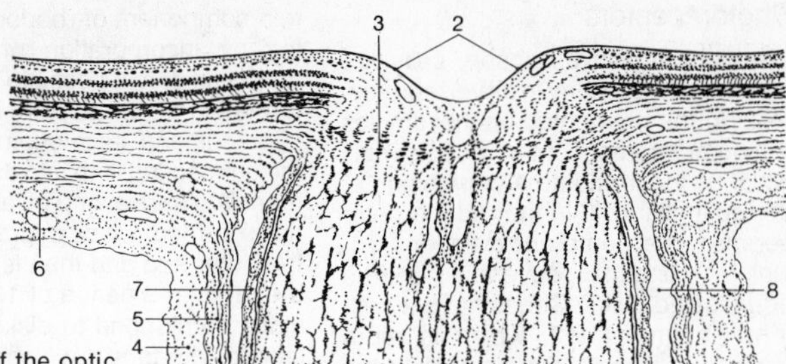

A Papilla of the optic nerve

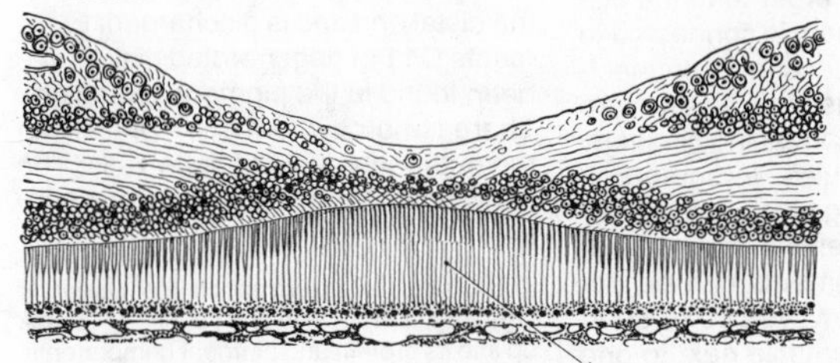

B Central fovea

Synapses in the retina

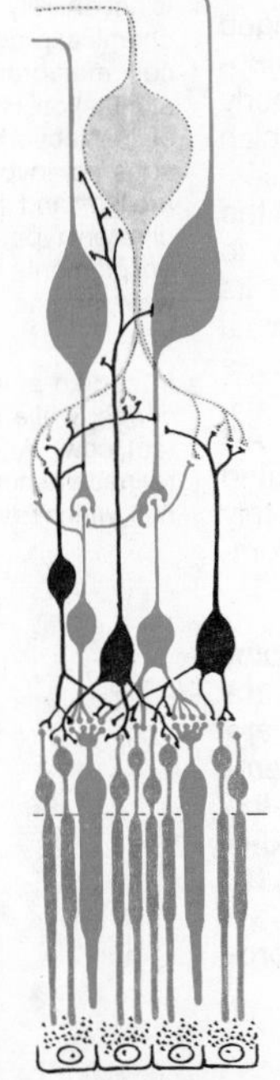

C Mixed rods and cones

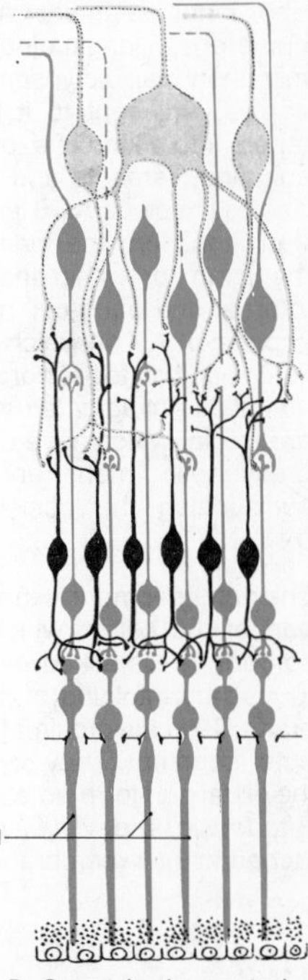

D Cones in the central fovea (after Polyak)

Photoreceptors

The light sensitive sensory cells are constructed similarly in all vertebrates. Next to the pigmented epithelium lies the outer segment of the receptor cell, which is inserted into a depression of the epithelial cell. In the rod the outer segment is a membrane cylinder **AB1,** which contains several hundred equal-sized disks. Cones have an external segment **B2,** which is more conical in shape and in which the proximal disks are larger than the distal. The inner segment is connected to the outer by a thin, eccentric connecting piece **A3.** This has a structure similar to a cilium containing circularly arranged microtubules. In some animals the connecting piece is relatively long, so that a definite interval between the two segments can be detected **A.** It may, however, be so short that the two segments lie next to one another without a visible interspace **B.** The inner segment **AB4** contains numerous, longitudinally arranged mitochondria, dictyosomes and ribosomes. Adjacent to it the cell body tapers into a kind of axon **AB5,** which contains filaments and microtubules. The cell nucleus **AB6** lies either at the transition from the inner segment to the axon, or within the axon. At its internal end each cell has a terminal expansion **A7** on which the synapses lie. In addition to the normal synapses, so-called *invaginated synapses* **A8** are found, which have an invaginated presynaptic membrane completely surrounding the postsynaptic complex.

The outer segment is the real receptor part of the cell in which light is absorbed. The stacked disks there are formed by infolding of the cell membrane **C9** in the proximal portion of the outer segment. They separate and in the distal part form isolated disks **C10.** *Rhodopsin,* the *visual purple,* is attached to their membranes. If the protein component of rhodopsin is labelled by incorporation of radioactive amino acid, it is possible to follow the formation of rhodopsin in the inner segment and its movement through the connecting limb into the outer segment **D.** There a strip of labelled material is formed, which migrates as far as the outer end and then is expelled (in the rat over a period of 10 days). The strips correspond to disks containing labelled rhodopsin, which is constantly being formed in the rods, migrates to the distal end and is discharged. Fragments **C11** of degenerated disks have been found in the pigment epithelium. There is no new formation of disks in the outer segments of cone cells. The infoldings of the membrane are constant.

The absorption of light affects the molecular structure of rhodopsin and causes it to break up into its protein and pigment components. From these components rhodopsin is constantly reformed *(rhodopsin-retinin cycle).* It is not entirely known in what way this photochemical process produces changes in the cell membrane and this leads to nerve stimulation. Rhodopsin is only present in rod cells. It absorbs light of all wave lenghts and so is not involved in color vision. The rods are light and dark receptors. There are three different types of cones which contain different pigments that only absorb light of certain wavelengths. The cones are color receptors.

In certain animals the retina contains only cones, while in others there are only rods (cat, cow). Animals with only rod-containing retinas are not able to distinguish color. The bull, which reacts to red, is in fact color blind.

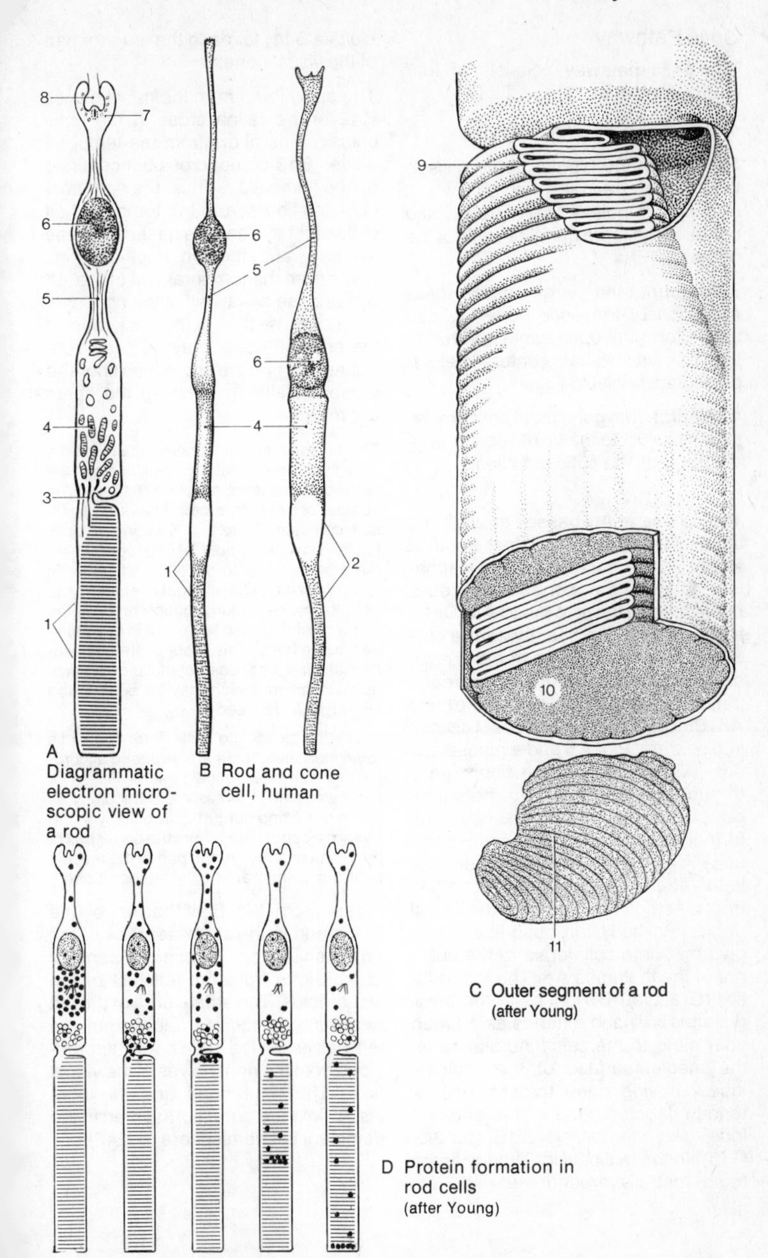

A Diagrammatic electron microscopic view of a rod

B Rod and cone cell, human

C Outer segment of a rod (after Young)

D Protein formation in rod cells (after Young)

Optic Pathway

The optic pathway consists of four neurons in series:

1. Neuron: the photoreceptors.

2. Neuron: the bipolar nerve cells in the retina *(retinal ganglia),* which transmit impulses from rods and cones to the large ganglion cells of the retina.

3. Neuron: the large nerve cells *(ganglion of the optic nerve)* whose axons form the optic nerve and run to the primary visual centers (lateral geniculate nucleus).

4. Neuron: the geniculate cells whose axons project to the visual cortex *(striate area)* as the optic radiation.

The **optic nerve A1** passes through the canal of the optic nerve into the cranial fossa. At the base of the diencephalon, it joins the contralateral optic nerve to form the **optic chiasm A2.** Distal to the chiasm, the fiber system is called the **optic tract A3.** The tracts of each side extend around the cerebral peduncle to the lateral geniculate body **A4.** Before that level each tract divides into a *lateral root* **A5** and a *medial root* **A6.** Whilst most of the fibers pass through the lateral root to the lateral geniculate body, the medial fibers run further beneath the medial geniculate body **A7,** to the superior colliculi. The latter fibers contain the optic reflex tracts. At their termination in the lateral geniculate body the optic fibers also give off some collaterals to the pulvinar of the thalamus **A8.** The **optic radiation** (Gratiolet) **B9** arises in the **lateral geniculate body** and extends as a broad fiber plate to the calcarine fissure on the medial surface of the occipital lobes, having gone through an anteriorly directed genu in the temporal lobe *genu temporale* **B10** (p. 242 C16). In the occipital lobe many fibers bend rostrally around the *genu occipitale* **B11,** to reach the anterior part of the visual cortex.

The optic fibers from the nasal halves **B12** of the retina cross in the optic chiasm. The fibers from the temporal halves **B13** do not cross but continue on the same side. Thus, the right tract contains fibers from the temporal half of the right eye and the nasal half of the left eye, and the left tract contains fibers from the temporal half of the left eye and the nasal half of the right eye. On a cross section of the tract, most of the crossed fibers may be seen ventrolaterally and the uncrossed fibers lie dorsomedially: in between the fibers are mixed.

The crossed and uncrossed fibers of the optic tract pass to different cell layers of the lateral geniculate body (pp. 178, 240). The number of geniculate cells at around 1 million corresponds to the number of optic fibers. However, most of the optic fibers terminate on five to six cells, which lie in different layers. Corticofugal fibers from the occipital cortex, which probably regulate the inflow of stimuli, also terminate in the lateral geniculate body. This is suggested by axoaxonal synapses, characteristic of presynaptic inhibition, which may be seen in the lateral geniculate body.

Axons of geniculate cells form the optic radiation. Their fibers are arranged according to the different retinal regions. Fibers for the lower half of the retina, particularly its periphery swing out furthest rostrally in the temporal genu. Fibers from the upper half of the retina and the central part of the macula only form a small arch in the temporal lobes.

Fibers from the right halves of the retina end in the **striate area B14** of the right hemisphere; it receives sensory impressions from the left half of the visual fields. The striate area of the left hemisphere receives fibers from the left halves of the retina with impressions from the right halves of the visual field. The right hand and the right visual field are both represented in the dominant left hemisphere (p. 246).

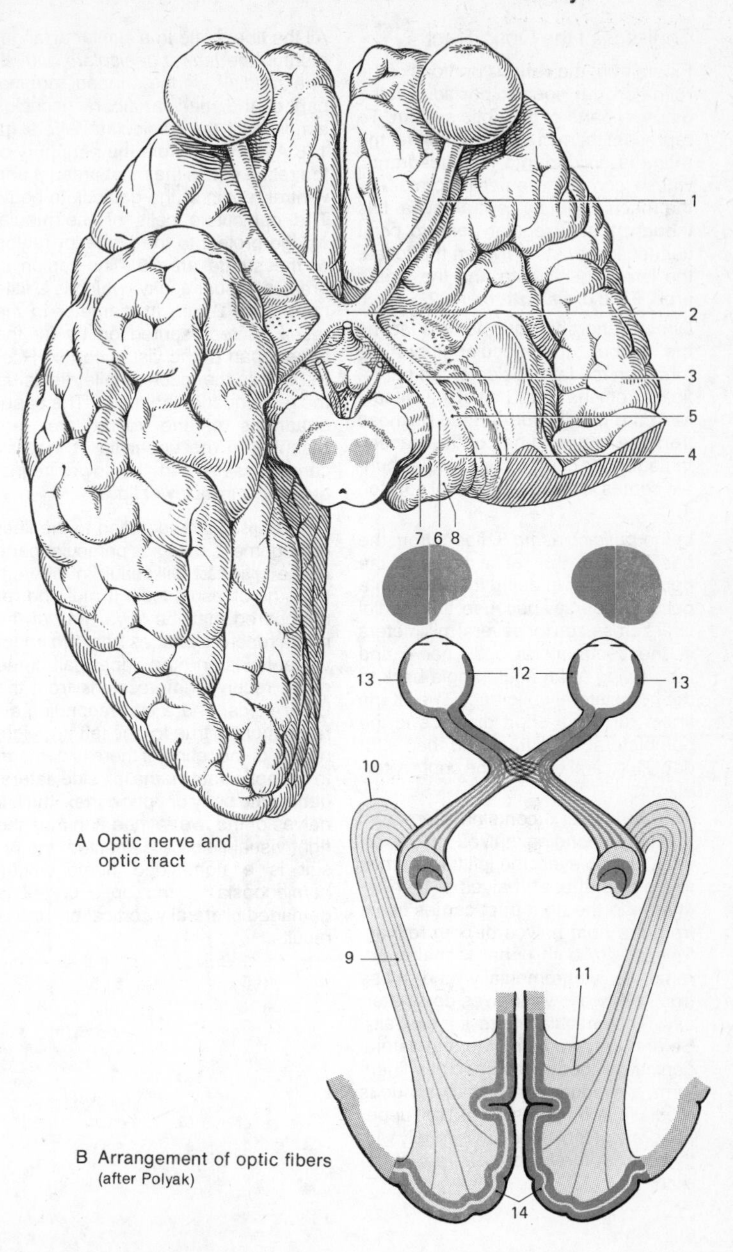

A Optic nerve and optic tract

B Arrangement of optic fibers (after Polyak)

Topistics of the Optic Tract

Fibers from the various portions of the retina occupy specific positions in the different parts of the optic system. To represent them in a simple manner the retina is divided into four quadrants, whose common center is formed by the macula with its central fovea. It is thought that a regular point to point connection exists between the fovea, the lateral geniculate and the striate area. Field of vision **1,** retina **2.**

Directly after the *optic nerve* **3** leaves the eyeball, the macular fibers are found on the lateral side of the nerve, fibers from the nasal half of the macula lie in the middle surrounded by those from the temporal half of the macula. In its further course the macular bundle comes to occupy a central position **4.**

In the *optic chiasma* **5,** fibers from the nasal half of the retina cross to the opposite side. In doing this they run a peculiar course, because the medial fibers cross, run for several millimeters in the contralateral optic nerve and then swing at an acute angle back to the homolateral optic tract. The lateral fibers run for a short distance in the homolateral optic tract and then suddenly turn and cross to the contralateral tract.

The *optic tract* **6** contains fibers from the corresponding halves of the retinas of both eyes: the left tract carries fibers from the left halves of both retinae and the right tract carries fibers from the right halves of both retinae. Fibers from both upper retinal quadrants lie ventromedially and those from the two lower halves dorsolaterally. The macula fibers lie centrally. Before they radiate into the lateral geniculate body, the macular fibers form a wedge which is continuous medially with fibers from the upper quadrant of the retina and laterally with those from the lower quadrant of the retina **7.**

All the fibers end in a similar arrangement in the *lateral geniculate body* **8.** The medial, wedge-shaped terminal part of the macular fibers occupies almost half the geniculate layers (p. 178 A9). Fibers from the periphery of the retina end in the most anterior and ventral region of the geniculate body. The geniculate cells of the medial wedge project to the posterior region of the *striate area* **9.** The region of greatest visual acuity, which is a little more than 2 mm in diameter in the retina, is represented on by far the largest part of the visual cortex. Rostral to it lie the much smaller fields for the remainder of the retina. The upper quadrants of the retina are represented in the upper lip of the calcarine fissure and the lower retinal quadrants in its lower lip.

Clinical Tips. Corresponding to the fiber arrangement, **damage to** particular parts of the **optic tract** will result in different patterns of visual loss. It must be remembered that the lower half of the retina receives images from the upper visual fields and the upper half of the retina receives impressions from the lower fields, and a corresponding arrangement is true for the left and right halves of the retina. If there is damage to the optic tract on the left side, lateral geniculate body or optic cortex, the left halves of the two retinae and thus the right visual fields are affected: the result is a right-sided homonymous hemianopsia. If the optic cortex is damaged bilaterally cortical blindness results.

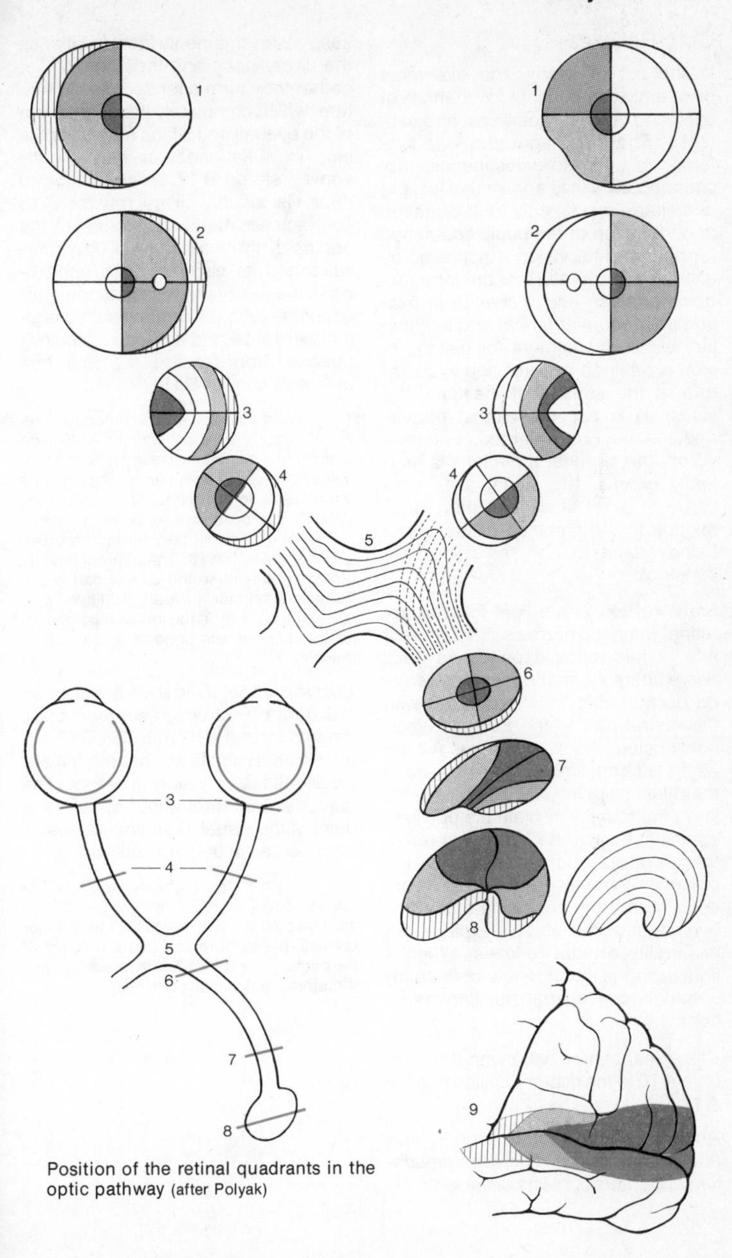

Position of the retinal quadrants in the optic pathway (after Polyak)

Optic Reflexes

In the act of seeing the eye must constantly compensate for changes of light and shade, nearness and distance. The diaphragm and lens systems must therefore continuously accommodate. While adjustment for light and shade results only from dilatation or contraction of the pupil, adaptation for near and far vision is achieved by altering the curvature of the lens *(accommodation)* and a change in fixation point *(convergence)* and in pupillary width. Adjustment for distant vision is attained by reduction in curvature of the surface of the lens, the visual axes become almost parallel and the pupil is enlarged. For near vision, the surface of the lens is markedly curved, the visual axes converge to meet at a point that corresponds to the distance of the object being gazed at and the pupils are constricted.

Light reflex. When light falls on the retina, the pupil narrows. The afferent arm of this reflex depends on optic nerve fibers **A1** that run to the *pretectal nucleus* **A2.** This is coupled with the rostral part of the Edinger-Westphal nucleus **A3,** whose fibers **A4** run as the efferent limb of the reflex arc to the ciliary ganglion **A5.** The postganglionic fibers **A6** innervate the pupillary sphincter muscle **A7.** The two pretectal nuclei are interconnected via the posterior commissure **A8.** In addition, optic fibers from each side end in the two pretectal nuclei. This explains the bilaterality of light reflexes. When a light is shone into one eye both pupils contract (consensual pupillary reaction).

Ciliospinal center **A9,** sympathetic fibers **A10** to the dilator pupillae muscle **A11.**

Accommodation (adaptation to near and distant vision). The accommodation apparatus consists of the lens, the suspensory ligaments (zonula ciliaris), the ciliary body and the choroid. Together they form a tensed elastic system, which completely lines the sclera of the eyeball and which maintains the lens in a flattened, or only slightly convex shape **B12.** In adaptation to near vision, the ciliary muscle **B13** contracts; its meridional fibers pull the origins of the long zonular fibers forward and its circular fibers approximate the ciliary processes to the margin of the lens. This relaxes the zonular fibers **B14** and the lens capsule is released from tension, the lens then becomes convex **B15.**

The fiber tracts for the accommodation reflex are not known with certainty. As fixation of an object is a prerequisite for accommodation, the optic nerve can be regarded as the afferent limb. Probably the reflex runs through the optic cortex to the pretectal nuclei, and possibly also through the two superior colliculi **A16.** The efferent limb, in any case, begins in the caudal part of the Edinger-Westphal nucleus. Its fibers synapse in the ciliary ganglion with postganglionic fibers which innervate the ciliary muscle.

Convergence. If the eyes are fixed on a distant object which comes closer, the two medial recti muscles **C17** adduct both eyeballs so that the initially parallel lines of vision intersect. The object is held throughout at the focal point of the visual axes and it remains represented in both maculae.

The fixation reflex probably runs via the optic tract to the occipital cortex, and thence via corticofugal fibers **A18** to the anterior colliculi, pretectal region and the nuclei of the ocular muscles **A19.** The reflex center is therefore the occipital cortex.

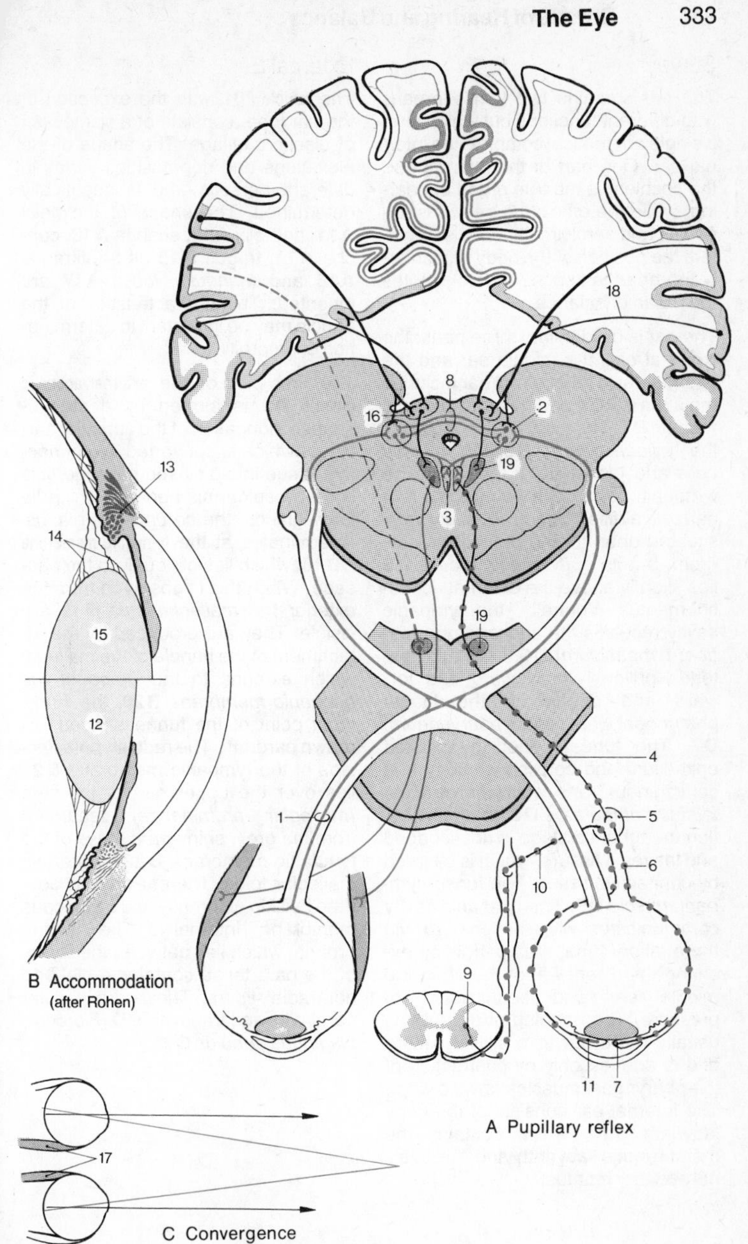

A Pupillary reflex

B Accommodation
(after Rohen)

C Convergence

Survey

The ear contains two sense organs with different functions but which form a single anatomical complex, the internal ear. One part of the internal ear, the cochlea, is the true organ of hearing, while the other (the sacculus, utriculus and semicircular canals) registers the position of the body particularly the head in space, and constitutes the organ of balance.

The ear is divided into three parts: the external ear, the middle ear and the internal ear. The external ear includes the auricle **AD1** and external acoustic meatus **D2.** The middle ear consists of the *tympanic cavity* **D3,** the *mastoid cells* and the *auditory tube* **D4.** The tympanic cavity with its ossicles is a narrow, airfilled space. It is not only situated deep to and behind the eardrum, but its *epitympanic recess* **D5** lies slightly above the external acoustic meatus. Ventrally, the tympanic cavity receives the auditory tube *(ostium tympanicum)* **D6.** The latter extends obliquely downward and forward and opens on the lateral pharyngeal wall *(ostium pharyngeum)* **D7.** The tube is lined by ciliated epithelium and consists of bony and cartilaginous parts, which merge at the *isthmus of the tube* **D8.** In cross section the tubal cartilage is hookshaped and leaves a fissure, which is covered by connective tissue. The *tensor tympani muscle* **D9.** The tympanic cavity communicates with the pharynx via the auditory tube, thus permitting exchange and renewal of the air in the middle ear and equalization of pressure. The opening of the tube is usually narrowed to a slitlike fissure and is opened only by contraction of the pharyngeal muscles (swallowing). The internal ear consists of the bony labyrinth **D10,** which contains the membranous labyrinth and the internal auditory meatus.

External Ear

The **auricle AD1,** with the exception of the ear lobe, consists of a framework of elastic cartilage. The shape of the elevations and depressions varies in different persons and is genetically determined. The shape of the helix **A11,** anthelix **A12,** scapha **A13,** concha **A14,** tragus **A15** and antitragus **A16** and *triangular fossa* **A17** are inherited. The characteristics of the auricle may be important in determination of paternity.

The first part of the **external acoustic meatus D2** is formed by a groove-shaped elongation of the auricular cartilage, which is converted by connective tissue into a closed passage. It is lined by epidermis beneath which lie large glands, the *ceruminous glands.* It terminates at the **tympanic membrane BD18,** which lies obliquely in the passage. When this is observed from the exterior the *malleolar stria* **B19** are visible. They are produced by the attachment of the handle of the malleus, which extends to the *umbo of the tympanic membrane* **B20,** the innermost point of the funnel-shaped, indrawn eardrum. The reddish *pars flaccida* of the tympanic membrane **B21** lies over the upper part of the stria *(malleolar prominence)* separated from the grey, shiny *pars tensa* of the tympanic membrane **B22** by the two malleolar folds. The ear drum is covered by skin externally and by mucous membrane internally. The lamina propria, which lies between the layers of the pars tensa contains radial and non-radial fibers. These are circular parabolic, and transverse **D.** *Fibrocartilaginous annulus* **C23.**

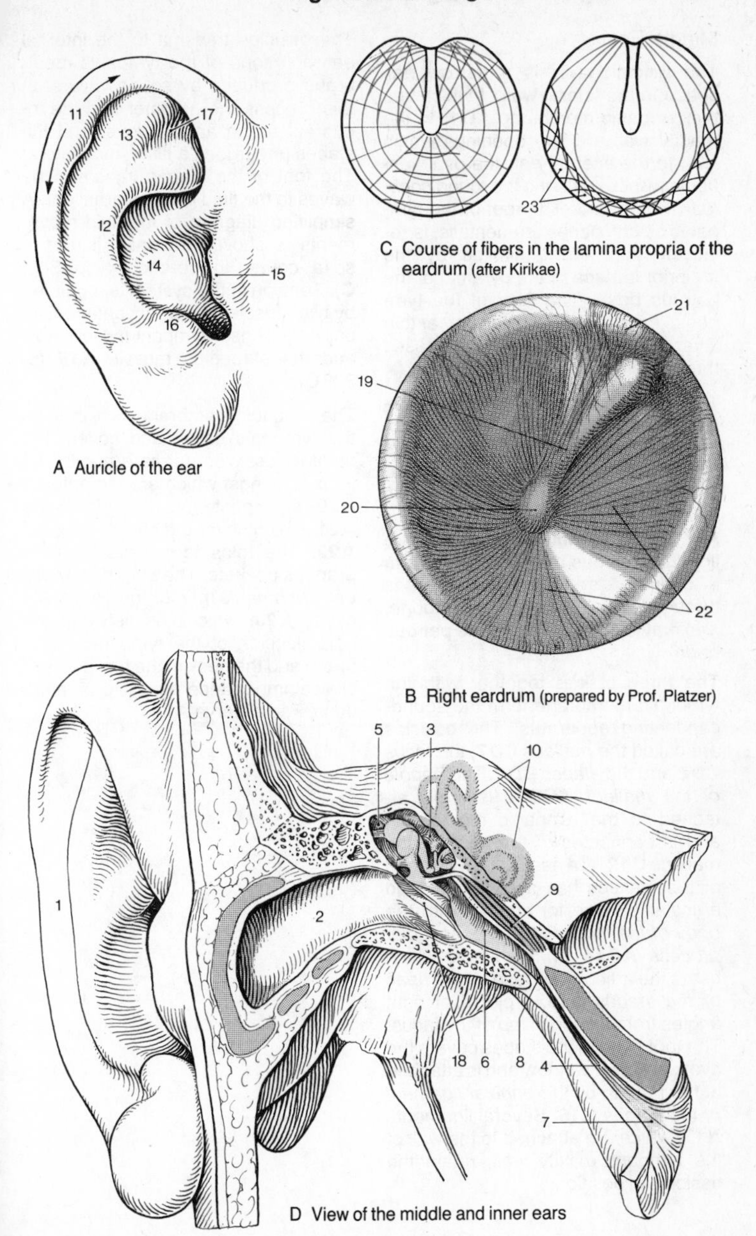

A Auricle of the ear

C Course of fibers in the lamina propria of the eardrum (after Kirikae)

B Right eardrum (prepared by Prof. Platzer)

D View of the middle and inner ears

Middle Ear

The **tympanic cavity** is a narrow, tall space in the lateral wall of which lies the tympanic membrane **AD1.** In the medial wall are two openings which lead to the internal ear, the oval *vestibular window* **D2** and the round *cochlear window* **D3.** The roof of the tympanic cavity, paries tegmentalis, is relatively thin and is limited by the superior surface of the pyramid of the petrous bone. The floor of the tympanic cavity also consists of a thin layer of bone, beneath which passes the jugular vein.

The tympanic cavity continues anteriorly as the auditory tube **A4** (p. 334). Posteriorly, the upper part opens into the *mastoid antrum* **A5,** a circular space into which open numerous small cavities, the *mastoid air cells* **A6.** These aircontaining cavities are lined by mucous membrane and form a chamber system which extends throughout the entire mastoid bone, and may even extend into the petrous bone.

The three **ossicles** together with the tympanic membrane form the sound-conducting apparatus. The ossicles are called the *malleus* **CD7,** the *incus* **CD8** and the *stapes* **CD9.** The *handle of the malleus* **ACD10** is firmly attached to the tympanic membrane, and is connected to the head of the malleus **C12** by a neck **C11.** The head of the malleus has a saddle-shaped articular surface for contact with the *body of the incus* **C13.** The *lenticular process* **AC14** of the latter, which bears the articular surface for the *head of the stapes* **C15,** projects at right angles from the long limb of the incus. The foot plate of the stapes covers the oval vestibular window and is attached at the margin by the *annular ligament of the stapes* **D16.** Several ligaments **A17,** which are attached to the wall of the tympanic cavity wall, retain the ossicles in position.

The ossicles transmit to the internal ear vibrations of the tympanic membrane produced by sound waves. In this process the hammer and the incus act as an angular lever and the stapes undergoes a tilting movement. The foot of the stapes transmits the waves to the fluid in the internal ear. A simplified diagram of the fluid movements is shown; in reality it runs a spiral course in the cochlea (p. 343, **C**). Tension in the system is regulated by two muscles which act antagonistically: the tensor tympani muscle **A18** and the stapedius muscle **A19** (p. 338C).

The mucous membrane, which lines the tympanic cavity and covers the auditory ossicles, forms a number of folds, amongst which are the anterior **A20** and posterior malleolar folds **A21,** which cover the chorda tympani **A22.** The folds form several membranous pockets. The *superior recess of the tympanic membrane,* Prussak's cavity **A23,** which lies between the pars flaccida of the tympanic membrane and the neck of the malleus is of clinical importance as a site of infections. Facial nerve **A24.**

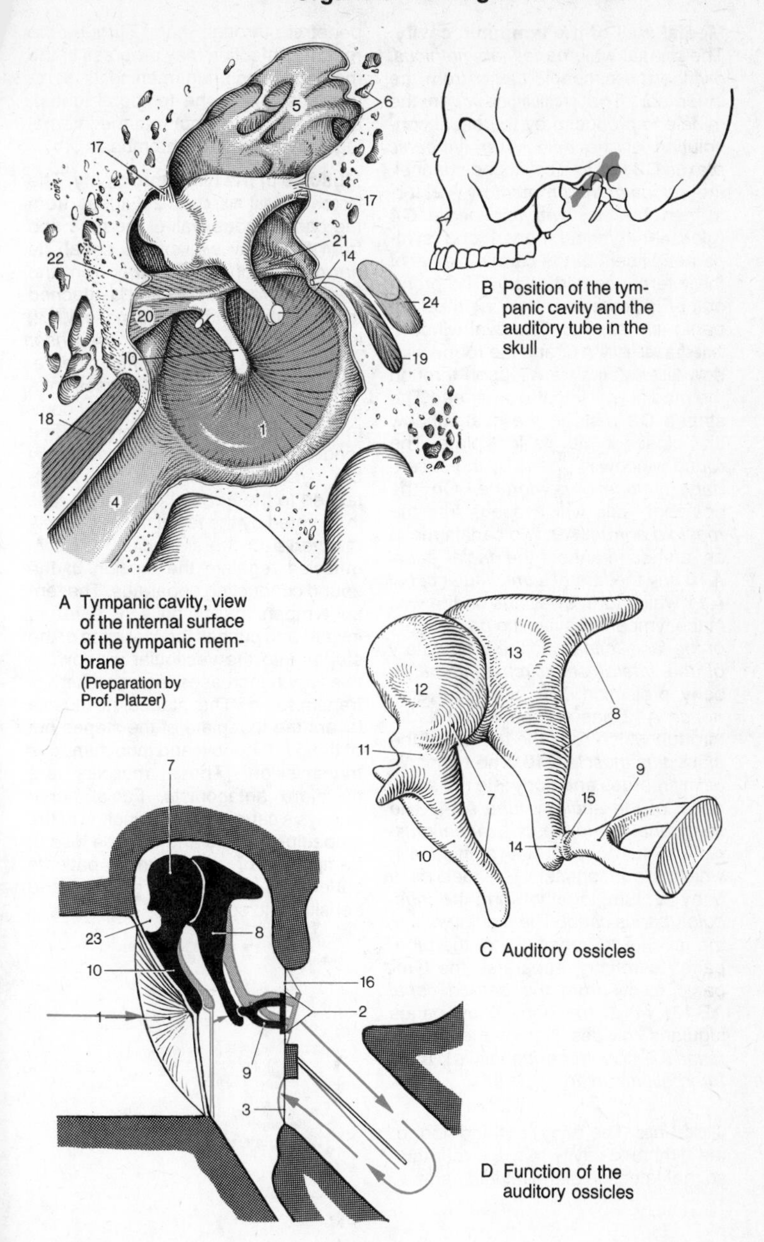

A Tympanic cavity, view of the internal surface of the tympanic membrane
(Preparation by Prof. Platzer)

B Position of the tympanic cavity and the auditory tube in the skull

C Auditory ossicles

D Function of the auditory ossicles

Medial wall of the tympanic cavity. The medial wall, *paries labyrinthicus,* divides the tympanic cavity from the inner ear. The **promontory A1** in the middle is produced by the basal convolution of the helix. The *tympanic plexus* **C3** lies in a branched channel, the *sulcus of the promontory* **A2.** It is formed by the *tympanic nerve* **C4** (glossopharyngeal nerve) and sympathetic fibers of the carotid plexus of the internal carotid artery. The promontory is limited ventrally by the *tympanic air cells* **A5.** The oval window, **fenestra vestibuli A6,** and the round window **fenestra cochleae A7** open through the medial wall into the inner ear. The stapes **C8** rests in the oval window and closes it with its foot plate. The round window is closed by the *secondary tympanic membrane.* On the posterior wall, which opens into the *mastoid antrum* **A9,** two canals run in an arched manner, the *facial canal* **A10** and the *lateral semicircular canal* **A11** which form swellings on the wall of the tympanic cavity, the *promontory of the facial canal* and the *promontory of the lateral semicircular canal.* A bony projection, the *pyramidal eminence* **A12** has an opening at its tip through which runs the tendon of the *stapedius muscle* **C13.** The tympanic cavity merges anteriorly into the *semicanal of the auditory tube* **A14.** The *semicanal of the tensor tympani muscle* **A15** lies above it. Both semicanals, which are incompletely divided by a bony septum, together form the *musculotubarius canal.* The medial wall, at the level of the openings of the tube, paries caroticus, separates the tympanic cavity from the *carotid canal* **AB16,** whilst the bony floor, paries jugularis, divides it from the *jugular fossa* **AB17.** *Internal jugular vein* **B18.** *Internal carotid artery* **B19.**

Clinical Tips. The bony roof and floor of the tympanic cavity may be very thin, so that infection of the middle ear may penetrate through them. Through the roof, the infection may progress to the meninges and brain (meningitis, cerebral abscess in the temporal lobes), and through the floor into the internal jugular vein (jugular thrombosis).

Muscles of the tympanic cavity. The **tensor tympani muscle C20** arises from the cartilaginous wall of the tube and from the bony wall of the canal. Its narrow tendon bends away from the *cochlear process* **C21** and is attached to part of the handle of the malleolus. it is innervated by the *nerve to the tensor tympani* from the mandibular nerve. The **stapedius muscle C22** arises in a small bony canal, which usually communicates with the facial canal. Its fine tendon passes through the opening in the pyramidal eminence and is attached to the head of the stapes. It is innervated by the facial nerve **C23** via the *nerve to the stapedius.* The two muscles regulate the tension of the sound conducting apparatus. The tensor tympani muscle pulls the eardrum inward and pushes the foot plate of the stapes into the vestibular window: in this way it increases the sensitivity of transmission. The stapedius muscle levers the foot plate of the stapes out of the oval window and thus dampens transmission. These muscles are therefore antagonists. Facial nerve paralysis causes loss of function of the stapedius muscle and with the loss of dampening of sound stimuli, patients suffer from *hyperacusis,* an increased sensitivity to sound.

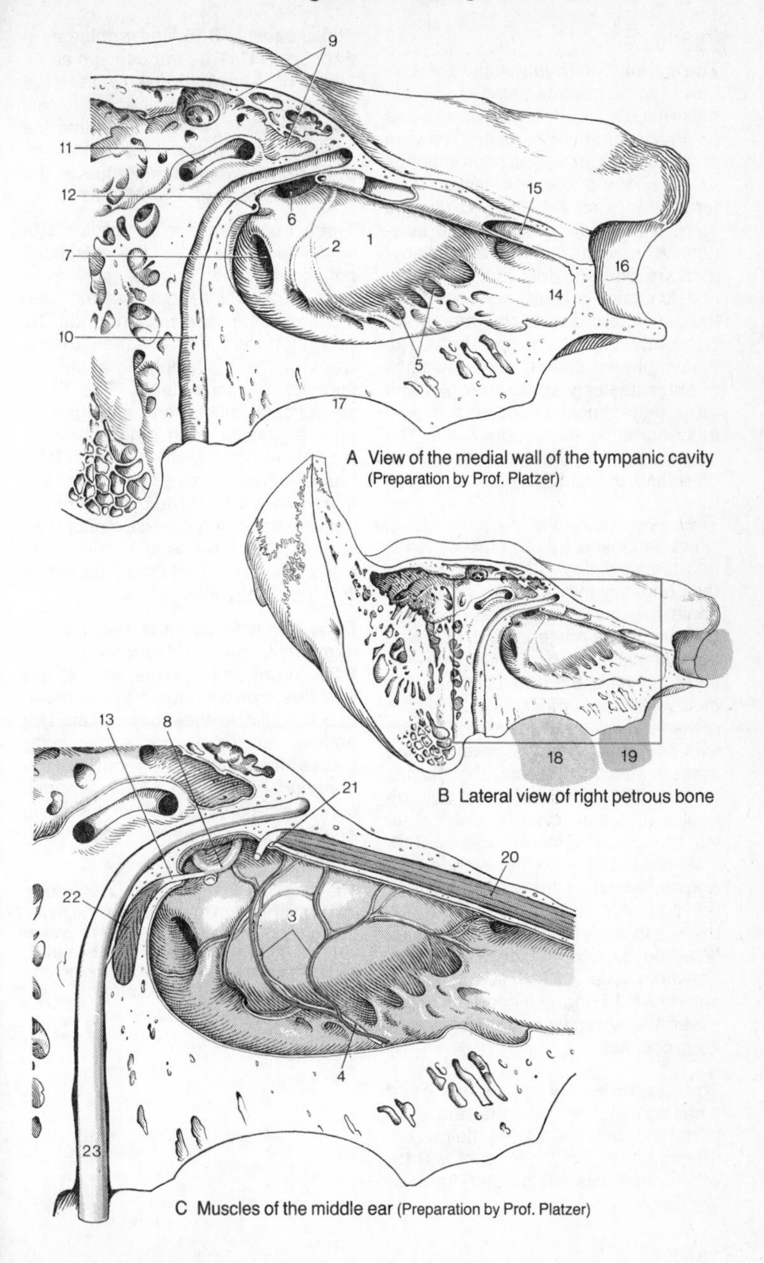

A View of the medial wall of the tympanic cavity (Preparation by Prof. Platzer)

B Lateral view of right petrous bone

C Muscles of the middle ear (Preparation by Prof. Platzer)

Internal Ear

The membranous labyrinth is a system of membranous vesicles and canals which are surrounded on all sides by a very hard bony capsule. The form of the cavities in the bone corresponds to the shape of the soft tissue structures, and a cast of them **C** gives a somewhat crude picture of the labyrinth. A bony and a membranous labyrinth are differentiated. The bony labyrinth contains a clear, aqueous fluid, the perilymph, in which the membranous labyrinth is suspended. The perilymphatic space communicates through the *perilymphatic duct* **A1** with the subarachnoid space at the posterior margin of the petrous bone. The membranous labyrinth contains a viscous fluid, the endolymph.

The oval vestibular window **AC2**, which is closed by the stapes, leads into the **vestibule AC3**, the midpart of the bony labyrinth. Anteriorly, the vestibule merges into the bony cochlea **C4**, and on its posterior wall open the bony *semicircular canals* **C5**. The vestibule contains the membranous *sacculus* **AB6** and *utriculus* **AB7**. In both of these a circumscribed region of their wall carries a sensory epithelium, the *macula sacculi* **AB8** and the *macula utriculi* **AB9**. The sacculus and utriculus communicate with each other via the *utriculosaccular duct* **AB10**. This canal gives off the slender *endolymphatic duct* **A11**, which runs to the posterior surface of the petrous bone and ends beneath the dura as flattened expansion, the *saccus endolymphaticus* **A12**. The *ductus reuniens* **AB13**, forms a connection between the sacculus and the membranous cochlea.

The bony **cochlea C4** has about two and a half convolutions. Its canal, the *spiral cochlear canal* **C14**, contains the membranous *cochlear duct* **AB15**, which begins as a blind end, the *vestibular caecum* **B16**, and terminates in the *cupula* **C17** as the *caecum cupulare* **B18**. Above and below it are the perilymphatic spaces: above the *scala vestibuli* **AB19**, which opens into the vestibule, and below it the *scala tympani* **AB20**, which terminates at the round *cochlear window* **ABC21**.

Three bony semicircular canals **C5**, which emanate from the vestibule, contain the membranous **semicircular ducts A22**, which communicate with the utriculus. These are surrounded by perilymph and are attached to the walls of the perilymphatic space by connective tissue fibers. The three semicircular ducts are arranged at right angles to each other: the (anterior) *superior semicircular duct* **B23** has its convexity directed toward the superior surface of the pyramid of the petrous bone, the *posterior duct* **B24** is parallel to the posterior surface of the petrous bone, and the *lateral duct* **B25** runs horizontally.

Each semicircular duct has an enlargement, the *membranous ampulla* **B26**, at one of its connections to the utriculus, corresponding to the osseous ampulla in the bony canal. The anterior and posterior semicircular canals join to form a common limb, the *common crus* **AB27**. Each ampulla contains sensory epithelium, the *crista ampullaris.* Tympanic membrane **C28**.

The courses taken by the semicircular ducts do not correspond to the axes of the body: the superior and posterior ducts diverge from the medial and frontal planes at 45°, and the lateral canal is tilted 30° posterolaterally from the horizontal plane.

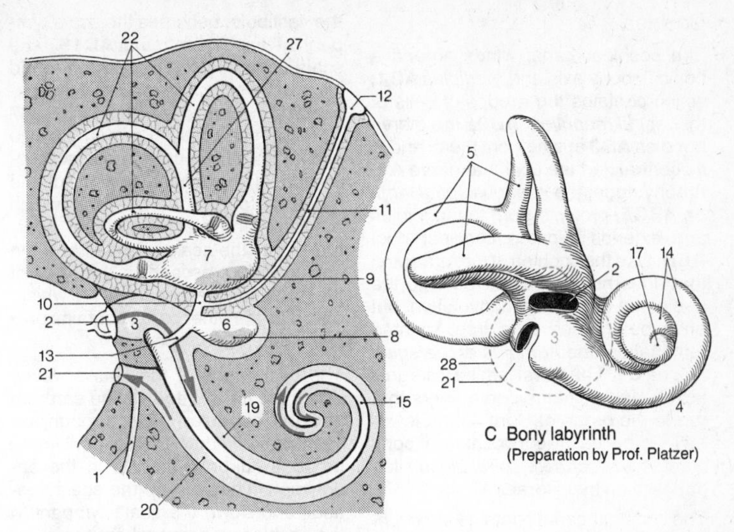

C Bony labyrinth
(Preparation by Prof. Platzer)

A Schematic view of the internal ear

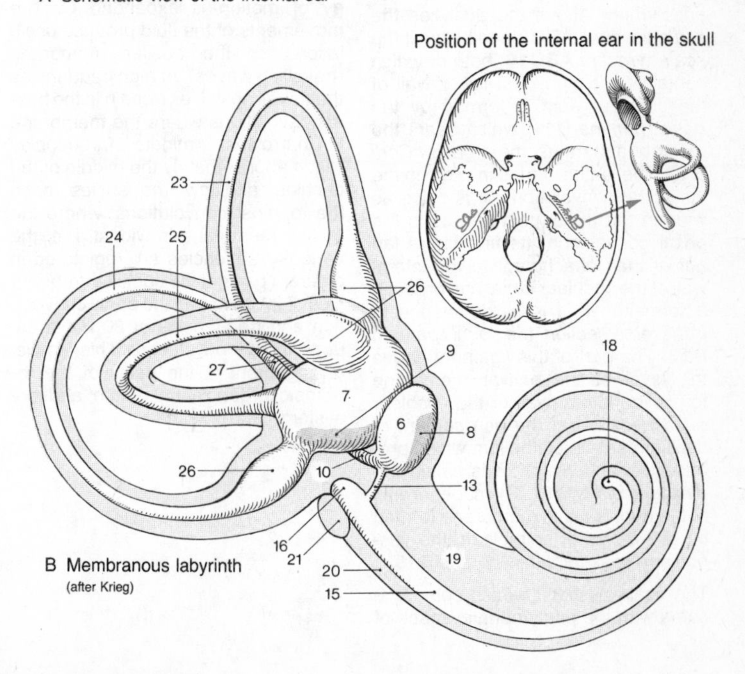

B Membranous labyrinth
(after Krieg)

Cochlea

The cochlear canal winds around a conical, bony axis, the *modiolus* **AC1,** which contains the groups of cells of the *spiral ganglion* **AB2,** the nerve bundles **AB3** arising from them, and in its central part the cochlear nerve **A4.** A bony ridge, the *osseous spiral lamina* **ABC5,** projects from the modiolus and extends into the cochlear duct **A6B.** Like the cochlea, it forms a spiral that does not, however, reach to the end of the uppermost convolution, but ends beforehand in a free, pointed termination, the *hamulus of the spiral lamina* **C7.** The spiral lamina is largely hollow and contains nerve fibers that run to the organ of Corti. In the lower half of the basal convolution a bony crest, the *secondary spiral lamina* lies opposite on the lateral wall.

The cochlear canal contains the *cochlear duct* **ABC8,** which is filled with endolymph. Above the duct lies the *scala vestibuli* **ABC9** and below it the *scala tympani* **ABC10,** both of which contain perilymph. The lower wall of the cochlear duct is formed by the *basilar lamina* **B11,** which bears the receptor apparatus, the *organ of Corti* **B12.** The lamina varies in width in the individual convolutions; it is twice as wide in the uppermost turn as in the basal convolution. Its fine fibers fan out at their attachment to the lateral wall of the cochlear canal and produce a formation that is sickle-shaped in transverse section, the *spiral ligament* **B13.** The part of this ligament above the level of the basilar membrane forms the lateral wall of the cochlear duct. Because of the rich content of capillaries in its epithelium which produce the endolymph, it is called the *stria vascularis* **B14.** The superior wall of the duct is a thin membrane formed by two layers of epithelium, the *vestibular (Reissner's) membrane* **B15.**

The scala vestibuli, which communicates with the perilymphatic space of the vestibule, becomes the scala tympani at the *helicotrema* **AC16.** The scala tympani descends to the round window, which is closed by the *secondary tympanic membrane* (p. 337 D). Communication between the two scalae is made possible by the separation of the lamina spiralis of the modiolus to form the hamulus, which results in the medial cochlear aperture. Only the scala vestibuli and the cochlear duct ascend to the apex of the cochlea, the cupula **A17,** which, unlike all the other parts, contains only two membranous spaces.

The oscillations of the sound waves which are transmitted via the eardrum and the ossicular chain through the vestibular window to the perilymph, produce fluid movements in the endolymph. They run up the scala vestibuli and down the scala tympani to the cochlear window, where the energy of motion is expended **C.** The movements of the fluid produce oscillations of the basilar membrane (travelling waves). At high frequencies the travelling waves remain in the basal convolutions where the membrane is narrowest, middle frequencies reach approximately the middle of the cochlea, and low frequencies reach the topmost convolutions, where the basilar membrane is widest, i. e. the various frequencies are registered in different parts of the cochlea: frequencies of 20,000 Hz in the basal convolution and frequencies of 20 Hz in the uppermost convolution. This spatial arrangement is the basis of the tonotopic organization of the auditory system.

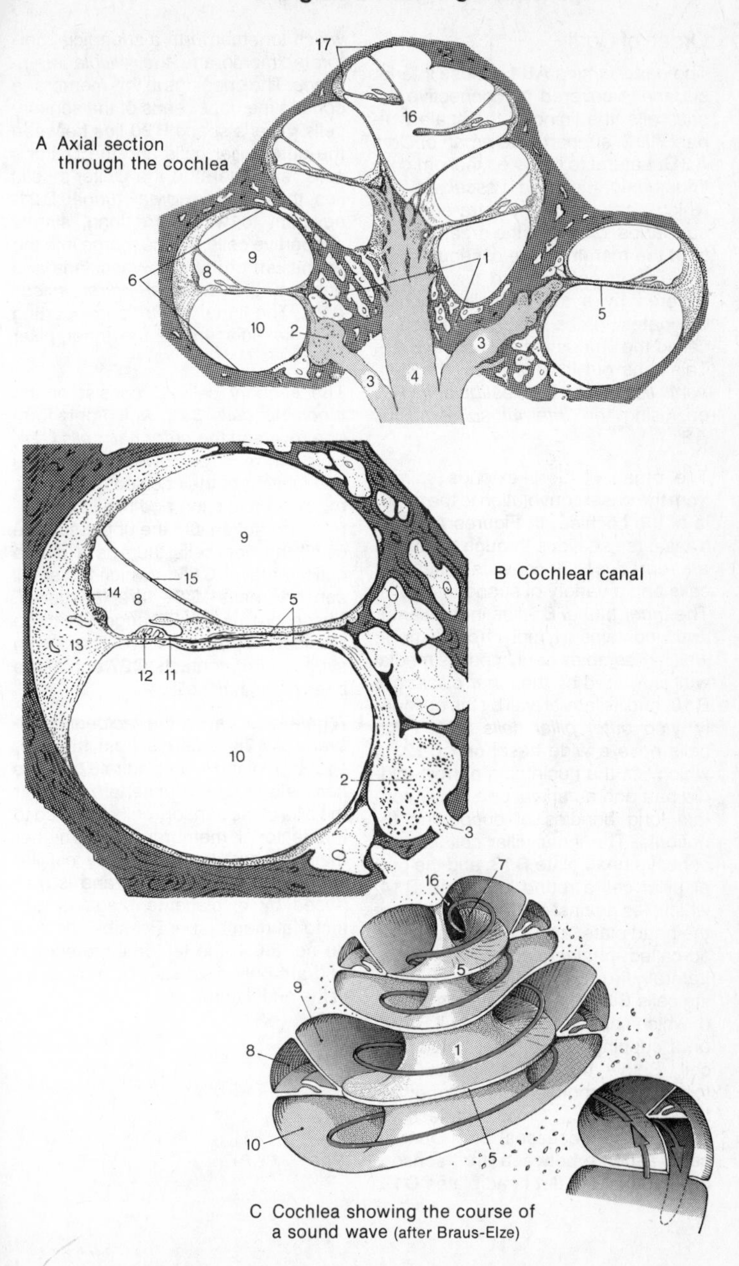

A Axial section through the cochlea

B Cochlear canal

C Cochlea showing the course of a sound wave (after Braus-Elze)

Organ of Corti

The *basal lamina* **AB 1,** whose inferior surface is covered by connective tissue cells, the lining of the scala tympani **AB 2,** supports the *organ of Corti* **A 3 B.** Lateral to it, the epithelium continues into the *stria vascularis* **A 4,** which contains many intra-epithelial capillaries. Medial to the organ of Corti, at the margin of the osseous spiral lamina, arises a dense tissue layer covered by a tall epithelium, which originates in the periosteum and is called the *limbus laminae spiralis* **A 5.** This runs out into two lips, the *tympanic lip* **A 6** and the *vestibular lip* **A 7,** enclosing the *internal spiral sulcus* **A 8.**

The organ of Corti extends spirally from the basal convolution to the cupula of the cochlea. In Figures **AB** only transverse sections through the organ are reproduced. It consists of sensory cells and a variety of supporting cells. The *inner tunnel* **B 9** lies in the center and contains lymph (cortilymph), which resembles perilymph. Its medial wall is formed by the *inner pillar cells* **B 10,** and its lateral wall by the obliquely lying *outer pillar cells* **B 11.** Pillar cells have a wide basal part **B 12,** in which lies the nucleus, a narrow middle part and an apical part. They contain long bundles of supporting tonofibrils. The inner pillar cell forms a concave head plate **B 13,** and the outer pillar cell a round head part **B 14,** which lies against the lower surface of the head plate, and a flat process, the so-called phalangeal process **B 15.** Laterally lie groups of Deiter's supporting cells **B 16** (phalangeal cells), each of which support a sensory cell **B 17 C** on a cupping of the lower part of the cell. Below the sensory cells their tonofibrils arborize into supporting baskets **B 18.** Small projections (phalangeal processes) from Deiter's cells ascend between the sensory cells and end in flat head plates **C 19,** which together form a superficial perforated membrane, the *reticular membrane.* The openings in this membrane contain the upper ends of the sensory cells. *Nuel's space* **B 20** lies between the outer pillar cells and the Deiter's cells, and lateral to the Deiter's cells lies the small *external tunnel* **B 21,** adjacent to which are long, simple supportive cells. These merge into the epithelium of the stria vascularis and enclose the *external spiral sulcus* **A 22.** The inner Deiter's supporting cells lie adjacent to the inner pillar cells.

The sensory cells C consist of the *inner hair cells* **C 23,** which only form one row, and the *outer hair cells* **C 24,** which form three rows in the basal convolution of the cochlea, four in the middle convolution and five in the upper convolution. On the upper surface of all the hair cells there is a dense cuticular layer **C 25** in which the small sensory hairs **C 26** are anchored. These are usually arranged in three semicircular rows. Nerve fibers with synapse-like contacts **C 27** end at the base of the hair cells.

A gelatinous layer, the *tectorial membrane* **AB 28,** extends from the vestibular lip of the limbus spirale over the hair cells. It is still uncertain whether the cilia of the sensory cells are fixed to the tectorial membrane, or whether their tangential deflection by oscillation of the basilar membrane is produced by displacement against the tectorial membrane. Possibly the cilia do not touch the tectorial membrane and are only displaced by movement in the endolymph.

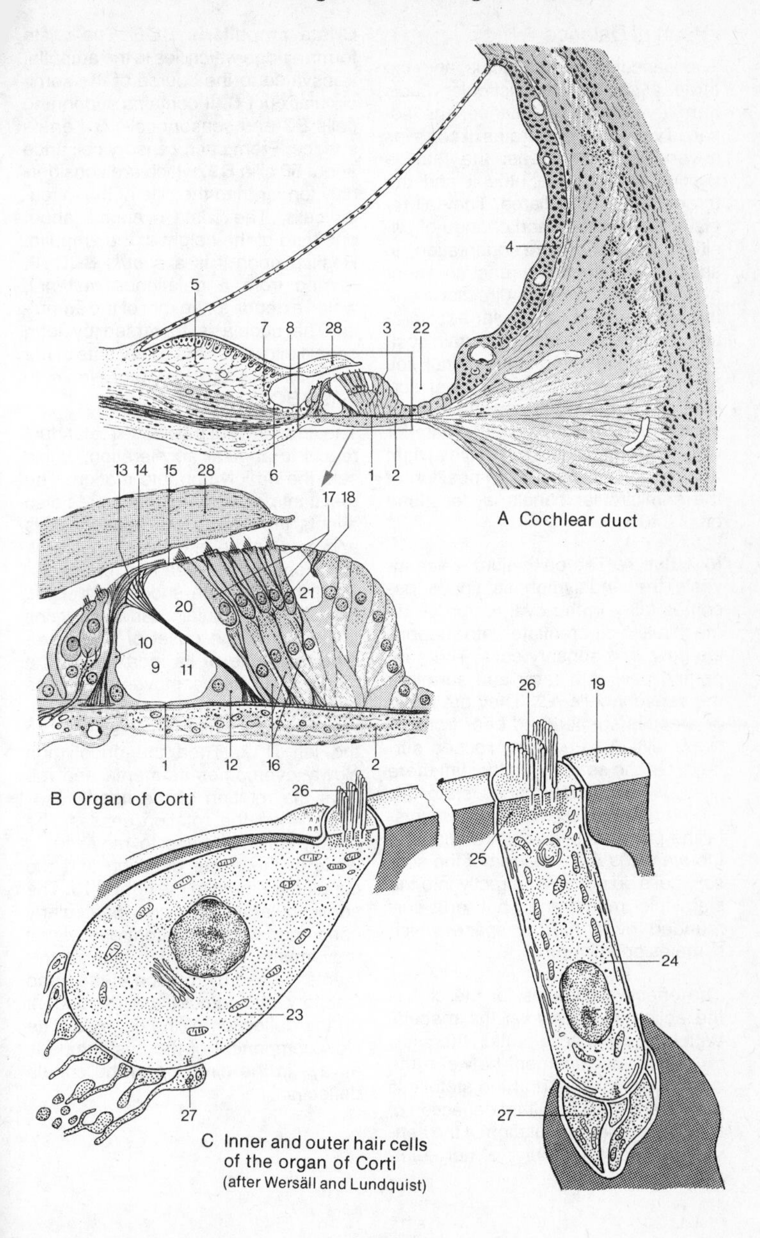

A Cochlear duct

B Organ of Corti

C Inner and outer hair cells of the organ of Corti (after Wersäll and Lundquist)

Organ of Balance

The sacculus, the utriculus and the three associated semicircular ducts form the organ of balance, the so-called vestibular apparatus. It contains several sensory areas: the macula sacculi, the macula utriculi and the three cristae ampullares. They all register acceleration and change of position, and therefore orientation in space. The maculae react to linear acceleration in various directions and the cristae react to angular acceleration. The maculae occupy certain positions in space; the macula utriculi lies almost horizontally on the lower surface of the utriculus, and the macula sacculi lies vertically on the anterior wall of the sacculus. The two lie at right angles to each other. The position of the semicircular canals is described on p. 340.

Maculae A. The epithelium which invests the endolymphatic space becomes taller in the oval region of the macula and differentiates into supporting cells and sensory cells. The supporting cells **A1** carry and surround the sensory cells **A2.** They are flask- or ampulla-shaped and bear from 70 to 80 cilia **A3** on their exposed surface. On the sensory epithelium there is a jellylike membrane, *statolithic membrane* **A4,** which contains crystalline particles of calcium carbonate, the *statoliths* **A5.** The cilia of the sensory cells do not enter directly into the statolithic membrane, but are surrounded by a narrow space which contains endolymph.

The effective stimulus for the cilia is the action of gravity on the macula. With increasing acceleration there is a tangential displacement between the sensory epithelium and the statolithic membrane. The resultant deflection of the cilia causes stimulation of the sensory cells and the firing of nerve impulses.

Crista ampullaris BC6. The crista forms a ridge which lies in the ampulla, transverse to the course of the semicircular duct **C.** It contains supporting cells **B7** and sensory cells **B8** on its surface. From each sensory cell arise about 50 cilia **B9,** which are considerably longer than the cilia of the macular cells. The crista occupies about one third of the height of the ampulla. Resting upon it is a *cupula* **BCD10,** formed from a gelatinous material, which extends to the roof of the ampulla. The cupula is traversed by long canals into which the ciliary tufts of the sensory cells are inserted. Nerve fibers **ABC11.**

The apparatus of the semicircular duct reacts to angular acceleration, which sets the endolymph into motion. The resultant deflection of the cupula also affects the cilia of the sensory cells and acts as the effective stimulus. If, for example, the head is rotated toward the right, the endolymph of the lateral semicircular canal remains stationary at the onset of the movement because of its inertia, and the result is a relative movement in the opposite direction (hydrodynamic inertia): both cupulae are displaced to the left **D12.** Then the endolymph slowly overcomes its inertia and follows the rotation of the head. However, when the rotation ceases, the endolymph continues to move for a while in the same direction and the cupulae are bent to the right **D13.** The semicircular ducts are particularly concerned with reflex control of visual movements. The jerky eye movements caused by rotation of the head (rotatory nystagmus) are dependent on the deflection of the cupulae. The slow component of the nystagmus is always in the direction of the cupula deflection.

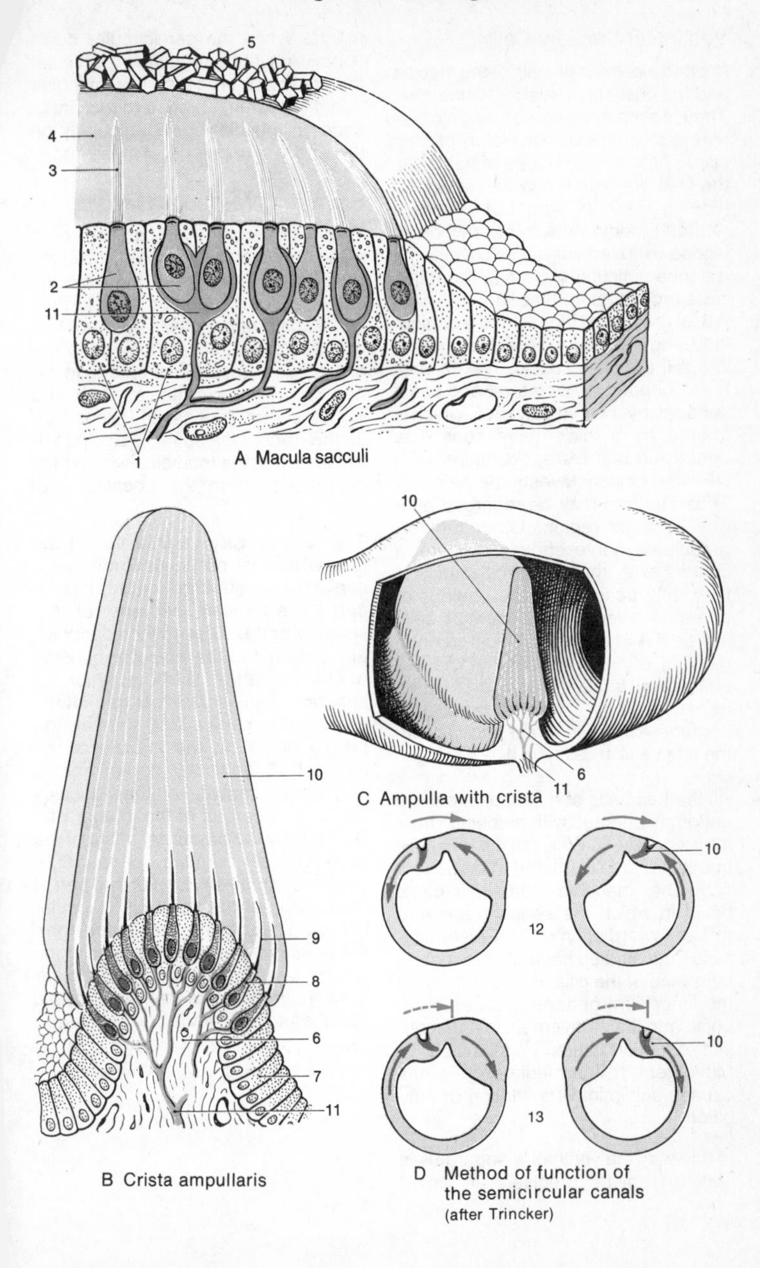

A Macula sacculi

B Crista ampullaris

C Ampulla with crista

D Method of function of the semicircular canals (after Trincker)

Vestibular Sensory Cells

In principle, the hair cells of the macula and the crista have a similar structure. They are mechanoreceptors which respond to tangential diversion of their cilia. There are two types of hair cells: the flasklike type I, and the cylindrical type II. Cells of type I **A1** have a rounded body with a narrow neck, whose exposed surface is covered by a dense cuticular plate **A2.** From this plate project about 60 cilia, *stereocilia* **A3,** of graduated length, and a particularly long *kinocilium* **A4,** at the origin of which there is no cuticle. The hair cell is surrounded on its lateral and basal surfaces by a nerve calyx **A5,** which is formed by a thick nerve fiber. The upper part of the calyx contains vesicles and closely invests the hair cell. Thus this part may be regarded as a true synaptic region. Other densely granulated nerve endings **A6** are in contact with the nerve calyces and they may possibly be the endings of efferent nerves. The slender hair cells of type II **A7** have an identical complement of cilia. At the base of the cell there are large, poorly granulated nerve endings. Heavily granulated boutons **A8** are in direct contact with the hair cells in cell type II.

All the hair cells of a sensory field are uniformly oriented with respect to their long kinocilia **B.** While all the hair cells in a crista are orientated in one direction, the maculae contain various fields in which the sensory cells are turned toward each other. Electrophysiological studies have shown that divergence of the cilia in the direction of the kinocilium generates a nerve impulse and their movement in the opposite direction leads to inhibition **C.** Movement in intermediate directions causes subliminal stimulation or inhibition.

In this way the vestibular apparatus is able to register precisely all movements. While the semicircular ducts (kinetic labyrinth) regulate in particular the direction of gaze, the maculae (tonic labyrinth) exert a direct influence on muscle tone, particularly on the extensor and cervical muscles.

Spiral and Vestibular Ganglia

The **spiral ganglion D9** consists of a chain of clusters of neurons lying in the modiolus, at the exit of the osseous spiral lamina. Together they form a spiral chain of ganglia. These contain true bipolar neurons, whose peripheral processes (dendrites) extend to the hair cells of the organ of Corti and whose central processes (axons) run as the *tract of the spiral foramen* **D10** to the axis of the modiolus where they combine to form the *cochlear root* **D11.**

The **vestibular ganglion D12** lies at the base of the internal acoustic meatus. It consists of a superior and an inferior part. The bipolar neurons of the superior part **D13** send their peripheral processes to the ampullary cristae of the anterior **D14** (anterior or superior ampullary nerve) and lateral semicircular canals **D15** (lateral ampullary nerve) to the macula of the utricle **D16** (ultricular nerve) and to part of the macula of the sacculus **D17.** The neurons of the inferior part **D18** supply the ampullary crista of the posterior semicircular canal **D19** (posterior ampullary nerve) and part of the macula of the sacculus (saccular nerve). The central processes form the *vestibular division* **D20,** which runs in a common sheath with the cochlear division, through the *internal acoustic meatus* into the middle cranial fossa.

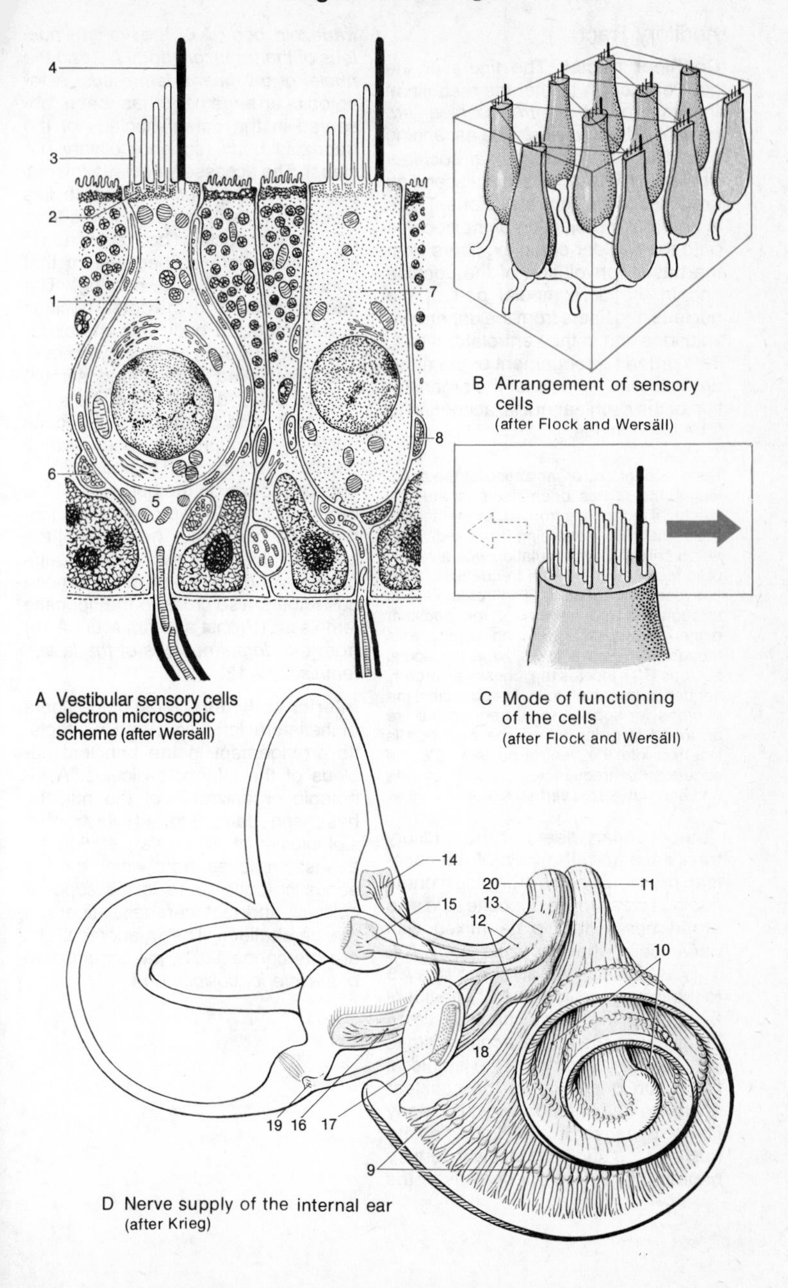

A Vestibular sensory cells electron microscopic scheme (after Wersäll)

B Arrangement of sensory cells (after Flock and Wersäll)

C Mode of functioning of the cells (after Flock and Wersäll)

D Nerve supply of the internal ear (after Krieg)

Auditory Tract

Cochlear nuclei. The fibers of the cochlear root **A1** enter the medulla at the level of the *ventral cochlear nucleus* **AB2** and divide. The ascending branches run to the *dorsal cochlear nucleus* **AB3** and the descending branches to the ventral nucleus. There is an orderly projection of the cochlea onto the nuclear complex: fibers from the basal convolution of the cochlea end in the dorsomedial part of the nucleus and fibers from the upper convolutions end in the ventrolateral part. This orderly arrangement of the afferent fibers is the basis for the organization of the cochlear nuclei according to tone frequencies.

Such a tonotopic organization of the cochlear complex has been demonstrated by electrical recordings from experimental animals (cat) **B.** Recording from individual nerve cells during stimulation with a variety of tones shows to which frequencies each cell responds optimally. The frontal section through the oral region of the cochlear nculei shows that when an electrode is moved from above to below in the dorsal nucleus **B3** it locates the tones in an orderly sequence from high to low frequencies, the neurons for specific tonal frequencies are arranged in order. When the electrode is inserted into the ventral nucleus **B2,** the sequence of frequencies abruptly ceases and only varies in a certain region.

The secondary fibers of the auditory tract arise from the cells of the cochlear nuclei. Bundles from the ventral nucleus cross to the opposite side as a broad nerve fiber plate mixed with nerve cells, the *trapezoid body* **A4.** This ascends as the **lateral lemniscus A5** to the inferior colliculus **A6.** The fibers that leave the dorsal nucleus cross obliquely as the *dorsal acoustic striae* **A7.** A large proportion of the lemniscal fibers run directly from the cochlear nuclei to the inferior colliculi. Many fibers synapse in the intermediate nuclei of the auditory tract with tertiary neurons: in the *dorsal nucleus of the trapezoid body* **A8,** the *ventral nucleus of the trapezoid body* **A9** and the *nuclei of the lateral lemniscus.* A tonotopic arrangement has been observed in the dorsal nucleus of the trapezoid body (superior olivary nucleus). The accessory nucleus *(medial superior olivary nucleus),* which lies medial to it **A10,** receives fibers from the cochlear nuclei of both sides and is interpolated into a fiber system that subserves directional hearing. The fiber connections of the dorsal nucleus of the trapezoid body with the abducens nucleus **A11** (reflex eye movements resulting from sound stimuli) are still under dispute. The fibers are said to extend around the abducens nucleus, through the contralateral cochlear nucleus, and to end as efferent conductors on the hair cells of the organ of Corti. They probably regulate the inflow of stimuli. The nuclei of the lateral lemniscus are scattered groups of cells along the course of the lemniscus. Fibers also cross to the opposite lemniscus (Probst's commissure **A13**) from the *dorsal nucleus of the lateral lemniscus* **A12.**

Inferior colliculus. Most of the fibers of the lateral lemniscus end in a topistic arrangement in the principal nucleus of the inferior colliculus. A tonotopic organization of the nucleus has been demonstrated electrophysiologically. It is a relay station for acoustic reflexes, from which run the acoustico-optic fibers to the superior colliculi, and tectocerebellar fibers to the cerebellum. The inferior colliculi are interconnected by the *commissure of the inferior colliculi* **A14.**

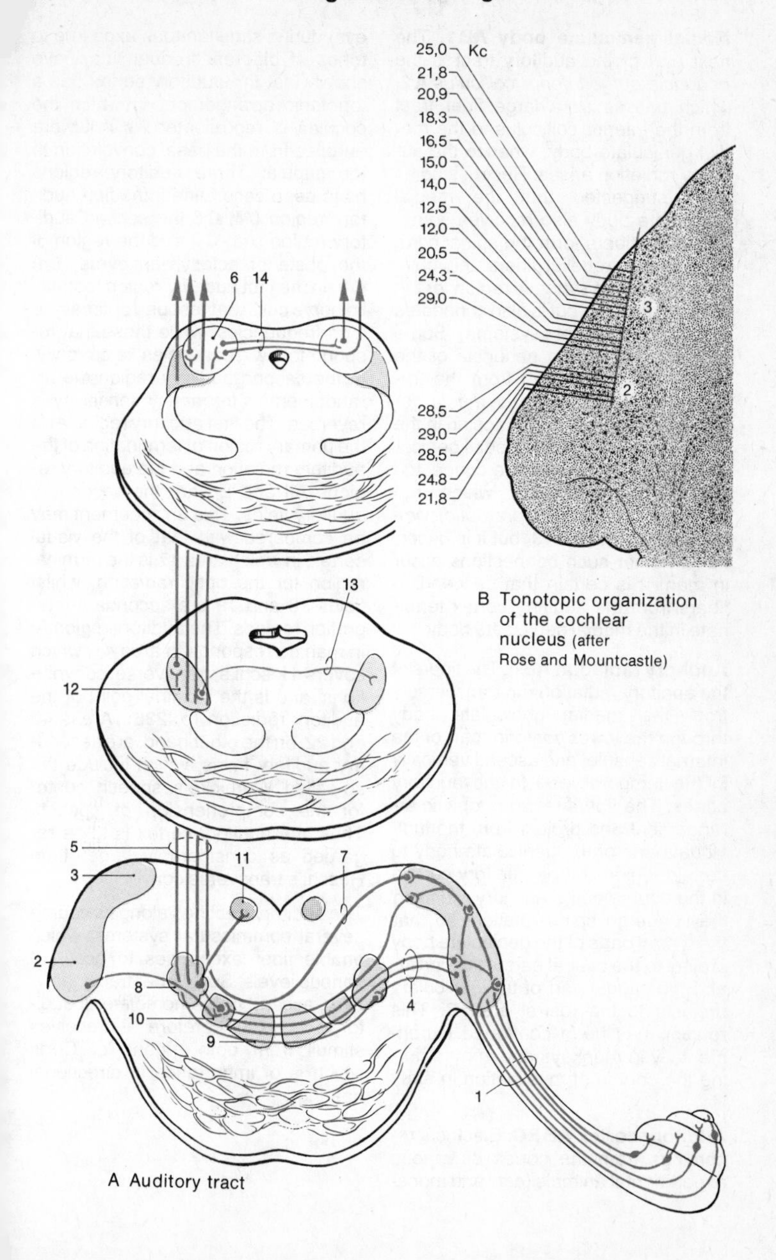

B Tonotopic organization of the cochlear nucleus (after Rose and Mountcastle)

A Auditory tract

Medial geniculate body AB1. The next part of the auditory tract is the *peduncle of the inferior colliculus* **A2,** which passes as a large fiber tract from the inferior colliculus to the medial geniculate body, whence the auditory radiation arises. Recent studies have suggested that the medial geniculate body also receives somatosensory fibers form the spinal cord, as well as cerebellar fibers. It is obviously not only a synaptic region for the acoustic system, but is also connected with various other systems. Some fiber bundles in the peduncle of the inferior colliculus stem from the trapezoid nuclei and reach the medial geniculate body without synapsing in the inferior colliculus. Both medial geniculate bodies are said to be connected by decussating fibers which run through the *inferior supra-optic commissure* (Gudden) **A3,** but it is uncertain whether such connections occur in man. It is certain that descending fibers from the auditory cortex terminate in the medial geniculate body.

Auditory radiation AB4. The fibers of the auditory radiation run transversely from the medial geniculate body through the lower posterior part of the internal capsule, and ascend vertically in the temporal lobe to the auditory cortex. The fibers retain a topistic arrangement and project from the individual parts of the geniculate body to certain regions of the auditory cortex. In the course of the auditory radiation they undergo spiral rotation, so that the rostral parts of the geniculate body project to the caudal part of the cortex, and the caudal part of the geniculate projects to the rostral area **B.** This rotation has been confirmed experimentally in monkeys and in man during the course of myelination in early life.

Auditory cortex AB5C. Electrical recordings from the cortex of various experimental animals (cats and monkeys) during simultaneous exposure to tones of different frequencies, have shown that the auditory cortex has a tonotopic arrangement in which the cochlea is represented as if it were unrolled from the basal convolution to the cupula. Three auditory regions have been ascertained: the first auditory region (AI) **C6** the second auditory region (AII) **C7** and the region of the posterior ectosylvian gyrus (Ep) **C8.** In the first auditory region rostrally lie nerve cells that respond optimally to high frequencies, while those that respond to low frequencies lie caudally. In the second auditory region the arrangement of frequency sensitivity is reversed. The first auditory region AI is the primary region of termination of the auditory radiation and the auditory regions AII and Ep are the secondary auditory fields. This arrangement may be compared with that of the visual cortex, in which area 17 is the terminal region for the optic radiation, whilst fields 18 and 19 are secondary integration regions. The auditory region AI in man corresponds to area 41, which covers Heschl's transverse convolutions, and is the terminal point of the auditory radiation (p. 236). Areas 42 and 22, on the other hand, are secondary auditory fields, which include the so-called Wernicke's speech center for the comprehension of speech. Thus, the auditory cortex is to be regarded as considerably larger than Heschl's transverse convolutions.

The auditory tract has along its course several commissural systems, which enable fiber exchanges to occur at various levels. Some fiber tracts, however, ascend to the homolateral auditory cortex. Therefore it receives stimuli from both organs of Corti, which is of importance in directional hearing.

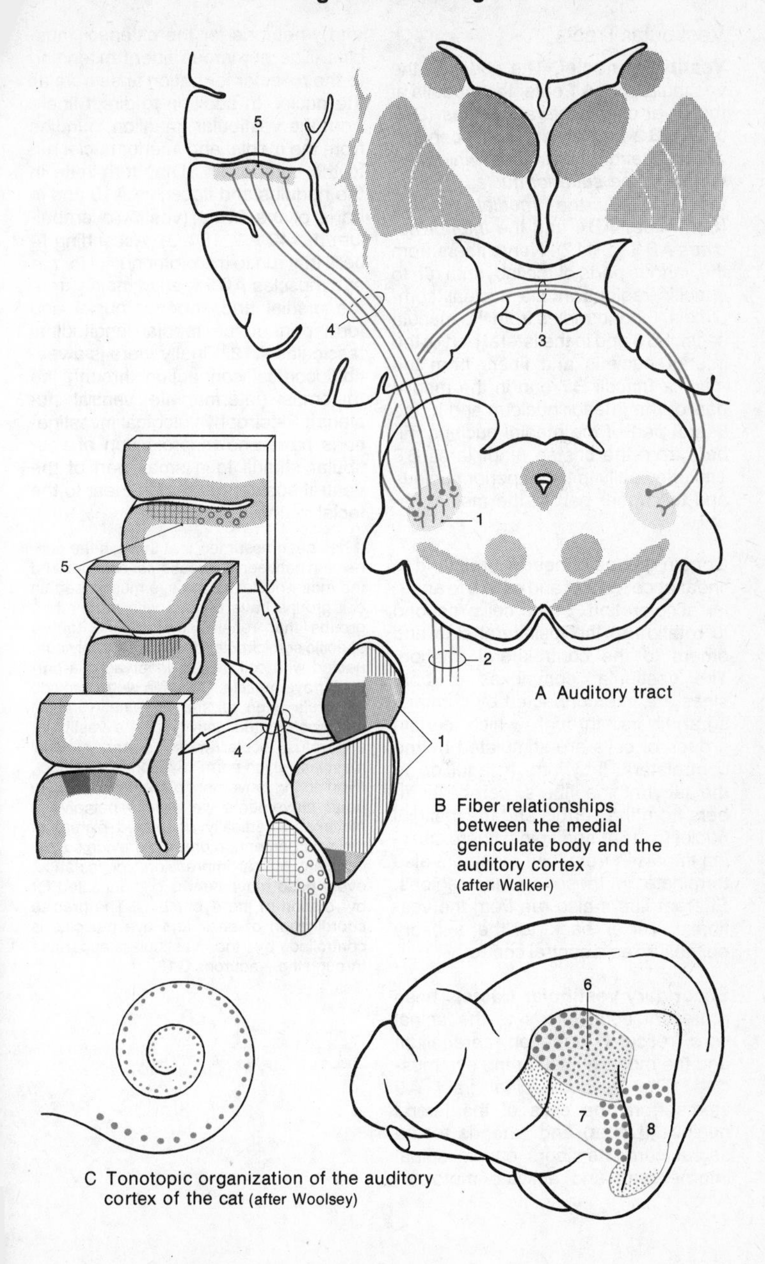

A Auditory tract

B Fiber relationships between the medial geniculate body and the auditory cortex (after Walker)

C Tonotopic organization of the auditory cortex of the cat (after Woolsey)

Vestibular Tracts

Vestibular nuclei. The fibers of the vestibular root **A1** enter the medulla at the level of the *lateral nucleus* (Deiters) **AB2** and divide into ascending and descending branches, which terminate in the *superior nucleus* (Bechterew) **AB3,** the *medial nucleus* (Schwalbe) **AB4** and the *inferior nucleus* **AB5** (p. 112). Nerve fibers from the various parts of the labyrinth run to specific regions in the nuclear complexes. Fiber bundles from the macula sacculi **B6** end in the lateral part of the inferior nucleus and fibers from the macula utriculi **B7** end in the medial part of the inferior nucleus and in the lateral part of the medial nucleus. Fibers from the cristae ampullares **B8** end principally in the superior nucleus and the upper part of the medial nucleus.

Certain groups of neurons respond to linear acceleration and others to angular acceleration. Some cells respond to rotation to the ipsilateral side and others to the contralateral rotation. The vestibular complexes of both sides are interconnected by commissural fibers, through which certain groups of cells are stimulated by the contralateral labyrinth. In addition to the labyrinthine fibers, cerebellar fibers from the vermis and the fastigial nuclei (p. 154) and spinal fibers carrying impulses from joint receptors also terminate in these nuclear regions. Efferent fibers also run from the vestibular nuclei back to the sensory epithelium as a central control.

Secondary vestibular tracts. These represent connections to the spinal cord, reticular formation, cerebellum and the motor nuclei for the eye muscles. The *vestibulospinal tract* **A9** stems from the cells of the lateral nucleus (Deiter) and extends to the sacral cord. Its fibers end in spinal interneurons and activate motor α- and γ-neurons for the extensor muscles. The numerous fibers extending to the reticular formation arise from all the nuclei. In addition to direct fibers from the vestibular ganglion, bundles from the medial and inferior nuclei run to the cerebellum. They terminate in the nodulus and flocculus **A10** and in parts of the uvula (vestibulocerebellum p. 142 A6, 154 C). Ascending fibers that run to the motor nuclei for the eye muscles **AC11,** arise mainly from the medial and superior nuclei and form part of the medial longitudinal fasciculus **A12.** Finally there is a vestibulocortical connection through the thalamus (intermediate ventral nucleus). Electrophysiological investigations have shown projection of vestibular stimuli to a small part of the ventral postcentral region, near to the facial region.

It has been assumed that through the connection between the vestibular complex and the motor nuclei for the eye muscle certain cell groups have contact with each other: groups that receive the stimuli from a specific semicircular duct are probably connected with cell groups innervating a particular eye muscle. Only this would provide an explanation for the extremely precise and coordinated activity of the vestibular apparatus, ocular muscles and cervical muscles, which permit fixation of an object, even during movement of the head. Despite head movements we always perceive a stationary, vertically orientated picture of the environment. In order to provide such a constant optical impression, for instance, every head movement is compensated for by rotation of the eyeballs **C.** The precise coordination of neck and eye muscles is controlled by the vestibular apparatus through the γ-neurons **C13.**

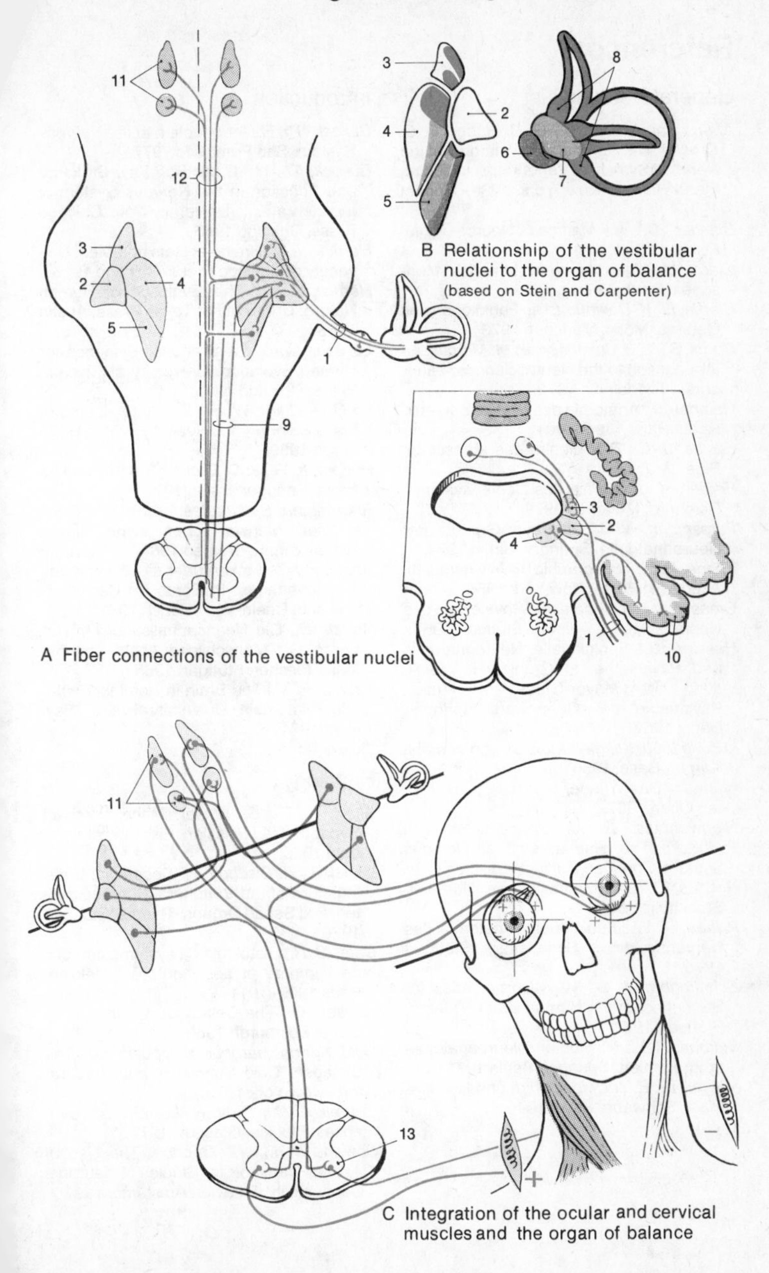

A Fiber connections of the vestibular nuclei

B Relationship of the vestibular nuclei to the organ of balance (based on Stein and Carpenter)

C Integration of the ocular and cervical muscles and the organ of balance

References

General Reading List

Ariëns Kappers, C. C., G. C. Huber, E. C. Crosby: The Comparative Anatomy of the Nervous System of Vertebrates, Including Man. Hafner, New York 1936, Reprint 1960

Biesold, D., H. Matthies: Neurobiologie. Fischer, Stuttgart 1977

Clara, M.: Das Nervensystem des Menschen. Barth, Leipzig 1959

Clarke, E., K. Dewhurst: Die Funktionen des Gehirns. Moos, München 1973

Curtis, B. A., S. Jakobson, E. M. Marcus: An Introduction to the Neurosciences. Saunders, Philadelphia 1972

Dejerine, J.: Anatomie des centres nerveux. Rueff, Paris 1895–1901

Eccles, J. C.: Das Gehirn des Menschen. Piper, München 1973

Ferner, H.: Anatomie des Nervensystems. Reinhard, München 1970

Forssmann, W. G., Ch. Heym: Grundriß der Neuroanatomie. Springer, Berlin 1974

Friede, R. L.: Topographic Brain Chemistry. Academic Press, New York 1966

Glees, P.: Morphologie und Physiologie des Nervensystems. Thieme, Stuttgart 1957

Hassler, R.: Funktionelle Neuroanatomie und Psychiatrie. In Gruhle, H. W., R. Jung, W. Mayer-Gross, M. Müller: Psychiatrie der Gegenwart. Springer, Berlin 1967

Ludwig E., J. Klingler: Atlas cerebri humani. Karger, Basel 1956

Mühr, A.: Das Wunder Menschenhirn. Walter, Olten 1957

Nieuwenhuis, R., J. Voogd, Chr. van Huizen: The Human Central Nervous System. Springer, Berlin 1978

Retzius, G.: Das Menschenhirn. Norstedt, Stockholm 1896

Rohen, J. W.: Funktionelle Anatomie des Nervensystems. Schattauer, Stuttgart 1975

Schaltenbrand, G., W. Wahren: Atlas for Stereotaxy of the Human Brain. Thieme, Stuttgart 1977

Sidman, R. L., M. Sidman: Neuroanatomie programmiert. Springer, Berlin 1971

Villiger, E., E. Ludwig: Gehirn und Rückenmark. Schwabe, Basel 1946

Introduction

Bullock, Th. H.: Introduction to the Nervous System. San Francisco 1977

Bullock, Th. H., G. A. Horridge: Structure and Function in the Nervous System of Invertebrates. University of Chicago Press, Chicago 1955

Herrick, J. C.: Brains of Rats and Men. University of Chicago Press. Chicago 1926

Herrick, J. C.: The Evolution of Human Nature. University of Texas Press, Austin 1956

Le Gros Clark, W. E.: Fossil Evidence for Human Evolution. University of Chicago Press, Chicago 1955

Le Gros Clark, W. E.: The Antecedents of Man. Edinburgh University Press, Edinburgh 1959

Popper, K. R., J. C. Eccles: The Self and Its Brain. Springer, Berlin 1977

Sherrington, Sir Charles: Körper und Geist – Der Mensch über seine Natur. Schünemann, Bremen 1964

Spatz, H.: Gedanken über die Zukunft des Menschenhirns. In Benz, E.: Der Übermensch. Rhein-Verl., Zürich 1961

Starck, D.: Die Neencephalisation. In Heberer, G.: Menschliche Abstammungslehre. Fischer, Stuttgart 1965

Tobias, P. V.: The Brain in Hominid Evolution. Columbia University Press, New York 1971

Histology

Akert, K., P. G. Waser: Mechanisms of Synaptic Transmission. Elsevier, Amsterdam 1969

Babel, J., A. Bischoff, H. Spoendlin: Ultrastructure of the Peripheral Nervous System and Sense Organs. Thieme, Stuttgart 1970

Cajal, S. R.: Histologie du système nerveux de l'homme et des vertébrés. Maloine, Paris 1909–1911

Causey, G.: The Cell of Schwann. Livingstone, Edinburgh 1960

Cold Spring Harbour Symposia 40: The Synapse. Cold Spring Harbour Laboratory, New York 1976

Cottrell, G. A., P. N. R. Usherwood: Synapses. Blackie, Glasgow 1977

Cowan, W. M., M. Cuenod: The Use of Axonal Transport for Studies of Neuronal Connectivity. Elsevier, Amsterdam 1975

De Robertis, E. D. P., R. Carrea: Biology of Neuroglia. Elsevier, Amsterdam 1965
Eränkö, O.: Histochemistry of Nervous Transmission. Elsevier, Amsterdam 1969
Friede, R. L., F. Seitelberger: Symposium über Axonpathologie und "Axonal Flow". Springer, Berlin 1971
Fuxe, K., L. Olson, Y. Zotterman: Dynamics of Degeneration and Growth in Neurons. Pergamon Press, Oxford 1973
Hild, W.: Das Neuron. In *Bargmann, W.:* Handbuch der mikroskopischen Anatomie, Suppl. to Vol. IV/1. Springer, Berlin 1959
Jones, D. G.: Synapses and Synaptosomes. Chapman & Hall, London 1975
Landon, D. N.: The Peripheral Nerve. Chapman & Hall, London 1976
Lehmann, H. J.: Die Nervenfaser. In *Bargmann, W.:* Handbuch der mikroskopischen Anatomie, Suppl. to Vol. IV/1. Springer, Berlin 1959
Nakai, J.: Morphology of Neuroglia. Igaku-Shoin. Osaka 1963
Pappas, G. D., D. P. Purpura: Structure and Function of Synapses. Raven Press, New York 1972
Penfield, W.: Cytology and Cellular Pathology of the Nervous System. Hoeber, New York 1932
Peters, A., S. L. Palay, H. F. Webster: The Fine Structure of the Nervous System. Harper & Row, New York 1970
Rapoport, St. J.: Blood-Brain Barrier in Physiology and Medicine. Raven Press, New York 1976
Roberts, E., T. N. Chase, D. B. Tower: GABA in Nervous System Function. Raven Press, New York 1976
Szentágothai, J.: Neuron Concept Today. Akadémiai Kiadó, Budapest 1977
Uchizono, K.: Excitation and Inhibition. Elsevier, Amsterdam 1975
Watson, W. E.: Cell Biology of Brain. Chapman & Hall, London 1976
Windle, W. F.: Biology of Neuroglia. Thomas, Springfield/Ill. 1958

Spinal Cord and Spinal Nerves

Bok, S. T.: Das Rückenmark. In *v. Möllendorff, W.:* Handbuch der mikroskopischen Anatomie, Vol. IV. Springer, Berlin 1928
Dyck, P. J., P. K. Thomas, E. H. Lambert: Peripheral Neuropathy. Vol. I: Biology of the Peripheral System. Saunders, Philadelphia 1975
Foerster, O.: Spezielle Anatomie und Physiologie der peripheren Nerven. In *Bumke, O., O. Foerster:* Handbuch der Neurologie, Suppl. II/1. Springer, Berlin 1928
Foerster, O.: Symptomatologie der Erkrankungen des Rückenmarks und seiner Wurzeln. In *Bumke, O., O. Foerster:* Handbuch der Neurologie, Vol. V. Springer, Berlin 1936
Hubbard, J. I.: The Peripheral Nervous System. Plenum Press, New York 1974
Jacobsohn, L.: Über die Kerne des menschlichen Rückenmarks. Abh. d. königl. preuß. Akad. d. Wiss., Phys.-math. Kl., Berlin 1908
Kadyi, H.: Über die Blutgefäße des menschlichen Rückenmarkes. Gubrynowicz & Schmidt, Lemberg 1886
Keegan, J. J., F. D. Garrett: The segmental distribution of the cutaneous nerves in the limbs of man. Anat. Rec. 102: 409–437, 1948
v. Lanz, T., W. Wachsmuth: Praktische Anatomie, Vol. I/2.–4. Springer, Berlin 1955–1972
Mumenthaler, M., H. Schliack: Läsionen peripherer Nerven, 4th Ed. Thieme, Stuttgart 1982
Noback, Ch. N., J. K. Harting: Spinal cord. In *Hofer, H., A. H. Schultz, D. Strack:* Primatologia, Vol. II/1. Karger, Basel 1971
Villiger, E.: Die periphere Innervation. Schwabe, Basel 1964

Brain Stem

Brodal, A.: The Cranial Nerves. Blackwell, Oxford 1954
Brodal, A.: The Reticular Formation of the Brain Stem. Oliver & Boyd, Edinburgh 1957
Clemente, C. D., H. W. Magoun: Der bulbäre Hirnstamm. In *Schaltenbrand, G., P. Bailey:* Einführung in die stereotaktischen Operationen mit einem Atlas des menschlichen Gehirns – Introduction to Stereotaxis with an Atlas of the Human Brain. Thieme, Stuttgart 1959
Crosby, E. C., E. W. Lauer: Anatomie des Mittelhirns. In *Schaltenbrand, G., P. Bailey:* Einführung in die stereotaktischen Operationen mit einem Atlas des menschlichen Gehirns – Introduction to Stereotaxis with an Atlas of the Human Brain. Thieme, Stuttgart 1959
Delafresnaye, J. F.: Brain Mechanisms and Consciousness. Blackwell, Oxford 1954.

Duvernoy, H. M.: Human Brain Stem Vessels. Springer, Berlin 1978

Feremutsch, K.: Mesencephalon. In *Hofer, H., A. H. Schultz, D. Starck:* Primatologia, Vol. II/2. Karger, Basel 1965

Gerhard, L., J. Olszewski: Medulla oblongata and Pons. In *Hofer, H., A. H. Schultz, D. Starck:* Primatologia, Vol. II/2. Karger, Basel 1969

Jasper, H., L. D. Proctor, R. S. Knighton, W. C. Noshay, R. T. Costello: Reticular Formation of the Brain. Churchill, Oxford 1958

Mingazzini, G.: Medulla oblongata und Brücke. In Handbuch der mikroskopischen Anatomie, Vol. IV. Springer, Berlin 1928

Olszewski, J., D. Baxter: Cytoarchitecture of the Human Brain Stem. Karger, Basel 1954

Pollak, E.: Anatomie des Rückenmarks, der Medulla oblongata und der Brücke. In *Bumke, O., O. Foerster:* Handbuch der Neurologie, Vol. I. Springer, Berlin 1935

Riley, H. A.: An Atlas of the Basal Ganglia, Brain Stem and Spinal Cord. Williams & Wilkins, Baltimore 1943

Spatz, H.: Anatomie des Mittelhirns. In *Bumke, O., O. Foerster:* Handbuch der Neurologie, Vol. I. Springer, Berlin 1935

Cerebellum

Angevine jr., J. B., E. L. Mancall, P. I. Yakovlev: The Human Cerebellum. Little, Brown, Boston 1961

Chan-Palay, V.: Cerebellar Dentate Nucleus. Springer, Berlin 1977

Dow, R. S., G. Moruzzi: The Physiology and Pathology of the Cerebellum. University of Minnesota Press, Minneapolis 1958

Eccles, J. C., M. Ito, J. Szentágothai: The Cerebellum as a Neuronal Machine. Springer, Berlin 1967

Fields, W. S., W. D. Willis: The Cerebellum in Health and Disease. Green, St. Louis 1970

Jakob, A.: Das Kleinhirn. In *v. Möllendorff, W.:* Handbuch der mikroskopischen Anatomie, Vol. IV. Springer, Berlin 1928

Jansen, J., A. Brodal: Das Kleinhirn. In *Bargmann, W.:* Handbuch der mikroskopischen Anatomie, Suppl. to Vol. IV/1. Springer, Berlin 1958

Larsell, O., J. Jansen: The Comparative Anatomy and Histology of the Cerebellum. University of Minnesota Press, Minneapolis 1972

Llinás, R.: Neurobiology of Cerebellar Evolution and Development. American Medical Association, Chicago 1969

Palay, S. L.: Cerebellar Cortex. Springer, Berlin 1974

Diencephalon

Akert, K.: Die Physiologie und Pathophysiologie des Hypothalamus. In *Schaltenbrand, G., P. Bailey:* Einführung in die stereotaktischen Operationen mit einem Atlas des menschlichen Gehirns – Introduction to Stereotaxis with an Atlas of the Human Brain. Thieme, Stuttgart 1959

Ariëns Kappers, J., J. P. Schadé: Structure and Function of the Epiphysis Cerebri. Elsevier, Amsterdam 1965

Bargmann, W., J. P. Schadé: Lectures on the Diencephalon. Elsevier, Amsterdam 1964

De Wulf, A.: Anatomy of the Normal Human Thalamus. Elsevier, Amsterdam 1971

Diepen, R.: Der Hypothalamus. In *Bargmann, W.:* Handbuch der mikroskopischen Anatomie, Vol. IV/7. Springer, Berlin 1962

Emmers, R., R. R. Tasker: The Human Somesthetic Thalamus. Raven Press, New York 1975

Feremutsch, K.: Thalamus. In *Hofer, H., A. H. Schultz, D. Starck:* Primatologia, Vol. II/2. Karger, Basel 1963

Frigyesi, T. L., E. Rinvik, M. D. Yahr: Thalamus. Raven Press, New York 1972

Harris, G. W., B. Donovan: The Pituitary Gland, Vol. III: Pars Intermedia and Neurohypophysis. Butterworths, London 1966

Hassler, R.: Anatomie des Thalamus. In *Schaltenbrand, G., P. Bailey:* Einführung in die stereotaktischen Operationen mit einem Atlas des menschlichen Gehirns – Introduction to Stereotaxis with an Atlas of the Human Brain. Thieme, Stuttgart 1959

Haymaker, W., E. Anderson, J. H. Nauta: Hypothalamus. Thomas, Springfield/Ill. 1969

Hess, W. R.: Das Zwischenhirn. Schwabe, Basel 1954

Kuhlenbeck, H.: The human diencephalon. Confin. neurol. (Basel), Suppl. 14 (1954)

Nir, I., R. J. Reiter, R. J. Wurtman: The Pineal Gland. Springer, Wien 1977

Paillas, J. E.: La journée du thalamus. Marseille 1969

Purpura, D. P.: The Thalamus. Columbia University Press, New York 1966

Richter, E.: Die Entwicklung des Globus pallidus und des Corpus subthalamicum. Springer, Berlin 1965

Wahren, W.: Anatomie des Hypothalamus. *In Schaltenbrand, G., P. Bailey:* Einführung in die stereotaktischen Operationen mit einem Atlas des menschlichen Gehirns – Introduction to Stereotaxis with an Atlas of the Human Brain. Thieme, Stuttgart 1959

Walker, A. E.: The Primate Thalamus. University of Chicago Press, Chicago 1938

Walker, A. E.: Normale und pathologische Physiologie des Thalamus. In *Schaltenbrand, G., P. Bailey:* Einführung in die stereotaktischen Operationen mit einem Atlas des menschlichen Gehirns – Introduction to Stereotaxis with an Atlas of the Human Brain. Thieme, Stuttgart 1959

Wolstenholme, G. E. W., J. Knight: The Pineal Gland. Ciba Foundation Symposium. Churchill-Livingstone, London 1971

Wurtman, R. J., J. A. Axelrod, D. E. Kelly: The Pineal. Academic Press, New York 1968

Telencephalon

Alajouanine P. Th.: Les grandes activitées du lobe temporal. Masson, Paris 1955

v. Bonin, G.: Die Basalganglien. In Schaltenbrand, G., B. Bailey: Einführung in die stereotaktischen Operationen mit einem Atlas des menschlichen Gehirns – Introduction to Stereotaxis with an Atlas of the Human Brain. Thieme, Stuttgart 1959

Brazier, A. B.: Architectonics of the Cerebral Cortex. Raven Press, New York 1978

Brodmann, K.: Vergleichende Lokalisationslehre der Großhirnrinde. Barth, Leipzig 1925

Bucy, P. C.: The Precentral Motor Cortex. University of Illinois Press, Urbana 1949

Ciba Foundation Symposium 58: Functions of the Septo-Hippocampal System. Elsevier, Amsterdam 1977

Critchley, M.: The Parietal Lobes. Arnold, London 1953

Denny-Brown, D.: The Basal Ganglia. Oxford University Press, London 1962

Dimond, St.: The Double Brain. Churchill-Livingstone, Edinburgh 1972

Eccles, J. C.: Brain and Conscious Experience. Springer, Berlin 1966

v. Economo, C., G. N. Koskinas: Die Cytoarchitektonik der Hirnrinde des erwachsenen Menschen. Springer, Berlin 1925

Eleftheriou, B. E.: The Neurobiology of the Amygdala. Plenum Press, New York 1972

Feremutsch, K.: Basalganglien. In *Hofer, H., A. H. Schultz, D. Starck:* Primatologia, Vol. II/2. Karger, Basel 1961

Gastaud, H., H. J. Lammers: Anatomie du rhinecéphale. Masson, Paris 1961

Hubel, D. H., T. N. Wiesel: Anatomical demonstration of columns in the monkey striate cortex. Nature (Lond.) 221: 747–750, 1969

Isaacson, R. L., K. H. Pribram: The Hippocampus. Plenum Press, New York 1975

Kahle, W.: Die Entwicklung der menschlichen Großhirnhemisphäre. Springer, Berlin 1969

Kennedy, C., M. H. Des Rosiers, O. Sakurada, M. Shinohara, M. Reivich, J. W. Jehle, L. Sokoloff: Metabolic mapping of the primary visual system of the monkey by means of the autoradiographic C^{14}deoxyglucose technique. Proc. nat. Acad. Sci. (Wash.) 73: 4230–4234, 1976

Kinsbourne, M., W. L. Smith: Hemispheric Disconnection and Cerebral Function. Thomas, Springfield/Ill. 1974

Passouant, P.: Physiologie de le hippocampe. Edition du Centre National de la Recherche Scientifique, Paris 1962

Penfield, W., H. Jasper: Epilepsy and the Functional Anatomy of the Human Brain. Little, Brown, Boston 1954

Penfield, W., T. Rasmussen: The Cerebral Cortex of Man. Macmillan, New York 1950

Penfield, W., L. Roberts: Speech and Brain Mechanisms. Princeton University Press, Princeton 1959

Ploog, D.: Die Sprache der Affen. In Gadamer, H. G., P. Vogler: Neue Anthropologie. Thieme, Stuttgart 1972

Rose, M.: Cytoarchitektonik und Myeloarchitektonik der Großhirnrinde. In *Bumke, O., O. Foerster:* Handbuch der Neurologie, Vol. I. Springer, Berlin 1935

Sanides, F.: Die Architektonik des menschlichen Stirnhirns. Springer, Berlin 1962

Stephan, H.: Allocortex. In Bargmann, W.: Handbuch der mikroskopischen Anatomie, Vol. IV/9. Springer, Berlin 1975

Valverde, F.: Studies on the Piriform Lobe. Harvard University Press, Cambridge/Mass. 1965

Cerebrospinal Fluid System

Hofer, H.: Circumventrikuläre Organe des Zwischenhirns. In *Hofer, H., A. H. Schultz, D. Starck:* Primatologia, Vol. II/2. Karger, Basel 1965
Lajtha, A., D. H. Ford: Brain Barrier System. Elsevier, Amsterdam 1968
Millen, J. W. M., D. H. M. Woollam: The Anatomy of the Cerebrospinal Fluid. Oxford University Press, London 1962
Schaltenbrand, G.: Plexus und Meningen. In *Bargmann, W.:* Handbuch der mikroskopischen Anatomie, Vol. IV/2. Springer, Berlin 1955
Sterba, G.: Zirkumventrikuläre Organe und Liquor. VEB Fischer, Jena 1969

Vascular System

Dommisee, G. F.: The Arteries and Veins of the Human Spinal Cord from Birth. Churchill-Livingstone, Edinburgh 1975
Hiller, F.: Die Zirkulationsstörungen des Rückenmarks und Gehirns. In *Bumke, O., O. Foerster:* Handbuch der Neurologie, Vol. III/11. Springer, Berlin 1936
Kaplan, H. A., D. H. Ford: The Brain Vascular System. Elsevier, Amsterdam 1966
Krayenbühl, H., M. G. Yasargil: Cerebral Angiography, 2nd Ed. by *P. Huber.* Thieme, Stuttgart 1982
Luyendijk, W.: Cerebral Circulation. Elsevier, Amsterdam 1968
Szilka, G., G. Bouvier, T. Hovi, V. Petrov: Angiography of the Human Brain Cortex. Springer, Berlin 1977

Autonomic Nervous System

Burnstock, G., M. Costa: Adrenergic Neurons. Chapman & Hall, London 1975
Csillik, B., S. Ariens Kappers: Neurovegetative Transmission Mechanisms. Springer, Berlin 1974
Gabella, G.: Structure of the Autonomous Nervous System. Chapman & Hall, London 1975
Kuntz, A.: The Autonomic Nervous System. Lea & Febiger, Philadelphia 1947
Mitchell, G. A. G.: Anatomy of the Autonomic Nervous System. Livingstone, Edinburgh 1953
Monnier, M.: Physiologie des vegetativen Nervensystems. Hippokrates. Stuttgart 1963
Müller, R. L.: Lebensnerven und Lebenstriebe. Springer, Berlin 1931
Newman, P. P.: Visceral Afferent Functions of the Nervous System. Arnold, London 1974
Pick, J.: The Autonomic Nervous System. Lippincott, Philadelphia 1970
White, J. C., R. H. Smithwick: The Autonomic Nervous System. Macmillan, New York 1948

Functional Systems

Adey, W. R., T. Tokizane: Structure and Function of the Limbic System. Elsevier, Amsterdam 1967
Andresi K. H., M. v. Dühring: Morphology of cutaneous receptors. In *Autrum, H., R. Jung, W. R. Loewenstein, D. M. MacKay, H. L. Teuber:* Handbook of Sensory Physiology, Vol. II. Springer, Berlin 1973
Barker, D.: The morphology of muscle receptors. In *Autrum, H., R. Jung, W. R. Loewenstein, D. M. MacKay, H. L. Teuber:* Handbook of Sensory Physiology, Vol. III/2. Springer, Berlin 1974
Campbell, H. J.: The Pleasure Areas. Eyre Methuen, London 1973
Couteaux, R.: Motor end plate structure. In Bourne, G. H.: The Structure and Function of Muscle. Academic Press, New York 1973
Douek, E.: The Sense of Smell and Its Abnormalities. Churchill-Livingstone, London 1974
Halata, Z.: The Mechanoreceptors of the Mammalien Skin, Advances in Anatomy, Embryology and Cell Biology, Vol. 50/5. Springer, Berlin 1975
Heppner, F.: Limbisches System und Epilepsie. Huber, Bern 1973
Isaacson, R. L.: The Limbic System. Plenum Press, New York 1974
Janzen, R., W. D. Keidel, A. Herz, C. Steichele: Pain. Thieme, Stuttgart 1972
Jung, R., R. Hassler: The extrapyramidal motor system. In *Field, J., H. W. Magoun, V. E. Hall:* Handbook of Physiology, Section 1, Vol. II. American Physiological Society, Washington 1960
Knight, J.: Mechanisms of Taste and Smell in Vertebrates. Ciba Foundation Symposium. Churchill, London 1970
Lassek, A. M.: The Pyramidal Tract. Thomas, Springfield/Ill. 1954
Monnier, M.: Functions of the Nervous System, Vol. III: Sensory Functions and Perception. Elsevier, Amsterdam 1975
Munger, B. L.: Patterns of organization of peripheral sensory receptors. In *Autrum,*

H., R. Jung, W. R. Loewenstein, D. M. MacKay, H. L. Teuber: Handbook of Sensory Physiology, Vol. I/1. Springer, Berlin 1971

de Reuk, A. V. S., J. Knight: Touch, Heat and Pain. Ciba Foundation Symposium. Churchill, London 1966

Wiesendanger, M.: The pyramidal tract. In: Ergebnisse der Physiologie, Vol. 61. Springer, Berlin 1969

Zacks, S. J.: The Motor Endplate Saunders, Philadelphia 1964

Zotterman, Y.: Olfaction and Taste. Pergamon Press, Oxford 1963

Zotterman, Y.: Sensory Mechanisms. Elsevier, Amsterdam 1966

Visual System

Carpenter, R. H. S.: Movements of the Eyes. Pion, London 1977

Fine, B. S., M. Yanoff: Ocular Histology. Harper & Row, New York 1972

Polyak, St.: The Vertebrate Visual System. University of Chicago Press, Chicago 1957

Rodieck, R. W.: Vertebrate Retina. Freeman, San Francisco 1973

Rohen, J. W.: Sehorgan. In *Hofer, H., A. H. Schultz, D. Starck:* Primatologia, Vol. II/1. Karger, Basel 1962

Rohen, J. W.: Das Auge und seine Hilfsorgane. In *Bargmann, W.:* Handbuch der mikroskopischen Anatomie, Suppl. to Vol. III/2. Springer, Berlin 1964

Straatsma, B. R., M. O. Hall, R. A. Allen, F. Crescitelli: The Retina Morphology, Function and Clinical Characteristics. University of California Press, Berkeley 1969

Walsh, F. W., W. F. Hoyt: Clinical Neuroophtalmology. Williams & Wilkins, Baltimore 1969

Warwick, R.: Wolff's Anatomy of the Eye and Orbit. Lewis, London 1976

Auditory and Vestibular Systems

Ades, H. W., H. Engström: Anatomy of the inner ear. In *Autrum, H., R. Jung, W. R. Loewenstein, D. M. MacKay, H. L. Teuber:* Handbook of Sensory Physiology, Vol. V/1. Springer, Berlin 1974

Beck, C.: Anatomie des Ohres. In *Berendes, J., R. Link, F. Zöllner:* Hals-Nasen-Ohren-Heilkunde, Vol. III/1. Thieme, Stuttgart 1965

Beck, C.: Histologie des Ohres. In Berendes, J., R. Link, F. Zöllner: Hals-Nasen-Ohren-Heilkunde, Vol. III/1. Thieme, Stuttgart 1965

Brodal, A., O. Pompeiano, F. Walberg: The Vestibular Nuclei and Their Connexions. Oliver & Boyd, Edinburgh 1962

Keidel, W. D.: Anatomie und Elektrophysiologie der zentralen akustischen Bahnen. In *Berendes, J., R. Link, F. Zöllner:* Hals-Nasen-Ohren-Heilkunde, Vol. III/3. Thieme, Stuttgart 1966

Kolmer, W.: Gehörorgan. In *v. Möllendorff:* Handbuch der mikroskopischen Anatomie, Vol. III/1. Springer, Berlin 1927

Precht, W.: Neuronal Operations in the Vestibular Systems. Springer, Berlin 1978

Rasmussen, G. L., W. F. Windle: Neural Mechanisms of the Auditory and Vestibular Systems. Thomas, Springfield/Ill. 1960

de Reuck, A. V. S., J. Knight: Hearing Mechanisms in Vertebrates. Ciba Foundation Symposium. Churchill, London 1968

Whitfield, I. C.: The Auditory Pathway. Arnold, London 1960

Index

Boldface page numbers indicate extensive coverage of the subject

U

V

W

Z